Second Edition

CULTURE OF ANIMAL CELLS

A Manual of Basic Technique

Second Edition

CULTURE OF ANIMAL CELLS
A Manual of Basic Technique

R. Ian Freshney

Department of Medical Oncology
Cancer Research Campaign Laboratories
University of Glasgow

Alan R. Liss, Inc., New York

Cover Illustrations. *From the top: Vero cells growing on microcarriers; suspension culture vessels; primary explant from human mammary carcinoma; human glioma cells.*

Address all Inquiries to the Publisher
Alan R. Liss, Inc., 41 East 11th Street, New York, NY 10003

Copyright © 1987 Alan R. Liss, Inc.

Printed in the United States of America

First Edition published 1983
Second Printing, April 1984
Third Printing, November 1984

Second Edition published 1987
Second Printing, February 1988

Library of Congress Cataloging-in-Publication Data

Freshney, R. Ian.
    Culture of animal cells.

    Bibliography: p.
    Includes index.
    1. Tissue culture.   2. Cell culture. I. Title.
[QH585.F74 1987]   591'.07'24   87-3313
ISBN 0-8451-4241-0

Contents

1   Introduction

2   Biology of the Cultured Cell

3   Design and Layout of the Laboratory

4   Equipping the Laboratory

5   Aseptic Technique

6   Laboratory Safety and Biohazards

7   The Culture Environment: Substrate, Gas Phase, Medium, and Temperature

8   Preparation and Sterilization

9   Disaggregation of the Tissue and Primary Culture

10   Maintenance of the Culture—Cell Lines

11   Cloning and Selection of Specific Cell Types

12   Physical Methods of Cell Separation

13   Characterization

14   Induction of Differentiation

15   The Transformed Phenotype

16   Contamination

17   Instability, Variation, and Preservation

18   Quantitation and Experimental Design

19   Measurement of Cytotoxicity and Viability

20   Culture of Specific Cell Types

21   Culture of Tumor Tissue

22   Three-Dimensional Culture Systems

23   Specialized Techniques

Preface to the Second Edition

In revising Culture of Animal Cells I have tried to keep the emphasis on the practical aspects of cell culture and have discussed the theoretical background only when it seemed necessary to the understanding of the technique or the status of the culture. For example, cell transformation and some of its implications are dealt with more fully to help the reader to appreciate the phenotypic properties that these cells might be expected to express and the roles that they might usefully play in experimental studies and commercial exploitation.

Major changes have been introduced in the presentation of serum free medium formulations as these have gained more general acceptance, and some have become commercially available, since publication of the first edition. In parallel with this, and in many cases as a direct consequence, the culture of specific cell types such as epidermal keratinocytes, melanocytes, and breast epithelium have become more feasible, so a number of protocols are included for such specialized cultures.

To enable these areas to be covered more effectively, I have enlisted the help of experts in each respective field to present protocols from their own experience where I feel my own expertise is insufficient. These specialist protocols cover areas of new technology, such as somatic hybridization and production of hybridomas as well as the culture of specific cell types, and have been presented in the same style as the previous protocols. I am very grateful to these new contributors and feel that they have extended the scope of the text more than I could have hoped to do alone. In some cases these protocols will be sufficient for readers to fulfill their needs without further recourse to the literature, but to satisfy those whose demands are greater, or where the technique is more complex, the appropriate references are provided.

A more extensive treatment has also been given to cytotoxicity assay and the culture of tumor cells, particularly from human tumors, in line with the emphasis that these techniques are currently being given in hospitals, basic research laboratories, and the biotechnology and drug industry.

In addition to the contributors of specialized protocols referred to above, I am again indebted to my colleagues in the Department of Medical Oncology including Jane Plumb, Stephen Merry, Carol McCormick, Alison Mackie, and Ian Cunningham, and a succession of graduate and undergraduate students including John McLean, Alison Murray, Jim Miller, Iain Singer, Barbara Christie, and Alan Beveridge who have provided data and ideas. While trying to answer their questions, I was stimulated into thinking more about the potential needs of the reader.

My thanks are also due to Mrs. Rae Fergusson for typing new material faster than I could generate it and handling my poor handwriting and illegible corrections with unbelievable accuracy.

Most of all I would like to thank my wife and family for their continuing help and encouragement. They provided much practical help, advice, and moral support. In particular my wife's many hours collating, referencing, and proof reading, have spared me many hours of often tedious work.

Preface to the First Edition

Tissue culture is not a new technique. It has been in existence since the beginning of this century and has passed through its simple exploratory phase, a later expansive phase in the 1950's, and is now in a phase of specialization concerned with control mechanisms and differentiated function. Matching the current trends, recent additions to the range of available tissue culture books have been concerned with specialized techniques and the result of this is that the basic procedures have become a little neglected.

It has been my objective in preparing this book to provide the novice to tissue culture with sufficient information to perform the basic techniques. It is anticipated that the reader will have a fundamental grasp of elementary anatomy, histology, cell physiology, and the basic principles of biochemistry, but will have had little or no experience in tissue culture. This book should prove useful at the advanced undergraduate level for technicians in training, for graduate studies, and at the postdoctoral level. It is intended as an introduction to the theory of the technique, and biology of cultured cells as well as a practical, step-by-step guide to procedures, and should be of value to anyone without any, or with little, prior experience in tissue culture. Of necessity, some of the more exciting developments in recent years, e.g., production of monoclonal antibodies by hybridoma cultures, can only be described briefly and references provided to further reading.

A list of reagents and commercial suppliers is located at the end of the book. Occasionally, a supplier's name is incorporated in the text but in most cases reference should be made to the trade index. Other reference materials included at the rear of the book are a glossary, a list of cell banks, a subject index, and the literature references cited in the text.

It is inevitable when preparing a text such as this that, in addition to my own experience, I have called upon the help and advice of many others both during the preparation of the book and in the twenty years or so since I was first introduced to the field. As with many other similar techniques, there is much of tissue culture that is never documented, but passed on by word of mouth at meetings, or, more often, in moments of conviviality after

meetings. Hence there may be occasions when I have reproduced advice or information as if it were my own, without due acknowledgment to published work, because I have been unable to trace a reference, or none exists. In all such cases I would like to thank those who have contributed consciously or unconsciously to my own accumulated experience in the field.

While it would be impossible to recall all of those with whom contact over the past two decades has influenced my current understanding of the field, there are those of whom I must make special mention. First among these is Dr. John Paul, who introduced me to the field and whose sound common sense and practicality were a good introduction to what can, in the correct hands, be a very precise discipline. I owe him my sincere gratitude, as his one-time student and now associate and friend.

In my years with the Beatson Institute I have had the privilege to work with many people, both resident and visitors, and share in their experience in the development of techniques to which I would otherwise not have been exposed. In some cases they are acknowledged in the text or figure legends, but I hope any who are not mentioned by name will still recognize my gratitude.

Among others who should be named are those who have worked most closely with me in recent years, helped in my own research activities, and generated some of the data that appear on these pages. They include Ms. Diana Morgan, Mrs. Elaine Hart, Mrs. Margaret Frame, Mr. Alistair McNab, Mrs. Irene Osprey, and Miss Sheila Brown. Although my wife and I do not work together usually, I have had the benefit of her skilled assistance at times, and, in addition, her experience in the field has added greatly to my own. Others who have worked with me for shorter periods, elements of whose work may be reported here in part, are Mohammad Hassanzadah, Peter Crilly, Fadik Akturk, Metyn Guner, Fahri Celik, Aileen Sherry, Bob Shaw and Carolyn MacDonald.

I have also been indebted to many people in Glasgow and elsewhere for helpful advice and collaboration. Among many others, these include David G.T. Thomas, David I. Graham, Michael Stack-Dunne, Peter Vaughan, Brian McNamee, David Doyle, Rona MacKie, Kenneth

C. Calman, and the late John Maxwell Anderson, with whom I had my first introduction to clinical collaboration.

I must also record my good fortune to have been able to spend time in other laboratories and learn from the approaches of others such as Robert Auerbach, Richard Ham, and Wally McKeehan.

I am also grateful to Flow Laboratories for their help and collaboration in running basic tissue culture courses and the resultant opportunity to broaden my knowledge of the field.

I would like to express my gratitude to Paul Chapple who first persuaded me that I should write a basic techniques book on tissue culture, and to numerous others, including Don Dougall, Wally and Kerstin McKeehan, Peter del Vecchio, John Ryan, Jim Smith, Rob Hay, Charity Waymouth, Sergey Federoff, Mike Gabridge, and Dan Lundin for help and advice during the preparation of the manuscript.

I would also like to thank Miss Donna Madore for converting my often illegible manuscript into typescript, Mrs. Marina LaDuke for expert photography, Miss Diane Leifheit for further help with the illustrations, and Ms. Jane Gillies for preparing the line drawings. These four ladies spent many hours on my behalf and their patience and skill is greatly appreciated. My thanks are also due to Mrs. Norma Wallace for completing the final retype quickly, efficiently, and at very short notice.

It would not be fitting for me to conclude this preface without further major acknowledgment to my wife, Mary, my daughter, Gillian, and son, Norman. Not only did I enjoy their sympathy and understanding at home, when I am sure, at times, I did not deserve it, but I also benefitted from the fruits of their labors during the day: drawing graphs, collecting references, researching and tabulating methods and information. My wife's experience in the field, plus countless hours of reading, revising, and collecting information, made her share in this work indispensible.

BACKGROUND

Tissue culture was first devised at the beginning of this century [Harrison 1907, Carrel, 1912] as a method for studying the behavior of animal cells free of systemic variations that might arise in the animal both during normal homeostasis and under the stress of an experiment. As the name implies, the technique was elaborated first with undisaggregated fragments of tissue, and growth was restricted to the migration of cells from the tissue fragment, with occasional mitoses in the outgrowth. Since culture of cells from such primary explants of tissue dominated the field for more than 50 years, it is not surprising that the name "tissue culture" has stuck in spite of the fact that most of the explosive expansion in this area since the 1950s has utilized dispersed cell cultures.

Throughout this book the term "tissue culture" will be used as the generic term to include organ culture and cell culture. The term "organ culture" will always imply a three-dimensional culture of undisaggregated tissue retaining some or all of the histological features of the tissue in vivo. "Cell culture" will refer to cultures derived from dispersed cells taken from the original tissue, from a primary culture, or from a cell line or cell strain, by enzymatic, mechanical, or chemical disaggregation. The term "histotypic culture" will imply that cells have been reassociated in some way to recreate a three-dimensional tissue-like structure, e.g., by perfusion and overgrowth of a monolayer, reaggregation in suspension, or infiltration of a three-dimensional matrix such as collagen gel.

Harrison chose the frog as his source of tissue presumably because it was a cold-blooded animal, and consequently incubation was not required. Furthermore, since tissue regeneration is more common in lower vertebrates, he perhaps felt that growth was more likely to occur than with mammalian tissue. Although his technique may have sparked off a new wave of interest in cultivation of tissue *in vitro*, few later workers were to follow his example in the selection of species. The stimulus from medical science carried future interest into warm-blooded animals where normal and pathological development are closer to human. The accessibility of different tissues, many of which grew well in culture, made the embryonated hen's egg a favorite choice; but the development of experimental animal husbandry, particularly with genetically pure strains of rodents, brought mammals to the forefront as favorite material. While chick embryo tissue could provide a diversity of cell types in primary culture, rodent tissue had the advantage of producing continuous cell lines [Earle et al., 1943].

The demonstration that human tumors could also give rise to continuous cell lines [e.g., HeLa: Gey et al., 1952], encouraged interest in human tissue, helped later by Hayflick and Moorhead's classical studies with normal cells of a finite life-span [1961].

For many years the lower vertebrates and the invertebrates have been largely ignored though unique aspects of their development (tissue regeneration in amphibia, metamorphosis in insects) make them attractive systems for the study of the molecular basis of development. More recently the needs of agriculture and pest control have encouraged toxicity and virological studies in insects, and the rapidly developing area of fish farming has required more detailed knowledge of normal development and pathogenesis in fish.

In spite of this resurgence of interest, tissue culture of lower vertebrates and the invertebrates remains a very specialized area, and the bulk of interest remains in avian and mammalian tissue. This has naturally influenced the development of the art and science of tissue culture, and much of what will be described in the ensuing chapters of this book reflect this, as well as my own personal experience. Hence advice on incubation and the physical and biochemical properties of media refers to homiotherms and guidance on the appropriate modification for poikilothermic animals will require recourse to the literature. This will be discussed in a little more detail in a later chapter.

Many of the basic techniques of asepsis, preparation and sterilization, primary culture, selection and cell separation, quantitation, and so on, apply equally to poikilotherms and will require only minor modification; on the whole the principles remain the same.

The types of investigation that lend themselves particularly to tissue culture are summarized in Figure 1.1.: (1) Intracellular activity, e.g., the replication and transcription of deoxyribonucleic acid (DNA), protein synthesis, energy metabolism; (2) intracellular flux, e.g., movement of ribonucleic acid (RNA) from the nucleus to the cytoplasm, translocation of hormone receptor complexes, fluctuations in metabolite pools; (3) "ecology," eg., nutrition, infection, virally or chemically induced transformation, drug action, response to external stimuli, secretion of specialized products; and (4) cell-cell interaction, e.g., embryonic induction, cell population kinetics, cell-cell adhesion.

The development of tissue culture as a modern sophisticated technique owes much to the needs of two major branches of medical research: the production of antiviral vaccines and the understanding of neoplasia. The standardization of conditions and cell lines for the production and assay of viruses undoubtedly provided much impetus to the development of modern tissue culture technology, particularly the production of large numbers of cells suitable for biochemical analysis.

This and other technical improvements made possible by the commercial supply of reliable media and sera, and by the greater control of contamination with antibiotics and clean air equipment, has made tissue culture accessible to a wide range of interests.

In addition to cancer research and virology, other areas of research have come to depend heavily on tissue culture techniques. The introduction of cell fusion techniques [Barski, et al., 1960; Soreuil and Ephrussi, 1961; Littlefield, 1964; Harris and Watkins, 1965] and genetic manipulation established somatic cell genetics as a major component in the genetic analysis of higher animals including man, and contributed greatly, via the monoclonal antibody technique, to the study of immunology, already dependent on cell culture for assay techniques and production of hemopoietic cell lines.

The insight into the mechanism of action of antibodies, and the reciprocal information that this provided about the structure of the epitope, derived from monoclonal antibody techniques [Kohler and Milstein, 1975] was, like the technique of cell fusion itself, a prologue to a whole new field of studies in genetic manipulation. This has supplied much basic information on the control of gene transcription and a vast new technology has grown from the ability to insert exploitable genes into prokaryotic cells. Cell products such as human

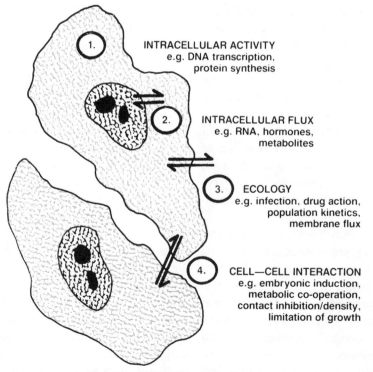

INTRACELLULAR ACTIVITY
e.g. DNA transcription,
protein synthesis

INTRACELLULAR FLUX
e.g. RNA, hormones,
metabolites

ECOLOGY
e.g. infection, drug action,
population kinetics,
membrane flux

CELL—CELL INTERACTION
e.g. embryonic induction,
metabolic co-operation,
contact inhibition/density,
limitation of growth

Fig. 1.1. *Areas of interest in tissue culture.*

growth hormone, insulin, and interferon have been genetically engineered, but the absence of post-transcriptional modifications, such as glycosylation, in bacteria suggest that mammalian cells may provide more suitable vehicles. The insertion of the appropriate genes into normal human cells (1) to make them continuous cell lines (see Chapter 2) and (2) to make them produce pharmaceutically viable drugs will have profound effects on the drug industry, which can only be overshadowed by radical innovations in organic chemical synthesis that are, as yet, not apparent. Other areas of major interest include the study of cell interactions and intracellular control mechanisms in cell differentiation and development [Auerbach and Grobstein, 1958; Cox, 1974; Finbow and Pitts, 1981] and attempts to analyze nervous function [Bornstein and Murray, 1958; Minna et al., 1972; Kingsbury et al., 1985]. Progress in neurological research has, however, not had the benefit of working with propagated cell lines as propagation of neurons has not so far been possible *in vitro* without resorting to the use of transformed cells (see Chapter 15).

Tissue culture technology has also been adopted into many routine applications in medicine and industry. Chromosomal analysis of cells derived from the womb by amniocentesis (see Chapter 23) can reveal genetic disorders in the unborn child, viral infections may be assayed qualitatively and quantitatively on monolayers of appropriate host cells, (see Chapter 23), and the toxic effects of pharmaceutical compounds and potential environmental pollutants can be measured in colony-forming and other *in vitro* assays (see Chapter 19).

Further developments in the application of tissue culture to medical problems may follow from the demonstration that cultures of epidermal cells form functionally differentiated sheets in culture [Green et al., 1979], and endothelial cells may form capillaries [Folkman and Haudenschild, 1980], suggesting possibilities in homografting and reconstructive surgery using an individual's own cells [Pittelow and Scott, 1986].

It is clear that the study of cellular activity in tissue culture may have many advantages; but in summarizing these, below, considerable emphasis must also be placed on its limitations, in order to maintain some sense of perspective.

ADVANTAGES OF TISSUE CULTURE

Control of the Environment

The two major advantages, as implied above, are the control of the physiochemical environment (pH, temperature, osmotic pressure, O_2, CO_2 tension), which may be controlled very precisely, and the physiological conditions, which may be kept relatively constant but cannot always be defined. Most cell lines still require supplementation of the medium with serum or other poorly-defined constituents. These supplements are prone to batch variation [Olmsted, 1967; Honn et al., 1975] and contain undefined elements such as hormones and other regulatory substances. Gradually the essential components of serum are being identified, making replacement with defined constituents more practicable [Birch and Pirt, 1971; Ham and McKeehan, 1978; Barnes and Sato, 1980; Barnes et al., 1984; Maurer, 1986] (see also Chapter 7).

Characterization and Homogeneity of Sample

Tissue samples are invariably heterogeneous. Replicates even from one tissue vary in their constituent cell types. After one or two passages, cultured cell lines assume a homogeneous, or at least uniform, constitution as the cells are randomly mixed at each transfer and the selective pressure of the culture conditions tends to produce a homogeneous culture of the most vigorous cell type. Hence, at each subculture each replicate sample will be identical, and the characteristics of the line may be perpetuated over several generations, or indefinitely if the cell line is stored in liquid N_2. Since experimental replicates are virtually identical, the need for statistical analysis of variance is reduced.

Economy

Cultures may be exposed directly to a reagent at a lower and defined concentration, and with direct access to the cell. Consequently, less is required than for injection *in vivo* where >90% is lost by excretion and distribution to tissues other than those under study.

Screening tests with many variables and replicates are cheaper, and the legal, moral, and ethical questions of animal experimentation are avoided.

DISADVANTAGES

Expertise

Culture techniques must be carried out under strict aseptic conditions, because animal cells grow much less rapidly than many of the common contaminants such as bacteria, molds, and yeasts. Furthermore, unlike microorganisms, cells from multicellular animals do not exist in isolation, and consequently, are not able to sustain independent existence without the provision of a complex environment, simulating blood plasma or interstitial fluid. This implies a level of skill and understanding to appreciate the requirements of the system and to diagnose problems as they arise. Tissue culture should not be undertaken casually to run one or two experiments.

Quantity

A major limitation of cell culture is the expenditure of effort and materials that goes into the production of relatively little tissue. A realistic maximum per batch for most small laboratories (two or three people doing tissue culture) might be 1–10 g of cells. With a little more effort and the facilities of a larger laboratory, 10–100 g is possible; above 100 g implies industrial pilot plant scale, beyond the reach of most laboratories, but not impossible if special facilities are provided.

The cost of producing cells in culture is about ten times that of using animal tissue. Consequently, if large amounts of tissue (>10 g) are required, the reasons for providing them by tissue culture must be very compelling. For smaller amounts of tissue ($\leqslant 10$ g), the costs are more readily absorbed into routine expenditure, but it is always worth considering whether assays or preparative procedures can be scaled down. Semimicro- or micro-scale assays can often be quicker due to reduced manipulation times, volumes, centrifuge times, etc. and are often more readily automated (see under Microtitration, Chapter 19).

Instability

This is a major problem with many continuous cell lines resulting from their unstable aneuploid chromosomal constitution. Even with short-term cultures, although they may be genetically stable, the heterogeneity of the cell population, with regard to cell growth rate, can produce variability from one passage to the next. This will be dealt with in more detail in Chapters 10 and 17.

MAJOR DIFFERENCES *IN VITRO*

Many of the differences in cell behavior between cultured cells and their counterparts *in vivo* stem from the dissociation of cells from a three-dimensional geometry and their propagation on a two-dimensional substrate. Specific cell interactions characteristic of the histology of the tissue are lost, and, as the cells spread out, become mobile and, in many cases, start to proliferate, the growth fraction of the cell population increases. When a cell line forms it may represent only one or two cell types and many heterotypic interactions are lost.

The culture environment also lacks the several systemic components involved in homeostatic regulation *in vivo*, principally those of the nervous and endocrine systems. Without this control, cellular metabolism may be more constant *in vitro* than *in vivo*, but may not be truly representative of the tissue from which the cells were derived. Recognition of this fact has led to the inclusion of a number of different hormones in culture media (see Chapter 7) and it seems likely that this trend will continue.

Energy metabolism *in vitro* occurs largely by glycolysis, and although the citric acid cycle is still functional it plays a lesser role.

It is not difficult to find many more differences between the environmental conditions of a cell *in vitro* and *in vivo* (see also Chapter 19) and this has often led to tissue culture being regarded in a rather skeptical light. Although the existence of such differences cannot be denied, it must be emphasized that many specialized functions are expressed in culture and as long as the limits of the model are appreciated, it can become a very valuable tool.

Origin of Cells

If differentiated properties are lost, for whatever reason, it is difficult to relate the cultured cells to functional cells in the tissue from which they were derived. Stable markers are required for characterization (see Chapter 13); and in addition, the culture conditions may need to be modified so that these markers are expressed (see next chapter and Chapter 14).

DEFINITIONS

There are three main methods of initiating a culture [Schaeffer, 1979] (see Glossary and Fig. 1.2): (1) *Organ culture* implies that the architecture characteristic of the tissue *in vivo* is retained, at least in part, in the culture (see Chapter 22). Toward this end, the tissue is cultured at the liquid gas interface (on a raft, grid, or gel) which favors retention of a spherical or three-dimensional shape. (2) In *primary explant culture* a fragment of tissue is placed at a glass (or plastic)/liquid interface where, following attachment, migration is promoted in the plane of the solid substrate (see Chapter 9). (3) *Cell culture* implies that the tissue, or outgrowth from the primary explant is dispersed (mechanically or enzymatically) into a cell suspension, which may then be cultured as an adherent monolayer on a solid substrate, or as a suspension in the culture medium (see Chapters 9 and 10).

Organ cultures, because of the retention of cell interactions as found in the tissue from which the culture was derived, tend to retain the differentiated properties of that tissue. They do not grow rapidly (cell proliferation is limited to the periphery of the explant and is restricted mainly to embryonic tissue) and hence cannot be propagated; each experiment requires fresh explantations and this implies greater effort and poorer sample reproducibility than with cell culture. Quantitation is, therefore, more difficult and the amount of material that may be cultured is limited by the dimensions of the explant ($\leqslant 1$ mm^3) and the effort required for dissection and setting up the culture.

However, it must be emphasized that organ cultures do retain specific histological interactions without which it may be difficult to reproduce the characteristics of the tissue.

Cell cultures may be derived from primary explants or dispersed cell suspensions. Because cell proliferation is often found in such cultures, propagation of cell

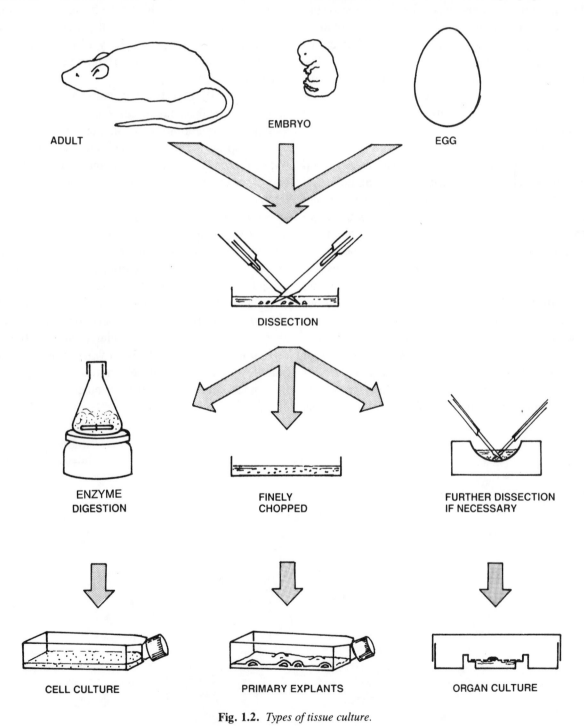

ADULT   EMBRYO   EGG

DISSECTION

ENZYME
DIGESTION

FINELY
CHOPPED

FURTHER DISSECTION
IF NECESSARY

CELL CULTURE   PRIMARY EXPLANTS   ORGAN CULTURE

Fig. 1.2. *Types of tissue culture.*

lines becomes feasible. A monolayer or cell suspension, with a significant growth fraction (see Chapter 18), may be dispersed by enzymatic treatment or simple dilution and reseeded, or subcultured, into fresh vessels. This constitutes a "passage" and the daughter cultures so formed are the beginnings of a "cell line."

The formation of a cell line from a primary culture implies: (1) an increase in total cell number over several generations; (2) that cells or cell lineages with similar high growth capacity will predominate; resulting in (3) a degree of uniformity in the cell population. The line may be characterized, and those characteristics will apply for most of its finite lifespan. The derivation of "continuous" (or "established" as they were once known) cell lines usually implies a phenotypic change or "transformation" and will be dealt with in Chapters 2 and 15.

When cells are selected from a culture, by cloning or some other method, the subline is known as a "cell strain." Detailed characterization is then implied. Cell lines, or cell strains, may be propagated as an adherent monolayer or in suspension. *Monolayer* culture signifies that the cells will attach to the substrate given the opportunity and that normally the cells will be propagated in this mode. *Anchorage dependence* means that attachment to (and usually some degree of spreading on) the substrate is a prerequisite for cell proliferation. Monolayer culture is the mode of culture common to most normal cells with the exception of mature hemopoietic cells. *Suspension* cultures are derived from cells which can survive and proliferate without attachment (*anchorage independent*); this ability is restricted to hemopoietic cells, transformed cell lines, or cells

from malignant tumors. It can be shown, however, that a small proportion of cells that are capable of proliferation in suspension exists in many normal tissues (see Chapter 15). The identity of these cells remains unclear, but a relationship to the stem cell or uncommitted precursor cell compartment has been postulated. This concept implies that some cultured cells represent precursor pools within the tissue of origin; the generality of this observation will be discussed more fully in the next chapter. Are cultured cell lines more representative of precursor cell compartments *in vivo* than of fully differentiated cells, bearing in mind that most differentiated cells do not normally divide?

Because they may be propagated as a uniform cell suspension or monolayer, cell cultures have many advantages in quantitation, characterization, and replicate sampling, but lack the potential for cell-cell interaction and cell-matrix interaction afforded by organ cultures. For this reason many workers have attempted to reconstitute three-dimensional cellular structures using aggregated cell suspension ("spheroids") or perfused high-density cultures (such as Vitafiber, Amicon) (see Chapter 22). In many ways some of the more exciting developments in tissue culture arise from recognizing the necessity of specific cell interaction in homogeneous or heterogeneous cell populations in culture. This may mark the transition from an era of fundamental molecular biology, where the regulatory processes have been worked out at the cellular level, to an era of cell or tissue biology where this understanding is applied to integrated populations of cells.

THE CULTURE ENVIRONMENT

The validity of the cultured cell as a model of physiological function *in vivo* has frequently been criticized. There are problems of characterization due to the alteration of the cellular environment; cells proliferate *in vitro* which would not normally *in vivo*, cell-cell and cell-matrix interactions are reduced because purified cell lines lack the heterogeneity and three-dimensional architecture found *in vivo*, and the hormonal and nutritional milieu is altered. This creates an environment which favors the spreading, migration, and the proliferation of unspecialized cells rather than the expression of differentiated functions. The provision of the appropriate environment, nutrients, hormones, and substrate is fundamental to the expression of specialized functions (see Chapter 14). Before considering such specialized conditions, let us examine the events accompanying the formation of a primary cell culture and a cell line derived from it (Fig. 2.1).

INITIATION OF THE CULTURE

Primary culture techniques are described in detail in Chapter 9. Briefly, a culture is derived either by outgrowth of migrating cells from a fragment of tissue, or by enzymatic or mechanical dispersal of the tissue. Regardless of the method employed, this is the first in a series of selective processes (Table 2.1) which may ultimately give rise to a relatively uniform cell line. In primary explantation (see Chapter 9) selection occurs by virtue of the cells' capacity to migrate from the explant, while with dispersed cells, only those cells which (1) survive the disaggregation technique and (2) adhere to the monolayer or survive in suspension will form the basis of a primary culture.

If the primary culture is maintained for more than a few hours, a further selection step will occur. Cells capable of proliferation will increase, some cell types will survive but not increase, and yet others will be unable to survive under the particular conditions used.

Hence, the distribution of cell types will change and continue to do so until, in the case of monolayer cultures, all the available culture substrate is occupied. After confluence is reached (i.e., all the available growth area is utilized and the cells make close contact with one another), the proportion of density-limited cells gradually decreases, and the proportion of cells which are less sensitive to density limitation of growth (see Chapters 9 and 15) increases. Virally or spontaneously transformed cells will overgrow their normal counterparts. Keeping the cell density low, e.g., by frequent subculture, helps to preserve the normal phenotype in cultures such as mouse fibroblasts, where spontaneous transformants tend to overgrow at high cell densities [Todaro and Green, 1963; Brouty-Boyé et al., 1979, 1980].

Some aspects of specialized function are expressed more strongly in primary culture, particularly when the culture becomes confluent. At this stage the culture will show its closest morphological resemblance to the parent tissue.

EVOLUTION OF CELL LINES

After the first subculture—or passage (see Fig. 2.1)—the primary culture becomes a cell line (Chapter 10) and may be propagated and subcultured several times. With each successive subculture, the component of the population with the ability to proliferate most rapidly will gradually predominate, and nonproliferating or slowly proliferating cells will be diluted out. This is most strikingly apparent after the first subculture, where differences in proliferative capacity are compounded with varying abilities to withstand the trauma of trypsinization and transfer (see Chapter 10).

Although some selection and phenotypic drift will continue, by the third passage the culture becomes more stable, typified by a rather hardy, rapidly proliferating cell. In the presence of serum and without specific selection conditions (see Chapters 7 and 20)

TABLE 2.1. Elements of Selection in the Evolution of Cell Lines

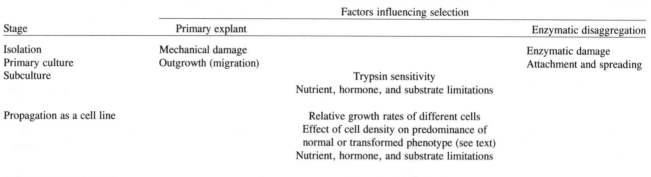

	Factors influencing selection		
Stage	Primary explant		Enzymatic disaggregation
Isolation	Mechanical damage		Enzymatic damage
Primary culture	Outgrowth (migration)		Attachment and spreading
Subculture		Trypsin sensitivity	
		Nutrient, hormone, and substrate limitations	
Propagation as a cell line		Relative growth rates of different cells	
		Effect of cell density on predominance of normal or transformed phenotype (see text)	
		Nutrient, hormone, and substrate limitations	
Senescence, transformation		Normal cells die out	
		Transformed cells overgrow	

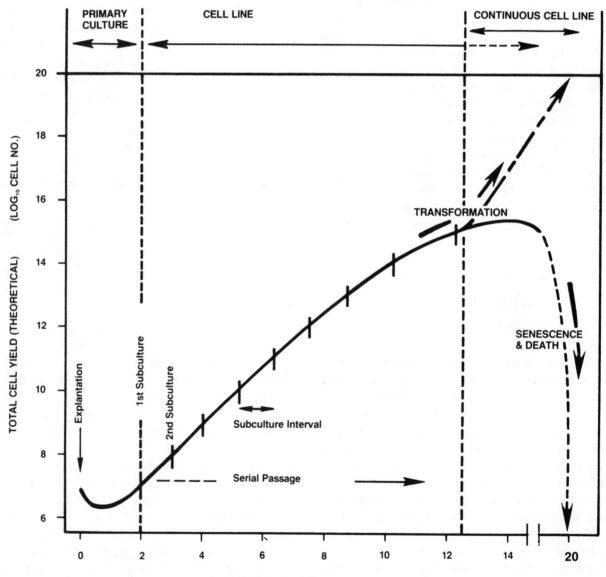

Fig. 2.1. *Evolution of a cell line. The vertical axis represents total cell growth (assuming no reduction at passage) on a log scale, the horizontal axis, time in culture on a linear scale, for a hypothetical cell culture. Although a continuous cell line is depicted as arising at 12½ wk it could, with different cultures, arise at any time. Likewise, senescence may arise at any time, but for human diploid fibroblasts, it is most likely between 30 and 60 cell doublings or ten to 20 wk depending on the doubling time.*

mesenchymal cells derived from connective tissue fibroblasts or vascular elements frequently overgrow the culture. While this has given rise to some very useful cell lines—e.g., WI38 human embryonic lung fibroblasts [Hayflick and Moorhead, 1961], BHK21 baby hamster kidney fibroblasts [MacPherson and Stoker, 1962] (see Table 10.2), and perhaps the most famous of all, the L-cell, a mouse subcutaneous fibroblast treated with methylcholanthrene [Earle et al., 1943; Sanford et al., 1948]—it has presented one of the major challenges of tissue culture since its inception: namely, how to prevent the overgrowth of the more fragile or slower-growing specialized cells such as hepatic parenchyma or epidermal keratinocytes. Inadequacy of the culture conditions is largely to blame for this problem and considerable progress has now been made in the use of selective media and substrates for the maintenance of many specialized cell lines (see Chapter 20).

"CRISIS" AND THE DEVELOPMENT OF CONTINUOUS CELL LINES

Most cell lines may be propagated in an unaltered form for a limited number of cell generations, beyond which they may either die out or give rise to continuous cell lines (Fig. 2.1). The ability of a cell line to grow continuously probably reflects its capacity for genetic variation allowing subsequent selection. Human fibroblasts remain predominantly euploid throughout their culture life-span and never give rise to continuous cell lines [Hayflick and Moorhead, 1961], while mouse fibroblasts and cell cultures from a variety of human and animal tumors often become aneuploid in culture and give rise to continuous cultures with fairly high frequency. The alteration in a culture giving rise to a continuous cell line is commonly called "in vitro transformation" (see Chapter 15) and may occur spontaneously or be chemically or virally induced.

Continuous cell lines are usually aneuploid and often have a chromosome complement between the diploid and tetraploid value (Fig. 2.2). There is also considerable variation in chromosome number and constitution among cells in the population (heteroploidy) (see also Chapter 16). It is not clear whether the cells that give rise to continuous lines are present at explantation in very small numbers or arise later as a result of transformation of one or more cells. The second would seem to be more probable on cell kinetic grounds as continuous cell lines can appear quite late in a culture's

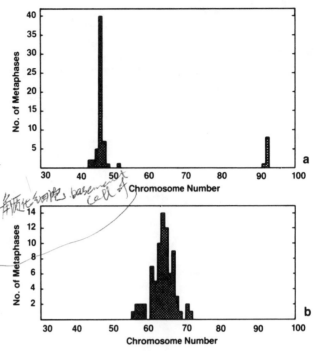

Fig. 2.2. *Chromosome numbers of finite and continuous cell lines. a. A normal human glial cell line. b. A continuous cell line from human metastatic melanoma.*

life history, long after the time it would have taken for even one preexisting cell to overgrow. The possibility remains, however, that there is subpopulation in such cultures with a predisposition to transform not shared by the rest of the cells.

The term "transformation" has been applied to the process of formation of a continuous cell line partly because the culture undergoes morphological and kinetic alterations, but also because the formation of a continuous cell line is often accompanied by an increase in tumorigenicity. A number of the properties of continuous cell lines are also associated with malignant transformations (see Chapter 10) such as reduced serum requirement, reduced density limitation of growth, growth in semisolid media, and aneuploidy (see also Table 10.3), etc. These will be reviewed in more detail in Chapter 15. Similar morphological and behavioral changes can also be observed in cells which have undergone virally or chemically induced transformation.

Many (if not most) normal cells do not give rise to continuous cell lines. In the classic example [Hayflick and Moorhead, 1961] normal human fibroblasts remain euploid throughout their life-span and at crisis (usually around 50 generations) will stop dividing, though they may remain viable for up to 18 months

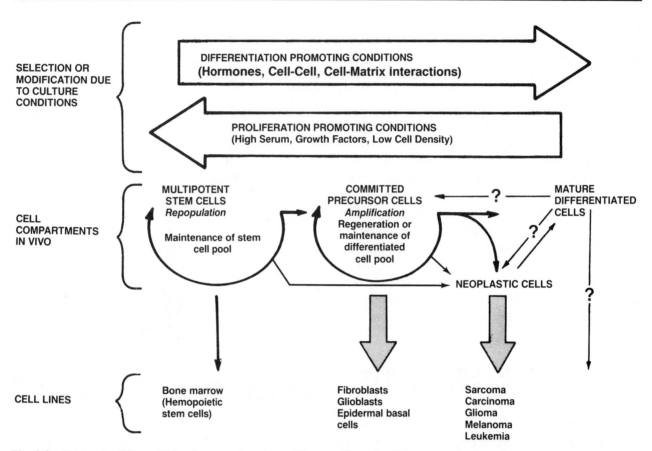

Fig. 2.3. *Origin of cell lines. With a few exceptions (e.g., differentiated tumor cells) culture conditions select for the proliferating cell compartment of the tissue or induce cells that are partially differentiated to revert to a precursor status. While neoplastic cells, and cell lines, may be derived from differentiated cells, it seems more likely that they arise from malignant precursor cells, some of which retain the capability to divide while continuing to differentiate.*

thereafter. Human glia [Pontén and Westermark, 1980] and chick fibroblasts [Hay and Strehler, 1967] behave similarly. Epidermal cells, on the other hand, have shown gradually increasing life-spans with improvements in culture techniques [Green et al., 1979] and may yet be shown capable of giving rise to continuous growth. This may be related to the self-renewal capacity of the tissue *in vivo* (see below, this chapter). Continuous culture of lymphoblastoid cell is also possible [Moore et al., 1967], although in this case, transformation with Epstein Barr virus may be implicated.

It is possible that the condition that predisposes to the development of a continuous cell line is inherent genetic variation, so it is not surprising to find genetic instability perpetuated in continuous cell lines. A common feature of many human continuous cell lines is the development of a subtetraploid chromosome number (see Fig. 2.2).

For a further discussion of variation and instability see Chapter 17.

DEDIFFERENTIATION

This implies that differentiated cells lose their specialized properties *in vitro*, but it is often unclear whether (1) undifferentiated cells of the same lineage (Fig. 2.3) overgrow terminally differentiated cells of reduced proliferative capacity or (2) the absence of the appropriate inducers (hormones: cell or matrix interaction) causes deadaptation (see Chapter 14). In practice both may occur. Continuous proliferation may select undifferentiated precursors, which, in the absence of the correct inductive environment, do not differentiate.

An important distinction should be made between dedifferentiation, deadaptation, and selection. Dedif-

TABLE 2.2. Examples of Cultured Cell Lines and Strains That Express Differentiated Properties *In Vitro*

Origin	Cell line	Species	Marker*	Reference
Finite cell lines				
Pigmented retina	Retina	Chick	Pigmentation	[Coon and Cahn, 1966]
Calvaria	Cartilage	Chick	Cartilage synthesis	[Coon and Cahn 1966]
Epidermis	Keratinocytes	Mouse	Cornification	[Fusenig et al., 1972]
Epidermis	Keratinocytes	Human	Cornification	[Rheinwald and Green, 1975b]
Skeletal muscle		Chick	Myogenesis, CK	[Richler and Yaffe, 1970]
Hypothalamus	C7	Mouse	Neurophysin	
			Vasopressin	[De Vitry et al., 1974]
Continuous cell lines				
Spleen	Friend	Mouse	Hemoglobin	[Scher et al., 1971]
Hepatoma	H-4-11-E-C3	Rat	Tyrosine aminotransferase	[Pitot et al., 1964]
Myeloid leukemia	K562	Human	Hemoglobin	[Anderson et al., 1979a, b]
Myeloid leukemia	HL60	Human	Phagocytosis NTB reduction	[Olsson and Ologson, 1981]
Glioma	C6	Rat	GFAP, GPDH	[Benda et al., 1968]
Glioma	MOG-CCM	Human	GFAP, GPDH	[Balmforth et al., 1986]
Pituitary tumor	GH2, GH3	Rat	Growth hormone	[Buonassisi et al., 1962]
Adrenal cortex tumor		Rat	Steroids	[Buonassisi et al., 1962]
Melanoma	B16	Mouse	Melanin	[Nils and Makarski, 1978]
Neuroblastoma	C1300	Rat	Neurites	[Lieberman and Sachs, 1978]
Kidney	MDCK	Dog	Domes, transport	[Rindler et al., 1979]
Kidney	LLCV-PK1	Pig	Na$^+$ dependent glucose uptake	[Hull, 1976; Saier, 1984]
Placenta		Human	hCG	[Yang, 1978]
Teratocarcinoma	Various	Mouse	Various	[Martin, 1975]
Myeloma	Various	Mouse	IGG	[Horibata and Harris, 70]
Pulmonary artery	CPAE	Cow	Factor VIII, ACE	[Del Vecchio and Smith, 1981]
Foreskin	Melanocytes	Human	Melanin	[Gilchrest, 1984, 1985 (see Chapter 20)]
Marrow	WEHI-3B D+	Mouse	Morphology	[Metcalf and Nicola, 1982]
Adrenal cortex		Cow	Steroids	[Simonian and White, 1984]
Breast	MCF-7	Human	Domes, α-lactalubumin.	[Soule et al., 1973]
Spleen	CT11-2, HT-1	Mouse	IL-2	[Gillis and Smith, 1977a,b]

*CK, creatinine phosphokinase; NTB, neotetrazolium blue; GFAP, glial fibrillary acidic protein; GPDH, glycerol phosphate dehydrogenase; hCG, human chorionic gonadotropin; IGG, immunogammaglobulin; ACE, angiotensin II converting enzyme; IL-2, interleukin 2.

ferentiation implies that the specialized properties of the cell are lost irreversibly, e.g., a hepatocyte would lose its characteristic enzymes (arginase, aminotransferases, etc.), could not store glycogen or secrete serum proteins, and these properties could not be reinduced once lost. Deadaptation, on the other hand, implies that synthesis of specific products, or other aspects of specialized function, are under regulatory control by hormones, cell-cell interaction, cell-matrix interaction, etc., and can be reinduced, given that the correct conditions can be recreated. Taking the hepatocyte again as an example, Michalopoulos and Pitot [1975] and Sattler et al. [1978] have shown that induction of tyrosine aminotransferase in normal rat liver cells requires hormones (insulin, hydrocortisone) and the correct matrix interaction (collagen). It is gradually becoming apparent that, given the correct culture conditions, differentiated functions can be expressed by a number of different cell types (Table 2.2), and the concept of dedifferentiation is now regarded as an unlikely explanation for the loss of specialized functions.

For correct inducing conditions to act, the appropriate cells must be present. In early attempts at liver cell culture, lack of expression of hepatocyte properties was due partly to overgrowth of the culture by connective tissue fibroblasts or endothelium from blood vessels or sinusoids. By using the correct disaggregation technique [Berry and Friend, 1969] and the correct culture conditions [Michalopoulos and Pitot, 1975;

Guguen-Guillouzo et al., 1983], hepatocytes can be selected preferentially. Similarly, epidermal cells can be grown either by using a confluent feeder layer [Rheinwald and Green, 1975] or selective medium [Peehl and Ham, 1980; Tsao et al., 1982]. The appearance of other examples, e.g., feeder selection for breast and colonic epithelium [Freshney et al., 1981], D-valine for the isolation of kidney epithelium [Gilbert and Migeon, 1975], and the use of cytotoxic antibodies [Edwards et al., 1980] (selection procedures reviewed in Chapters 11, 12, and 20), clearly demonstrate that the selective culture of specialized cells is not the insuperable problem that it once appeared.

WHAT IS A CULTURED CELL?

The question remains open, however, as to the exact nature of the cells that grow in each case. Expression of differentiated markers under the influence of inducing conditions may either mean that the cells being cultured are mature and only require induction to maintain synthesis of specialized proteins, or that the culture is composed of precursor or stem cells which are capable of proliferation but remain undifferentiated until the correct inducing conditions are applied, whereupon some or all of the cells mature to differentiated cells. It may be useful to think of a cell culture as being in equilibrium between multipotent stem cells, undifferentiated but committed precursor cells, and mature differentiated cells (see Fig. 2.3) and that the equilibrium may shift according to the environmental conditions. Routine serial passage at relatively low cell densities would promote cell proliferation and little differentiation while high cell densities, low serum, and the appropriate hormones would promote differentiation and inhibit cell proliferation.

The source of the culture will also determine which cellular components may be present. Hence cell lines derived from the embryo may contain more stem cells and precursor cells and be capable of greater self-renewal than cultures from adults. In addition, cultures from tissues which are undergoing continuous renewal *in vivo* (epidermis, intestinal epithelium, hemopoietic cells) will still contain stem cells, which, under the appropriate culture conditions, may survive indefinitely, while cultures from tissue which renew only under stress (fibroblasts, muscle, glia) may only contain committed precursor cells with a limited culture life-span.

Thus, the identity of the cultured cell is not only defined by its lineage *in vivo* (hemopoietic, hepato-

cyte, glial, etc.) but also by its position in that lineage (stem cell, committed precursor cell, or mature differentiated cell). With the exception of mouse teratomas and one or two other examples from lower vertebrates, it seems unlikely that cells will change lineage (transdifferentiate), but they may well change position in the lineage, and may even do so reversibly in some cases.

When cells are cultured from a neoplasm, they need not adhere to these rules. Thus a hepatoma from rat may proliferate *in vitro* and still express some differentiated features, but the closer they are to the normal phenotype, the more induction of differentiation may inhibit proliferation. Although the relationship between position in the lineage and cell proliferation may become relaxed (though not lost; B16 melanoma cells still produce more pigment at high cell density and at a low rate of cell proliferation than at a low cell density and a high rate of cell proliferation), transfer between lineages has not been clearly established (see also Chapter 14).

FUNCTIONAL ENVIRONMENT

Since the inception of tissue culture as a viable technique, culture conditions have been adapted to suit two major requirements: (1) production of cells by continuous proliferation and (2) preservation of specialized functions. The upsurge of interest in cellular and molecular biology and virology in the 1950s and 1960s concentrated mainly on fundamental intracellular processes such as the regulation of protein synthesis, often requiring large numbers of cells. Later, the development of such techniques as molecular hybridization and gene transfer allowed the emphasis to shift to the study of the regulation of specialized functions. While the need for bulk cultures remains, more attention has been directed to the creation of an environment which will permit the controlled expression of differentiation.

It has been recognized for many years that specific functions are retained for longer where the three-dimensional structure of the tissue is retained, as in organ culture (see Chapter 22). Unfortunately, organ cultures cannot be propagated, must be prepared *de novo* for each experiment, and are more difficult to quantify than cell cultures. For this reason there have been numerous attempts to recreate three-dimensional structures by perfusing monolayer cultures [Kruse et al., 1970; Whittle and Kruse, 1973; Knazek et al., 1972; Knazek, 1974; Gullino and Knazek, 1979] and to reproduce elements of the environment *in vivo* by culturing cells on or in special matrices like collagen gel [Michalopoulos and Pitot, 1975; Yang et al., 1981; Burwen and Pitelka, 1980], cellulose [Leighton, 1951] or gelatin sponge [Douglas et al., 1976], or matrices from other natural tissue matrix glycoproteins such as fibronectin, chondronectin, and laminin [Gospodarowicz et al., 1980; Kleinman et al., 1981; Reid and Rojkind, 1979] (see Chapter 7). These techniques present some limitations, but with their provision of homotypic cell interactions, cell matrix interactions, and the possibility of introducing heterotypic cell interactions, they may hold considerable promise for the examination of tissue-specific functions.

The development of normal tissue functions in culture would facilitate investigation of pathological behavior such as demyelination and malignant invasion. But, from a fundamental viewpoint, it is only when cells *in vitro* express their normal functions that any attempt can be made to relate them to their tissue of origin. Expression of the differentiated phenotype need not be complete, since the demonstration of a single cell type-specific cell surface antigen may be sufficient to place a cell in the correct lineage. More complete functional expression may be required, however, to place a cell in its correct position in the lineage, and to reproduce a valid model of its function *in vivo*.

The major requirement that distinguishes tissue culture from most other laboratory techniques is the need to maintain asepsis. This is accentuated by the much slower growth of cultured animal cells relative to most of the major potential contaminants. The introduction of laminar flow cabinets has greatly simplified the problem and allows the utilization of unspecialized laboratory accommodation (see below and Chapters 4 and 5).

There are six main functions to be accommodated: sterile handling, incubation, preparation, wash-up, sterilization, and storage (Table 3.1). The clean area for sterile handling should be located at one end of the room and wash-up and sterilization at the other, with preparation, storage, and incubation in between. The preparation area should be adjacent to the wash-up and sterilization areas, and storage and incubators should be readily accessible to the sterile working area.

STERILE HANDLING AREA

This should be located in a quiet part of the laboratory, its use should be restricted to tissue culture, and there should be no through traffic or other disturbance likely to cause dust or drafts. Use a separate room or cubicle if laminar flow cabinets are not available. The work area, in its simplest form, should be a plastic laminate-topped bench, preferably plain white or neutral gray, to facilitate observation of cultures, dissection, etc., and enable accurate reading of pH when using phenol red as an indicator. Nothing should be stored on this bench and any shelving above should only be used in conjunction with sterile work, e.g., for holding pipette cans and instruments. The bench should either be freestanding (away from the wall) or sealed to the wall with a plastic sealing strip.

Laminar Flow

The introduction of laminar flow cabinets (or "hoods") with sterile air blown over the work surface (Fig. 3.1) (see Chapters 4 and 5) affords greater control of sterility at a lower cost than providing a separate sterile room. Individual cabinets are preferable as they separate operators and can be moved around, but laminar flow wall or ceiling units in batteries can be used (Fig. 3.2). With cabinets, only the operator's arms enter the sterile area, while with laminar flow wall or ceiling units, there is no cabinet and the operator is part of the work area. While this may give more freedom of movement, particularly with large pieces of apparatus (roller bottles, fermentors), greater care must be taken by the operator not to disrupt the laminar flow, and it may prove necessary to wear caps and gowns to avoid contamination.

Select cabinets that suit your accommodation—freestanding or bench top— and allow plenty of leg room underneath with space for pumps, aspirators, etc. Freestanding cabinets should be on casters so they can be moved if necessary. Select chairs which are of a suitable height, preferably with adjustable seat height and back angle, and make sure that they can be drawn up close enough to the front edge to allow comfortable working well within the cabinet. A small section of bench or folding flap (300–500 mm minimum) should be provided beside each hood for depositing apparatus or reagents not in immediate use.

It is a good principle, made easier by the introduction of laminar flow, to create a "sterility gradient" in the tissue culture laboratory. Hence, a single room housing all the necessary functions of a tissue culture laboratory should have its sterile cabinets located at one end, furthest from the door, while wash-up, preparation of glassware or reagents, centrifugation, etc., would be best performed at the opposite end of the room (Fig. 3.3, 3.4). This principle still applies where laminar flow is not available, in fact, even more so; but the introduction of laminar flow, particularly horizontal laminar flow, makes the gradient easier to maintain.

In addition, you may wish to provide a small room or cubicle for use as a containment area (Fig. 3.5).

15

TABLE 3.1. Tissue Culture Facilities

Minimum requirements (essential)	Desirable features (beneficial)	Useful additions
Sterile area: clean and quiet area, no through traffic	Filtered air (air conditioning	Piped CO_2 and compressed air
Separate from animal house and microbiological labs	Hot room with temperature recorder	Storeroom for bulk plastics
Preparation area	Microscope room	Containment room for biohazard work
Storage areas: liquids—ambient, 4°C, −20°C; glassware (shelving); plastics (shelving); small items (drawers); specialized equipment (slow turnover), cupboard(s); chemicals— ambient, 4°C, −20°C (share with liquids but keep chemicals in sealed container over desiccant)	Dark-room Service bench adjacent to culture area Separate prep room Separate sterilizing room Cylinder store	Liquid N_2 storage tank (~500 l)
Space for incubator(s)		
Space for liquid N_2 freezer(s)		
Wash-up area (not necessarily within tissue culture laboratory, but adjacent)		

This must be separated by a door or air lock from the rest of your suite, and will need its own incubators, freezer, refrigerator, centrifuge, etc. It will also require a biohazard cabinet or pathogen hood with separate extract and pathogen trap (for fuller description of containment facilities see Chapter 6).

INCUBATION

The requirement for cleanliness is not as stringent as with sterile handling, but clean air, low disturbance level, and no through traffic will endow your incubation area with a better chance of avoiding dust, spores, and drafts that carry them.

Incubation may be carried out in separate incubators or in a thermostatically controlled hot room. Incubators, bought singly, are inexpensive and economic in space; but as soon as you require more than two, their cost is more than a simple hot room and their use less convenient. Incubators also lose more heat when opened and are slower to recover than a hot room. As a rough guide, you will need 0.2 m³ (200 l, 6 ft³) of incubation space (0.5 m², 6 ft² shelf space) per person. Extra provision may need to be made for a humid incubator(s) with a controlled CO_2 level in the atmosphere (see Chapter 4).

Hot Room

If you have the space within the laboratory area or have an adjacent room readily available and accessible, it may be possible to convert this into a hot room (Fig. 3.6). It need not be specifically constructed but should be sufficiently well insulated not to allow "cold

spots" on the walls. If insulation is required, line with plastic laminate-veneered board, separated from the wall by about 5 cm (2 in) fiberglass, mineral wool, or fire-retardant plastic foam. Mark the location of the straps or studs carrying the lining panel to identify anchorage points for shelving. Shelf supports should be spaced at 500–600 mm (21 in) to support shelving without sagging.

Do not underestimate the space that you will require in the lifetime of the hot room. It costs very little more to equip a large hot room than a small one. Calculate on the basis of the amount of shelf space you will require; if you have just started, multiply by five or ten; if you have been working for some time, by two or four. Allow 200–300 mm (9 in) between shelves and use wider shelves (450 mm, 18 in) at the bottom, and narrower (250–300 mm, 12 in) above eye level. Slatted shelving (mounted on adjustable brackets) should be used to allow for air circulation. It must be flat and perfectly horizontal with no bumps or irregularities.

Wooden furnishings should be avoided as much as possible as they warp in the heat and can harbor infestations.

A small bench, preferably stainless steel or solid plastic laminate, should be provided at some part of the hot room. This should accommodate a microscope, its transformer, and the flasks that you wish to examine. If you contemplate doing cell synchrony experiments or having to make any sterile manipulations at 36.5°C, you should also allow space for a small laminar flow unit (300 × 300 or 450 × 450 mm, 12–18 in

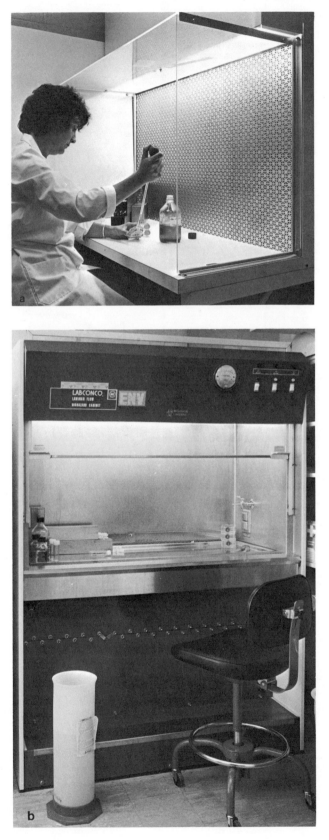

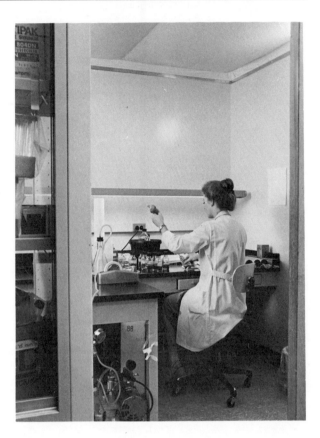

Fig. 3.2. *Aseptic room. Filtered air is supplied from the ceiling, and the whole room is regarded as a sterile working area.*

square, filter size) either wall mounted or on a stand over part of the bench. Alternatively a small laminar flow cabinet (not more than 1,000 mm, 3 ft, wide) could be located in the room. The fan motor should be tropically wound and should not run continuously. Apart from wear of the motor, it will generate heat in the room and the motor may burn out.

Incandescent lighting is preferable to fluorescent, which can cause degradation of constituents of the medium. Furthermore, some fluorescent tubes have difficulty in striking in a hot room.

The temperature should be controlled within $\pm 0.5°C$ at any point and at any time. This depends on: (1) the sensitivity and accuracy of the control gear; (2) the siting of the thermostat sensor; (3) the circulation of air in the room; (4) correct insulation; and (5) the evolution of heat by other apparatus (stirrers, etc.) in the room

Heaters. Heat is best supplied via a fan heater, domestic or industrial, depending on the size of the room. Approximately 2–3 kW per 20 m^3 (700 ft^3) will be required (or two 1.0–1.5 kW) depending on the insulation. The fan on the fan heater should run continuously, and the power to the heating element should come from a proportional controller (see below).

Fig. 3.1. *Laminar flow cabinets. a. horizontal flow. b. vertical flow (biohazard type).*

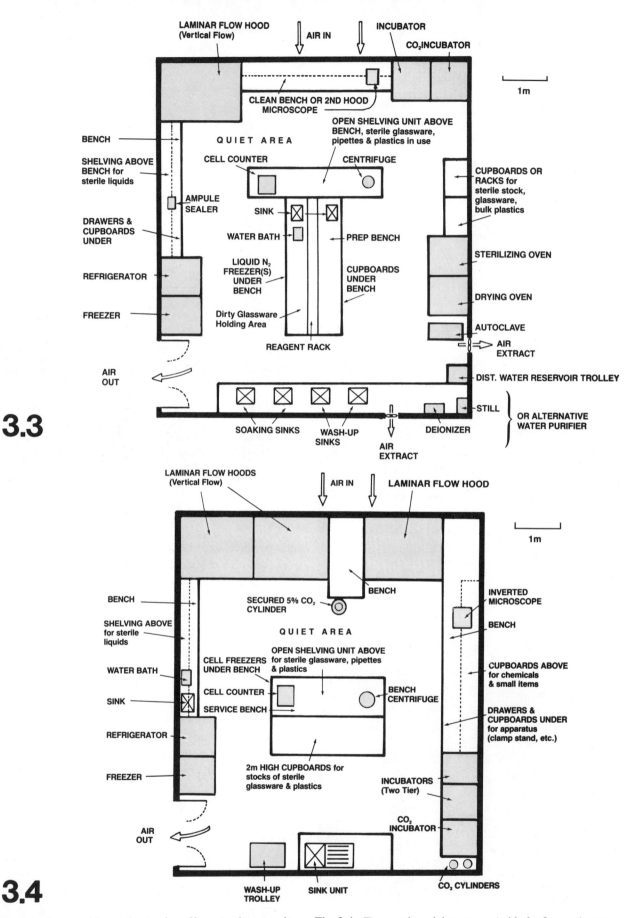

Fig. 3.3. *Suggested layout for simple, selfcontained tissue culture laboratory for use by two or three persons. Shaded areas represent movable equipment.*

Fig. 3.4. *Tissue culture laboratory suitable for five or six persons with washing-up and preparation facility located elsewhere. Shaded areas represent movable equipment.*

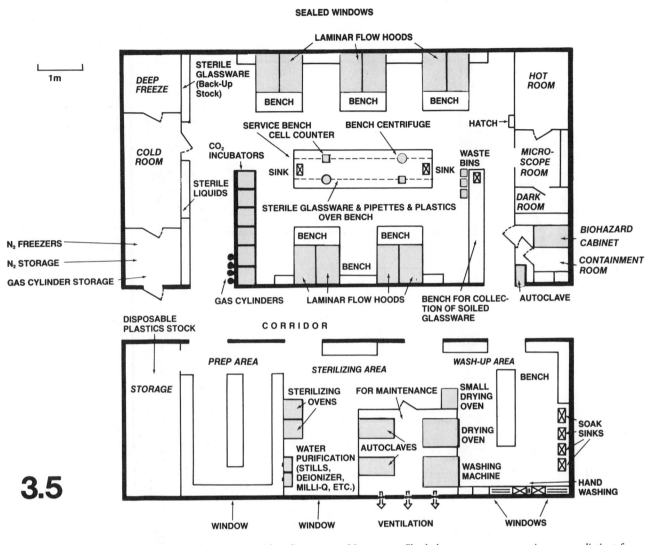

Fig. 3.5. *Large scale tissue culture laboratory with adjacent washing-up, sterilization, and preparation area. Suitable for 20 to 30 persons. Shaded areas represent equipment as distinct from furniture.*

Air circulation. A second fan, positioned on the opposite side of the room, and with the air flow opposing that of the fan heater, will ensure maximum circulation. If the room is more than 2 m × 2 m (6 ft × 6 ft) some form of ducting may be necessary. Blocking off the corners as in Figure 3.6 is often easiest and most economical in space in a square room. In a long rectangular room, a false wall may be built at either end, but be sure to insulate it from the room and make it strong enough to carry shelving.

Thermostats. Thermostats should be of the "proportional controller" type acting via a relay to supply heat at a rate proportional to the difference between the room temperature and the set point. So, when the door opens and the room temperature falls, recovery is rapid; but the temperature does not overshoot, as the closer the room temperature approaches the set point, the less heat is supplied.

If possible, dual thermostats in parallel, but preferably controlling separate heaters, should be installed so that if one fails it is overridden by the other (Fig. 3.7). One thermostat ("regulating") is set at the required temperature, 36.5°C, with a narrow fluctuation range, say 0.4°C (±0.2°C). The second, or safety, thermostat is set slightly below the first so that if the temperature falls below the range controlled by the first, the second will be activated. Pilot lights in series with each circuit will indicate when they are activated. Finally, there should be an overriding thermostat in series (on both heaters if two are installed) set at 38.5°C, so that if a thermostat locks on, the override thermostat will cut out at 38.5°C and illuminate a warning light (Fig. 3.8).

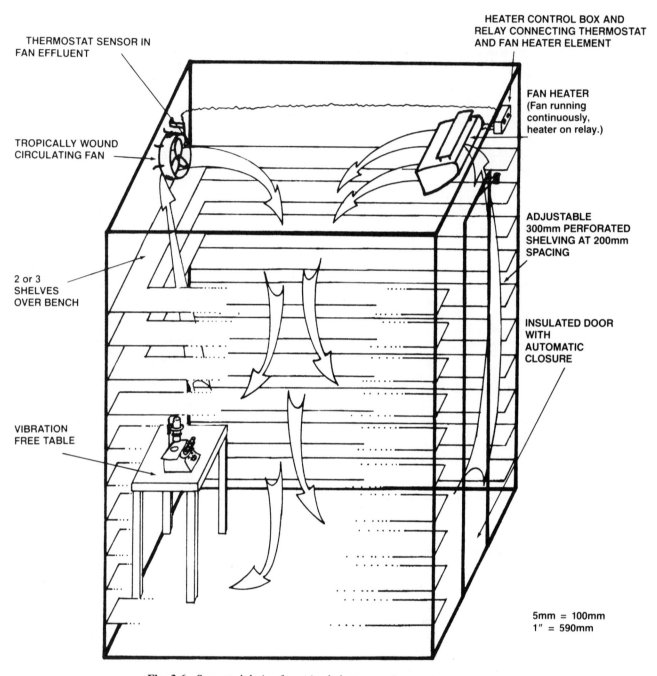

HEATER CONTROL BOX AND
RELAY CONNECTING THERMOSTAT
AND FAN HEATER ELEMENT

THERMOSTAT SENSOR IN
FAN EFFLUENT

FAN HEATER
(Fan running
continuously,
heater on relay.)

TROPICALLY WOUND
CIRCULATING FAN

ADJUSTABLE
300mm PERFORATED
SHELVING AT 200mm
SPACING

2 or 3
SHELVES
OVER BENCH

INSULATED DOOR
WITH
AUTOMATIC
CLOSURE

VIBRATION
FREE TABLE

5mm = 100mm
1″ = 590mm

Fig. 3.6. *Suggested design for a simple hot room. Arrows represent air circulation (based on an original design by Dr. John Paul).*

The thermostat sensors should be located in an area of rapid air flow close to the effluent from the second, circulating, fan for greatest sensitivity. A rapid response, high thermal conductivity sensor (thermistor or thermocouple) should be used in preference to a pressure bulb type.

Overheating. Since so much care is taken to provide heat and replenish its loss rapidly, another problem is often forgotten—namely, unwanted heat gain. This can arise (1) because of a rise in ambient temperature in the laboratory in hot weather or (2) due to heat produced from within the hot room by apparatus such as stirrer motors, roller racks, laminar flow units, etc.

Try to avoid heat-producing equipment in the hot-room, and arrange for heat dissipation either by a thermostatically controlled fan extract (and inlet) or an

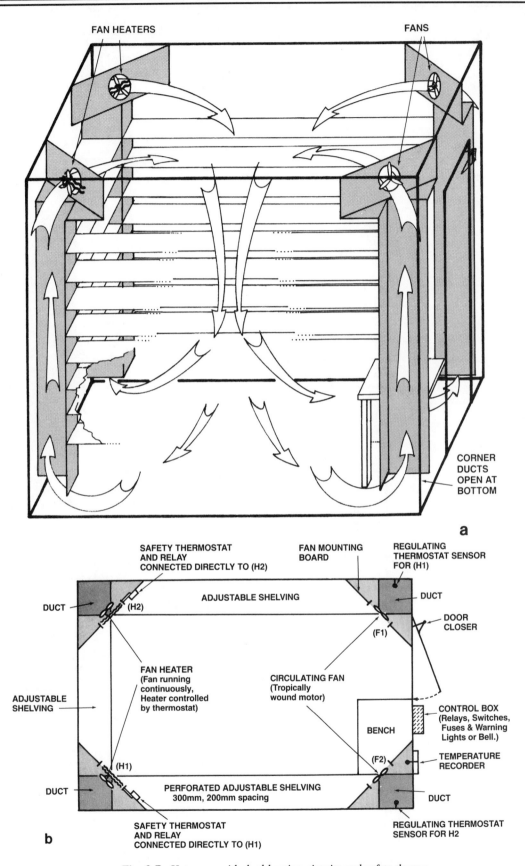

FAN HEATERS

FANS

CORNER
DUCTS
OPEN AT
BOTTOM

a

SAFETY THERMOSTAT
AND RELAY
CONNECTED DIRECTLY TO (H2)

FAN MOUNTING
BOARD

REGULATING
THERMOSTAT SENSOR
FOR (H1)

DUCT

(H2)

ADJUSTABLE SHELVING

DUCT

(F1)

DOOR
CLOSER

FAN HEATER
(Fan running
continuously,
Heater controlled
by thermostat)

CIRCULATING FAN
(Tropically
wound motor)

ADJUSTABLE
SHELVING

CONTROL BOX
(Relays, Switches,
Fuses & Warning
Lights or Bell.)

BENCH

(H1)

(F2)

TEMPERATURE
RECORDER

DUCT

PERFORATED ADJUSTABLE SHELVING
300mm, 200mm spacing

DUCT

b

SAFETY THERMOSTAT
AND RELAY
CONNECTED DIRECTLY TO (H1)

REGULATING THERMOSTAT
SENSOR FOR H2

Fig. 3.7. *Hot room with dual heating circuits and safety thermostats. a. Oblique view. b. Plan view. Arrows represent air circulation.*

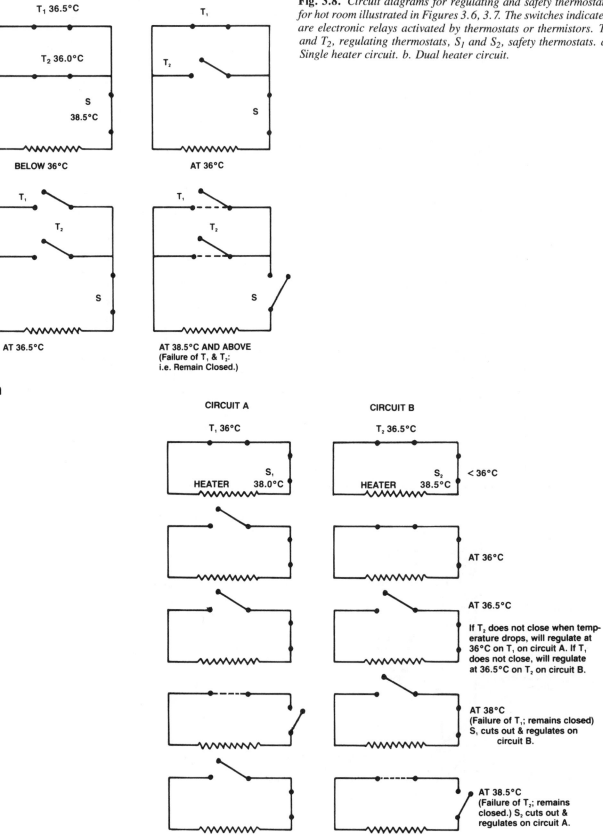

Fig. 3.8. *Circuit diagrams for regulating and safety thermostats for hot room illustrated in Figures 3.6, 3.7. The switches indicated are electronic relays activated by thermostats or thermistors. T_1 and T_2, regulating thermostats, S_1 and S_2, safety thermostats. a. Single heater circuit. b. Dual heater circuit.*

T_1 36.5°C

T_2 36.0°C

S
38.5°C

BELOW 36°C

T_1

T_2

S

AT 36°C

T_1

T_2

S

AT 36.5°C

T_1

T_2

S

AT 38.5°C AND ABOVE
(Failure of T_1 & T_2:
i.e. Remain Closed.)

a

CIRCUIT A

T_1 36°C

S_1
HEATER   38.0°C

CIRCUIT B

T_2 36.5°C

S_2
HEATER   38.5°C   < 36°C

AT 36°C

AT 36.5°C

If T_2 does not close when temperature drops, will regulate at
36°C on T_1 on circuit A. If T_1
does not close, will regulate
at 36.5°C on T_2 on circuit B.

AT 38°C
(Failure of T_1; remains closed)
S_1 cuts out & regulates on
circuit B.

AT 38.5°C
(Failure of T_2; remains
closed.) S_2 cuts out &
regulates on circuit A.

T_1 & T_2: REGULATING THERMOSTATS
S_1 & S_2: CUT OUTS

b

air conditioner. In either case, set the thermostat well below ($>2°C$) the heater thermostats so that the latter will regulate the temperature.

Access. If a proportional controller, good circulation, and adequate heating are provided, an air lock will not be required. The door should still be well insulated (foam plastic or fiberglass filled), light, and easily closed, preferably self-closing. It is also useful to have a hatch leading into the tissue culture area, with a shelf on both sides, so that cultures may be transferred easily into the room. The hatch door should also have an insulated core. If the hatch is located above the bench, this will avoid any risk of a "cold spot" on the shelving.

A temperature recorder should be installed such that the chart is located in easy and obvious view of the people working in the tissue culture room. A weekly change of chart is convenient and still has sufficient resolution. If possible, one high-level and one low-level warning light should be placed beside the chart or at a different, but equally obvious, location.

SERVICE BENCH

It may be convenient to position a bench to carry cell counter, microscope, etc., close to the sterile handling area, either dividing the area, or separating it from the other end of the lab (see Figs. 3.3, 3.4, 3.5). The service bench should also have provision for storage of sterile glassware, plastics, pipettes, screw caps, syringes, etc., in drawer units below and shelves above. This bench may also house other accessory equipment such as an ampule sealing device and a small bench centrifuge. The bench should provide a close supply of all the immediate requirements.

PREPARATION

The need for extensive media preparation in small laboratories can be avoided if there is a proven source of reliable commercial culture media. While a large laboratory (~ 50 people doing tissue culture) may still find it more economic to prepare their own media, most smaller enterprises may prefer to purchase ready-made media. This reduces preparation to reagents, such as salt solutions, EDTA, etc., bottling these and water, and packaging screw caps and other small items for sterilization. While this area should still be clean and quiet, sterile handling is not necessary as all the items will be sterilized.

If there is difficulty in obtaining reliable commercial media, a larger area should be allocated for prepara-

tion to accommodate a coarse and fine balance, pH meter, and, preferably, an osmometer. Bench space will be required for dissolving and stirring solutions, and for bottling and packaging. If possible, an extra horizontal laminar flow cabinet should be provided in the sterile area for filtering and bottling sterile liquids, and incubator space must be allocated for quality control of sterility, i.e., incubation of samples of media in broth and after plating out.

Heat-stable solutions and equipment can be autoclaved or dry-heat sterilized at the nonsterile end of the area. Both streams then converge on the storage areas (see below).

WASH-UP

If possible wash-up and sterilization facilities should be situated outside the tissue culture lab as the humidity and heat that they produce may be difficult to dissipate without increasing air flow above desirable limits. Autoclaves ovens, and distillation apparatus should be located in a separate room, if possible (see Fig. 3.5), with an efficient extraction fan. The wash-up area should have plenty of space for soaking glassware, and space for an automatic washing machine, should you require one. There should also be plenty of bench space for handling baskets of glassware, sorting pipettes, packaging and sealing sterile packs, and a pipette washer and drier. If the sterilization facilities must be located in the tissue culture lab, site them where greatest ventilation is possible and furthest from the sterile handling area.

Trolleys are often useful for collecting dirty glassware and redistributing fresh sterile stocks, but remember to allocate parking space for them.

STORAGE

Storage must be provided for: (1) sterile liquids (a) at room temperature (salt solutions, water, etc.), (b) at $4°C$ (media), and (c) at $-20°C$ or $-70°C$ (serum, trypsin, glutamine, etc.); (2) sterile glassware: (a) media bottles, glass culture flasks and (b) pipettes; (3) sterile disposable plastics: (a) culture flasks and petri dishes, (b) centrifuge tubes and vials, and (c) syringes; (4) screw caps, filter tubes, stoppers, etc.; (5) apparatus: filters and large receiver flasks; and (6) gloves, disposal bags etc. All should be within easy reach of the sterile working area. Refrigerators and freezers should be located toward the nonsterile end of the lab as the doors and compressor fans create dust and drafts and they may harbor fungal spores. Also, they require maintenance and periodic defrosting, which creates a level and kind of activity best separated from your sterile working area.

The keynote of storage areas is ready access both for withdrawal and replenishment. Double-sided units are useful because they may be restocked from one side and used from the other. Storage boxes or trays which can be taken away and filled and replaced when full are also useful.

Remember to allocate sufficient space for storage as this will allow you to make bulk purchases, and thereby save money, and at the same time reduce the risk of running out of valuable stocks at times when they cannot be replaced. As a rough guide, you will need 200 l (~ 8 ft^3) of 4°C storage and 100 l (~ 4 ft^3) of -20°C storage per person. The volume per person increases with fewer people. Thus, one person may need a 250-l (10-ft^3) fridge and a 150-l (6-ft^3) freezer. This refers to storage space only; and allowance must be made for access for working in the cold room where walk-in cold rooms and deep freezers are planned.

In general, separate -20°C freezers are better than a walk-in cold room. They are easier to clean out and maintain, and they provide better backup if one unit fails.

CONSTRUCTION AND LAYOUT

The rooms should be supplied with air filtered to usual industrial or office standards and be designed for easy cleaning. Furniture should fit tight to the floor or be suspended from the bench allowing space to clean underneath. Cover the floor with vinyl or other dust-proof finish and allow a slight fall in the level toward a floor drain located in the center of the room. This allows liberal use of water if the floor has to be washed, but, more important, it protects equipment from damaging floods if stills, autoclaves or sinks overflow.

If the tissue culture lab and preparation, wash-up, and sterilization areas can be separated, so much the better. Adequate floor drainage should still be provided in both areas although clearly the wash-up and sterilization area will be most important. If you have a separate wash-up and sterilization facility, it will be convenient to have this on the same floor and adjacent, with no steps to negotiate, so that trolleys may be used. Across a corridor is probably ideal (see Fig. 3.5) (see next chapter for sinks, soaking baths, etc.).

Try to imagine the flow of traffic—people, reagents, wash-up, trolleys, etc.—and arrange for minimum conflict, easy and close access to stores, and easy withdrawal of soiled items. Make sure your doors are wide enough and high enough to allow entry of all the equipment you want, particularly laminar flow units, and allow space for maintenance.

Inevitably space will be the first problem and some degree of compromise will be inevitable, but a little thought ahead of time can save much space and ultimately people's tempers.

Chapter 4
Equipping the Laboratory

The specific needs of a tissue culture laboratory, like most labs, can be divided into three categories: (1) essential—you cannot perform a job without them; (2) beneficial—the work would be done better, more efficiently, quicker, or with less labor; and (3) useful—it would make life easier, improve working conditions, reduce fatigue, enable more sophisticated analyses to be made, or generally make your working environment more attractive.

Equipment that might be used in tissue culture is listed in Table 4.1 in the grouping suggested above. Remember two main points: assuming you need it and can afford it, you must be able to get it into the room (access) and you must have space for it (accommodation). Suppliers of items of equipment will be found in the "Trade Index—Sources of Material" at the end of the book.

ESSENTIAL EQUIPMENT

Incubator

This should be large enough, probably 200 1 (6 ft^3) per person, have forced air circulation, temperature control $\pm$ 0.5°C, and a safety thermostat which cuts out if the incubator overheats or, better, which regulates it if the first fails. It should be corrosion resistant, e.g., stainless steel (anodized aluminum is acceptable for a dry incubator), and easily cleaned. A double cabinet, one above the other, independently regulated, gives you more accommodation with the added protection that if one-half fails the other can still be used. This is also useful when you need to clean out one compartment.

Incubation Temperature

The optimal temperature for cell culture is dependent on: (1) the body temperature of the animal from which the cells were obtained; (2) any regional variation in temperature (e.g., skin may be lower); and (3) the incorporation of a safety factor to allow for minor errors in incubator regulation. Thus, the temperature

recommended for most human and warm-blooded animal cell lines is 36.5°C, close to body heat but set a little lower for safety.

Avian cells, because of the higher body temperature in birds, should be maintained at 38.5°C for maximum growth but will grow quite satisfactorily, if more slowly, at 36.5°C.

Cultured cells will tolerate considerable drops in temperature, can survive several days at 4°C and can be frozen and cooled to -196°C (see Chapter 17), but they cannot tolerate more than about 2°C above normal (39.5°C) for more than a few hours, and will die quite rapidly at 40°C and over.

Epidermal cells from the mouse may grow better at a slightly lower temperature, e.g., 33°C.

In general the cells of poikilothermic animals have a wide temperature tolerance but should be maintained at a constant level within the normal range of the donor species. This requires incubators with cooling as well as heating as the incubator temperature may need to be below ambient (e.g., for fish). Cooling capacity should be sufficient to lower the temperature about 2°C, or more, below ambient so that regulation is performed by the heater circuit, which is more sensitive.

If necessary, poikilothermic animal cells can be maintained at room temperature, but the variability of the ambient temperature in laboratories makes this undesirable.

Regulation of temperature should be kept within ±0.5°C; consistency is more important than accuracy. Cells will grow quite well between 33° and 39°C but will naturally vary in growth rate and metabolism. The incubation temperature should be kept constant both in time and at different parts of the incubator. Water baths give the most accurate control of temperature, but present problems of contamination, particularly since the flasks need to be immersed for proper temperature control. They are, therefore, seldom used and incubators are preferable. The air should be circulated

TABLE 4.1. Tissue Culture Equipment

Minimum requirements (essential)	Desirable features (beneficial)	Useful additions
Incubator	Laminar flow hood(s), vertical, horizontal, biohazard	$-70°C$ freezer
Sterilizer (autoclave, pressure cooker, oven)	Cell counter	Glassware washing machine
Refrigerator	Vacuum pump	Closed-circuit TV for inverted microscope(s)
Freezer (for $-20°C$ storage)	CO_2 incubator	Colony counter
Inverted microscope	Coarse and fine balance	High-capacity centrifuge (6 × 1 liter)
Soaking bath or sink	pH meter	Cell sizer (e.g., Coulter ZB series)
Deep washing sink	Osmometer	Time-lapse cinemicrographic equipment
Pipette cylinder(s)	Phase-contrast and fluorescence microscope(s)	Interference-contrast microscope
Pipette washer	Portable temperature recorder	Polythene bag sealer (for packaging sterile items for long-term storage)
Still or water purifier	Permanent temperature recorders on sterilizing oven and autoclave	Controlled-rate cooler (for cell freezing)
Bench centrifuge	Roller racks for roller bottle culture	Filing for freezer records and catalogues
Liquid N_2 freezer (~35 1, 1,500–3,000 ampules)	Magnetic stirrer racks for suspension cultures	Centrifugal elutriator centrifuge and rotor
Liquid N_2 storage flask (~25 1)	Pipette drier	Fluorescence-activated cell sorter
	Pipette plugger	Densitometer
	Trolleys for collecting soiled glassware and redistributing fresh supplies	Density meter (for density gradient cell separation)
	Pipette aid	
	Autopipette or other form of automatic dispensor, dilutor	
	Separate sterilizing oven and drying oven	

by a fan to give even temperature distribution, and cultures should be placed on perforated shelves and not on the floor or touching the sides of the incubator. Further discussion of temperature control in hot rooms is given in Chapter 3.

Sterilizer

The simplest of these is a domestic pressure cooker which will generate 1 atm (15 lb/in^2) above ambient. More complex autoclaves exist, but the main consideration is the capacity; will it accommodate all you want to do? A simple bench top autoclave (Fig. 4.1a) may be sufficient but a larger model with a timer and a choice of presterilization and poststerilization evacuation (Fig. 4.1b) will give more capacity and greater flexibility in use. A "wet" cycle (water, salt solutions, etc.) is performed without evacuation before or after sterilization. Dry items (instruments, swabs, screw caps, etc.) require the chamber to be evacuated before sterilization, to allow efficient access of hot steam, and should be evacuated after sterilization to remove steam and promote subsequent drying; otherwise the articles will emerge wet, leaving a trace of contamination from the condensate on drying. To minimize this risk always use deionized or reverse osmosis water to supply the autoclave.

If you require a high sterilization capacity (300 l, 9 ft^3, or more), buy two smaller autoclaves rather than one large one, so that during routine maintenance and accidental breakdowns you still have one functioning machine. Furthermore, a smaller machine will heat up and cool more quickly and can be used more economically for small loads. Leave sufficient space around them for maintenance and ventilation and provide adequate air extraction to remove heat and steam.

Refrigerators and Freezers

Usually a domestic item will be found to be quite efficient and cheaper than special laboratory equip-

Fig. 4.1. *Autoclaves. a. Bench-top model. b. Large, recessed model.*

ment. Domestic refrigerators are available with no ice box ("larder refrigerators") giving more space and eliminating the need for defrosting. However, if you require a lot of accommodation (400 l, 12 ft³, or more; see Chapter 3), a large hospital (blood bank) or catering freezer may be better. While autodefrost freezers may be bad for some reagents (enzymes, antibiotics, etc.), they are very useful for most tissue culture stocks where their bulk and nature precludes severe cryogenic damage. Conceivably, serum could deteriorate during oscillations in the temperature of an autodefrost freezer, but in practice it does not seem to. Many of the essential constituents of serum are small proteins, polypeptides, and simpler organic and inorganic compounds which may be insensitive to cryogenic damage.

Microscope

It cannot be overstressed that, in spite of considerable and highly desirable progress toward quantitative analysis of cultured cells, it is still vital to look at them regularly. A morphological change is often the first sign of deterioration in a culture (see Chapter 10) and the characteristic pattern of microbiological infection (see Chapter 16) is easily recognized.

A simple inverted microscope is essential (Fig. 4.2). Make certain that the stage is large enough to accommodate large roller bottles (see Chapter 23) in case you should require them. There are many simple and inexpensive inverted microscopes on the market; but if you foresee the need for photography of living cultures, then you should invest in one with high-quality optics, a long working distance phase-contrast condenser and objectives, with provision to take a camera (e.g., Nikon, Zeiss).

Washing-up Equipment

Soaking baths or sinks. Soaking baths or sinks should be deep enough so that all your glassware (except pipettes and large aspirators) can be totally immersed in detergent during soaking, but not so deep that the weight of glass is sufficient to break smaller

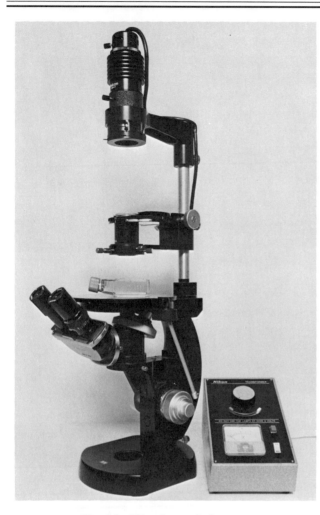

Fig. 4.2. *Nikon inverted microscope.*

items at the bottom, e.g., 400 mm (15 in) wide × 600 mm (24 in) long × 300 mm (12 in) deep.

If you are designing a lab from scratch, then you can get sinks built in of the size that you want. Stainless steel or polypropylene are best, the former if you plan to use radioisotopes and the latter for hypochlorite disinfectants.

Washing sinks should be deep enough (450 mm, 18 in) to allow manual washing and rinsing of your largest items without having to stoop too far to reach into them, and about 900 mm (3 ft) from floor to rim (Fig. 4.3). It is better to be too high than too low; a short person can always stand on a raised step to reach a high sink but a tall person will always have to bend down if the sink is too low. There should be a raised edge around the top of the sink to contain spillage and prevent the operator getting wet when bending over the sink. The raised edge should go around behind the taps at the back.

Each washing sink will require four taps: a single cold water, combined hot/cold mixer, a cold hose connection for a rinsing device, and a nonmetallic tap for deionized water, from a reservoir above the sink (see Fig. 4.3). A centralized supply for deionized water should be avoided as the pipework can build up dirt and algae and is difficult to clean.

Pipette cylinders. These should be made from polypropylene and be freestanding, distributed around the lab, one per work station.

Pipette washer. Following an overnight soak in detergent, reusable pipettes are easily washed in a standard siphon-type washer (see Chapter 8 and Fig. 8.4). This should be placed at floor rather than bench level to avoid awkward lifting of the pipettes and connected to the deionized water supply so that the final rinse can be done in deionized water. If possible a simple changeover valve should be incorporated into the deionized water feed line (see Fig. 4.3).

Pipette drier. If a stainless steel basket is used in the washer, this may then be transferred directly to an electric drier. Alternatively, pipettes can be dried on a rack or in a regular drying overn.

Sterilizing and Drying Oven

Although all sterilizing can be done in an autoclave, it is preferable to sterilize pipettes and other glassware by dry heat, avoiding the possibility of chemical contamination from steam condensate or corrosion of pipette cans. This will require a high-temperature (160–180°C) fan-powered oven to ensure even heating throughout the load. As with autoclaves, do not get an oven that is too big for the size of glassware that you use. It is better to use two small ovens than one big one; heating is easier, more uniform, quicker, and more economical when only a little glassware is being used. You are also better protected during breakdowns.

Water Purification

Pure water is required for rinsing glassware, dissolving powdered media, or diluting concentrates. The first of these is usually satisfied by deionized water but the second and third require a higher degree of purity, demanding a three- or four-stage process (Fig. 4.4). The important principle is that each stage is qualitatively different; reverse osmosis may be followed by charcoal filtration, deionization, and micropore filtration (e.g., Millipore) or distillation (with a silica sheathed cement) may be substituted for the first stage. Reverse osmosis is cheaper if you pay the fuel bills

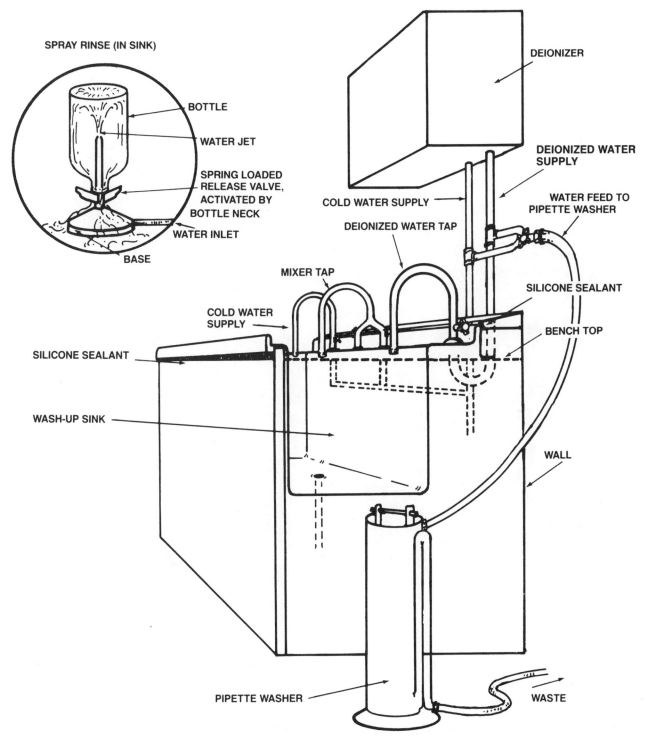

SPRAY RINSE (IN SINK)

BOTTLE

WATER JET

SPRING LOADED RELEASE VALVE, ACTIVATED BY BOTTLE NECK

WATER INLET

BASE

DEIONIZER

DEIONIZED WATER SUPPLY

COLD WATER SUPPLY

WATER FEED TO PIPETTE WASHER

DEIONIZED WATER TAP

SILICONE SEALANT

MIXER TAP

BENCH TOP

COLD WATER SUPPLY

SILICONE SEALANT

WASH-UP SINK

WALL

PIPETTE WASHER

WASTE

SCALE 1cm = 100mm

Fig. 4.3. *Washing-up sink and pipette washer, drawn to scale (bench height, 900 mm). Inset: bottle rinsing device, located in sink.*

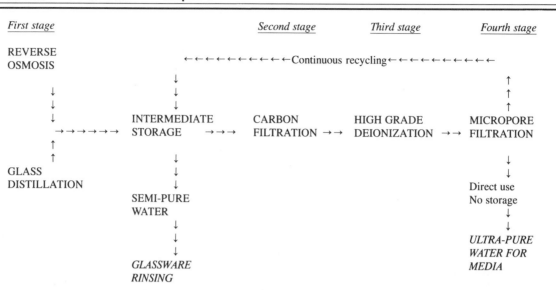

First stage   *Second stage*   *Third stage*   *Fourth stage*

REVERSE
OSMOSIS
    ←←←←←←←←←Continuous recycling←←←←←←←←←←←

INTERMEDIATE   CARBON   HIGH GRADE   MICROPORE
STORAGE   →→→   FILTRATION →→   DEIONIZATION →→   FILTRATION

GLASS
DISTILLATION

SEMI-PURE
WATER
                                                    Direct use
                                                    No storage

GLASSWARE   *ULTRA-PURE*
RINSING   *WATER FOR*
                                                   MEDIA

Fig. 4.4A. *Preparation of ultra-pure water. First stage can be either reverse osmosis or distillation. Ultra-pure water is not stored but continuously recycled to intermediate store. Water collected first thing in the morning will be maximum purity.*

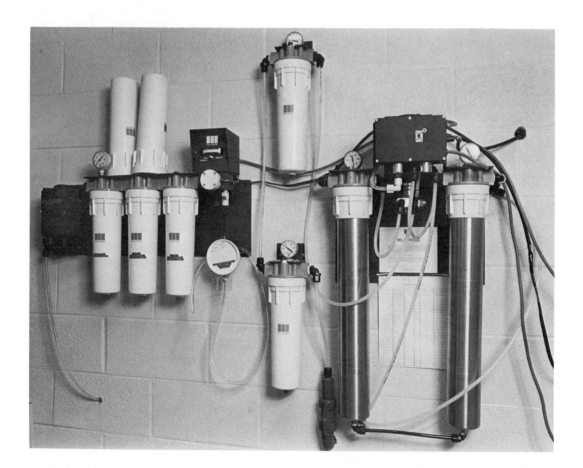

Fig. 4.4B. *Suggested layout for high purity water supply. Tap water is fed through reverse osmosis or glass distillation to a storage container. This semi-purified water is then recycled via carbon filtration, deionization, and micropore filtration back to the storage container. Reagent quality water is available at all times from storage; media quality water is available from the micropore filter supply (right of diagram). If the apparatus recycles continuously then the highest purity water will be collected first thing in the morning for the preparation of medium.*

but if not, distillation is better and more likely to give a sterile product. If reverse osmosis is used, the type of cartridge should be chosen to suit the pH of the water supply.

The deionizer should have a conductivity meter monitoring the effluent to indicate when the cartridge must be changed. Other cartridges should be dated and replaced according to the manufacturer's instructions.

Purified water should not be stored but recycled through the apparatus continually to preclude algal infection. Any tubing or reservoirs in the system should be checked regularly (every 3 months or so) for algal infection, cleaned out with hypochlorite and detergent (e.g., Chloros), and thoroughly rinsed in purified water before reuse.

Water is the simplest but probably the most critical constituent of all media and reagents, particularly with serum-free media (see Chapter 7).

Centrifuge

Periodically, cell suspensions require centrifugation to increase the concentration of cells or to wash off a reagent. A small bench-top centrifuge, preferably with proportionally controlled braking, is sufficient for most purposes. Cells sediment satisfactorily at 80–100g; higher g may cause damage and promote agglutination of the pellet. A large-capacity refrigerated centrifuge, 4 × 1 liters or 6 × 1 liters, will be required, if large scale suspension cultures (see Chapter 23) are contemplated.

Cell Freezing

The procedures for cell freezing will be dealt with in detail elsewhere (Chapter 17), but the basic facilities should be considered here. The freezing process can be carried out satisfactorily without sophisticated equipment, but storage requires a properly constructed liquid N_2 freezer and storage dewar (see Fig. 17.2). Freezers range in size from around 25 l to 500 l, i.e., 250–15,000 1-ml ampules. It is best to freeze a minimum of five ampules for each cell strain, 20 for a commonly used strain, and 100 for one in continuous use. A capacity of 1,200–1,500 is appropriate for most small laboratories.

The choice of freezer is determined by three factors: (1) capacity (no. of ampules); (2) economy and static holding time (the time taken for all the liquid N_2 to evaporate)—both of which are governed by the evaporation rate, which in turn is dependent on the frequency of access; and (3) convenience of access. Gen-

erally speaking, 2 is inversely proportional to 1 and 3. There are two main types of freezer: narrow-necked with slow evaporation but with more difficult access, and wide-necked with easier access but three times the evaporation rate (see Fig. 17.8). If the cost of liquid nitrogen and its supply presents no problem, then a wide-necked freezer may be more convenient (e.g., Union Carbide LR40 or equivalent, 3,000 ampules capacity, 3–5 l/d evaporation) although the holding time will only be about 1 wk to 10 d. A narrow-necked freezer, on the other hand, will be more economical and last up to 2 months if N_2 supplies run out (e.g., L'aire Liquide 35-1, 1,500 ampules capacity, 0.5 l/d evaporation).

If you require bulk storage (~ 10,000 ampules), then you will need to consider a vessel of around 300 l capacity. Wide-necked freezers are most common in this size because of the mechanical difficulties in operating narrow-necked freezers of high capacity, but the latter are available and will save a considerable amount in expenditure on liquid N_2. At 300 l the evaporation rate is approximately 10 l/d in a wide-necked freezer.

The advantages of gas-phase and liquid-phase storage will be discussed elsewhere (Chapter 17), but one major implication of storing in the gas phase is that the liquid phase is necessarily reduced to the space below your ampule storage area, usually 20–30% of the full volume. Hence, the static holding time is reduced to one-third or one-fifth of that of the filled freezer, filling must be carried out more regularly, and the chances of accidental thawing are increased. Where the investment is higher (many ampules or rare cell strains) automatic alarm systems should be fitted and, for the high-capacity freezers, an automatic filling system is recommended. However, automatic systems can fail and a twice-weekly check of liquid levels with a dipstick should be maintained and a record kept.

An appropriate storage vessel should also be purchased to enable a backup supply of liquid N_2 to be held. The size of this depends: (1) on the size of the freezer; (2) the frequency and reliability of delivery of liquid N_2; and (3) the rate of evaporation. A 40 l, wide-necked freezer will require about 20–30 l/wk, so a 50 l dewar flask (or two 25 l flasks, which are easier to handle) is advisable. A 35 l, narrow-necked freezer, on the other hand, using 5–10 l/wk will only require a 25 l dewar. Larger freezers are best supplied on line from a dedicated storage tank, e.g., a 160 l storage vessel linked to a 320 l freezer with automatic filling and alarm.

BENEFICIAL EQUIPMENT

The above describes the essential equipment for a modest tissue culture facility; but there are several items of equipment, which, if your budget will stretch to them, will make your laboratory easier to use and more efficient.

Laminar Flow Hood

Usually one hood is sufficient for two to three people (see Chapters 4 and 5). A horizontal flow hood is cheaper and gives best sterile protection to your cultures, but for potentially hazardous materials (radioisotopes, carcinogenic or toxic drugs, virus-producing cultures, or any primate (including human) cell lines), a Class II biohazard cabinet should be used (Figs. 3.1, 5.3). It is important to consult local biohazard regulations before equipping, as legal requirements and recommendations vary (see Chapter 6).

Choose a hood that is: (1) large enough (usually 1,200 mm (4 ft) wide × 600 mm (2 ft) deep); (2) quiet (noisy hoods are more fatiguing); (3) easily cleaned both inside the working area and below the work surface in the event of spillage; and (4) comfortable to sit at (some cabinets have awkward ducting below the work surface, which leaves no room for your knees, or have screens which obscure your vision). The front screen should be able to be raised, lowered, or removed completely to facilitate cleaning and handling bulky culture apparatus. Remember, however, that a biohazard cabinet will not give you, the operator, the required protection if you remove the front screen.

Insist that you be allowed to examine and sit at a hood, as if using it, before committing yourself to purchase. Can you get your knees under it while sitting comfortably and close enough to work? Is there a foot rest in the correct place? Is your head conveniently placed to see what you are doing? Is the work surface perforated and will this give you trouble with spillage (a solid work surface vented at front and back may be better). If the work surfaces are lifted, are the edges sharp or rounded so that you will not cut yourself when cleaning out the hood? Is the lighting convenient and adequate?

Cell Counter

A cell counter (see Fig. 18.2) is a great advantage when more than two or three cell lines are carried and is essential for precise quantitative growth kinetics. Several companies now market models ranging in sophistication from simple particle counting up to automated cell size analysis. For routine counting, the Coulter "D Industrial" is more than adequate and much less expensive than equipment with cell sizing facilities (see also Cell Counting, Chapter 18).

Vacuum Pump

A vacuum pump or simple tap siphon saves a lot of time and effort when handling large numbers of cultures or large fluid volumes. Tap siphons require a minimum of 6 m (20 ft) head of water to create sufficient suction, but are by far the cheapest, simplest, and most efficient way to dispose of nonhazardous tissue culture effluent. If you do not have sufficient water pressure, or are handling potentially hazardous material, use a vacuum pump similar to that supplied for sterile filtration. If necessary, the same pump will serve both duties. The effluent should be collected in a reservoir into which a sterilizing agent such as glutaraldehdye or hypochlorite may be added when work is finished and at least 30 min before the reservoir is emptied (Fig. 4.5). A drying agent, hydrophobic filter (Gelman), or second trap placed in the line to the pump prevents fluid being carried over. Do not draw air through a pump from a reservoir containing hypochlorite as the free chlorine will corrode the pump and could be toxic. Avoid vacuum lines; if they become contaminated with fluids, they can be very difficult to clean out.

A peristaltic pump may be used instead of a vacuum pump. No trap is required, effluent can be collected directly into disinfectant, and there is less chance of discharging aerosol into the atmosphere. However, the pump tubing should be checked regularly for wear and the pump operated by a self cancelling foot switch.

Always switch on the pump before inserting a pipette in the tubing to avoid effluent running back.

CO$_2$ Incubator

Although incubations can be performed in sealed flasks in a regular dry incubator or hot room, some vessels, e.g., petri dishes or multiwell plates, require a controlled atmosphere with high humidity and elevated CO$_2$ tension (Fig. 4.6). The cheapest way of controlling the gas phase is to place the cultures in a plastic box, a desiccator, or an anaerobic jar (Fig. 4.7). Gas the container with the correct CO$_2$ mixture and then seal. If the container is not full, include an open dish of water to increase the humidity. Making the culture medium about 10% hypotonic will also help to counteract evaporation.

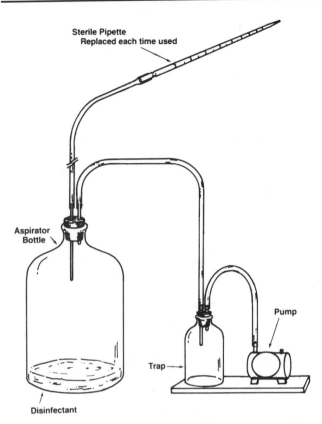

Sterile Pipette
Replaced each time used

Aspirator
Bottle

Pump

Trap

Disinfectant

Fig. 4.5. *Vacuum pump assembly for withdrawing spent medium, etc.*

Fig. 4.6. *Automatic CO₂ incubator (Napco). This is a dual-chamber model although only the top chamber is shown. Dual controls are located in the top panel, temperature regulation on the left, and a CO₂ controller on the right (see also Fig. 4.7).*

CO_2 incubators are rather expensive, but their ease of use and the superior control of CO_2 tension and temperature (anaerobic jars and desiccators take longer to warm up) justify the expenditure. A controlled atmosphere is achieved by blowing air over a humidifying tray (Fig. 4.8) and controlling the CO_2 tension with a CO_2 monitoring device. Alternatively, CO_2 tension may be controlled by mixing air and CO_2 in the correct ratio; but CO_2 controllers, although they add to the capital cost of the incubator, reduce CO_2 consumption considerably and give better control and recovery after opening the incubator. They function by drawing air from the incubator into the sample chamber, determining the concentration of CO_2, and injecting pure CO_2 into the incubator to make up any deficiency. Air is circulated around the incubator to keep both the CO_2 level and the temperature uniform. Most CO_2 controllers require calibration every few months, although some manufacturers produce self-calibrating detectors (Heraeus).

Since humid incubators require regular cleaning, the interior should dismantle readily without leaving inaccessible crevices or corners.

Preparation and Quality Control

A coarse and a fine balance and a simple pH meter are useful additions to the tissue culture area for the preparation of media and special reagents. Although a phenol red indicator is sufficient for monitoring pH in most solutions, a pH meter will be required when phenol red cannot be used, e.g., in preparation of cultures for fluorescence assays.

One of the most important physical properties of culture medium, and one that is often difficult to predict, is the osmolality. An osmometer (see Fig. 7.10) is, therefore, a useful accessory to check solutions as they are made up, to adjust new formulations, or to compensate for the addition of reagents to the medium. They usually work by freezing-point depression or elevation of vapor pressure. Choose one with a low sample volume ($\leqslant$ 1 ml), since on occasion you may want to measure a valuable or scarce reagent.

Fig. 4.7. *Becton Dickinson Anaerobic Jar. This type of jar or a desiccator (preferably plastic) can be used to maintain a regulated atmosphere in the absence of a CO_2 incubator.*

Upright Microscope

An upright microscope may be required, in addition to an inverted microscope, for chromosome analysis, mycoplasma detection, and autoradiography. Select a high-grade research microscope, such as the Reichert Polyvar, with regular brightfield optics up to × 100 objective magnification, phase-contrast up to at least × 40 objective magnification, and preferably × 100, and fluorescence optics with epi-illumination and × 40 and × 100 objectives. Leitz supplies a × 50 water immersion objective, which is particularly useful for observation of routine mycoplasma preparations with Hoechst stain (see Chapter 16). An automatic camera should also be fitted.

Temperature Recording

Ovens, incubators, and hot rooms should be monitored regularly for uniformity and stability of temperature control. A recording thermometer with ranges from below −50°C to about +200°C will enable you to monitor cell freezing, incubators, and sterilizing ovens with one instrument fitted with a resistance thermometer or thermocouple with a long Teflon-coated lead.

Ideally, in addition to the above, recording thermometers should be permanently fixed into your hot room, sterilizing oven and autoclave, and a regular check kept for abnormal behavior.

Magnetic Stirrer

There are certain specific requirements for magnetic stirrers. A rapid stirring action for dissolving chemicals is available with any stirrer but to be used for disaggregation (Chapter 9) or suspension culture (Chapters 10 and 22): (1) the stirrer motor should not heat the culture (use the rotating field type of drive or belt drive from an external motor); (2) the speed must be controlled down to 50 rpm; (3) the torque at low rpm should still be capable of stirring up to 10 l of fluid; (4) it should have more than one place if several cultures are to be maintained simultaneously; (5) each stirrer position should be individually controlled; and (6) there should be a readout of rpm at each position.

Roller Racks

Roller racks are used to scale up monolayer culture (see Chapter 23). The choice of apparatus is determined by the scale, i.e., the size and number of bottles to be rolled. This may be calculated from the number of cells required, the maximum attainable cell density, and the surface area of the bottles. A large number of small bottles gives the highest surface area but tends to be more labor intensive in handling, so a usual compromise is around 125 mm (5 in) diameter and various lengths from 150–500 mm (6–20 in) long. The length of the bottle will determine the maximum yield but is limited by the size of the rack; the height of the rack will determine the number of tiers, i.e., rows of bottles. Although it is cheaper to buy a larger rack than several small ones, the latter alternative: (1) allows you to build up gradually (having confirmed that the system works); (2) can be easier to locate in a hot room; and (3) will still provide accommodation if one rack requires maintenance. The New Brunswick has been found to be reliable and a good size as a starter module if larger scale production is required. Otherwise the Bellco or Luckhams bench-top models may be satisfactory for smaller scale activities.

Pipette Aids and Automatic Pipetting

If a large number of cultures is to be made, an automatic pipette such as the Compu-pet or Watson-Marlow (Fig. 4.9) will be found to be an advantage. (Pipette aids and automatic pipetting are reviewed below). For smaller numbers, Gilson-type pipettors are good for small volumes; Bellco supplies an automated pipette aid which takes regular pipettes (Fig. 4.10). Only the disposable pipette tips of micropipettes need

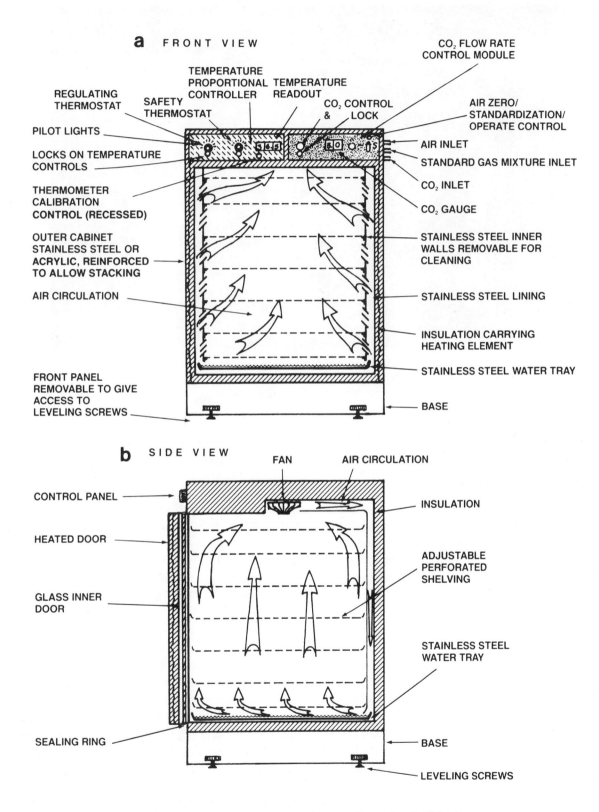

Fig. 4.8. *Components of a typical CO$_2$ incubator. a. Front view. b. Side view.*

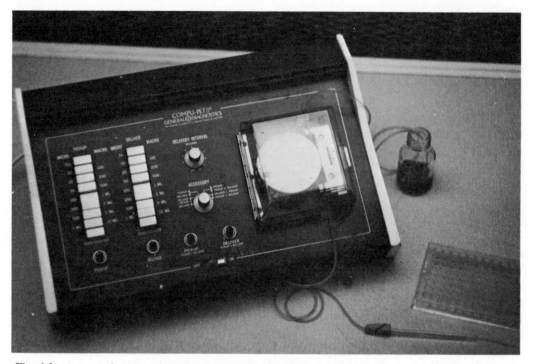

Fig. 4.9. *Automated pipetting device. The Compu-pet, suitable for repetitive dispensing and dilution in the 5 µl–10 ml range. Only the delivery tube requires sterilization.*

be sterile; the Bellco pipette aid uses regular sterile plugged pipettes.

Mechanical Aids and Automation

Repetitive pipettes have been designed in a variety of patterns and those most suited to tissue culture are the syringe types. They operate either by alternately drawing up and expressing liquid through a two-way valve (Cornwall Syringe) or by incremental movement of the syringe piston (Hamilton, Flow).

Repetitive dispensers can also be mounted on reagent bottles (Boehringer/Oxford), in which case the culture flask is taken to the pipette rather than vice versa.

All of these repeating pipettes have problems in use resulting from the necessity to autoclave glass syringes, two-way valves, etc. The valves tend to stick (though making these of Teflon helps), syringe pistons deform, or, if Teflon, may contract due to compression during autoclaving. It is preferable to have a nonsterile metering and repeating mechanism so that only the dispensing element need be sterile. The Bellco automated pipette handle, though it has no facility for repetitive pipetting, conforms to this requirement, as does the Tridak Stopper (Bellco) syringe dispenser.

Automated pipetting can also be provided by a peristaltic pump controlled in small increments (e.g., Compu-pet) (see Fig. 4.9). In these, only the delivery tube is autoclaved, and accuracy and reproducibility can be maintained to high levels over ranges from 10 µl up to 10 ml. In addition, a number of delivery tubes may be sterilized and held in stock, allowing a quick changeover in the event of accidental contamination or change in cell type or reagent. Larger pumps (10–50 ml) are also available (e.g., Watson-Marlow). In this volume range, it is also possible to use a simple transfusion device with a graduated reservoir (Fig. 4.11). Graduated reservoirs are less convenient where smaller volumes or greater accuracy is required, although a burette, preferably with a two-way valve,can be used.

The introduction of microtitration trays (see Fig. 7.4), has brought with it many automated dispensers, diluters (Fig. 4.12), and other accessories. Transfer devices using perforated trays or multipoint pipettes make it easier to seed from one plate to another, and there are also plate mixers and centrifuge carriers available. The range of equipment is so extensive that it cannot be covered here and the appropriate trade catalogues should be consulted (Flow, Microbiological Associates, Dynatech, Gibco). Two items worthy of

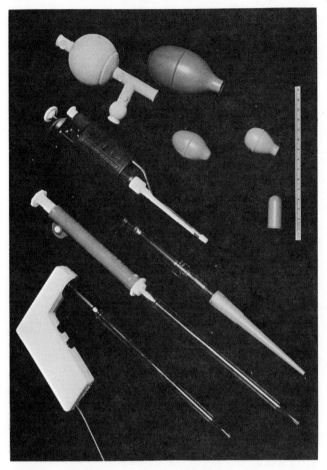

Fig. 4.10. *Pipetting aids. Top to bottom: Bellco, Pi-pump (standard pipettes); Finn-pipette, Gilson (micropipettors, take special plastic tips); various plain bulbs and bulb with inlet and outlet valves (see also Figs. 5.2, 5.5, 4.12).*

note, however, are the Rainin programmable single or multitip micropipette and the Costar Transtar media transfer and replica plating device (Fig. 4.15).

See Figure 6.1 for the proper method of inserting a pipette into a pipetting device.

USEFUL ADDITIONAL EQUIPMENT

Low-Temperature Freezer

Most tissue culture reagents can be stored at 4°C or −20°C, but occasionally some drugs, reagents, or derivatives from cultures may require a temperature of −70°C where most, if not all the water is frozen and most chemical and radiolytic reactions are severely limited. Such a freezer is also a useful accessory for cell freezing (see Chapter 17). The chest type is more efficient at maintaining a low temperature for minimum power consumption, but vertical cabinets are much less extravagant in floor space. If you do

choose a cabinet type, make sure that it has individual compartments (six to eight in a 400 l (15 ft³) freezer) with separate close-fitting doors, and expect to pay 20% more than for a chest type.

Glassware Washing Machine

A reliable person doing your washing-up is probably the best way of producing clean glassware; but when the amount gets to be too great, or reliable help is not readily available, it may be worth considering an automatic washing machine (Fig. 4.13). There are several of these currently available which are quite satisfactory. You should look for the following principles of operation:

(1) Choice of racks with individual spigots over which you can place bottles, flasks, etc. Open vessels such as petri dishes and beakers will wash satisfactorily in a whirling arm spray, but narrow-neck vessels need individual jets. Each jet should have a cushion at its base to protect the neck of the bottle from chipping.

(2) The water pump which pumps the water through the jets should have a high delivering pressure, requiring around 2–5 hp, depending on the size of machine.

(3) Washing water should be heated to 90°C.

(4) There should be a facility for a deionized water rinse at the end of the cycle. This should be heated to 50–60°C; otherwise the glassware may crack after the hot wash and rinse, and should be delivered as a continuous flush and not recycled. If recycling is unavoidable, a minimum of three separate deionized rinses will be required.

(5) Preferably, rinse water from the end of the previous wash cycle should be discarded and not retained for the prerinse of your next wash. This reduces the risk of cross contamination when the machine is used for chemical and radioisotope wash-up.

(6) The machine should be lined with stainless steel and plumbed in stainless steel or nylon pipework.

(7) If possible, a glassware drier should be chosen that will accept the same racks (see Fig. 4.13), so that they may be transferred directly via a suitably designed trolley without unloading.

Betterbuilt makes such machines of different sizes with compatible drying ovens.

Closed-Circuit TV

Since the advent of cheap microcircuits, television cameras and monitors have become a valuable aid to the discussion of cultures and the training of new staff or students (Fig. 4.14). Choose a high-resolution, but

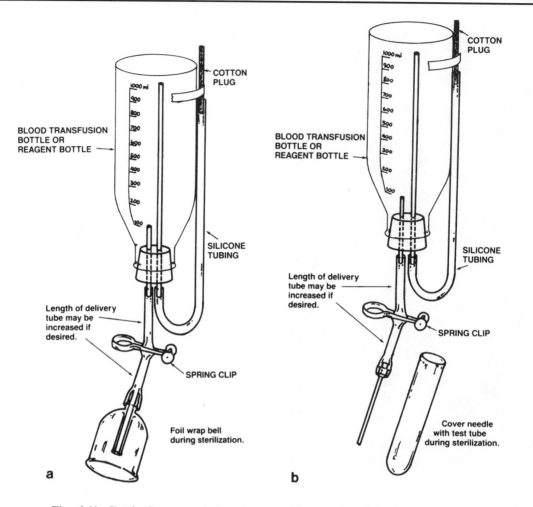

Fig. 4.11. *Simple dispensing devices for use with a graduated bottle. a. With bell used in conjunction with open bottle. b. With needle for slower delivery via a skirted cap or membrane type closure. (From a design by Dr. John Paul.)*

not high-sensitivity, camera, as the standard camera sensitivity is usually sufficient, and high sensitivity may lead to problems of over-illumination. Black and white usually gives better resolution and is quite adequate for phase-contrast observation of living cultures. Color is preferable for fixed and stained specimens. If you will be discussing cultures with a technician or one or two associates, a 12- or 15-in monitor is adequate and gives better definition, but if you are teaching a group of ten or more students then go for a 19- or 21-in monitor. Addition of a video recorder will enable time-lapse films to be made.

Colony Counters

Monolayer colonies are easily counted by eye or on a dissecting microscope with a felt-tip pen to mark off the colonies, but if a lot of plates are to be counted, then an automated counter will help.

There are three levels of sophistication in colony counters. The simplest use an electrode-tipped marker pen, which counts when you touch down on a colony. They often have a magnifying lens to help visualize the colonies (see Fig. 18.6). From there, a large increase in sophistication and cost takes you to an electronic counter employing a fixed program, which counts colonies using a hard-wired program. These counters are very rapid and can discriminate between colonies of different diameters (though this is not necessarily proportional to cell number per colony; see also Chapter 19).

At the highest level of sophistication, image-analysis equipment may be used for colony counting but will

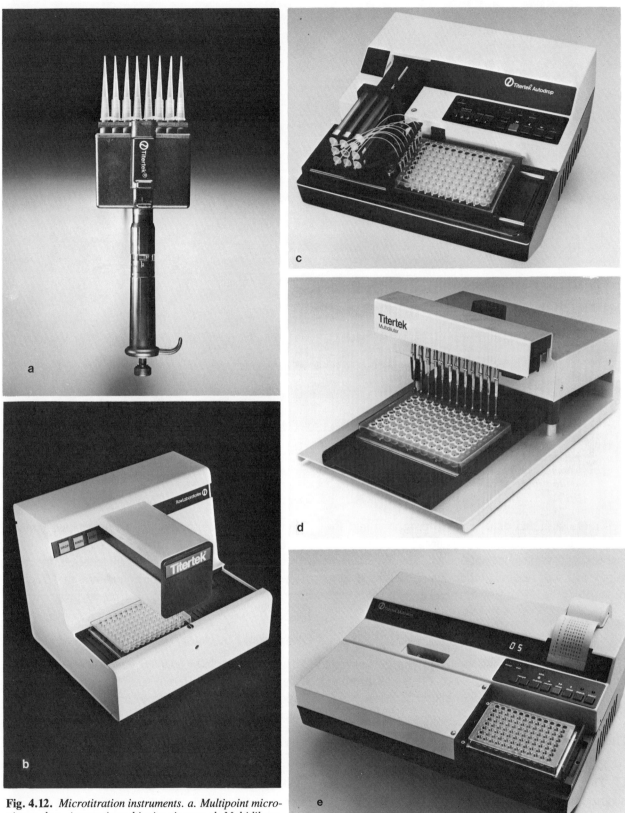

Fig. 4.12. *Microtitration instruments. a. Multipoint micro-pipette. b,c. Automatic multipoint pipettor. d. Multidiluter. e. Densitometer. Photographs reproduced by permission of Flow Laboratories Ltd.*

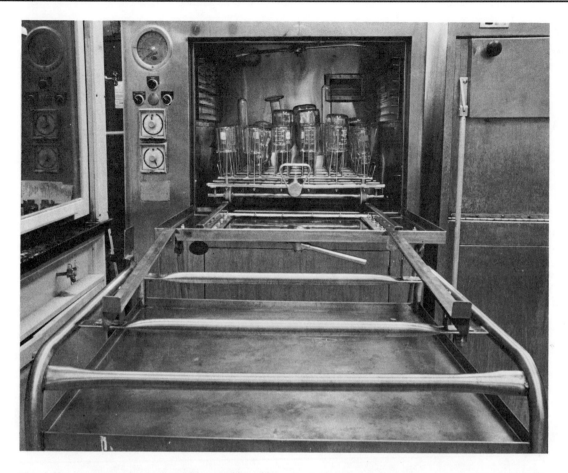

Fig. 4.13. *Automatic glassware washing machine. Glassware is placed on individual jets, which ensures thorough washing and rinsing. After washing, glassware is withdrawn on the rack onto the trolley (front) and transferred to drier (right) fitted with same rails as washing machine and drier (Betterbuilt).*

need more skill and experience in programming. Because of this programmable feature, however, image analysis will cope with almost any size or shape of colony and will perform other complex tasks such as measuring area of outgrowth round an explant.

Cell Sizing

A dual-threshold cell counter (e.g., the Coulter multisizer, see Fig. 18.2) with the facility for pulse-height analysis scans a cell population at a range of threshold settings simultaneously and prints out cell size distributions automatically.

Time-Lapse Cinemicrography

This technique is discussed in more detail in Chapter 23. The apparatus may be added to most good-quality inverted microscopes.

Controlled-Rate Cooler

While cells may be frozen by simply placing them in an insulated box at $-70°C$, some cells may require

different cooling rates or differently shaped cooling curves [Mazur et al., 1970; Leibo and Mazur, 1971]. A programmable freezer enables the cooling rate to be varied by blowing liquid nitrogen into the freezing chamber, under the control of a preset program (see Fig. 17.6).

Centrifugal Elutriator

This is a specially adapted centrifuge suitable for separating cells of different sizes (see under Cell Separation, Chapter 12). They are costly but very effective.

Flow Cytophotometer

This instrument, also known as a fluorescence-activated cell sorter (impulse cytophotometer, or cytofluorimeter) can analyze cell populations and separate them according to a variety of criteria (see Chapters 12 and 18). It has almost unlimited potential but is too expensive to come within most tissue culture laboratory equipment budgets.

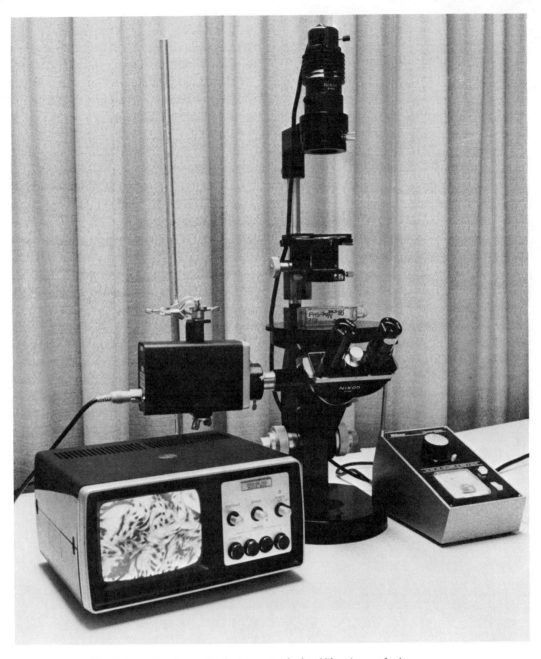

Fig. 4.14. *Closed-circuit television attached to Nikon inverted microscope.*

It is always very tempting to purchase new pieces of equipment as they appear on the market, but weigh the advantages that they may offer against the space that they will occupy and what they will cost. Try to be sure also that (1) they will be of lasting benefit and (2) you and others will want to use them.

CONSUMABLE ITEMS

This includes general items such as pipettes, culture flasks, ampules for freezing, centrifuge tubes (10–15 ml, 50 ml, 250 ml; Sterilin, Corning), universal containers (Sterilin, Nunclon), disposable syringes and needles (21–23 g for withdrawing fluid from vials, 18 g for dispensing cells), filters of various sizes (see Chapter 8) for sterilization of fluids, surgical gloves, and paper towels.

Pipettes

These should be "blow out" and wide tipped for fast delivery, graduated to the tip with the maximum point

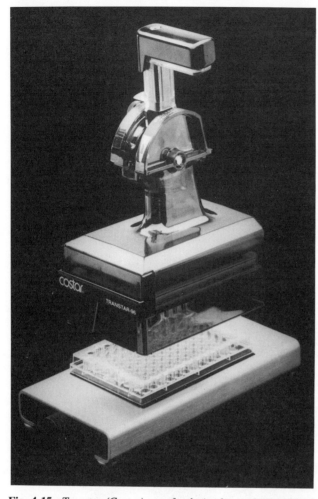

Fig. 4.15. *Transtar (Costar) transfer device for seeding, transferring medium, replica plating, and other similar manipulations with microtitration plates, enabling simultaneous handling of all 96 wells (Reproduced by permission of Northumbria Biologicals Ltd).*

of the scale at the top rather than at the tip. Disposable pipettes can be used but are expensive and may need to be reserved for holidays or crises in wash-up or sterilization. Pipettes to be re-used are collected in pipette cylinders or hods, one per work station.

Pasteur pipettes are best regarded as disposable and should not be discarded into pipette cylinders but into secure glassware waste.

Culture Vessels

Choice of culture vessels is determined by: (1) the yield (cell number) required (See Table 7.1); (2) whether the cell is grown in monolayer or suspension; and (3) the sampling regime, i.e., are the samples to be collected simultaneously or at intervals over a period of time.

"Shopping around" will often result in a cheaper price, but do not be tempted to change too often and always test a new supplier's product before committing yourself.

Care should be taken to label "sterile," "non-sterile," "tissue culture" grade, and "non-tissue culture" grade plastics clearly, and preferably they should be stored separately. Glass bottles with flat sides can be used instead of plastic provided a suitable wash-up and sterilization service is available (see Chapter 8).

Chapter 5

In spite of the introduction of antibiotics, contamination by microorganisms remains a major problem in tissue culture. Bacteria, mycoplasma, yeasts, and fungal spores may be introduced via the operator, atmosphere, work surfaces, solutions, and many other sources (see Table 16.1). Contaminations can be minor and confined to one or two cultures, can spread between several and infect a whole experiment, or can be widespread and wipe out your, or even the whole laboratory's, entire stock. Catastrophes can be minimized if: (1) cultures are checked on the microscope, preferably by phase contrast, every time that they are handled; (2) they are kept antibiotic-free for at least part of the time to reveal cryptic contaminations (see Chapters 10 and 16); (3) reagents are checked for sterility before use (by yourself or the supplier); (4) bottles of media, etc., are not shared with other people or used for different cell lines; and (5) the standard of sterile technique is kept high at all times.

Mycoplasmal infection, invisible under regular microscopy, presents one of the major threats. Undetected, it can spread to other cultures around the laboratory. It is, therefore, essential to back up visual checks with a mycoplasma test, particularly if cell growth appears abnormal. (For a more detailed account of contamination see Chapter 16.)

OBJECTIVES OF ASEPTIC TECHNIQUE

Correct aseptic technique should provide a barrier between microorganisms in the environment outside the culture and the pure uncontaminated culture within its flask or dish. Hence, all materials which will come into direct contact with the culture must be sterile and manipulations designed such that there is no direct link between the culture and its nonsterile surroundings.

It is recognized that the sterility barrier cannot be absolute without working under conditions which would severely hamper most routine manipulations. Since testing the need for individual precautions would be an extensive and lengthy controlled trial, procedures are adopted largely on the basis of common sense and experience. Aseptic technique is a combination of procedures designed to reduce the probability of infection, and the correlation between the omission of a step and subsequent contamination is not always absolute. The operator may abandon several precautions before the probability rises sufficiently that a contamination occurs. By then, the cause becomes multifactorial and consequently no simple solution is obvious. If, once established, all precautions are maintained consistently, breakdown will be rarer and more easily detected.

Although laboratory conditions have improved in some respects (air conditioning and filtration, laminar flow facilities, etc.), the modern laboratory is often more crowded and accommodation may have to be shared. However, with reasonable precautions, maintenance of sterility is not difficult.

QUIET AREA

In the absence of laminar flow, a separate sterile room should be used if possible (see Fig. 3.2). If not, pick a quiet corner of the laboratory with little or no traffic and no other activity (see Chapter 3). With laminar flow, an area should be selected which is free from drafts and traffic should still be kept to a minimum. Animals and microbiological culture should be excluded from the tissue culture area. It should be kept clean and dust free and should not contain equipment other than that connected with tissue culture.

WORK SURFACE

One of the more frequent examples of bad technique is the failure to keep the work surface clean and tidy. The following rules should be observed:
(1) Start with a completely clear surface.
(2) Swab down liberally with 70% alcohol.

(3) Bring on to it only those items you require for a particular procedure and swab bottles, cans, etc., with 70% alcohol beforehand.

(4) Arrange your apparatus (a) to have easy access to all of it without having to reach over one item to get at another and (b) to leave a wide, clear space in the center of the bench (not just the front edge!) to work on (Fig. 5.1). If you have too much equipment too close to you, you will inevitably brush the tip of a sterile pipette against a nonsterile surface.

(5) Work within your range of vision, e.g., insert a pipette in a bulb with the tip of the pipette pointing away from you so that it is in your line of sight continuously and not hidden by your arm.

(6) Mop up any spillage immediately and swab with 70% alcohol.

(7) Remove everything when you have finished and swab down again.

PERSONAL HYGIENE

There has been much discussion about whether hand-washing encourages or reduces the bacterial count on the skin. Regardless of this debate, washing will moisten the hands and remove dry skin likely to blow onto your culture and reduce loosely adherent microorganisms which are the greatest risk to your cultures. Surgical gloves may be worn and swabbed frequently, but it may be preferable to work without (where no hazard is involved) and retain the extra sensitivity that this allows.

Caps, gowns, and face masks are often worn but are not always strictly necessary, particularly when working with laminar flow. However, if you have long hair, tie it back. When working on the open bench, do not talk while working aseptically; and if you have a cold, wear a face mask, or, better still, do not do any tissue culture during the height of the infection. Talking is permissible when working in vertical laminar flow with a barrier between you and the culture but should still be kept to a minimum.

PIPETTING

Standard glass or disposable plastic pipettes are still the easiest form of manipulating liquids. Syringes are often used, but regular needles are too short to reach into most bottles. Syringing may produce high shearing forces when dispensing cells and increase the risk of self-inoculation.

Pipettes of a convenient size range should be selected—1 ml, 2 ml, 5 ml, 10 ml, and 25 ml cover most requirements. If you only require a few of each, make

up mixed cans and save space. Mouth pipetting, even with plugged pipettes or a filter tube/mouthpiece should be avoided, as it has been shown to be a contributory factor in mycoplasmal infection and may introduce an element of hazard to the operator, e.g., with virus-infected cell lines and human biopsy or autopsy specimens and biohazards (see Chapter 6). Inexpensive bulbs and pipetting devices are available; try a selection of these to find one that suits you (see Fig. 4.10). They should accept securely all the sizes of pipette that you use without forcing them and without the pipette falling out. The regulation of flow should be easy and rapid, but at the same time, capable of fine adjustment. You should be able to draw liquid up and down repeatedly (e.g., to disperse cells) and there should be no fear of carry-over. The device should fit comfortably in your hand and should be easy to operate with one hand.

The Marburg-type pipette (see Fig. 4.10) (Gilson, Oxford, Eppendorf, etc.) is particularly useful for small volumes (1 ml and less) though there can be some difficulty in reaching down into larger vessels with most of them. They are best used in conjunction with a shallow vial or bottle and are particularly useful when dealing with microtitration assays and other multiwell dishes but should not be used for serial propagation. Multipoint pipettors (4, 6, or 12 point) are available for microtitration dishes (see Fig. 4.10).

It is necessary to insert a cotton plug in the top of a glass pipette before sterilization to maintain sterility in the pipette during use. If this becomes wet in use, discard the pipette into the wash-up. Plugging pipettes for sterile use is a very tedious job, as is the removal of plugs before washing. Automatic pipette pluggers are available, and although expensive, speed up the process and reduce the tedium (see Fig. 8.5). Alternatively, a sterile filter tube (Fig. 5.2) may be attached to the bulb, eliminating the need to plug pipettes. It is important that the filter tube is changed between handling of different cell lines to avoid the risk of cross contamination.

Automatic pipetting devices and repeating dispensers are discussed in Chapter 4.

STERILE HANDLING

Swabbing

Swab bottles, particularly from the cold room, before using for the first time each day.

Capping

Deep screw caps should be used in preference to stoppers although care must be taken when washing

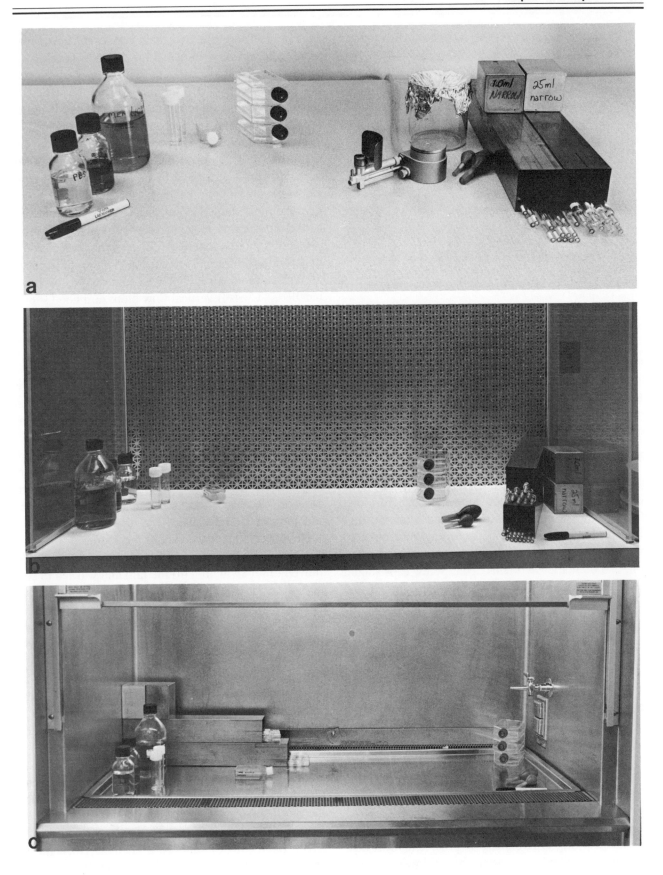

Fig. 5.1. *Suggested layout of work area. a. Open bench. b. Horizontal laminar flow. c. Vertical laminar flow.*

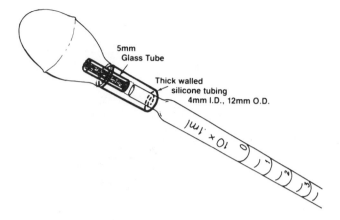

Fig. 5.2. *Filter tube. Interposed between bulb and pipettes, avoids the necessity to plug pipettes. Filter tube must be renewed between cell lines or if wetted (developed at the Beatson Institute from an original idea by Dr. John Paul).*

caps to ensure that all detergent is rinsed from behind rubber liners. Wadless polypropylene caps should be used if possible. The screw cap should be covered with aluminum foil to protect the neck of the bottle from sedimentary dust.

Flaming

When working on the open bench, the necks of bottles and screw caps should be flamed before and after opening a bottle and before and after closing. Pipettes should be flamed before use. Work close to the flame where there is an up-current due to convection, and do not leave bottles open. Screw caps should be placed open side down on a clean surface and flamed before replacing on the bottle.

Pouring

Whenever possible, do not pour from one sterile container into another unless the bottle you are pouring from is to be used once only and, preferably, is to deliver all its contents (premeasured) in one single delivery. The major risk in pouring lies in the generation of a bridge of liquid between the outside of the bottle and the inside which may permit infection to enter the bottle.

LAMINAR FLOW

The major advantage of working in laminar flow is that the working environment is protected from dust and contamination by a constant stable flow of filtered air passing over the work surface (Fig. 5.3) (see also

Fig. 3.1). There are two main types: (1) horizontal, where the air flow blows from the side facing you, parallel to the work surface, and is not recirculated; and (2) vertical, where the air blows down from the top of the cabinet on to the work surface and is drawn through the work surface and either recirculated or vented. In recirculating hoods, 20% is vented and made up by drawing in air at the front of the work surface. This is designed to minimize overspill from the work area of the cabinet. Horizontal flow hoods give the most stable airflow and best sterile protection to the culture and reagents; vertical flow gives more protection to the operator. If potentially hazardous material (radioisotopes, mutagens, human- or primate-derived cultures, virally infected cultures, etc.) is being handled, a Class II vertical flow biohazard hood should be used (see Fig. 6.3a).

If known human pathogens are handled, a Class III pathogen cabinet with a pathogen trap on the vent is obligatory (see Fig. 6.3b and Chapter 6).

Laminar flow hoods depend, for their efficiency, on a minimum pressure drop across the filter. When filter resistance builds up, the pressure drop increases and the flow rate of air in the cabinet falls. Below 0.4 m/s (80 ft/min), the stability of the laminar air flow is lost and sterility can no longer be maintained. The pressure drop can be monitored with a manometer fitted to the cabinet, but direct measurement of airflow with an anemometer is preferable.

Routine maintenance checks are required (every 3–6 months) of the primary filters, which may be removed (after switching off the fan) and washed in soap and water, as they are usually made of polyurethane foam. Every 6 months the main filter should be checked for air flow and holes (detectable by locally increased air flow and an increased particulate count). This is best done on a contract basis.

Regular weekly checks should be made below the work surface and any spillage mopped up and the area sterilized. Spillages should, of course, be mopped up when they occur; but occasionally they go unnoticed, so a regular check is imperative.

Laminar flow hoods are best left running continuously because this keeps the working area clean. Should any spillage occur, either on the filter or below the work surface, it dries fairly rapidly in sterile air, reducing the chance of growth of microorganisms.

Ultraviolet (uv) lights are used to sterilize the air and exposed work surfaces in laminar flow cabinets between use. The effectiveness of this is doubtful be-

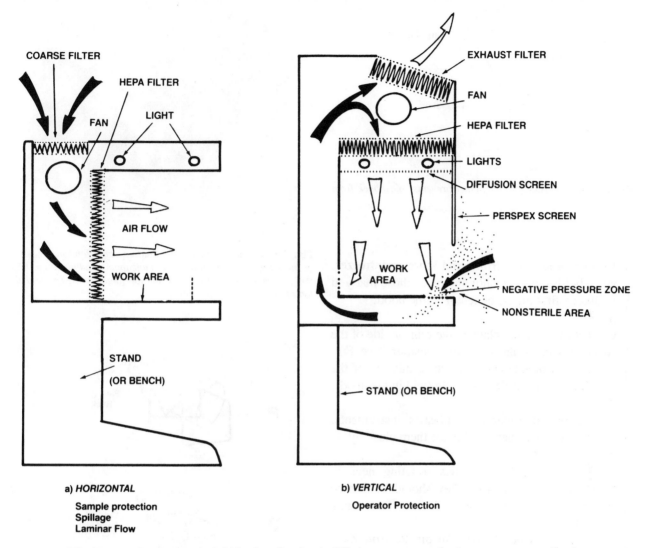

a) *HORIZONTAL*

Sample protection
Spillage
Laminar Flow

b) *VERTICAL*

Operator Protection

Fig. 5.3. *Horizontal (a) and vertical (b) laminar flow hoods. Filled arrows, nonsterile air; open arrows, sterile air.*

cause crevices are not reached, and these are treated more effectively with alcohol or other sterilizing agents, which will run in by capillarity. Ultraviolet irradiation will also lead to crazing of some clear plastic panels (e.g., Perspex) after 6 months to 1 year.

STANDARD PROCEDURE

Emphasis is being placed here on aseptic technique. Media preparation and other manipulations will be discussed in more detail under the appropriate headings.

Outline

Clean and prepare work area, with bottles, pipettes, etc. Carry out preparative procedures first before culture work. Flame articles as necessary and keep the work surface clean and clear. Finally, tidy up and wipe over surface with 70% alcohol.

Materials

70% alcohol
swabs
bunsen (not in laminar flow) and
    lighter
pipette-aid or bulb (see Fig. 4.10)
waste beaker (Fig. 5.4) or aspiration
    pump (see Fig. 4.5)
scissors
marker pen
media, stocks, etc

Protocol

1.

Swab down bench surface or all inside surfaces of laminar flow hood with 70% alcohol.

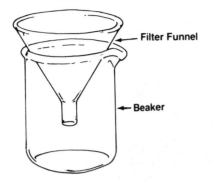

Fig. 5.4. *Waste beaker. Filter funnel prevents splashback from beaker.*

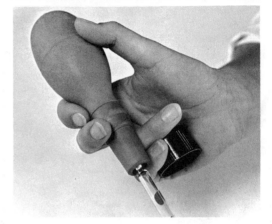

Fig. 5.5. *Uncapping bottle and holding cap. Hand may need to be moved up or down bulb between uncapping and pipetting. With this particular bulb (Aspirette), the forefinger is used to seal the top of the bulb when pipetting.*

2.
Bring media, etc., from cold store and freezer, swab bottles with alcohol and place those that you will need first on the bench or in the hood.
3.
Collect pipettes and place at the rear or side of the work surface in an accessible position (see Fig. 5.1). Open pipette cans and place lids out of the way but still in sterile work area (within the hood).
4.
Collect any other glassware, plastics, instruments, etc. that you will need and place them close by.
5.
 (a) Flame necks of bottles, rotating neck in flame, and slacken caps—they should be flamed outside the hood, if one is used—and flame again after slackening.
 (b) On open bench: (i) take pipette from can, touching other pipettes as little as possible, particularly at the tops; (ii) flame top to burn off any cotton protruding from the pipette; (iii) insert in bulb, pointing pipette away from you and holding it well above the graduations.
 ◊ Note. Take care not to exert too much pressure as pipettes can break when being forced into a bulb.
 (iv) flame pipette by pushing lengthwise through flame, rotate 180°, and pull back through flame. This should only take 2–3 s or the pipette will get too hot. You are not attempting to sterilize the pipette, merely to fix any dust which may have settled on it. If you have touched anything or contaminated the pipette in any other way, discard it into the wash-up; do not attempt to resterilize it by flaming; (v) holding the pipette still pointing away from you, remove the cap of your first bottle into the crook formed between your little finger

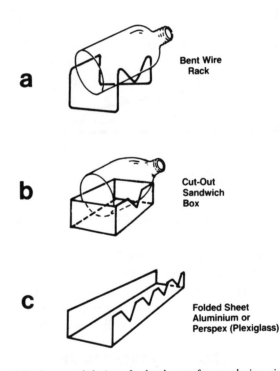

Fig. 5.6. *Suggested designs for bottle rest for use during pipetting. a. Wire rack, suggested by M. Stack-Dunne. b. V-cut in a plastic storage box. c. Folded aluminium or plexiglas suggested by A.C. McKirdy.*

and the heel of your hand (Fig. 5.5); (vi) flame the neck of the bottle; (vii) withdraw the requisite amount of fluid and hold; (viii) flame bottle neck and recap; (ix) remove caps of receiving bottle, flame neck, insert fluid, reflame, and replace cap; (x) when finished, tighten caps, flame thoroughly, and replace foil. Work with the bottles tilted so that your hand does not come over the open neck. If you have difficulty in holding the cap in your hand while you pipette, leave the foil in place and place the cap on the bench resting on the skirt of the foil. If bottles are to be left open, they should be sloped as close to horizontal as possible, laying them on the bench or on a bottle rest (Fig. 5.6).

(c) In laminar flow, proceed as for open bench but omit flaming during manipulations. Bottles may be left open more safely but should still be closed if you leave the hood for more than a few minutes. In vertical laminar flow, do not work immediately above an open bottle or dish. In horizontal laminar flow, do not work behind an open bottle or dish.

6.
On completion of the operation, remove stock solutions from work surface keeping only the bottles that you will require.

7.
Check cultures, decide what they require, and bring to sterile work area.

8.
Swab bottles, flame necks (plastic very briefly), and place on work surface, preferably one cell strain at a time with its own bottle of medium and other solutions.

9.
For fluid change, proceed as follows:

(a) Take sterile pipette, flame (if not in hood), and insert bulb.

(b) Flame neck of culture bottle, open, withdraw medium and discard into waste beaker (see Fig. 5.4), or preferably, via a suction pump with a collection trap, or a tap siphon, with a suction line into the hood (see Fig. 4.5).

(c) Transfer fresh medium to culture flask as in (5) above.

(d) Tighten caps, flame necks, replace foil.

10.
Return flasks to incubator and media to cold-room.

11.
Clear away all pipettes, glassware, etc., and swab down the work surface.

The essence of good sterile technique is similar to good laboratory technique anywhere. Keep a clean clear space to work and have on it only what you require at one time. Prepare as much as possible in advance so that cultures are out of the incubator for the shortest possible time and the various manipulations can be carried out quickly, easily, and smoothly. Keep everything in direct line of sight and develop an awareness of accidental contacts between sterile and nonsterile surfaces. Leave the area clean and tidy when you finish.

Chapter 6
Laboratory Safety and Biohazards

There is increasing concern about safety in the laboratory. At one time, scientific laboratories were exempt from health and safety legislation, but this is no longer the case. Accidents happen in the laboratory as in other work places and a greater understanding of the biological or medical consequences offer no greater protection; perhaps the converse, as familiarity often leads to a more casual approach in dealing with regular, biological, and radiological hazards just as it does in the factory dealing with equally hazardous engineering tools.

To draw attention to items in the text which refer specifically to safety, they will be identified with this diamond signal ◊.

A major problem which arises constantly in establishing safe practices in a biology laboratory is the disproportionate concern given to the more esoteric and poorly understood risks, such as those arising from genetic manipulation, relative to the known proven hazards of chemicals, toxins, fire, ionizing radiation, electrical shock, and broken glass. No one should ignore potential biohazards [Barkley, 1979], but they should not displace the recognition of everyday safety problems.

The following typical examples should not be interpreted as a code of practice, but rather as advice which might help in compiling safety regulations. Local and national regulations should be consulted before making up your own laboratory code of practice.

GENERAL SAFETY

Glassware and Sharp Items

◊ The most common form of injury in tissue culture results from accidental handling of broken glass and syringes, e.g., broken pipettes in a wash-up cylinder when too many pipettes, particularly Pasteur pipettes, are forced into too small a container. Pasteur pipettes should be discarded or if reused handled separately and with great care. Avoid syringes unless they are needed for loading ampules or withdrawing fluid from a vial. When disposable needles are discarded, the point should be bent over and trapped inside the sheath or taped. Provide separate receptacles for the disposal of sharp items and broken glass and do not use them for general waste.

◊ Take care when fitting a bulb or pipetting device onto a pipette. Choose the correct size to guard against the risk of the pipette breaking at the neck and lacerating your hand. Check the neck is sound, hold the pipette as near the end as possible, and apply gentle pressure with the pipette pointing away from your knuckles (Fig. 6.1).

Chemical Toxicity

Relatively few major toxic substances are used in tissue culture, but when they are, the conventional precautions should be taken paying particular attention to the distribution of aerosols by laminar flow cabinets (see Biohazards, this chapter). Detergents, particularly those used in automatic machines, are usually caustic; even when they are not they can cause irritation to the skin, eyes, and lungs. Use dosing devices where possible, wear gloves, and avoid procedures which cause the detergent to spread as dust. Liquid detergent concentrates are more easily handled but are often more expensive.

◊ Chemical disinfectants such as hypochlorite should also be used cautiously and with a dispenser. Hypochlorite disinfectants will bleach clothing and cause skin irritations and will even corrode stainless steel.

◊ Specific chemicals used in tissue culture requiring special attention are (1) Dimethyl sulphoxide (DMSO), which is a powerful solvent and skin penetrant and can, therefore, carry many substances through the skin [Horita and Weber, 1964] and even through protective gloves, and (2) mutagens and carcinogens, which should be handled only in a safety cabinet (see below) and are sometimes dissolved in DMSO.

51

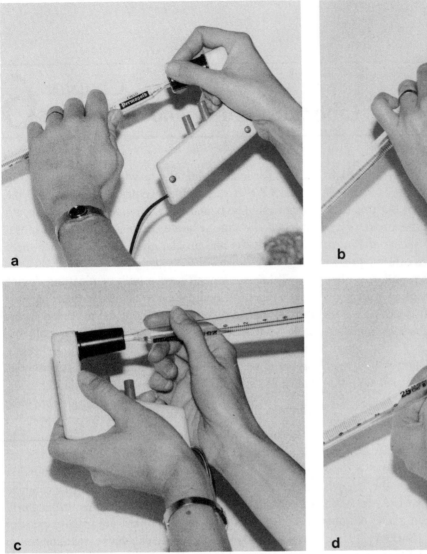

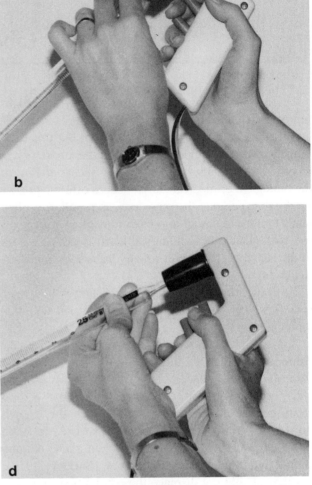

Fig. 6.1. *Inserting a pipette into a pipetting device. a. Wrong position. Left hand too far down pipette, risking contamination of the pipette and exerting too much leverage, which might break the pipette; right hand too far over and exposed to end of pipette or splinters should the pipette break at the neck during insertion. b,c,d. Correct positions. Left hand further up pipette, right hand clear of top of pipette.*

Gases

◊ Most gases used in tissue culture (CO_2, O_2, N_2) are not harmful in small amounts but are, nevertheless, dangerous if handled improperly. They are contained in pressurized cylinders which must be properly secured (Fig. 6.2). When a major leak occurs, there is a risk of asphyxiation from CO_2 and N_2 and of fire from O_2. Evacuation and maximum ventilation are necessary in each case and for O_2, call the fire department.

◊ Ampule sealing is usually performed in a gas-oxygen flame, so great care must be taken both to guard the flame and to prevent unscheduled mixing of the gas and oxygen. A one-way valve should be incorporated in the gas line so that oxygen cannot blow back.

Liquid N_2

◊ There are three major risks associated with liquid N_2: frostbite, asphyxiation, and explosion. Since the temperature of liquid N_2 is $-196°C$, direct contact with it (splashes, etc.), or with anything, particularly metallic, submerged in it, presents a serious hazard. Gloves thick enough to act as insulation but flexible enough to allow manipulation of ampules should be

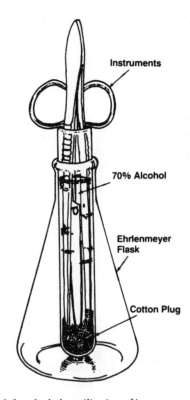

Instruments

70% Alcohol

Ehrlenmeyer
Flask

Cotton Plug

Fig. 6.2. *Cylinder clamp. Clamps onto edge of bench and secures gas cylinder with fabric strap. Fits different sizes of cylinder and can be moved from one position to another if necessary. Available from most laboratory suppliers.*

Fig. 6.3. *Flask for alcohol sterilization of instruments. Wide base prevents tiping and center tube reduces the amount of alcohol required so that spillage, if it occurs, is minimized (from an original idea by M.G. Freshney).*

worn. When liquid N_2 boils off during routine use of the freezer, regular ventilation is sufficient to remove excess nitrogen; but when nitrogen is being dispensed, or a lot of material is being inserted in the freezer, extra ventilation will be necessary.

◇ When ampules are submerged in liquid N_2, a high-pressure difference results between the outside and the inside of the ampule. If they are not perfectly sealed, this results in inspiration of liquid N_2 which will cause the ampule to explode violently when thawed. This can be avoided by storing in the gas phase (see Chapter 17) or by ensuring that the ampules are perfectly sealed. Thawing from storage under liquid N_2 should always be performed in a container with a lid, such as a plastic bucket (see Chapter 17), and a face shield or goggles must be worn.

FIRE

◇ Particular fire risks associated with tissue culture stem from the use of bunsen burners for flaming, together with alcohol for swabbing or sterilization. Keep the two separate; always ensure that alcohol for

sterilizing instruments is kept in a narrow-necked bottle or flask which is not easily upset, and with the minimum volume of alcohol (Fig. 6.3). Alcohol for swabbing should be kept in a plastic wash bottle. When instruments are sterilized in alcohol and the alcohol subsequently burnt off, care must be taken not to return the instruments to the alcohol while they are still alight.

RADIATION

◇ There are three main types of radiation hazard from tissue culture associated procedures. The first is from ingestion of radiolabelled compounds. Tritiated nucleotides, if accidentally ingested, will become incorporated into DNA, and, due to the short path length of the low energy β-emission from ^3H, will cause radiolysis within the DNA. Radioactive isotopes of iodine will concentrate in the thyroid and may also cause local damage.

◇ The second type of risk is from irradiation from higher energy β and γ emitters such as ^{32}P, ^{125}I, ^{131}I, and ^{51}Cr. Protection can be obtained by working be-

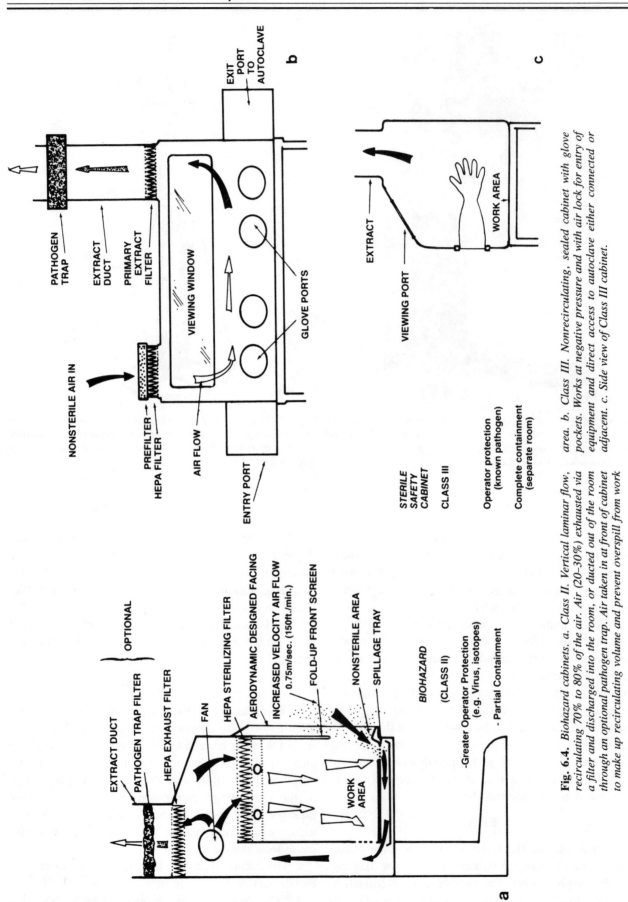

Fig. 6.4. *Biohazard cabinets. a. Class II. Vertical laminar flow, recirculating 70% to 80% of the air. Air (20–30%) exhausted via a filter and discharged into the room, or ducted out of the room through an optional pathogen trap. Air taken in at front of cabinet to make up recirculating volume and prevent overspill from work area. b. Class III. Nonrecirculating, sealed cabinet with glove pockets. Works at negative pressure and with air lock for entry of equipment and direct access to autoclave either connected or adjacent. c. Side view of Class III cabinet.*

hind a 2mm thick lead shield and storing the concentrated isotope in a lead pot. Perspex screens (5mm) can be used with ^{32}P at low concentrations for short periods.

◊ In both cases, work in a biohazard cabinet, to contain aerosols, and wear gloves. The items that you are working with should be held in a shallow tray to contain any accidental spillage. Clean up carefully when finished and monitor regularly for any spillage.

◊ The third type of irradiation risk is from X-ray machines, high energy sources such as ^{60}Co or uv sources used for sterilizing apparatus or stopping cell proliferation in feeder layers (see Chapters 11, 23). Since the energy, particularly from X-rays or ^{60}Co is high, these sources are usually located in specially designed accommodation and subject to strict control. UV sources can cause burns to the skin and can damage the eyes. They should be carefully screened to prevent direct irradiation of the operator and barrier filter goggles should be worn.

Consult your local radiological officer and code of practice before embarking on radioisotopic experiments. Local rules vary but most places have strict controls on the amount of radioisotopes that can be used, stored, and discarded. The advice given above is general and should not be construed as satisfying any legal requirement.

BIOHAZARDS

The need for protection against biological hazards [see also Barkley, 1979] is defined (1) by the source the material and (2) by the nature of the operation being carried out. It is also governed by the conditions under which culture is performed. Standard microbiological technique on the open bench has the advantage that the techniques in current use have been established as a result of many years of accumulated experience. Problems arise when new techniques are introduced or when the number of people sharing the same area increases. With the introduction of horizontal laminar flow cabinets, the sterility of the culture was protected more effectively, but the exposure of the operator to aerosols was increased. This led to the development of vertical laminar hoods with an air curtain at the front (see Chapters 4 and 5) to minimize overspill from within the cabinet.

We can define three levels of handling: (1) a sealed pathogen cabinet with filtered air entering and leaving via a pathogen trap filter (Class III); (2) a vertical laminar flow cabinet with front protection in the form of an air curtain (Class II); (Fig. 6.4, National Sanitation Foundation Standard 49, NIH specification NIH 03-112, British Standard BS5726); and (3) open bench, depending on good microbiological technique. Table 6.1 lists common procedures with suggested levels of

TABLE 6.1 Biohazard Procedures and Suggested Levels of Containment

Procedure	Level of protection
Media preparation	Open bench, standard microbiological practice, or horizontal laminar flow
Cell lines other than human and other primates	Open bench, standard microbiological practice, or horizontal or vertical laminar flow
Primary culture and serial passage of human and other primate cells	Vertical laminar flow cabinet with air curtain protection at front and filtered extract (Class II)
Interspecific hybrids or other recombinants between human cells and animal tumor cells	Vertical laminar flow cabinet with air curtain protection at front and filtered extract (Class II)
Virus-producing human cell lines	Pathogen cabinets with glove pockets, filtered air entering and pathogen trap on vented air (Class III). Located in a separate room with separate provision for incubation, centrifugation, cell counting, etc. No access except to designated personnel. All waste, soiled glassware, etc., to be sterilized as it leaves the room and extracted air to be filtered
Tissue samples and cultures carrying known human pathogens	Pathogen cabinet with glove pockets, filtered air entering and pathogen trap on vented air (Class III). Located in separate room with separate provision for incubation, centrifugation, cell counting, etc. No access except to designated personnel. All waste, soiled glassware, etc., to be sterilized as it leaves the room and extracted air to be filtered. Shower facilities and change of clothing on entering and leaving.

containment. These are suggestions only, however, and you should seek the advice of your local safety committee for legal requirements.

◊ There is often less doubt when known classified pathogens are being used since the regulations are laid down by the Howie Report (U.K.) [1978] and Center for Disease Control (U.S.A.) or when harmless, sterile solutions are being prepared. It is the "gray area" in the middle that causes concern, as development of new techniques such as interspecific cell hybridization and new facilities such as laminar flow, introduces putative risks for which there are no epidemiological data available for assessment. Transforming viruses, transformed human cell lines, and human-mouse hybrids, for example, should be treated cautiously until data accumulates that they carry no risk.

◊ Risks which are more easily recognized are those associated with biopsy and autopsy specimens from human and primate tissue. When infection has been confirmed, the type of organism will determine the degree of containment, but where there is no known infection, the possibility remains that the sample may yet carry hepatitis B, tuberculosis, or other pathogens, as yet undiagnosed. Such samples should be handled with caution (Class II biohazard cabinet, no sharp instruments used in handling, discard into disinfectant or autoclave) until they can be shown to be uninfected by the appropriate clinical diagnostic tests.

◊ Potentially biohazardous materials must be sterilized before disposal. They may be placed in autoclavable sacks (unsealed) and autoclaved, or immersed in a sterilizing agent such as hypochlorite or glutaraldehyde. Various proprietary preparations are available, e.g., Chloros liquid concentrate (I.C.I.). Recommended concentrations vary according to local rules but a rough guide can be obtained from the manufacturer's instructions. Hypochlorite is often used at 300 ppm available chlorine but some authorities demand 2500 ppm as recommended in the Howie Report [1978]. Hypochlorite is effective and easily washed off items to be reused but is highly corrosive particularly in alkaline solutions. It will bleach clothing and even corrodes stainless steel, so gloves and a lab coat or apron should be worn when handling hypochlorite and soaking baths and cylinders should be made of polypropylene.

The Culture Environment: Substrate, Gas Phase, Medium, and Temperature

The influence of the environment on the culture is expressed via four routes: (1) the nature of the substrate or phase on or in which the cells grow — this may be solid, as in monolayer growth on plastic, semisolid as in a gel such as collagen or agar, or liquid as in suspension culture; (2) the physicochemical and physiological constitution of the medium; (3) the constitution of the gas phase; and (4) the incubation temperature. It is, perhaps, useful to think of the four elements of the ancient alchemists in remembering these routes: "air"—the gas phase, "earth"—the substrate, "fire"—the temperature, and "water"—the medium.

THE SUBSTRATE

The majority of vertebrate cells cultured *in vitro* have been grown as monolayers on an artificial substrate. Spontaneous growth in suspension is restricted to hemopoietic cell lines, rodent ascites tumors, and a few other selected cell lines. From the earliest attempts, glass has been used as the substrate, initially because of its optical properties, but subsequently because it appears to carry the correct charge for cells to attach and grow. With the exception of the above-mentioned cells and other transformed cell lines, most cells need to spread out on a substrate in order to proliferate [Fisher and Solursh, 1979; Folkman and Moscona, 1978]. Inadequate spreading due to poor adhesion or overcrowding will inhibit cell proliferation. Cells shown to require attachment for growth are said to be "anchorage dependent" (see also Chapter 15). Cells which have undergone transformation frequently become anchorage independent and can grow in suspension when stirred or held in suspension with semisolid media such as agar.

This assumes, however, that cell proliferation is the principal objective. It may not be; cells which are anchored only to each other as spheroids in suspension or which are growing as a secondary layer on top of a confluent monolayer may proliferate more slowly but may still reflect more accurately behavior *in vivo*.

Glass

Glass is often used as a substrate. It is cheap, easily washed without losing its growth-supporting properties, can be sterilized readily by dry or moist heat, and is optically clear. Treatment with strong alkali (e.g., NaOH or caustic detergents) renders glass unsatisfactory for culture until it is neutralized by an acid wash.

Disposable Plastic

Single-use polystyrene flasks provide a simple, reproducible substrate for culture. They are usually of good optical quality, and the growth surface is flat, providing uniform and reproducible cultures. Polystyrene, as manufactured, is hydrophobic, and does not provide a suitable surface for cell growth, so tissue culture plastics are treated by γ-irradiation, chemically, or with an electric arc to produce a charged surface which is then wettable. As the resulting product varies in quality from one manufacturer to another, samples from a number of sources should be tested by determining the plating efficiency and growth rate of your cells (See Chapters 11 and 19), in medium containing the normal and half-normal concentration of serum (high serum concentrations may mask imperfections in the plastic).

While polystyrene is by far the most common and cheapest plastic substrate, cells may also be grown on polyvinylchloride, polycarbonate (PVC), polytetrafluorethylene (PTFE), thermanox (TPX), and a number of other plastics. If you need to use a different plastic, it is worth trying to grow a regular monolayer and then attempting to clone cells on it (see Chapters 11 and 19), with and without pretreatment of the surface (see below).

Teflon (PTFE) is available in a charged (hydrophilic) and uncharged (hydrophobic) form; the charged form can be used for regular monolayer cells and the uncharged for macrophages and some transformed cell lines. TFE films are available as disposable petri dishes ("Petriperm," Heraeus), or as membranes to be incorporated in an autoclavable reusable culture vessel ("Chamber/Dish," Bionique, see Fig. 13.2). These dishes have two other advantages: (1) the substrate is permeable to O_2 and CO_2 and (2) the plastic is thin and, therefore, well suited to histological sectioning for light or electron microscopy.

Permeable substrates have been in use for many years. In 1965, Sandström suggested that hepatocytes survived better in the higher oxygen tension provided by growth in a cellophane sandwich. Growth of cells on floating collagen [Michalopouloos and Pitot, 1975; Lillie et al., 1980] and cellulose nitrate membranes [Savage and Bonney, 1978] have been used to improve the survival of epithelial cells and promote terminal differentiation (see Chapters 14 and 20).

It is possible that growth of cells on a permeable substrate contributes more than increased diffusion of oxygen, CO_2, and nutrients. Attachment to a natural substrate such as collagen may exert some biological control of phenotypic expression due to the interaction of receptor sites on the cell surface with specific sites in the extracellular matrix. Permeability of the surface to which the cell is anchored may, in itself, signify polarity to the cell by simulating the basement membrane underlying an epithelial cell layer or between tissue cells and endothelium surrounding the vascular space. Such polarity may be vital to full functional expression in secretory epithelia and many other cell types. This prompted Reid and Rojkind [1979], Gospodarowicz [Vlodavsky et al., 1980], and others to explore the growth of cells on natural substrates related to basement membrane (See below and Chapter 14).

Microcarriers

Polystyrene (Nunclon, GIBCO), Sephadex (Flow Laboratories and Pharmacia), and polyacrylamide (Biorad) are available in bead form for propagation of anchorage-dependent cells in suspension (see Chapter 23).

Sterilization of Plastics

Disposable plasticware is usually supplied sterile and cannot be reused as washing in detergents renders the surface unsuitable for monolayer culture. For the sterilization of other plastics, see Chapter 8.

Alternative Artificial Substrates

Although glass and plastic are employed for more than 90% of all cell propagation, there are alternative substrates which can be used for specialized applications. Westermark [1978] developed a method for the growth of fibroblasts and glia on palladium. Using electron microscopy shadowing equipment, he produced islands of palladium on agarose, which does not allow cell attachment in fluid media. The size and shape of the islands was determined by masks made in the manner of electronic printed circuits, and the palladium was applied by "shadowing" under vacuum, as used in electron microscopy.

Cells may be grown on stainless steel discs [Birnie and Simons, 1967], or other metallic surfaces [Litwin, 1973]. Observation of the cells on an opaque substrate requires surface interference microscopy, unless very thin metallic films are used, as with Westermark's palladium islands.

Treated Surfaces

Cell attachment and growth can be improved by pretreating the substrate in a variety of ways [Barnes et al., 1984a]. It is a well-established piece of tissue culture lore that used glassware supports growth better than new. This may be due to etching of the surface or minute traces of residue left after culture. Growth of cells in a flask also improves the surface for a second seeding and this type of conditioning may be due to collagen [Hauschka and Konigsberg, 1966] or fibronectin [Thom et al., 1979] released by the cells. The substrate can be conditioned by treatment with spent medium for another culture [Stampfer et al., 1980], or by purified fibronectin (1ng/ml) added to the medium [Gilchrest et al., 1980], or collagen [Elsdale and Bard, 1972; Kleinman et al., 1979, 1981]. Treatment with denatured collagen improves the attachment of many cells such as epithelial cells [Lillie et al., 1980; Freeman et al., 1976] and muscle cells [Hauschka and Konigsberg, 1966], and it may be necessary for the expression of differentiated functions by these cells (see Chapters 14 and 20).

Coating with denatured collagen may be achieved using rat tail collagen or commercially supplied alternatives (Vitrogen, Flow) and simply pouring the collagen solution over the surface of the dish, draining off the excess, and allowing the residue to dry. As this sometimes leads to detachment of the collagen layer during culture, the following protocol has been suggested to ensure the collagen remains firmly anchored to the substrate. The following protocol on coating surfaces with cross-linked collagen was contributed by Jeffrey O. Macklis, Department of Neuroscience, Children's Hospital, and Department of Neuropathology, Harvard Medical School, Boston, Massachusetts, 02115.

Principle

A new type of collagen surface for culture of nervous system cells was described by Macklis et al., 1985, which allowed extended culture survival, improved microscopy, and dry storage of coated culture dishes. Collagen was derivatized to plastic culture dishes by a cross-linking reagent, 1-cyclohexyl-3-(2-morpholinoethyl)-carbodiimide-metho-p-toluenesulfonate (carbodiimide), and comparison to conventional ammonia-polymerized or adsorbed surfaces showed superior culture viability and improved optical characteristics. Simple covalent bonding of collagen fibrils to active groups on tissue culture plastic is described below. A large supply of coated dishes can be prepared in a single five-h session and stored for later use.

Outline

Prepare stock collagen solution, then dilute into an aqueous solution of carbodiimide. Coat dishes, incubate, wash, air dry, sterilize under UV, and use or store dry.

Materials

Collagen solution in dilute acetic acid at protein concentration of approximately 500 μg/ml. This can be purchased commercially or prepared by extraction from rat tails by the method of Bornstein (1958) carbodiimide (Aldrich Chemical Co., Milwaukee, WI) tissue culture plates (Falcon Plastics; Becton, Dickenson & Co.) double distilled water, sterilized by autoclaving

Protocol

1.
Place approximately 2 μg of carbodiimide in each of several 15 ml sterile medium tubes. Seal and store at 4°C until used.
2.
Add 14 ml of sterile, double distilled water at room temperature to each tube containing carbodiimide to be used (each prepared tube will coat fifteen 35 mm culture dishes).

3.
Vortex each tube for approximately ten seconds and set aside.
4.
Add 1 ml of stock collagen solution to one tube containing carbodiimide solution (approximately 130 μg/ml) and rapidly vortex until uniform.
5.
Rapidly transfer collagen-carbodiimide solution to dishes, generously covering the bottoms (approximately 1 ml in a 35 mm dish). The rapidity of transfer minimizes derivitization to the solution tube and maximizes early contact with the dish.
6.
Incubate dishes at 25°C for three h.
7.
Wash three times with sterile, double-distilled water.
8.
Air dry at room temperature for one h.
9.
Sterilize under ultraviolet irradiation for one h.
10.
Use dishes immediately or store dry for later use.

Collagen may also be applied as an undenatured gel (see Chapter 14), and this type of substrate has been shown to support neurite outgrowth from chick spinal ganglia [Ebendal, 1976], morphological differentiation of breast [Yang et al., 1981], and other epithelia [Sattler et al., 1978] and to promote expression of tissue-specific functions of a number of other cells *in vitro* [Meier and Hay, 1974, 1975; Kosher and Church, 1975]. In this case the collagen is diluted 1:10 with culture medium and neutralised to pH 7.4. This causes the collagen to gel so the dilution and dispensing must be rapid. It is best to add the growth medium to the gel for a further 4–24 h to ensure the gel equilibrates with the medium before adding cells. At this stage fibronection (25–50 μg/ml) and/or laminin (1–5 μg/ml) may be added to the medium.

Evidence is gradually accumulating that specific treatment of the substrate with biologically significant compounds can induce specific alterations in attachment or behavior of specific cell types. For example, chondronectin enhances chondrocyte adherence [Varner et al., 1984] and laminin epithelial cells [Kleinman et al., 1981]. Reid and Rojkind [1979] described methods for preparing reconstituted "basement membrane rafts" from tissue extracts for optimization of culture conditions for cell differentiation.

Gelatin coating has been found to be beneficial for the culture of muscle [Richler and Yaffe, 1970] and endothelial cells [Folkman et al., 1979] (see Chapter 20), and it is necessary for some mouse teratomas. McKeehan and Ham [1976] found that it was necessary to coat the surface of plastic dishes with 1 mg/ml poly-D-lysine before cloning in the absence of serum (see Chapter 11).

This raises the interesting question of whether the cell requires at least two components of interaction with the substrate: (1) adhesion to allow the attachment and spreading necessary for cell proliferation [Folkman and Moscona, 1978] and (2) specific interactions, reminiscent of the interaction of an epithelial cell with basement membrane, with other extracellular matrix constituents, or with adjacent tissue cells [Auerbach and Grobstein, 1958]. The second type of interaction may be less critical to sustained proliferation of undifferentiated cells but may be required for the expression of some specialized functions (see Chapters 2 and 14).

While inert coating of the surface may suffice, it may yet prove necessary to use a monolayer of an appropriate cell type to provide the correct matrix for maintenance of some specialized cells. Gospodarowicz et al. [1980] were able to grow endothelium on confluent monolayers of 3T3 cells which had been extracted with Triton × 100, leaving a residue on the surface of the substrate. This so-called extracellular matrix (ECM) has also been used to promote differentiation in ovarian granulosa cells [Gospodarowicz, 1980] and in studying tumor cell behavior [Vlodavsky et al., 1980].

Outline

Extract a postconfluent monolayer of matrix forming cells with detergent, wash and seed required cells on to matrix so produced.

Materials

3T3 mouse fibroblasts, MRC-5 human fibroblast or CPAE bovine arterial endothelial cells (or any other cell line shown to be suitable for producing extracellular matrix)

1% Triton X 100 in sterile distilled deionized water

Sterile distilled deionized water

Protocol

1.

Set up matrix producing cultures, and grow to confluence.

2.

After 3–5 days at confluence, remove the medium and add an equal volume of sterile 1% Triton X 100 in distilled water to the cell monolayer.

3.

Incubate for 30 min at 37°C.

4.

Remove Triton X solution and wash residue three times with the same volume of sterile distilled water.

5.

Flasks may be used directly or stored at 4°C for up to 3 weeks.

Feeder Layers

Cultures of mouse embryo fibroblasts, or other cells, have been used for many years to enhance growth particularly at low cell densities (see Chapter 11) [Puck and Marcus, 1955]. This action is due partly to supplementation of the medium but may also be due to conditioning of the substrate by cell products. Feeder layers grown as a confluent monolayer may make the surface suitable for attachment for other cells (see Chapters 20 and 21). We have shown selective growth of breast and colonic epithelium, and of glioma, on confluent feeder layers of normal fetal intestine [Freshney et al., 1982].

The survival and neurite extension by central and peripheral neurons can be enhanced by culturing the neurons on a monolayer of glial cells, although in this case the effect is due to a diffusible factor rather than direct cell contact [Lindsay, 1979].

After a monolayer culture reaches confluence subsequent proliferation causes cells to detach from the artificial substrate and migrate over the surface of the monolayers. Their morphology may change (Fig. 7.1), and the cells are less well spread, more densely staining, and may be more highly differentiated. Apparently, the interaction of a cell with a cellular underlay is different from the interaction with a synthetic substrate. This can cause the change in morphology and reduce proliferative potential.

Three-Dimensional Matrices

It has long been realized that while growth in two dimensions is a convenient way of preparing and observing a culture and allows a high rate of cell prolif-

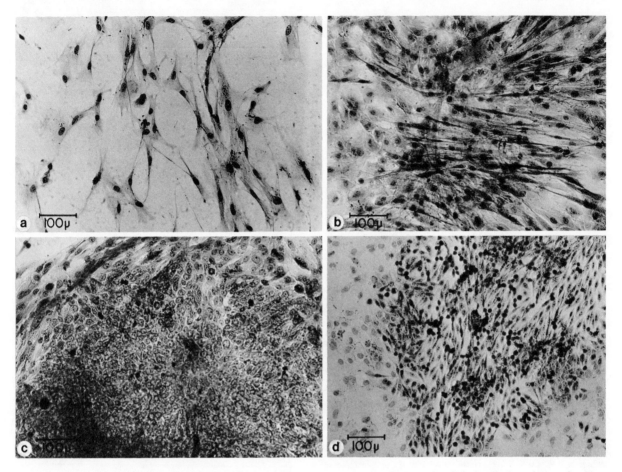

Fig. 7.1. *Morphological alteration in cells growing on feeder layers. a. Fibroblasts from human breast carcinoma growing on plastic and (b) growing on a confluent feeder layer of fetal human intestinal cells (FHI). c. Epithelial cells from human breast carcinoma growing on plastic and (d) on same confluent feeder layer as in b.*

eration, it lacks the cell-cell and cell-matrix interaction characteristic of whole tissue *in vivo*. The very first attempts to culture animal tissues [Harrison, 1907; Carrel, 1912] were performed with gels formed of clotted lymph or plasma on glass. In these cases, however, the cells migrated along the glass/clot interface rather than within the gel and tissue architecture and cell-cell interaction was gradually lost. Migration was often accompanied by proliferation of cells in the outgrowth, leading, in later studies, to the development of propagated cell lines. It gradually became apparent that many functional and morphological characteristics were lost during serial subculture, as discussed in Chapter 2.

These deficiencies encouraged the exploration of three-dimensional matrices such as collagen gel [Douglas et al., 1980]; cellulose sponge, alone [Leighton et al., 1951], or collagen-coated [Leighton et al.,

1968]; or Gelfoam (see Chapter 22). Many different cell types can be shown to penetrate such matrices and establish a tissue-like histology. Breast epithelium, seeded within collagen gel, displays a tubular morphology, while breast carcinoma grows in a more disorganized fashion, confirming the correlation between this mode of growth and the condition *in vivo* [Yang et al., 1981].

Neurite outgrowth from sympathetic ganglia neurons growing on collagen gels follows the orientation of the collagen fibers in the gel [Ebendal, 1976] (see further discussion of three-dimensional cultures, Chapter 22).

Nonadhesive Substrates

There are situations where attachment of the cell is undesirable. The selection of virally transformed colonies, for example, can be achieved by plating cells in

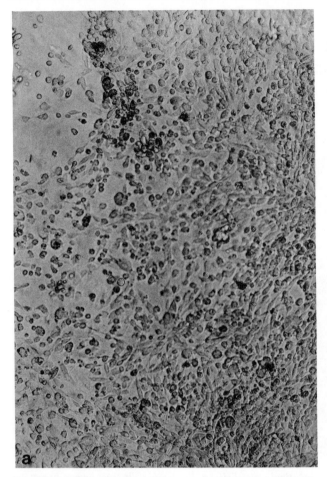 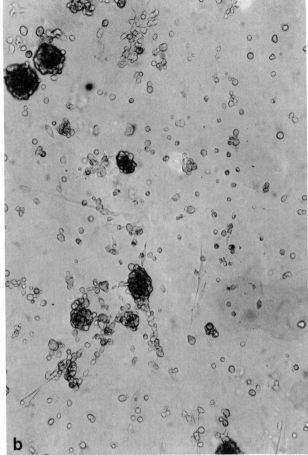

Fig. 7.2. *Cell growth at interface between Methocel-containing medium and agar gel. Methocel concentration, 1.5%; agar, 1.25%. Human metastatic melanoma. In a, 2.5 × 10^5 cells per ml, cloned alone. In b, 5 × 10^4 cell cloned with 2 × 10^5 homologous feeder cells per ml.*

agar [Macpherson and Montagnier, 1964], as the untransformed cells do not form colonies readily in this matrix.

There are two principles involved in this system: (1) prevention of attachment at the base of the dish where spreading would occur and (2) immobilization of the cells such that daughter cells remain associated with the colony even if nonadhesive. The usual agents employed are agar, agarose, or Methocel (Methylcellulose viscosity 4,000 cps). The first two are gels and the third is a high-viscosity sol. Because Methocel is a sol, cells will sediment slowly through it. It is, therefore commonly used with an underlay of agar (see Chapter 11). Non-tissue culture grade dishes can be used without an agar underlay, but some attachment and spreading may occur.

Liquid-Gel or Liquid-Liquid Interfaces

While the Methocel-over-agar system usually gives rise to discrete colonies at the interface of the agar and the Methocel, some cells can migrate across the gel surface and form monolayers or cords of cells (Fig. 7.2). The reason for this remains obscure, although the concentrations of Methocel and agar in this example are higher than normal and may have contributed to the effect. Rosenberg [1965] observed cell spreading and monolayer formation with HeLa cells at the liquid-liquid interface between various fluorinated hydrocarbons (FC43, FC73) and aqueous culture media. The occurrence of spreading and locomotion on nonrigid substrates conflicts somewhat with current concepts of cell adhesion and locomotion unless denatured serum protein or some other substance forms a layer at the interface sufficient to permit anchorage. Methocel, particularly, often contains particulate debris which may help to promote this.

Perfused Microcapillary Bundles

Knazek et al. [1972] developed a technique for the growth of cells on the outer surface of bundles of

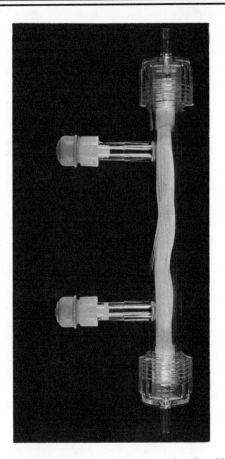

Fig. 7.3. *Vitafiber Chamber (Amicon). A bundle of hollow fibers of permeable plastic is enclosed in a transparent plastic outer chamber, accessible via either of the two side arms for seeding cells. During culture, the chamber is perfused down the center of the hollow fibers through connections attached to either end of the chamber (see also Fig. 21.3).*

plastic microcapillaries (Fig. 7.3) (see Chapter 22). The plastic allows the diffusion of nutrients and dissolved gases from medium perfused through the capillaries. Cells will grow up to several cells deep on the outside of the capillaries and an analogy with whole tissue is suggested.

Culture Vessels

Some typical culture vessels are listed in Table 7.1. The anticipated yield of HeLa cells is quoted for each vessel; the yield from a finite cell line, e.g., diploid fibroblasts, would be about one-fifth of the HeLa figure. Several factors govern the choice of culture vessel including: (1) the cell yield; (2) whether the cells grow in suspension or as a monolayer; (3) whether the culture should be vented to the atmosphere or sealed; (4) what form of sampling and analysis is to be performed; and (5) the anticipated cost.

Cell yield. For monolayer cultures, the cell yield is proportional to the available surface area of the flask. Small volumes and multiple replicates are best performed in multiwell dishes (Fig. 7.4) which range from Teresaki plates (60–72 wells, 10 μl culture volume) up to four wells, 50 mm in diameter. The most popular are microtitration dishes (96 or 144 wells, 0.1–0.2 ml, 0.25 cm^2 growth area) and 24-well "cluster dishes" (1–2 ml each well, 1.75 cm^2 (see Table 7.1). The middle of the size range embraces both petri dishes (Fig. 7.5) and flasks ranging from 20 cm^2–175 cm^2 (Fig. 7.6) Flasks are usually designated by their surface area, e.g., No. 25 or No. 120. Glass bottles are more variable since they are usually drawn from standard pharmaceutical supplies (Fig. 7.7). They should have: (1) one reasonably flat surface; (2) a deep screw cap with a good seal and nontoxic liner; and (3) shallow sloping shoulders to facilitate harvesting monolayer cells after trypsinization and to improve the efficiency of washing.

If you require large cell yields (e.g., ~ 10^9 HeLa cervical carcinoma cells or 2 × 10^8 MCR-5 diploid human fibrolast), then increasing the size and number of conventional bottles becomes cumbersome and special vessels are required. These are described in Chapter 23. Increasing the yield of cells growing in suspension requires only that the medium volume be increased, as long as cells in deep culture are kept agitated and sparged with 5% CO_2 in air (see Chapter 23).

Venting. Multiwell dishes and petri dishes chosen for replicate sampling or cloning have loose-fitting lids to give easy access to the dish. Consequently, they are not sealed and will require a humid atmosphere with control of the CO_2 tension (see below and Chapter 4). Because a thin film of liquid may form around the inside of the lid, partially sealing some dishes, loose-fitting lids should be provided with supports to raise them off the rim of the dish (Fig. 7.8). If a perfect seal is required, some multiwell dishes can be sealed with self-adhesive Mylar film (Flow). Flasks may be vented by slackening the caps. Again, because of variable sealing due to liquid inside the cap, the cap must be slackened one full turn. Flasks are vented in this way to allow CO_2 to enter (in a CO_2 incubator) or to allow excess CO_2 to escape in excessive acid-producing cell lines.

Sampling and analysis. Multiwell plates are ideal for replicate culture if all samples are to be removed simultaneously and processed in the same way. If, on the other hand, samples need to be withdrawn at differ-

TABLE 7.1 Culture Vessel Characteristics*

Culture vessel	Plastic or glass	No. of replicates	Vol.	Surface area	Approx. cell yield	Supplier
Microtest (Terasaki)	P	60, 72	0.01 ml	0.78 mm^2	2.5×10^3	F, N
Microtitration plate	P	96, 144	0.1 ml	32 mm^2	10^5	C, F, N, L
Multiwell plate†	P	4 round	1.0 ml	2 cm^2	5×10^5	C, F, N
Multiwell plate	P	12 round	2.0 ml	4.5 cm^2	10^6	F
Multiwell plate	P	24 round	1.0 ml	2 cm^2	5×10^5	C, F, L
Multiwell plate	P	8 rectangular	2.0 ml	7.8 cm^2	2×10^6	Lu
Multiwell plate	P	4 rectangular	3.0 ml	16.1cm^2	4×10^6	Lu
Multiwell plate	P	6 round	2.5 ml	9.6 cm^2	2.5×10^6	C, L
Multiwell plate	P	4 round	5.0 ml	28 cm^2	7×10^6	F, N, L
Petri dishes†						
30 mm	P		2.0 ml	6.9 cm^2	1.7×10^6	S
35 mm	P		3.0 ml	8.0 cm^2	2.0×10^6	C, Cg
50 mm	P		4.0 ml	17.5 cm^2	4.4×10^6	S, F, N
60 mm	P		5.0 ml	21 cm^2	5.2×10^6	C, Cg, S, N
90 mm	P		10.0 ml	49 cm^2	12.2×10^6	S, F
100 mm	P		10.0 ml	55 cm^2	13.7×10^6	C, Cg, F, N
100 mm, square	P		15.0 ml	100 cm^2	20×10^6	S
Tissue culture tubes						
Leighton†	P & G		1.0 ml	4.00 cm^2	10^6	Be
One side flattened	P		2.0 ml	5.50 cm^2	10^6	N
Round with screw cap	P		2.0 ml			N
Flasks						
25	P		5.0 ml	25 cm^2	5×10^6	C, Cg, F, N
50	G		10–20 ml	50 cm^2	10^7	
75	P & G		15–30 ml	75 cm^2	2×10^7	C, Cg, F, N
120	G		40–100 ml	120 cm^2	5×10^7	
150	P		75 ml	150 cm^2	6×10^7	C, Cg
175	P		50–100 ml	175 cm^2	7×10^7	F, N
Roller bottles						
2500 ml	G		100–250 ml	700 cm^2	2.5×10^8 (~ 1 g)	N.B
Roller disposable	P		100–250 ml	850 cm^2	3.0×10^8	F, Cg
Large	G		100–500 ml	1,585 cm^2	6.0×10^8	N.B
Nunc cell factory	P		1,800 ml	6,000 cm^2	2.0×10^9	N
Spiral	P		1,600 ml	8,500 cm^2	2.5×10^9 (~ 10 g)	S
Microcarriers	See Chapter 21		See Stirrer bottles, below			Ph, B, N, F
Stirrer bottles						
Reagent bottle, round (500 ml)	G		200 ml		3×10^8	Cg, Be
Reagent bottle, round (1,000 ml)	G		400 ml		8×10^8	Cg, Be
Aspirator (2,000 ml)	G		600 ml		10^9	P
Aspirator (5,000 ml)	G		4,000 ml	Gas with 5% CO_2	6×10^9	P
Aspirator (10,000 ml)	G		8,000 ml	Gas with 5% CO_2	8×10^9	P

*Abbreviations: B, Bio Rad; Be, Bellco; Cg, Corning; C, Costar; F, Falcon; L, Linbro (Flow); Lu, Lux; N, Nunc (GIBCO); N.B, New Brunswick; P, Pyrex; Ph, Pharmacia; S, Sterilin.

†Dishes and Leighton Tubes can be used on their own or with a coverslip, e.g., glass, TPX (Lux), Polystyrene (Lux), Melinex (I.C.I.). Non-tissue culture grade dishes may be used with coverslips. Petri dish sizes often refer to the outside diameter of the base or lid. Surface area must be calculated from the inside diameter of the base.

ent time intervals and processed immediately, it may be preferable to use separate vessels (flasks, test tubes, etc.) (Fig. 7.9) Individual wells in microtitration plates can be sampled by cutting and removing only that part of the adhesive plate sealer overlying the wells to be sampled. Alternatively, microtitration plates are avail-able with removable wells for individual processing, although these are only suitable for suspension cells. They are not tissue culture treated as yet.

If processing of the sample involves extraction in acetone, toluene, ethyl acetate, or certain other organic solvents, then a problem will arise with polystyrene.

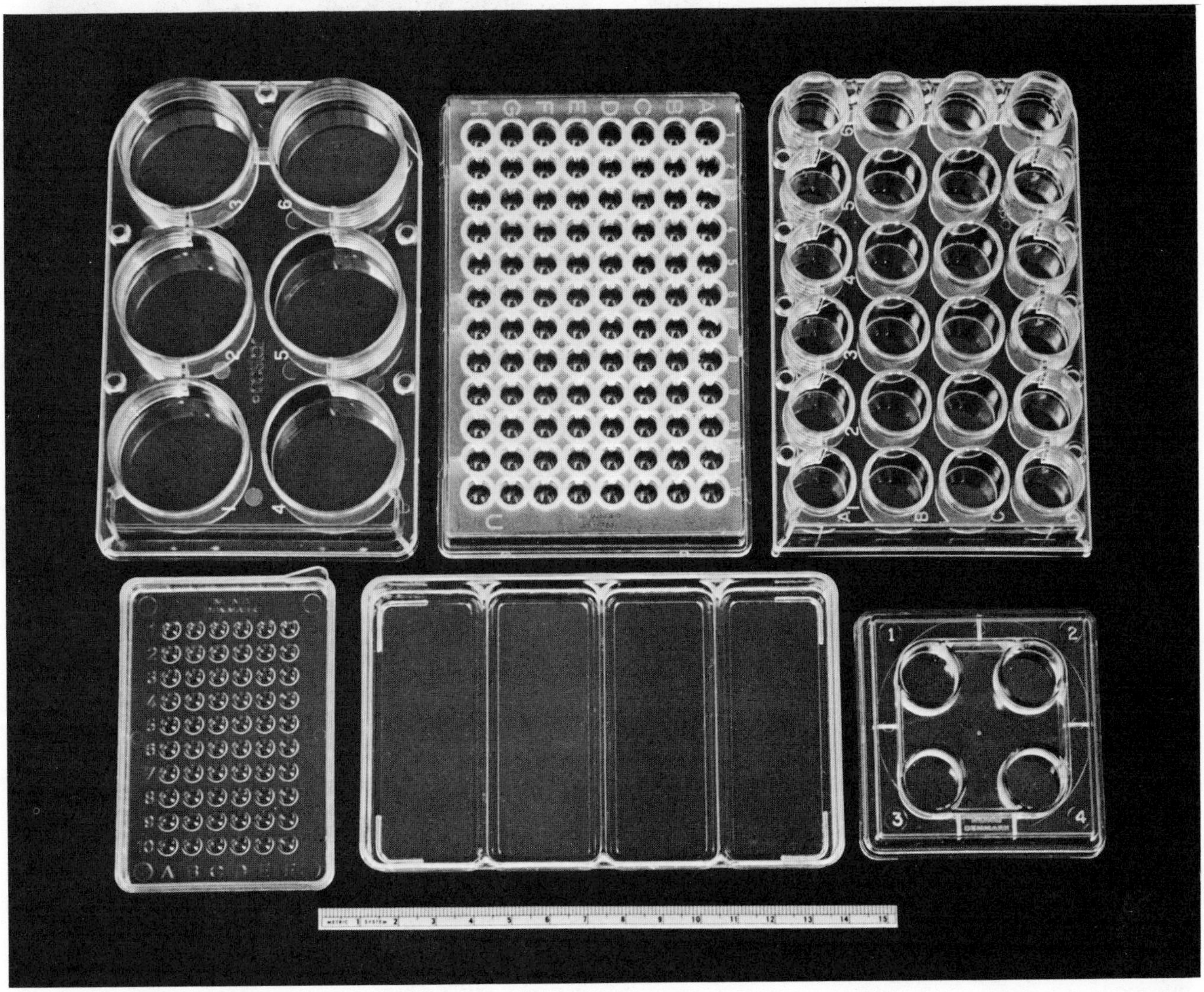

Fig. 7.4. *Multiwell plates (see Table 7.1 for sizes and capacities).*

Since this problem is often associated with histological procedures, Lux supplies Thermanox (TPX) plastic coverslips, suitable for histology, to fit into regular multiwell dishes (which need not be tissue culture grade). However, they are of poor optical quality and should be mounted cells uppermost with a conventional glass coverslip on top.

Glass vessels are required for procedures such as hot perchloric acid extractions of DNA. Plain-sided test tubes or Erlenmeyer flasks (no lip) used in conjunction with sealing tape or Oxoid caps are quick to use and are best kept in a humid CO_2-controlled atmosphere. Regular glass scintillation vials, or "minivials," are also good culture vessels as they are flat-bottomed and have a screw closure. Once used with scintillant, however, they should not be reused for culture.

Cost. Cost always has to be balanced against convenience—e.g., petri dishes are always cheaper than

Fig. 7.5. *Some common sizes of disposable plastic petri dishes. Sizes range from 35 mm to 90 mm diameter, circular, and 9 cm × 9 cm, square. Larger dishes are available but are seldom used for cell culture. A grid pattern can be provided to help in scanning the dish—to count colonies, for example.*

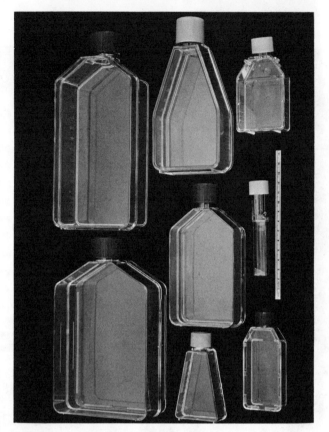

Fig. 7.6. *Disposable plastic culture vesels (Falcon, Costar and Corning). The triangular bottles (Costar) are designed to improve access to all of the growth surface when dispersing a monolayer (see Table 7.1 for sizes and capacities).*

flasks of an equivalent surface area but require humid CO_2-controlled conditions and are more prone to infection.

Cheap soda glass bottles, though not always of good optical quality, are often better for culture than higher grade Pyrex or optically clear glass, which usually contains lead.

A major disadvantage of glass is that it is labor intensive in preparation as it must be carefully washed and resterilized before it can be reused. The cost of this will depend on your existing staff and the number of flasks used. To employ a new member of staff to wash and sterilize glassware will cost about half the amount per flask, relative to disposable plastic, for an annual output of 10,000 flasks. If your usage is substantially less than this, it will be better to use disposable plastic, particularly for smaller flasks (25 cm^2). If you do not need to meet the cost of employing washing-up staff, glass will be found much cheaper than plastic.

THE GAS PHASE

Oxygen

The significant constituents of the gas phase are oxygen and carbon dioxide. Cultures vary in their oxygen requirement; the major distinction lying between organ and cell cultures. While atmospheric, or lower, oxygen tensions [Cooper et al., 1958; Balin et al., 1976] are preferable for most cell cultures, some organ cultures, particularly from late stage embryo, newborn, or adult, require up to 95% O_2 in the gas phase [Trowell, 1959; DeRidder and Mareel, 1978]. This may be a problem of diffusion related to the geometry of organ cultures (see Chapter 22) rather than a distinct cellular requirement, since most dispersed cells prefer lower oxygen tensions, and some systems, e.g., human tumor cells in clonogenic assay [Courtney et al., 1978], and human embryonic lung fibroblasts [Balin et al., 1976], do better in less than the normal atmospheric oxygen tension. It has been

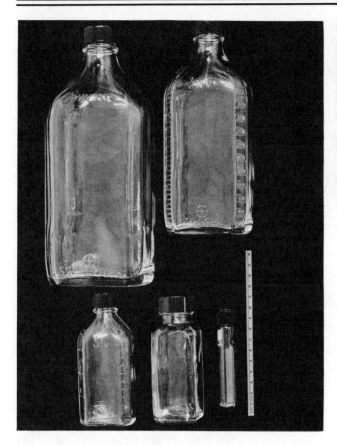

Fig. 7.7 *Examples of standard glass bottles which may be used as culture flasks.*

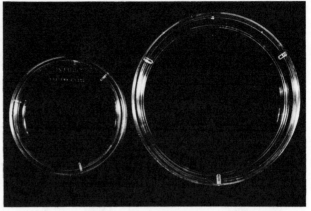

Fig. 7.8. *"Vented" dishes (9 cm and 6 cm diameter).*

As HCO^-_3 has a fairly low dissociation constant with most of the available cations, it tends to reassociate, leaving the medium acid. The net result of increasing atmospheric CO_2 is to depress the pH, so the effect of elevated CO_2 tension is neutralized by increasing the bicarbonate concentration:

$$NaHCO_3 \rightleftharpoons Na^+ + HCO_3^- \qquad (2)$$

The increased HCO^-_3 concentration pushes equation 1 to the left until equilibrium is reached at pH 7.4. If another alkali, e.g., NaOH, is used instead, the net result is the same.

$$NaOH + H_2CO_3 \rightleftharpoons NaHCO_3 + H_2O \qquad (3)$$
$$\rightleftharpoons Na^+ + HCO_3^- + H_2O$$

The equivalent $NaHCO_3$ concentrations commonly used with different CO_2 tensions are listed in Table 7.2 and 7.4.

Intermediate values of CO_2 and HCO^{3-} may be used provided the concentration of both are varied simultaneously. As many media are made up in acid solution and may incorporate a buffer, it is difficult to predict how much bicarbonate to use when other alkali may also end up as bicarbonate, as in equation 3 above. When preparing a new medium for the first time, add the specified amount of bicarbonate and then sufficient 1 N NaOH such that the medium equilibrates to the desired pH after incubation at 36.5°C overnight. When dealing with media already at working strength, vary the amount of HCO_3^- to suit the gas phase (Table 7.2) and leave overnight to equilibrate at 36.5°C. Each medium has a recommended bicarbonate concentra-

suggested [McKeehan et al., 1976] that the requirement for selenium in medium is related to oxygen tension and that this element helps to remove free radicals of oxygen. This requirement may only arise in the absence of serum proteins and, as there is a trend toward serum-free media, the role of dissolved O_2 may become more important in the future, requiring, perhaps, controlled O_2 tension during incubation. As the depth of the culture medium can influence the rate of oxygen diffusion to the cells, it is advisable to keep the depth of medium within the range 2–5 mm (0.2–0.5 ml/cm²) in static culture.

Carbon Dioxide

Carbon dioxide has a rather complex role to play, and because many of its actions are interrelated, e.g., dissolved CO_2, pH, and HCO^-_3 concentration, it is difficult to determine its major direct effect. The atmospheric CO_2 tension will regulate the concentration of dissolved CO_2 directly, as a function of temperature. This in turn produces H_2CO_3 which dissociates:

$$H_2O + CO_2 \rightleftharpoons H_2CO_3 \rightleftharpoons H^+ + HCO_3^- \qquad (1)$$

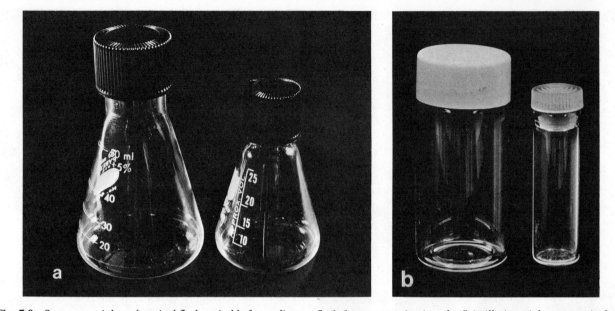

Fig. 7.9. *Screw-cap vials and conical flasks suitable for replicate cultures or sample storage. a. Screw caps are preferable to stoppers as they are less likely to leak and protect the neck of the flask from contamination. b. Scintillation vials are particularly useful for isotope incorporation studies but should not be reused for culture after containing scintillation fluid.*

TABLE 7.2. Variations in HCO_3^- and CO_2 Concentrations in Variants of Eagle's Minimum Essential Medium (MEM)

	Eagle's MEM Hanks' salts	Eagle's MEM Earle's salts	Dulbecco's modification of Eagle's MEM
$NaHCO_3$	4 mM	26 mM	44 mM
CO_2	Atmospheric & CO_2 evolved from culture	5%	10%

TABLE 7.3. Relationship Between HCO_3^-, CO_2 and HEPES Concentrations

Gas phase CO_2	Liquid phase	
	HCO_3^-	HEPES
Atmospheric	4 mM	10 mM
2%	8 mM	20 mM
5%	24 mM	50 mM

tion and CO_2 tension to achieve the correct pH and osmolality but minor variations will occur in different methods of preparation.

With the introduction of Good's buffers (e.g., Hepes, Tricine) [Good et al., 1966] into tissue culture, there was some speculation that since CO_2 was no longer necessary to stabilize the pH, it could be omitted. This has since proved to be untrue [Itagaki and Kimura, 1974], at least for a large number of cell types, particularly at low cell concentrations. Although 20 mM HEPES buffer can control pH within the physiological range, the absence of atmospheric CO_2 allows equation 1 to move to the left, eventually eliminating dissolved CO_2, and ultimately HCO_3^-, from the medium. This appears to limit cell growth, although whether the cells require the dissolved CO_2 or the HCO_3^- (or both) is not clear. Recommended HCO_3^-, CO_2, and HEPES concentrations are given in Table 7.3.

The inclusion of pyruvate in the medium enables cells to increase their endogenous production of CO_2, making them independent of exogenous CO_2 and HCO_3^-. Leibovitz L15 medium [Leibovitz, 1963] contains a higher concentration of sodium pyruvate (550 mg/1) but no $NaHCO_3$ and does not require CO_2 in the gas phase. Sodium-glycerophosphate can also be used to buffer autoclavable media lacking CO_2 and HCO_3^- [Waymouth, 1979].

In summary, cultures at low cell concentration in an open vessel need to be incubated in an atmosphere of CO_2, the concentration of which is in equilibrium with the sodium bicarbonate in the medium. At very low cell concentrations (e.g., during cloning), it is necessary to add CO_2 to the gas phase of sealed flasks for most cultures. At high cell concentrations, it will not be necessary to add CO_2 to the gas phase in sealed flasks, but it may yet be necessary in open dishes. Where the culture produces a lot of acid, and the endogenous production of CO_2 is high, it may be desirable to slacken the cap of a culture flask and allow excess CO_2 to escape. In these cases it is advisable to incorporate HEPES (20 mM) in the medium to stabilize the pH.

TABLE 7.4. Balanced Salt Solutions

Component	Earle's balanced salt solution gm/l	Dulbecco's phosphate buffered saline (solution A) (PBSA) gm/l	Hanks' balanced salt solution gm/l	Spinner salt solution (Eagle) gm/l
Inorganic salts				
$CaCl_2$ (anhyd.)	0.02	—	0.14	—
$CaCl_2 \cdot 2H_2O$	—	—	—	—
KCl	0.04	0.20	0.40	0.40
KH_2PO_4	—	0.20	0.06	—
$MgCl_2 \cdot 6H_2O$	—	—	0.10*	—
$MgSO_4 \cdot 7H_2O$	0.20	—	0.10	0.20
NaCl	6.68	8.00	8.00	6.80
$NaHCO_3$	2.20	—	0.35	2.20
$Na_2HPO_4 \cdot 7H_2O$	—	2.16	0.09**	—
$NaH_2PO_4 \cdot H_2O$	0.14†	—	—	1.40
Other components				
D-glucose	1.00	—	1.00	1.00
Phenol red	0.01	—	0.01‡	0.01

*$MgCl_2 \cdot 6H_2O$ added to original formula.

†Original formulation calls for 150.0 mg/l $NaH_2PO_4 \cdot 2H_2O$, In Vitro, 9, #6 (1974).

**Original formulation calls for 0.06 gm/l $Na_2HPO_4 \cdot H_2O$, Proc. Soc. Exp. Biol. and Med., 71 [1949].

‡Original formulation calls for 0.02 gm/l, Proc. Soc. Exp. Biol. and Med., 71 [1949].

MEDIA AND SUPPLEMENTS

The discovery that cells from explants could be subcultured and propagated *in vitro* led to attempts to provide more defined media to sustain continuous cell growth and replace the "natural" media like embryo extract, protein hydrolysates, lymph, etc. Basal media of Eagle [1955, 1959] and the more complex media 199 of Morgan et al. [1950] and CMRL 1066 of Parker et al. [1957], although "defined", are usually supplemented with 5–20% serum; it was the desire to eliminate this remaining undefined constituent that led to the evolution of such complex media as NCTC 109, Evans et al. [1956], and 135, Evans and Bryant [1965], Waymouth's MB 572/1 [1959], Ham's F10 [1963] and F12 [1965], Birch and Pirt [1971], the MCDB series [Ham and McKeehan, 1978], and Sato's hormone-supplemented media [Barnes and Sato, 1980].

One approach to developing a medium is to start with a rich medium such as Ham's F12 [1965] or medium 199 supplemented with a high concentration of serum (say 20%) and gradually attempt to reduce the serum concentration by manipulating the concentrations of existing constitutents and by adding new ones. This is a very laborious procedure but it has resulted in a number of different formulations for the culture of human fibroblasts and other cell types either in low serum concentrations or in its complete absence (see below).

Even after many years of exhaustive research into matching particular media to specific cell types and culture conditions, the choice of medium is not obvious and is often empirical. No all purpose medium has been developed for the more demanding requirements of specialized cells, and even transformed cells, cultured from spontaneous tumors, have highly specific requirements, differing among tumors, even of one type, and often differing from the normal cells of the same tissue. Hence serum has been retained for many cell types and is only gradually being eliminated after many years of careful and painstaking work.

PHYSICAL PROPERTIES

pH

Most cell lines will grow well at pH 7.4. Although the optimum pH for cell growth varies relatively little among different cell strains, some normal fibroblast lines perform best at pH 7.4–7.7, and transformed cells may do better at pH 7.0–7.4 [Eagle, 1973]. There have been reports that epidermal cells may be maintained at pH 5.5 [Eisinger et al., 1979]. It may prove advantageous to do a brief growth experiment (see Chapter 11 and 18) or special function analysis (e.g. Chapter 14) to determine the optimum pH.

Phenol red is commonly used as an indicator. It is red at pH 7.4, becoming orange at pH 7.0, yellow at pH 6.5, slightly bluish red at pH 7.6, and purple at pH 7.8. Since the assessment of color is highly subjective, it is useful to make up a set of standards using sterile balanced salt solution (BSS) and phenol red at the correct concentration and in the same type of bottle that you normally use for preparing medium.

Preparations of pH Standards

Materials

Hanks' Balanced Salt Solution, × 10 concentrate or powder, without bicarbonate or glucose, with 20 mM HEPES

nine bottles of a size closest to your standard medium bottles or culture flasks

distilled deionized water

sterile 0.1 N NaOH (make up in distilled deionized water and filter sterilize; see Chapter 8)

pH meter

Protocol

1.

Make up the BSS at pH 6.5, dispense into nine bottles (two extra to allow for breakage) of the appropriate size and autoclave.

2.

Adjust the pH to 6.5, 6.8, 7.0, 7.2, 7.4, 7.6, 7.8, with sterile 0.1 N NaOH, checking the pH of a sample of each bottle on a pH meter after allowing each bottle to equilibrate with the atmosphere.

3.

Keep sterile and sealed.

Buffering

Culture media require to be buffered under two sets of conditions: (1) open dishes, where evolution of CO_2 causes the pH to rise, and (2) overproduction of CO_2 and lactic acid in transformed cell lines at high cell concentrations, when the pH will fall. A buffer may be incorporated in the medium to stabilize the pH but in (1) exogenous CO_2 may still be required by some cell lines, particularly at low cell concentrations, to prevent total loss of dissolved CO_2 and bicarbonate from the medium (see above). In (2) it is usually preferable to leave the cap slack (shrouded in aluminum foil) or to use a CO_2-permeable cap (Camlab) to promote the release of CO_2.

A bicarbonate buffer is still used more frequently than any other, in spite of its poor buffering capacity at physiological pH, because of its low toxicity, low cost, and and nutritional benefit to the culture. HEPES is a much stronger buffer in the pH 7.2–7.6 range and is now used extensively at 10 or 20 mM. When HEPES is used with exogenous CO_2, it has been found that the

HEPES concentration must be more than double that of the bicarbonate for adequate buffering (see Table 7.3). A variation of Ham's F12 with 20 mM HEPES, 8mM bicarbonate, and 2% CO_2 has been used successfully in the author's laboratory for the culture of a number of different cell lines.

Osmolality

Most cultured cells have a fairly wide tolerance for osmotic pressure [see also Waymouth, 1970]. Since the osmolality of human plasma is about 290 mOsm/kg, it is reasonable to assume that this is the optimum for human cells *in vitro*, although it may be different for other species (e.g., around 310 mOsm/kg for mice [Waymouth, 1970]). In practice, osmolalities between 260 mOsm/kg and 320 mOsm/kg are quite acceptable for most cells. Slightly hypotonic medium may be better for petri dish culture to compensate for evaporation during incubation.

Osmolality is usually measured by freezing-point depression (Fig. 7.10) or elevation of vapor pressure. The measurement of osmolality is a useful quality control step if you are making up medium yourself as it helps to guard against errors in weighing and dilu-

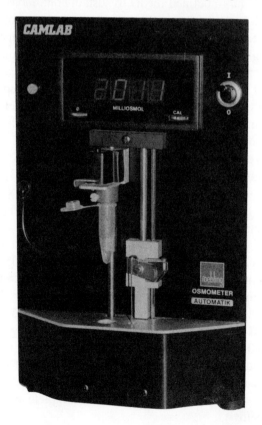

Fig. 7.10. *Roebling osmometer (Camlab, Cambridge, England). This model accepts samples of 50 μl.*

tion, etc. It is particularly important to monitor osmolality if alterations are made in the constitution of the medium. Addition of HEPES, drugs dissolved in strong acids and bases, and subsequent neutralization, can all markedly affect the osmolality.

Temperature

Apart from the direct effect of temperature on cell growth, it will also influence pH due to the increased solubility of CO_2 at lower temperatures, and possibly, due to changes in ionization and the pK_a of the buffer. The pH should be adjusted to 0.2 units lower at room temperature than at 36.5°C. It is best to make up the medium complete with serum and incubate a sample overnight at 36.5°C under the correct gas tension to check the pH when making up a medium for the first time.

Viscosity

The viscosity of culture medium is influenced mainly by the serum content, and in most cases will have little effect on cell growth. It becomes important, however, whenever a cell suspension is agitated, e.g., when a suspension culture is stirred, or when cells are dissociated after trypsinization. If there is cell damage under these conditions, then this may be reduced by increasing the viscosity of the medium with carboxy methyl cellulose or polyvinyl pyrolidone (see Reagent Appendix) [Birch and Pirt, 1971]. This becomes particularly important in low serum concentrations and in the absence of serum.

Surface Tension and Foaming

The surface tension of medium may be used to promote adherence of primary explants to the substrate (see Chapter 9) but is seldom controlled in any way. In suspension cultures, where 5% CO_2 in air is bubbled through medium containing serum, foaming may result. The addition of a silicone antifoam (Dow Chemical, Co.) helps to prevent this by reducing surface tension.

The effects of foaming have not been clearly defined. The rate of protein denaturation may increase, and the risk of contamination increases if the foam reaches the neck of the culture vessel.

CONSTITUENTS OF MEDIA

Balanced Salt Solutions

It is useful to distinguish balanced salt solutions from media for the present discussion. A balanced salt solution (BSS) is composed of inorganic salts, usually including sodium bicarbonate, and is supplemented with glucose, although glucose and bicarbonate are often omitted. The compositions of some common examples are given in Table 7.4. HEPES buffer (5–20 mM) may be added to these if necessary and the equivalent weight of NaCl omitted to maintain the correct osmolality.

BSS is used as a diluent for more complete media, as a washing or dissection medium, for short incubations up to about 4 h, and for a variety of other purposes which require an isotonic solution which is not necessarily nutritionally complete.

The choice of BSS is dependent on: (1) the CO_2 tension (see above and Tables 7.2, 7.4). The bicarbonate concentration must be such that equilibrium is reached at pH 7.4 at 36.5°C, e.g., Earle's BSS (EBSS) is commonly used as a diluent for Eagle's media for equilibration with 5% CO_2 while Hanks' BSS [Hanks and Wallace, 1949] (HBSS) (see Reagent Appendix) is used with air; (2) its use for tissue disaggregation, or monolayer dispersal. In these cases Ca^{2+} and Mg^{2+} are usually omitted as in Moscona's [1952] calcium- and magnesium-free saline (CMF), or Dulbecco and Vogt's [1954] phosphate-buffered saline, solution A (PBSA) (see Table 7.4); and (3) its use for suspension culture of adherent cells. MEM(S) is a variant of Eagle's [1959] minimum essential medium, deficient in Ca^{2+}, to reduce cell aggregation and attachment (see Table 7.4).

HBSS, EBSS, and PBS rely on the relatively weak buffering of phosphate, which is not at its most effective at physiological pH. Paul [1975] constructed a tris-buffered BSS which is more effective, but for which the cells sometimes require a period of adaptation. HEPES is currently the most effective buffer in the pH 7.2–7.8 range and TRICINE in the pH 7.4–8.0 range, although these tend to be expensive if used in large quantities. Sodium β-glycerophosphate (pK_a 6.6) also works well [Waymouth, personal communication].

DEFINED MEDIA

Defined media vary in complexity from Eagle's minimum essential medium (MEM) [Eagle, 1959], which contains essential amino acids, vitamins, and salts, to complex media such as 199 [Morgan et al., 1950], CMRL 1066 [Parker et al., 1957], RPMI 1640 [Moore et al., 1967], and F12 [Ham, 1965] (Table 7.5). The complex media contain a larger number of different amino acids and vitamins and are often supplemented with extra metabolites (e.g., nucleosides) and minerals. Nutrient concentrations are, on the whole, low in F12 and high in Dulbecco's modification of Eagle's MEM (DMEM) [Dulbecco and Freeman, 1959; Morton, 1970], although the latter has fewer constitutents. Barnes and Sato [1980] employed a 1:1 mixture of DMEM and F12 as the basis for their serum-free formulations to combine the richness of F12 and the higher nutrient concentration of DMEM.

The common constituents of medium may be grouped as follows (Table 7.5).

Amino Acids

The essential amino acids, i.e., those which are not synthesized in the body, are required by cultured cells with, in addition, cysteine and tyrosine, although individual requirements for amino acids will vary from one cell to another. Other nonessential amino acids are often added to compensate either for a particular cell type's incapacity to make them or because they are made but lost into the medium. The concentration of amino acids usually limits the maximum cell concentration attainable, and the balance may influence cell survival and growth rate. Glutamine is required by most cells although some cell lines will utilize glutamate. Recent evidence suggests that glutamine is used in cultured cells as an energy and carbon source [Reitzer et al., 1979].

Vitamins

Eagle's MEM contains only the B group vitamins (see Table 7.5), other requirements presumably being derived from the serum. The requirement for extra vitamins is most apparent where the serum concentration is reduced, but there are other cases (e.g., low cell densities for cloning) where they may be essential even in the presence of serum. Vitamin limitation is usually expressed in terms of cell survival and growth rate rather than maximum cell density.

Salts

The salts are chiefly those of Na^+, K^+, Mg^{2+}, Ca^{2+}, Cl^-, SO_4^{2-}, PO_4^{3-}, and HCO_3^- and are the major components contributing to the osmolality of the medium. Calcium is reduced for suspension cultures to minimize cell aggregation and attachment (see above). The sodium bicarbonate concentration is determined by the concentration of CO_2 in the gas phase (see above).

Glucose

Glucose is included in most media as an energy source. It is metabolized principally by glycolysis to form pyruvate which may be converted to lactate or to acetoacetate and enter the citric acid cycle to form CO_2. The accumulation of lactic acid in the medium, particularly evident in embryonic and transformed cells, implies that the citric acid cycle may not function entirely as *in vivo* and recent data have shown that much of its carbon is derived from glutamine rather than glucose. This may explain the exceptionally high requirement of some cultured cells for glutamine or glutamate.

Organic Supplements

A variety of other compounds including nucleosides, citric acid cycle intermediates, pyruvate, and lipids appear in complex media. Again these constituents have been found to be necessary when the serum concentration is reduced and may help in cloning and in maintaining certain specialized cells.

Hormones and Growth Factors

See below under Serum.

SERUM

The sera used most in tissue culture are calf serum, fetal bovine serum, horse serum, and human serum. The choice of serum will be discussed below in more detail. Calf serum is the most widely used, fetal bovine second, usually for more demanding cell lines, and human serum in conjunction with some human cell lines. Horse serum is preferred to calf serum by some workers as it can be obtained from a closed herd and is often more consistent from batch to batch.

Although most cell lines still require the supplementation of the medium with serum, there are now many instances where cultures may be maintained and may proliferate serum-free (see Table 7.6). Continuous cell strains such as the L929 and HeLa were among the first to be grown serum-free [Evans et al., 1956; Waymouth, 1959; Birch and Pirt, 1971; Higuchi, 1977] and a degree of selection may have been involved. However, results from the laboratories of Ham [Ham and McKeehan, 1978], Sato [Barnes and Sato, 1980], and others [Carney et al., 1981] have demonstrated that serum may be reduced or omitted without cellular adaptation if nutritional and hormonal modifications are made to the media appropriate to the cell line being studied [Sato et al., 1982]. This has provided indirect evidence for the constitution of serum which will be considered before further discussion of serum-free media.

Protein

Although proteins are a major component of serum, the functions of many of these *in vitro* remain obscure; and it may be that relatively few proteins are required other than as carriers for minerals, fatty acids, and hormones, or as hormones themselves. Those proteins which have been found beneficial are albumin [Iscove and Melchers, 1978; Barnes and Sato, 1980] and globulins [Tozer and Pirt, 1964]. Fibronectin; cold-insoluble globulin) promotes attachment and α_2- macroglobulin inhibits trypsin [DeVonne and Mouray, 1978]. Fetuin in fetal serum enhances cell attachment [Fisher et al., 1958], and transferrin [Guilbert and Iscove, 1976] binds iron making it less toxic but available to the cell. There may be other proteins as yet uncharacterized, essential for cell attachment and growth.

Polypeptides

Natural clot serum stimulates cell proliferation more than serum from which the cells have been removed physically. This appears to be due to the release of a polypeptide from the platelets during clotting. This polypeptide, platelet-derived growth factor (PDGF) [Heldin et al., 1979; Antoniades et al., 1979] is one of a family of polypeptides with mitogenic activity and is probably the major growth factor in serum. PDGF stimulates growth in fibroblasts and glia, but platelet-derived factors may be inhibitory to growth, or promote differentiation in epithelial cells [Lechner et al., 1981]. Other growth factors such as fibroblast growth factor (FGF) [Gospodarowicz, 1974], epidermal growth factor (EGF) [Cohen, 1962; Carpenter and Cohen, 1977; Gospodarowicz et al., 1978], endothelial growth factor [Folkman et al., 1979; Maciag et al., 1979], multiplication-stimulating activity (MSA) [Dulak and Temin, 1973a,b], and insulin-like growth factors IGFI, IGFII (IGFII is homologous to MSA) [Greenstein et al., 1989], which have been isolated from whole tissue or released into the medium by cells in culture, have varying degrees of specificity [Hollenberg and Cuatrecasas, 1973] and are probably present in serum in small amounts [Gospodarowicz and Moran, 1974]. Many of these growth factors are available commercially (see "Growth Factors," Trade Index) in pure form.

Hormones

Hormones may exhibit a variety of different effects on cells, and it is often difficult to recognize the key pathway. Insulin promotes the uptake of glucose and amino acids [Hokin and Hokin, 1963; Segal, 1964; Fritz and Knobil, 1964; Kelly et al., 1978] and may owe its mitogenic effect to this property. Some growth factors (IGFI, IGFII) bind to the insulin receptor on the cell surface and may act similarly. Growth hormone may be present in sera, particularly fetal sera, and in conjunction with the somatomedins, may have a mitogenic effect. Hydrocortisone is also present in serum in varying amounts. It can promote cell attachment [Ballard and Tomkins, 1969; Fredin et al., 1979] and cell proliferation [Guner et al., 1977 McLean et al., 1986] (see also Chapter 20) but under certain conditions; (e.g., high cell density) may be cytostatic [Freshney et al., 1980a,b] and can induce cell differentiation [Moscona and Piddington, 1966; Ballard, 1979; McLean et al., 1986].

Serum replacement experiments suggest that other hormones required for culture may be present in serum (see below), making it necessary at present to retain serum for culture but underlining the desirability for its replacement.

Metabolites and Nutrients

Serum also contains amino acids, glucose, ketoacids, and a number of other nutrients and intermediary metabolites. These may be important in simple media but less so in complex media, particularly those with higher amino acid concentrations and other supplements.

Minerals

Serum replacement experiments have also suggested that trace elements and iron, copper, and zinc may be provided by serum [Ham and McKeehan, 1978], bound to serum protein. A requirement for selenium has also been demonstrated in the same way [McKeehan et al., 1976].

Inhibitors

Serum may also contain substances inhibiting cell proliferation [Harrington and Godman, 1980]. Some of these may be artifacts of preparation, e.g., bacterial toxins from contamination prior to filtration; the γ-globulin fraction may contain antibodies cross-reacting with the culture. Heat inactivation removes comple-

TABLE 7.5. Media*

Component	Eagle's MEM mg/l	Dulbecco's modification mg/l	Ham's F12 mg/l	CMRL 1066 mg/l	RPMI 1640 mg/l	199 mg/l	L15 mg/l	Fischer's mg/l	Waymouth's MB 752/1 mg/1
Amino acids									
L-alanine	—	—	8.90	25.0	—	25.0	225	—	—
L-arginine (free base)	—	—	—	—	200	—	500	—	—
L-arginine·HCl	126	84.0	211	70.0	—	70.0	—	15.0	75.0
L-asparagine	—	—	—	—	50.0	—	—	—	—
L-asparagine·H_2O	—	—	15.0	—	—	—	260	11.4	—
L-aspartic acid	—	—	13.3	30.0	20.0	30.0	—	—	60.0
L-cysteine (free base)	—	—	—	—	—	—	120	—	61.0
L-cystine	24	48.0	—	20.0	50.0	—	—	—	15.0
L-cystine, 2Na	—	—	—	—	—	23.7	—	23.7	—
L-cysteine·HCl·H_2O	—	—	35.1	260	—	0.0987	—	—	—
L-glutamic acid	—	—	14.7	75.0	20.0	66.8	—	—	150
L-glutamine	292	584	146	100	300	100	300	200	350
Glycine	—	30.0	7.50	50.0	10.0	50.0	200	—	50.0
L-histidine (free base)	—	—	—	—	15.0	—	250	—	128
L-histidine HCl·H_2O	42.0	42.0	21.0	20.0	—	21.9	—	81.1	—
L-hydroxy-proline	—	—	—	10.0	20.0	10.0	—	—	—
L-isoleucine	52.0	105	3.94	20.0	50.0	20.0	125	75.0	25.0
L-leucine	52.0	105	13.1	60.0	50.0	60.0	125	30.0	50.0
L-lysine HCl	73.1	146	36.5	70.0	40.0	70.0	93	50.0	240
L-methionine	15.0	30.0	4.48	15.0	15.0	15.0	75.0	100	50.0
L-phenylalanine	33.0	66.0	4.96	25.0	15.0	25.0	125	67.0	50.0
L-proline	—	—	34.5	40.0	20.0	40.0	—	—	50.0
L-serine	—	42.0	10.5	25.0	30.0	25.0	200	15.0	—
L-threonine	48.0	95.0	11.9	30.0	20.0	30.0	300	40.0	75.0
L-tryptophan	10.0	16.0	2.04	10.0	5.00	10.0	20.0	10.0	40.0
L-tyrosine	36.0	72.0	5.40	40.0	20.0	—	—	—	40.0
L-tyrosine 2Na	—	—	—	—	—	49.7	373	74.6	—
L-valine	47.0	94.0	11.7	25.0	20.0	25.0	100	70.0	65.0
Vitamins									
L-ascorbic acid	—	—	—	50.0	—	0.050	—	—	17.5
Biotin	—	—	0.0073	0.010	0.200	0.010	—	0.010	0.02
D-Ca pantothenate	1.00	4.00	0.480	0.010	0.250	0.010	1.00	0.500	1.00
Calciferol	—	—	—	—	—	0.100	—	—	—
Choline chloride	1.00	4.00	14.0	0.500	3.00	0.500	1.00	1.50	250
Folic acid	1.00	4.00	1.30	0.010	1.00	0.010	1.00	10.0	0.40
i-inositol	2.00	7.20	18.0	0.050	35.0	0.050	2.00	1.50	1.00
Nicotinamide	1.00	4.00	0.04	0.025	1.00	0.025	1.00	0.50	1.00
Pyridoxal HCl	1.00	4.00	0.062	0.025	—	0.025	—	0.50	—
Riboflavin	0.10	0.40	0.038	0.010	0.20	0.010	—	0.50	1.00
Thiamin HCl	1.00	4.00	0.34	0.010	1.00	0.010	—	1.00	10.0
Vitamin B_{12}	—	—	1.36	—	0.005	—	—	—	0.20
Pyridoxine HCl	—	—	0.062	0.025	1.00	0.025	—	—	1.00
Cholesterol	—	—	—	0.200	—	—	—	—	—
Para-aminobenzoic acid	—	—	—	0.050	1.00	0.050	—	—	—
Nicotinic acid	—	—	—	—	—	0.025	—	—	—
Menaphthone sodium bisulphite 3H_2O	—	—	—	—	—	0.019	—	—	—
Dl-α tocopherol PO_4·2Na	—	—	—	—	—	0.01	—	—	—
Vitamin A acetate	—	—	—	—	—	0.115	—	—	—
Riboflavin PO_4, 2Na	—	—	—	—	—	—	0.10	—	—
Thiamin mono PO_4, 2H_2O	—	—	—	—	—	—	1.00	—	—
Inorganic salts									
$CaCl_2$ (anhyd.)	200	200	—	200	—	—	—	—	—
$CaCl_2$·2H_2O	—	—	44.0	—	—	186.00	186	91.0	120
$Fe(NO_3)_3$·9H_2O	—	0.10	—	—	—	0.10	—	—	—
KCl	400	400	224	400	400	400	400	400	150

(continued)

TABLE 7.5. Media* (continued)

Component	Eagle's MEM mg/l	Dulbecco's modification mg/l	Ham's F12 mg/l	CMRL 1066 mg/l	RPMI 1640 mg/l	199 mg/l	L15 mg/l	Fischer's mg/l	Waymouth's MB 752/1 mg/l
KH_2PO_4	—	—	—	—	—	60.0	60.0	—	80.0
$MgCl_2 \cdot 6H_2O$	—	—	122	—	—	—	—	—	240
$MgSO_4 \cdot 7H_2O$	200	200	—	200	100	200	400	121	200
NaCl	6,800	6,400	7,599	6,799	6,000	8,000	8,000	8,000	6,000
$NaHCO_3$	2,200	3,700	1,176	2,200	2,200	350	—	1,125	2,240
$NaH_2PO_4 \cdot H_2O$	140	125	—	140	—	—	—	78.0	—
Na_2HPO_4 (anhyd.)	—	—	—	—	—	47.5	190	60.0	—
$Na_2HPO_4 \cdot 7H_2O$	—	—	268	—	1,512	—	—	—	566
$CuSO_4 \cdot 5H_2O$	—	—	0.00249	—	—	—	—	—	—
$FeSO_4 \cdot 7H_2O$	—	—	0.834	—	—	—	—	—	—
$ZnSO_4 \cdot 7H_2O$	—	—	0.863	—	—	—	—	—	—
$CaNO_3 \cdot 4H_2O$	—	—	—	—	100	—	—	—	—
Other components									
D-glucose	1,000	4,500	1,802	1,000	2,000	1,000	—	1,000	5,000
D-galactose	—	—	—	—	—	—	9,000	—	—
Lipoic acid	—	—	0.21	—	—	—	—	—	—
Phenol red	10.0	15.0	12.0	20.0	5.00	17.0	10.0	5.00	10.0
Sodium pyruvate	—	110	110	—	—	—	550	—	—
Hypoxanthine	—	—	4.10	—	—	0.30	—	—	—
Linoleic acid	—	0.084	—	—	—	—	—	—	25.0
Putrescine 2HCl	—	—	0.161	—	—	—	—	—	—
Thymidine	—	—	0.73	10.0	—	—	—	—	—
Cocarboxylase	—	—	—	1.00	—	—	—	—	—
Coenzyme A	—	—	—	2.50	—	—	—	—	—
Deoxyadenosine	—	—	—	10.0	—	—	—	—	—
Deoxycytidine HCl	—	—	—	10.0	—	—	—	—	—
Deoxyguanosine	—	—	—	10.0	—	—	—	—	—
Diphosphopyridine nucleotide·$4H_2O$	—	—	—	7.00	—	—	—	—	—
Ethanol for solubilizing lipid components	—	—	—	16.0	—	—	—	—	—
Flavine adenine dinucleotide	—	—	—	1.00	—	—	—	—	—
Glutathione (reduced)	—	—	—	10.0	1.00	0.05	—	—	15.0
5-methyl-deoxycytidine	—	—	—	0.10	—	—	—	—	—
Sodium acetate·$3H_2O$	—	—	—	83.0	—	—	—	—	—
Sodium glucuronate·H_2O	—	—	—	4.20	—	—	—	—	—
Triphosphopyridine nucleotide	—	—	—	1.00	—	—	—	—	—
Tween 80	—	—	—	5.00	—	5.00	—	—	—
Uridine triphosphate·$4H_2O$	—	—	—	1.00	—	—	—	—	—
Adenine SO_4	—	—	—	—	—	10.0	—	—	—
5'AMP	—	—	—	—	—	0.20	—	—	—
ATP·2Na	—	—	—	—	—	10.0	—	—	—
Cholesterol	—	—	—	—	—	0.20	—	—	—
2-Deoxyribose	—	—	—	—	—	0.50	—	—	—
Guanine HCl	—	—	—	—	—	0.30	—	—	—
D-ribose	—	—	—	—	—	0.50	—	—	—
Na acetate	—	—	—	—	—	36.7	—	—	—
Thymine	—	—	—	—	—	0.30	—	—	—
Uracil	—	—	—	—	—	0.30	—	—	—
Xanthine	—	—	—	—	—	0.30	—	—	—
CO_2 (gas phase)	5%	10%	5% (pH 7.0) 2% (pH 7.4)	5%	5%	Ambient	Ambient	5% (pH 7.0) 2% (pH 7.4)	5%

*Note: For use in a gas phase of air in sealed containers Hanks' salts may be substituted in Eagle's MEM. Likewise to use 199 in 5% CO_2,, Substitute Earle's salts (see Table 7.4).

TABLE 7.6A. Examples of Serum-Free Media Basal Formulae*

Component	MCDB110	MCDB202	MCDB402	MCDB153	Iscove's	LHC
Amino acids						
(All as L-enantiomers)						
Alanine	1.0 E-4	1.0 E-4	—	1.0 E-4	2.8 E-4	1.0 E-4
Arginine·HCl	1.0 E-3	3.0 E-4	3.0 E-4	1.0 E-3	4.0 E-4	2.0 E-3
Asparagine	1.0 E-4	1.0 E-3	1.0 E-4	1.0 E-4	1.9 E-4	1.0 E-4
Aspartic acid	1.0 E-4	1.0 E-4	1.0 E-5	3.0 E-5	2.3 E-4	3.0 E-5
Cysteine·HCl	5.0 E-5	2.0 E-4	—	2.4 E-4	—	2.4 E-4
Cystine	—	2.0 E-4	4.0 E-4	—	2.9 E-4	—
Glutamic acid	1.0 E-4	1.0 E-4	1.0 E-5	1.0 E-4	5.1 E-4	1.0 E-4
Glutamine	2.5 E-3	1.0 E-3	5.0 E-3	6.0 E-3	4.0 E-3	6.0 E-3
Glycine	3.0 E-4	1.0 E-4	1.0 E-4	1.0 E-4	4.0 E-4	1.0 E-4
Histidine·HCl	1.0 E-4	1.0 E-4	2.0 E-3	8.0 E-5	2.0 E-4	1.6 E-4
Isoleucine	3.0 E-5	1.0 E-4	1.0 E-3	1.5 E-5	8.0 E-4	3.0 E-5
Leucine	1.0 E-4	3.0 E-4	2.0 E-3	5.0 E-4	8.0 E-4	1.0 E-3
Lysine·HCl	2.0 E-4	2.0 E-4	8.0 E-4	1.0 E-4	8.0 E-4	2.0 E-4
Methionine	3.0 E-5	3.0 E-5	2.0 E-4	3.0 E-5	2.0 E-4	6.0 E-5
Phenylalanine	3.0 E-5	3.0 E-5	3.0 E-4	3.0 E-5	4.0 E-4	6.0 E-5
Proline	3.0 E-4	5.0 E-5	—	3.0 E-4	3.5 E-4	3.0 E-4
Serine	1.0 E-4	3.0 E-4	1.0 E-4	6.0 E-4	4.0 E-4	1.2 E-3
Threonine	1.0 E-4	3.0 E-4	5.0 E-4	1.0 E-4	8.0 E-4	2.0 E-4
Tryptophan	1.0 E-5	3.0 E-5	1.0 E-5	1.5 E-5	7.8 E-5	3.0 E-5
Tyrosine	3.0 E-5	5.0 E-5	2.0 E-4	1.5 E-5	4.6 E-4	3.0 E-5
Valine	1.0 E-4	3.0 E-4	2.0 E-3	3.0 E-4	8.0 E-4	6.0 E-4
Vitamins						
d-Biotin	3.0 E-8	3.0 E-8	3.0 E-8	6.0 E-8	5.3 E-8	6.0 E-8
Folic acid	—	—	—	1.8 E-6	9.1 E-6	1.8 E-6
Folinic acid	1.0 E-9	1.0 E-8	1.0 E-6	—	—	—
DL-a-lipoic acid	1.0 E-8	1.0 E-8	1.0 E-8	1.0 E-6	—	1.0 E-6
Niacinamide	5.0 E-5	5.0 E-5	5.0 E-5	3.0 E-7	3.3 E-5	3.0 E-7
D-pantothenate ½Ca	1.0 E-6	1.0 E-6	5.0 E-5	1.0 E-6	1.7 E-5	1.0 E-6
Pyridoxal	—	—	—	—	2.0 E-5	—
Pyridoxine·HCl	3.0 E-7	3.0 E-7	1.0 E-4	3.0 E-7	—	3.0 E-7
Riboflavin	3.0 E-7	3.0 E-7	1.0 E-6	1.0 E-7	1.1 E-6	1.0 E-7
Thiamin·HCl	1.0 E-6	1.0 E-6	1.0 E-4	1.0 E-6	1.2 E-5	1.0 E-6
Vitamin B12	1.0 E-7	1.0 E-7	1.0 E-8	3.0 E-7	9.6 E-9	3.0 E-7
Other organic constituents						
Acetate	—	—	—	3.7 E-3	—	3.7 E-3
Adenine	1.0 E-5	1.0 E-6	1.0 E-6	1.8 E-4	—	1.8 E-4
Choline chloride	1.0 E-4	1.0 E-4	1.0 E-4	1.0 E-4	2.9 E-5	2.0 E-4
D-glucose	4.0 E-3	8.0 E-3	5.5 E-3	6.0 E-3	2.5 E-2	6.0 E-3
i-Inositol	1.0 E-4	1.0 E-4	4.0 E-5	1.0 E-4	4.0 E-5	1.0 E-4
Linoleic acid	—	2.0 E-7	3.0 E-7	—	—	—
Putrescine·2HCl	1.0 E-9	1.0 E-9	1.0 E-9	1.0 E-6	—	1.0 E-6
Na Pyruvate	1.0 E-3	5.0 E-4	1.0 E-3	5.0 E-4	1.0 E-3	5.0 E-4
Thymidine	3.0 E-7	3.0 E-7	1.0 E-6	3.0 E-6	—	3.0 E-6
Major inorganic salts						
$CaCl_2$	1.0 E-3	2.0 E-3	1.6 E-3	3.0 E-5	1.5 E-3	1.1 E-4
KCl	5.0 E-3	3.0 E-3	4.0 E-3	1.5 E-3	4.4 E-3	1.5 E-3
KNO_3	—	—	—	—	7.5 E-7	—
$MgCl_2$	—	—	—	6.0 E-4	—	2.2 E-2
$MgSO_4$	1.0 E-3	1.5 E-3	8.0 E-4	—	8.1 E-4	—
NaCl	1.1 E-1	1.2 E-1	1.2 E-1	1.2 E-1	7.7 E-2	1.0 E-1
Na_2HPO_4	3.0 E-3	5.0 E-4	5.0 E-4	2.0 E-3	1.0 E-3	2.0 E-3
Trace elements						
$CuSO_4$	1.0 E-9	1.0 E-9	5.0 E-9	1.1 E-8	—	1.0 E-8
$FeSO_4$	5.0 E-6	5.0 E-6	1.0 E-6	5.0 E-6	—	5.4 E-4
H_2SeO_3	3.0 E-8	3.0 E-8	1.0 E-8	3.0 E-8	1.0 E-7	3.0 E-8
$MnSO_4$	1.0 E-9	5.0 E-10	1.0 E-9	1.0 E-9	—	1.0 E-9

(continued)

TABLE 7.6A (continued)

Component	MCDB110	MCDB202	MCDB402	MCDB153	Iscove's	LHC
Na_2SiO_3	5.0 E-7	5.0 E-7	1.0 E-5	5.0 E-7	—	5.0 E-7
$(NH_4)_6Mo_7O_{24}$	1.0 E-9	1.0 E-9	3.0 E-9	1.0 E-9	—	1.0 E-9
NH_4VO_3	5.0 E-9	5.0 E-9	5.0 E-9	5.0 E-9	—	5.0 E-9
$NiCl_2$	5.0 E-10	5.0 E-12	3.0 E-10	5.0 E-10	—	5.0 E-10
$SnCl_2$	5.0 E-10	5.0 E-12	—	5.0 E-10	—	5.0 E-10
$ZnSO_4$	5.0 E-7	1.0 E-7	1.0 E-6	5.0 E-7	—	4.8 E-7
Buffers and indicators						
Hepes	3.0 E-2	3.0 E-2	—	2.8 E-2	2.5 E-2	2.3 E-2
$NaHCO_3$	—	—	1.4 E-2	1.4 E-2	3.6 E-2	1.2 E-2
Phenol red	3.3 E-6	3.3 E-6	3.3 E-5	3.3 E-6	4.0 E-5	3.3 E-6
CO_2	2%	2%	5%	5%	10%	5%

*Computer style notation used for concentrations, e.g., 3.0 E-2 = 30 mM. Sufficient NaOH is added to give pH 7.3–7.4 at 37°C in correct gas phase.

ment from the serum and reduces the cytotoxic action of immunoglobulins without damaging polypeptide growth factors, but it may remove some more labile constitutents, and is not always as satisfactory as untreated serum.

The presence of tissue-specific inhibitors of cell proliferation, or chalones, in serum has not been verified although some hormones may be inhibitory in a nonspecific fashion.

SERUM-FREE MEDIA

Ever since the observation in the 1950s that natural media could be replaced in part by synthetic media, attempts have been made to culture cells without serum. NCTC 109 [Evans et al., 1956], 135 [Evans and Bryant, 1965], Waymouth's MB752/1 [Waymouth, 1959], MB705/1 [Kitos et al., 1962], and the media of Birch and Pirt [1970, 1971] and Higuchi [1977] were all able to sustain the growth of L-cells without serum. Pirt and co-workers further modified their formulation by adding insulin and, with other minor modifications, were able to culture HeLa cells without serum [Blaker et al., 1971]. Ham [1963, 1965] was able to clone CHO cells serum free and, more recently, specific formulations, (e.g., MCDB110) have been derived to culture human fibroblasts [Ham, 1984], many normal and neoplastic murine and human cells [Barnes and Sato, 1980], lymphoblasts [Iscove and Melchers, 1978], and several different primary cultures [Mather and Sato, 1979a,b] in the absence of serum with, in some cases, traces of serum proteins added. Some examples of serum-free media are given in Table 7.6.

Culturing cells in the presence of serum has a number of serious disadvantages:

1. The constitution of serum is well-known in terms of its major constituents like albumin, transferrin etc., but it also contains a wide range of minor components which may have a considerable effect on cell growth. These include peptide growth factors, hormones, minerals, and lipids, the presence and action of which has not been fully determined.

2. Serum varies from batch to batch and at best a batch will last 1 year, perhaps deteriorating during that time. It then requires to be replaced with another batch which may be selected as similar but will never be identical.

3. Changing serum batches requires extensive testing to ensure that the replacement is as close as possible to the previous batch. This can involve several tests (growth, plating efficiency, special functions) (see above) and may involve several different cell lines.

4. If more than one cell type is used, each type may require a different batch of serum requiring several batches to be held on reserve simultaneously. Coculture of different cell types will present an even greater problem.

5. Periodically the supply of serum is restricted due to drought in the cattle rearing areas, spread of disease amongst the cattle or economic or political reasons. This can create problems at any time, restricting the amount available and number of batches to choose from, but can be particularly acute at times of high demand. Demand is currently increasing and will probably exceed supply unless the majority of commercial users are able to adopt serum free media.

6. For anyone interested in downstream processing of culture medium to recover cell products, the presence of serum creates a major obstacle to purification and may even limit the pharmaceutical acceptance of the product.

7. Serum is frequently contaminated with viruses, many of which may be harmless to cell culture but which represent an additional unknown factor outside

TABLE 7.6B. Examples of Serum-Free Media Supplements and Modifications*

Basal medium	MCDB 110 F MCDB110	MCDB 170MDS ME MCDB202	MCDB 153KDS K MCDB153	WAJC 404 PE MCDB151	Iscove L Iscove	HITES SCLC RPMI1640	Masui AL F12/DME	Lechner LHC-9 LE MCDB153
Modifications								
CaCl$_2$				1.3E-4				
Cysteine		7.0E-5		Delete				
Cystine				2.4E-4				
Glutamine		2.0E-3						
Pyruvate		1.0E-3		1.0E-4				
Trace elements				As in 110				
CuSO$_4$				1.0E-9				
ZnSO$_4$		5.0E-7						
Supplements								
Na$_2$SeO$_3$						3.0E-8	2.5E-8	
Dithiothreitol	6.5E-6							
Glutathione	6.5E-7							
Phosphoenolpyruvate	1.0E-5							
Phosphoethanolamine		1.0E-4						5.0E-7
Ethanolamine		1.0E-4						5.0E-7
PGE1	2.5E-8	2.5E-8						
Cholesterol	7.6E-6							
Soya lecithin	6 µg/ml							
Soybean lipid					50 µg/ml			
Sphingomyelin	1 µg/ml							
Retinoic acid								0.1 µg/ml
Vitamin E	1.4E-7							
Vitamin E acetate	4.2E-7							
Dexamethasone	5.0E-7			5.0E-7				
Hydrocortisone		1.4E-7	1.4E-7			1.0E-8		2.0E-7
Estradiol						1.0E-8		
EGF	30 ng/ml	10 ng/ml	25 ng/ml	25 ng/ml				5.0 ng/ml
Insulin	1 µg/ml	5 µg/ml	5 ng/ml	10 µg/ml		5 µg/ml	5 µg/ml	5.0 µg/ml
Glucagon							0.2 µg/ml	
Prolactin		5 µg/ml						
Triiodothyronine							5E-10	6.5 ng/ml
Epinephrine								0.5 µg/ml
Transferrin, Fe^{3+} saturated		5 µg/ml			30–300 µg/ml	100 µg/ml		10 µg/ml
BSA					0.5–10 mg/ml			
Bovine pituitary extract		70 µgP/ml†		25 µgP/ml				35 µg/ml
Ovine prolactin		1.0 µg/ml						
Dialysed FBS			1 mgP/ml					
Cholera toxin				2.0E-10				

*F, fibroblasts; ME, mammary epithelium; K, keratinocytes; PE, prostatic epithelium; L, lymphoblasts; SCLC, small lung cell cancer; AL, adenocarcinoma of the lung; LE, lung epithelium; PG, prostaglandin; BSA, bovine serum albumin. For preparation of lipid and peptides see Barnes et al., 1984. See footnote to Table 7.6 (a).

†Ovine prolactin may be substituted for bovine pituitary extract. Bovine pituitary extract is prepared according to Tsao et al., 1982.

the operator's control. Fortunately, improvements in serum sterilization techniques have virtually eliminated the risk of mycoplasma infection from sera from most reputable suppliers.

8. Cost has been cited as a disadvantage of serum supplementation. Certainly, serum constitutes the major part of the cost of a bottle of medium (more than ten times the cost of the chemical constituents), but if it is replaced by defined constituents the cost of these may be as high as the serum. However, it is to be hoped that as demand for such items as transferrin, selenium, insulin, etc. increases the cost will come down and serum-free media will become relatively cheaper.

9. As well as its growth-promoting activity serum contains growth-inhibiting activity and the net effect of the serum is the combination of both inhibition and stimulation of growth.

10. Serum growth factors such as platelet derived growth factor (PDGF) tend to stimulate fibroblastic overgrowth in mixed cultures. They may also induce differentiation in some epithelia removing them from the proliferative state.

11. Standardization of experimental and production protocols is difficult.

Selective Media

One of the major advantages of the control over growth-promoting activity afforded by serum-free media is in the ability to make a medium selective for a particular cell type (Table 7.7). The long-standing problem of overgrowth by stromal fibroblasts can now be tackled effectively in breast and skin cultures using MCDB 170 and 153, melanocytes can be cultivated in the absence of fibroblasts, and keratinocytes and separate lineages and even stages of development may be selected in hemopoietic cells by selecting the correct growth factor or group of growth factors (see Chapter 20). Add to this the possibility of switching from a growth factor, after necessary amplification of the culture, to a differentiation factor, or set of factors, and the amplified culture may then be made to perform one or more specialized functions.

Disadvantages

Unfortunately the transition to serum-free conditions, however desirable, is not as straightforward as it seems. Each cell type appears to require a different recipe and cultures from malignant tumors may vary in requirements from tumor to tumor even within one class of tumors. Removal of serum also requires that the degree of purity of reagents and water and the degree of cleanliness of all apparatus must be extremely high as the removal of serum also removes the protective, detoxifying action that some serum proteins may have. Some of the constituents of serum-free media are not commerically available as yet (see Chapter 20) and may require to be prepared in the lab.

Finally, growth is often slower in serum-free media, fewer generations are achieved with finite cell lines and some degree of selection may be involved in finite and continuous lines adapting to serum free medium.

Replacement of Serum

The essential factors in serum have been described above and include: (1) adhesion factors such as fibronectin; (2) peptides regulating growth and differentiation such as insulin, PDGF, and TGFβ (a tumor-derived growth factor also extractable from platelets); (3) essential nutrients such as minerals, vitamins, fatty acids, and intermediary metabolites; and (4) hormones regulating membrane transport, phenotypic status, and cell surface constitution such as insulin, hydrocortisone, estrogen, and triiodotyrosine.

Adhesion factors. When serum is removed it may be necessary to treat the plastic growth surface with fibronectin (25–50 μg/ml) or laminin (1–5 μg/ml) added directly to the medium [Barnes et al., 1984]. Pretreat the plastic with polysine of 1 mg/ml and wash off (McKeenan and Ham, 1976). (See also "Treated Surfaces" above and Barnes et al., 1984a).

Protease inhibitors. Following trypsin mediated subculture, the addition of serum inhibits any residual proteolytic activity. Consequently, protease inhibitors

TABLE 7.7 Examples of Selective Media

Cells or cell line	Medium	Reference
Fibroblasts	MCDB 202	[McKeehan and Ham, 1977]
Fibroblasts	MCDB 110	[Bettger, 1981]
Keratinocytes	MCDB 153	[Tsao et al., 1982]
Bronchial epithelium	LHC	[Lechner and Laveck, 1985]
Mammary epithelium	MCDB 170	[Hammond et al., 1984]
Prostate epithelium (rat)	WAJC 401	[McKeehan et al., 1982]
Prostate epithelium (human)	WAJC 404	[McKeehan et al., 1984]
Glial cells		[Michler-Stuke and Bottenstein, 1982]
Melanocytes		[Gilchrest (see Chapter 20)]
Small cell lung cancer	HITES	[Carney et al., 1981]
Adenocarcinoma of lung		[Brower et al., 1986]
Colon carcinoma		[Van der Bosch, et al., 1981]
Endothelium	MCDB 130	[Knedler and Ham, 1983]

such as soya bean trypsin inhibitor (Sigma) must be added to serum-free media after subculture [Rockwell et al., 1980]. Furthermore, because crude trypsin is a complex mixture of proteases some of which may require different inhibitors, it is preferable to use pure trypsin (e.g., Sigma Gr. III). Alternatively, cells may be washed by centrifugation to remove trypsin.

Special care may be required when trypsinizing cells from serum-free media as they are more fragile and may need to be chilled to reduce damage [McKeehan, 1977].

Hormones. Hormones which have been used include growth hormone (somatotropin), 50 ng/ml, insulin at 1–10 units/ml, which improves plating efficiency in a number of different cell types, and hydrocortisone, which improves the cloning efficiency of glia and fibroblasts (see Table 7.6B and Chapter 11) and which has been found necessary for the maintenance of epidermal keratinocytes and some other epithelial cells (see Chapter 20). Barnes and Sato [1980] describe the use of 10 pM 5-tri-iodo tyrosine (T_3) as a necessary supplement for MDCK (dog kidney) cells, and various combinations of estrogen, androgen, or progesterone with hydrocortisone and prolactin at around 10 nm can be shown to be necessary for the maintenance of mammary epithelium (see Chapter 20).

Other hormones with functions not usually associated with the cells they were tested on were found to be effective in replacing serum, e.g., FSH (follicle-stimulating hormone) with B16 murine melanoma [Barnes and Sato, 1980]. It is possible that sequence homologies exist between some growth-stimulating polypeptides and well-established peptide hormones. Alternatively, processing of some of the large proteins or polypeptides may release active peptide sequences with quite different functions.

Peptide growth factors. The family of polypeptides which have been found to be mitogenic *in vitro* is now quite extensive (see above) and includes FGF, EGF, PDGF, and MSA [Barnes et al., 1984 a,c] active in the 10 ng/ml range. Until recently it was felt that the bulk of these peptide growth factors had very low specificity, but recently a number of new factors have been isolated, mostly active in the hemopoietic system [Barnes et al. 1984d] (see Chapter 20), but also acting on non-hemopoetic cells such as melanocytes [Gilchrest et al., 1980] (see Chapter 20). Many of these factors have not been purified and few of the solid tissue factors are commercially available, but the hemopoietic factors are often highly purified, many are

genetically engineered, and several are commercially available. (BRL, Collaborative Research, Gibco).

Growth factors may act synergistically or additively with prostaglandin $F_2\alpha$, somatomedin C, and hydrocortisone. [Westermark and Wasteson, 1975; Gospodarowicz, 1974].

Nutrients. Iron, copper, and a number of minerals have been included in serum-free recipes although the evidence for a positive requirement for some of the rarer minerals is still lacking. Selenium (Na_2SeO_3) at around 20nM, is found in most formulae and there appears to be some requirement for lipids or lipid precursors such as choline, linoleic acid, ethanolamine, or phosphoethanolamine.

Proteins and polyamines The inclusion of proteins such as bovine serum albumin (BSA) or tissue extracts often increases growth and survival but adds undefined constituents to the medium. BSA, fatty acid free, is used at 1–10 mg/ml.

Transferrin, at around 10 ng/ml, is required as a carrier for iron and may also have a mitogenic role. Putrescine is used at 100 nM.

Selection and Development of Serum-Free Medium

In some cases recipes may already exist in the literature and may be used with only minor modification. Otherwise it will be necessary to develop a new formulation.

There are two general approaches to the development of serum-free medium for a particular cell line or primary culture [Maurer, 1986]. The first is to take a known recipe for a related cell type and alter the constituents individually until the medium is optimised for your own particular requirement. This has been the approach adopted by Ham and coworkers [Ham, 1984] and generally will provide the most optimal conditions. However, it is a very time consuming and laborious process, involving growth curves and clonal growth assays at each stage, and it is not unreasonable to expect to spend at least 3 years developing a new medium for a new cell type.

This has led to the second approach using existing media or combinations such as RPMI1640 [Carney et al., 1981] or Ham's F12 and DMEM [Barnes and Sato, 1980] and restricting the manipulation of the constituents to a shorter list of such substances as selenium, transferrin, albumin, insulin, hydrocortisone, estrogen, triiodotyrosine, ethanolamine, phosphoethanolamine, growth factors (EGF, FGF, PDGF, endothelial growth supplement, etc.), prostaglandins (PGE_1,

PGF$_2^\alpha$), and any others which may have special relevance to your own system. Among these, selenium, transferrin, and insulin will usually be found to be essential, the requirement for the others more variable.

So far there seem to be no clear guidelines to indicate which supplements may be required. Some may be fairly universal, like insulin, transferrin, and selenium, while others such as estrogens, androgens, and T$_3$ may be more specific for individual cell types (though not necessarily those cell types that would be traditionally associated with those hormones). As with medium and serum selection, trial and error may be the only method to select the correct supplements. If a group of compounds is found to be effective in reducing serum supplementation, the active constitutents may be identified by systematic omission of single components, and then their concentrations optimized. [Ham, 1984].

Preparation of Serum-Free Media

A number of recipes are now available for particular cell types (see Tables and 7.6 and 7.7 and Chapter 20). Some of these are available commercially, others will have to be made up. The procedure for making up serum-free recipes is similar to regular media (see above) [Waymouth, 1984]. Ultra pure reagents and water should be used and care taken with solutions of Ca^{2+} and Fe^{2+} or Fe^{3+} to avoid precipitation. It is often recommended that these be added last, immediately before use. Otherwise the constituents are generally made up as a series of stock solutions, minerals and vitamins 1,000 $\times$, tyrosine, tryptophan, and phenylalanine in 0.1 N HC1 at 50 $\times$, essential amino acids 100 $\times$ in water, salts 10 $\times$ in water, and any other special cofactors, lipids, etc. 1,000 $\times$ in the appropriate solvents. These are combined in the correct proportions diluted to the final concentration and the pH and osmolality checked.

Growth factors, hormones, and cell adhesion factors are best added separately just before use as these may need to be adjusted to suit particular experimental conditions.

Serum Substitutes

A number of products have been developed commercially to replace all or part of the serum in conventional media. These include SerXtend (NEN), Ventrex (Ventrex Laboratories Ltd), Nutricyte (Brooks Laboratories), Nu-serum (Collaborative Research), and CLEX (Dextran Products Ltd). While these may offer a degree of consistency not obtainable with regular sera, batch variations can still occur and their constitution is not fully defined. They may be useful as an *ad hoc* measure or to increase economy but are not a replacement for serum-free media. Nutridoma (BCL), ITS (Flow), SIT (Sigma), Selectakit (Gibco), and Ultraser-G (LKB) are defined supplements to replace serum, partially or completely (Maurer, 1986).

Conclusions

However desirable serum-free conditions may be, there is no doubt that the relative simplicity of retaining serum, the lack of a reliable source of most serum-free media, and the considerable investment in time, effort, and resources that must go into preparing new recipes or even existing ones all act as considerable deterrents to most laboratories to enter the serum-free arena. There is no doubt, however, that the need for consistent and defined conditions for the investigation of regulatory processes governing growth and differentiation, the pressure from biotechnology for easier product purification, and the gradually worsening situation world-wide in the supply of serum will eventually force the adoption of serum-free media on a more general scale. But first, recipes must be found which are less temperamental than some in current use and which can be used with equal facility and effectiveness in different laboratories. Unfortunately that time is not yet with us.

SELECTION OF MEDIUM AND SERUM

Unfortunately, there are few good guidelines for the selection of the appropriate medium for a given cell type. Information is usually available in the literature, or from the source of the cells, for cell lines currently available (Table 7.8; Chapter 20) failing this, the choice is either empirical or by comparative testing of several media (see below). Many continuous cell lines (e.g., HeLa, L-cells, BHK-21), primary cultures of human, rodent, and avian fibroblasts, and cell lines derived from them can be maintained on a relatively simple medium such as Eagle's MEM, supplemented with calf serum. More complex media are required where a specialized function is being expressed (e.g., aminotransferase activity in rat hepatoma [Pitot et al., 1964], cartilage and pigment secretion in chick embryo cells [Coon and Cahn, 1966] (see also Chapter 14), or when cells are passaged at low seeding density ($<$ 10^3/ml), as for cloning (see Chapter 11). Frequently the more demanding culture conditions that require complex media also require fetal bovine serum rather than calf or horse serum.

TABLE 7.8. Selecting a Suitable Medium*

| Cells or cell line | With serum | | Serum-free medium |
	Medium	Serum	
Chick embryo fibroblasts	Eagle's MEM	CS	MCDB 202
Chick embryo pigmented retina and cartilage	Ham's F10		
Chinese hamster ovary (CHO)	Eagle's MEM	CS	MCDB 301
Chondrocytes	F12	FB	[Adolphe et al., 1983]
Continuous cell lines	Eagle's MEM	CS	199, Waymouth, 1984 MB752/1, MD7505/1 [Kitos et al., 1962] CMRL1066
Endothelium	DME, 199, MEM	FB	MCDB 130
Fibroblasts	Eagle's MEM	CS	MCDB 110, 202, 402
Glial cells	MEM, SF12	FB	[Michler et al., 1984]
Glioma	MEM, SF12	FB	
HeLa cells	Eagle's MEM	CS	[Blaker et al., 1971]
Hemopoietic cells	RPMI 1640, Fischer's	FB	[Iscove, 1984] (see also chapter 20)
Human diploid fibroblasts	Eagle's MEM	CS	MCDB 110, 202
Human leukemia	RPMI 1640	FB	[Breitman et al., 1984] [Iscove, 1984]
Human tumors	SF12, L15, RPMI 1640, DME	FB	See Table 7.6
L cells (L929, LS)	Eagle's MEM	CS	[Birch and Pirt, 1970, 1971] [Higuchi, 1971]
Lymphoblastoid cell lines (human)	RPMI 1640	FB	[Iscove, 1984]
MDCK dog kidney epithelium	DME, F10/DME	FB	[Taub, 1984]
Melanocytes	See chapter 20		
Melanoma	MEM, SF12	FB	[Barnes and Sato, 1980]
Mouse embryo fibroblasts	Eagle's MEM	CS	MCDB 402
Mouse leukemia	Fischer's RPMI 1640	FB,HoS	[Murakami, 1984]
Mouse erythroleukemia	SF12, RPMI1640	FB,HoS	[Iscove, 1984]
Mouse myeloma	DME, RPMI 1640	FB	[Murakami, 1984] HB101 (Hana Biologics)
Mouse neuroblastoma	DME, F12/DME	FB	MCDB411
NRK rat kidney fibroblasts	MEM, DME	CS	
Rat minimal deviation hepatoma (HTC, MDH)	Swim S77, SF12	FB	
Skeletal muscle	DME, F12	FB,HoS	F12
Syrian hamster fibroblasts, e.g., BHK 21	MEM, GMEM, DME	CS	[Pardee et al., 1984]
3T3 cells	MEM, DME	CS	MCDB402

*Abbreviations as in Table 7.5. SF12 is Ham's F12 plus Eagle's essential amino acids and non-essential amino acids as in DME (available as 100 times stock from Flow Labs., Gibco, etc.). Further recommendations on the choice of media can be found in McKeehan, 1977, Barnes et al., 1984, Maurer, 1986.

Some examples are given in Table 7.8 of cell types and the media used for them, but this list is neither exhaustive nor binding. For a more complete list, see Maurer [1986] Barnes et al. [1984 a–d] Ham and McKeehan [1979], and Morton [1970]. If a clear indication of the correct culture conditions is not available, a simple cell growth experiment with commercially available media and mutliwell plates (see Chapter 18 and 19) can be carried out in about 2 weeks. Assaying for clonal growth (see Chapters 11 and 19) and measuring the expression of specialized functions may narrow the choice further. You may be surprised to find that your best conditions do not agree with the literature, and you will have to decide between the optimal growth and behavior of the cells as you find them, or reproducing the conditions found in another laboratory, which may be difficult due to variations in the impurities present in reagents and water and in batches of serum, if present. It is to be hoped that as serum, requirements are reduced and reagent purity increases, medium standardization will improve and the need for such a choice will not arise.

Finally, you may have to compromise in your choice of medium or serum because of cost. Autoclavable media are available from commercial suppliers (Flow, GIBCO). They are simple to prepare from powder and are suitable for many continuous cell strains. They may need to be supplemented with glutamine for some cells and will usually require serum. The cost of serum should be calculated on the basis of medium volume where cell yield is not important, but where the objective is to produce large quantities of cells, calculate serum costs on a per cell basis. If a culture grows to 10^6/ml in serum A and 2×10^6/ml in serum B, serum B becomes the less expensive by a factor of two.

If fetal bovine serum seems essential, try mixing it with calf serum. This may allow you to reduce the concentration of the more expensive fetal serum. If you can, leave out serum altogether, or reduce the concentration and use one of the serum-free formulations suggested above.

Batch Reservation

Serum standardization is difficult as batches vary considerably and one batch will only last about 1 year, stored at $-20°C$. Select the type of serum that is most appropriate and request batches to test from a number of suppliers. Most serum suppliers will normally reserve a batch until a customer can select the most suitable one (provided this does not take longer than 3 weeks or so). When a suitable batch has been selected, the supplier is requested to hold it for up to 1 year for regular dispatch. Other suppliers should also be informed so that they may return theirs to stock.

Testing Serum

The quality of a given serum is assured by the supplier, but their quality control is usually performed with one of a number of continuous cell lines. If your requirements are more discriminating, then you will need to do your own testing. There are four main parameters for testing serum:

Cloning efficiency. During cloning the cells are at a low density and hence are at their most sensitive, making this a very stringent test. Plate the cells out at ten to 100 cells/ml and look for colonies after 10d to 2 wk. Stain and count the colonies (see Chapters 11 and 19) and look for differences in cloning efficiency (sur-

vival) and colony size (cell proliferation). Each set should be tested at a range of concentrations from 2% up to 20%. This will reveal whether one serum is equally effective at a lower concentration, thereby saving money and prolonging the life of the batch, and will show up any toxicity at high serum concentration.

Growth curve. A growth curve should be performed in each serum (see Chapter 18) determining the lag period, doubling time, and saturation density (density at "plateau"). A long lag implies that the culture is having to adapt; short doubling times are preferable if you want a lot of cells quickly; and a high saturation density will provide more cells for a given amount of serum and will be more economical.

Preservation of cell culture characteristics. Clearly the cells must do what you require of them in the new serum, whether they are acting as host to a given virus, producing a certain differentiated cell product, or expressing a characteristic sensitivity to a given drug.

Sterility. Serum from a reputable supplier will have been tested and shown to be free of microorganisms, but occasionally odd bottles or parts of a batch may slip through even the most stringent sampling procedures. To be certain, you should grow cells in antibiotic-free medium supplemented with the serum, look for any microbiological contamination, and stain the cells for mycoplasmas (see Chapter 16).

The previous precautions and tests are advisable in selecting a new batch of serum partly because of the investment both in terms of cash and cell lines, but also because of the need for stringent control of your culture conditions. However, it is not always within the capacity of a small laboratory to cover all these requirements, and short cuts may have to be taken. If this is the case, you may be obliged to accept the assurances of the supplier regarding sterility and growth-promoting capacity, but you must still test the serum for any special functional requirements.

OTHER SUPPLEMENTS

Tissue extracts and digests have been used for many years as supplements to tissue culture media in addition to serum. Many are derived from microbiological culture techniques and autoclavable broths, e.g., bactopeptone, tryptose, and lactalbumin hydrolysate, which are proteolytic digests of beef heart or lactalbumin, and contain mainly amino acids and small peptides. Bactopeptone and tryptose may also contain nucleosides and other heat-stable tissue constitutents such as fatty acids and carbohydrates.

Embryo Extract. Embryo extract (or embryo juice) is a crude homogenate of 10 day chick embryo clarified by centrifugation (see Appendix). The crude extract was fractionated by Coon and Cahn [1966] to give high and low molecular weight fractions. The low molecular weight fraction promoted cell proliferation while the high molecular weight fraction promoted pigment and cartilage cell differentiation. Embryo extract was originally used as a component of plasma clots (see Chapter 1 and 9) to promote cell migration from the explant. It has been retained in some organ culture techniques (Chapter 22) and is still sometimes used in nerve and muscle culture (Chapter 20).

Conditioned Medium. Puck and Marcus [1955] found that the survival of low-density cultures could be improved by growing the cells in the presence of feeder layers (see Chapter 11). While part of this effect may have been due to conditioning of the substrate, the main effect was presumed to be conditioning of the medium by release of small molecular metabolites and macromolecules into the medium r[Takahashi and Okada, 1970]. Hauschka and Konigsberg [1966], showed that the conditioning of culture medium necessary for the growth and differentiation of myoblasts was due to collagen released by the feeder cells. Feeder layers and conditioning of the medium by embryonic fibroblasts or other cell lines remains a useful method of culturing difficult cells [e.g., Stampfer et al., 1980] (see Chapter 11).

INCUBATION TEMPERATURE

The optimal temperature for cell culture is dependent on (1) the body temperature of the animal from which the cells were obtained, (2) any regional variation in temperature (e.g., skin may be lower), and (3) the incorporation of a safety factor to allow for minor errors in incubator regulation. Thus, the temperature recommended for most human and warm-blooded animal cell lines is 36.5° C, close to body heat but set a little lower for safety.

Avian cells, because of the higher body temperature in birds, should be maintained at 38.5° C for maxi-mum growth but will grow quite satisfactorily, if more slowly, at 36.5° C.

Cultured cells will tolerate considerable drops in temperature, can survive several days at 4° C and can be frozen and cooled to −196° C (see Chapter 17), but they cannot tolerate more than about 2° C above normal (39.5° C) for more than a few hours, and will die quite rapidly at 40° C and over.

Epidermal cells from the mouse animals may grow better at a slightly lower temperature of 33° C.

Much of the preceding discussion has been based on observations with warm-blooded animals but is, nevertheless, applicable in principle to lower vertebrates, and perhaps to a lesser extent, to invertebrates. Temperature must be considered separately, however. In general the cells of poikilothermic animals have a wide temperature tolerance but should be maintained at a constant level within the normal range of the donor species. This requires incubators with cooling as well as heating as the incubator temperature may need to be below ambient (e.g., for fish). As for a hot-room, cooling capacity should be sufficient to lower the temperature about 2° C, or more, below ambient so that regulation is performed by the heater circuit which is more sensitive.

If necessary, poikilothermic animal cells can be maintained at room temperature, but the variability of the ambient temperature in laboratories makes this undesirable.

Regulation of temperature should be kept within ±0.5° C; consistency is more important than accuracy. Cells will naturally vary in growth rate and metabolism. The incubation temperature should be kept constant both in time and at different parts of the incubator. Water baths give the most accurate control of temperature, but present problems of contamination, particularly since the flasks need to be immersed for proper temperature control. They are, therefore, seldom used and incubators are preferable. The air should be circulated by a fan to give even temperature distribution, and cultures should be placed on perforated shelves and not on the floor or touching the sides of the incubator. Further discussion of temperature control in hot rooms is given in Ch. 3.

All stocks of chemicals and glassware used in tissue culture should be reserved for that purpose alone. Traces of heavy metals or other toxic substances can be difficult to detect other than by a gradual deterioration of your cultures. It also follows that separate stocks imply separate glassware washing. The requirements of tissue culture washing are higher than for general glassware; a special detergent may be necessary (see below) and cross-contamination from chemical glassware must be avoided.

All apparatus and liquids which come in contact with cultures or other reagents must be sterile. A summary of the procedures used is given in Tables 8.1 and 8.2.

PROCEDURES FOR THE PREPARATION AND STERILIZATION OF APPARATUS

Glassware

Items of glassware used for dispensing and storage of media, and for cell culture, must be cleaned very carefully to avoid traces of toxic materials, contaminating the inner surfaces, becoming incorporated into the medium (Fig. 8.1). Where the glass surface is to be used for cell propagation it must not only be clean but also carry the correct charge. Caustic alkaline detergents render the surface of the glass unsuitable for cell attachment and require subsequent neutralization with HCl or H_2SO_4, but many modern detergents do not alter the glass surface and can be removed completely.

For the most effective washing procedure: (1) Do not let soiled glassware dry out. A sterilizing agent, such as sodium hypochlorite, should be included in the water used to collect soiled glassware (a) to remove any potential biohazard and (b) to prevent microbial contamination growing up in the water; (2) select a detergent which is effective in the water of your area, rinses off easily, and is nontoxic (see below); (3) ensure that the glassware is thoroughly rinsed in tap water and deionized or distilled water, before drying; (4) dry inverted; and (5) sterilize by dry heat to mini-

mize the risk of depositing toxic residues from steam sterilization.

Plastic culture flasks are, on the whole, meant for single use, as washing detergent renders them unsuitable for cell propagation (in monolayer) and resterilization is difficult. Cells may be reseeded back into the same flask after subculture but this tends to increase the risk of contamination.

Sterilization procedures are designed not just to kill the bulk of microorganisms but to eliminate spores which may be particularly resistant. Moist heat is more effective than dry heat but does carry a risk of leaving a residue. Dry heat is preferable but at a minimum of 160°C for 1 hr. Moist heat (for fluids and perishable items) need only be maintained at 121°C for 15–20 min (Fig. 8.1). For moist heat to be effective, steam penetration must be assured and for this the sterilization chamber must be evacuated prior to steam injection. Insertion of Thermalog indicators monitors both temperature and humidity during sterilization.

Materials

Pipette cylinders (to collect used pipettes)
disinfectant (if required)
detergent
soaking baths
bottle brushes
stainless steel baskets (to collect washed and rinsed glassware for drying)
aluminum foil
sterility indicators (Browne's tubes, Thermalog Indicators)
glass petri dishes (for screw caps)
autoclavable plastic film (Portex, Cedanco) or paper sterilization bags
sterile-indicating autoclave tape

Collection and Washing
1.
Collect immediately after use into detergent containing a disinfectant such as sodium hypochlo-

TABLE 8.1. Sterlization of Equipment and Apparatus

Item	Sterilization
Apparatus containing glass & silicone tubing	Autoclave*
Disposable tips for micropipettes	Autoclave
Dispenser tubing for Compu-pet	Autoclave
Filters—Millipore, Sartorius	Autoclave—do not prevac. or postvac.
Glassware	Dry heat†
Glass bottles with screw caps	Autoclave
Glass coverslips	Dry heat
Glass slides	Dry heat
Instruments	Dry heat
Magnetic stirrer bars	Autoclave
Pasteur pipettes—glass	Dry heat
Pipettes—glass	Dry heat
Screw caps	Autoclave
Silicone tubing	Autoclave
Stoppers—rubber, silicone	Autoclave
Test tubes	Dry heat

*Autoclave—100 kPa (1 bar,/15 lb/in^2) 121°C for 20 min.
†Dry heat—160°C/1 hr.

TABLE 8.2. Sterilization of Liquids

Solution	Sterilization	Storage
Agar	Autoclave*	Room temperature
Amino acids	Filter†	4°C
Antibiotics	Filter	−20°C
Bacto-peptone	Autoclave	Room temperature
Bovine serum albumin	Filter—use stacked filters (see text)	4°C
Carboxylmethyl cellulose	Steam—30 min**	4°C
Collagenase	Filter	−20°C
DMSO	Self-sterilizing, aliquot into sterile tubes	Room temperature—keep dark, avoid contact with rubber or plastics
EDTA	Autoclave	Room temperature
Glucose—20%	Autoclave	Room temperature
Glucose—1–2%	Filter (low concentrations caramelize if autoclaved)	Room temperature
Glutamine	Filter	−20°C
Glycerol	Autoclave	Room temperature
HEPES	Autoclave	Room temperature
HC1 1 N	Filter	Room temperature
Lactalbumin hydrolsate	Autoclave	Room temperature
Methocel	Autoclave	4°C
NaHCO$_3$	Filter	Room temperature
NaOH 1 N	Filter	Room temperature
Phenol red	Autoclave	Room temperature
Salt solutions (without glucose)	Autoclave	Room temperature
Serum	Filter—use stacked filters (see text)	−20°C
Sodium pyruvate 100 mM	Filter	−20°C
Transferrin	Filter	−20°C
Tryptose	Autoclave	Room temperature
Trypsin	Filter	−20°C
Vitamins	Filter	−20°C
Water	Autoclave	Room temperature

*Autoclave—100 kPa (15lb/in^2) 121°C for 20 min.
†Filter—0.2-μm pore size.
**Steam—100°C for 30 min.

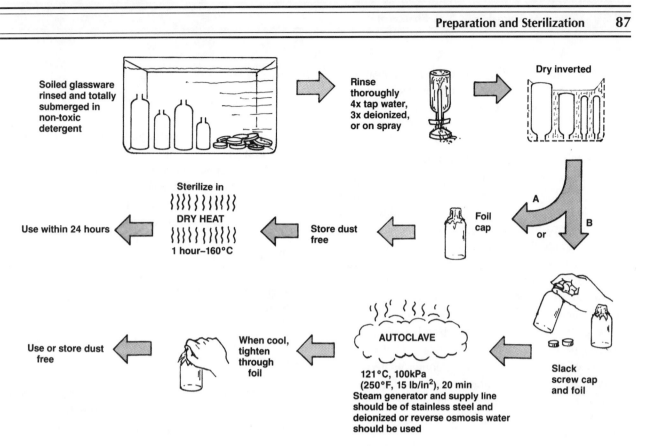

Fig. 8.1. *Wash-up and sterilization of glassware.*

rite. It is important that apparatus should not dry before soaking, or cleaning will be much more difficult.

2.

Soak overnight in detergent (see below).

3.

Machine wash or brush by hand or machine (Fig. 8.2) the following morning and rinse thoroughly in four complete changes of tap water followed by three changes of deionized water. If rinsing is done by hand, a sink spray (Fig. 4.3) is a useful accessory; otherwise bottles must be emptied and filled completely each time. Clipping bottles in a basket will help to speed up this stage.

4.

After rinsing thoroughly, invert bottles, etc., in stainless steel wire baskets and dry upside down.

5.

Cap with aluminum foil when cool and store.

Sterilization

1.

Place in an oven with fan-circulated air at 160°C.

2.

Check that temperature has returned to 160°C, seal the oven with a strip of tape with the time

Fig. 8.2. *Motorized bottle brushes. Care must be taken not to press down too hard on the brush lest the bottle break.*

recorded on it, and leave for 1 hr.; ensure that the center of the load achieves 160°C by using a sterility indicator or recording thermometer with the sensor in the middle of the load. Do not pack the load too tightly; leave room for hot air circulation.

3.

After 1 hr, switch off the oven and allow to cool with the door closed. It is convenient to put the oven on an automatic timer so that it can be left to switch off on its own.

4.

Use within 24–48 hr. Alternatively, bottles may be loosely capped with screw caps, autoclaved for 20 min at 120°C with prevacuum and postvacuum cycle (see Chapter 4), and the caps tightened when cool. Caps must be very slack (one complete turn) during autoclaving to allow entry of steam and to prevent the liner being sucked out of the cap and sealing the bottle. If the bottle seals during sterilization in an autoclave, sterilization will not be complete. Unfortunately, misting often occurs when bottles are autoclaved and a slight residue may be left when this evaporates. There is also a risk of the bottles becoming contaminated as they cool by drawing in nonsterile air before they are sealed. Dry-heat sterilization is better, allowing the bottles to cool down within the oven before removal.

Pipettes

1.

Place in water with detergent (e.g. 1% Decon) tip first and a sterilizing agent (hypochlorite or glutaraldehyde) immediately after use (Fig 8.3). Do not put pipettes which have been used with agar or silicones (water repellents, antifoams, etc.) in the same cylinder as regular pipettes. Use disposable pipettes for silicones and either rinse agar pipettes after use in hot tap water or use disposable pipettes.

2.

Soak overnight and remove plugs with compressed air the following morning.

3.

Transfer to pipette washer (Fig. 8.4; see also Fig. 4.3), tips uppermost.

4.

Rinse by siphoning action of pipette washer for a minimum of 4 hr or in an automatic washing machine with a pipette adaptor.

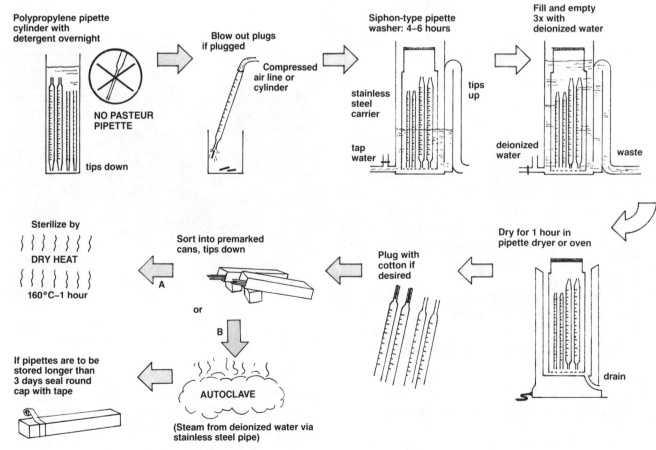

Fig. 8.3. *Wash-up and sterilization of pipettes.*

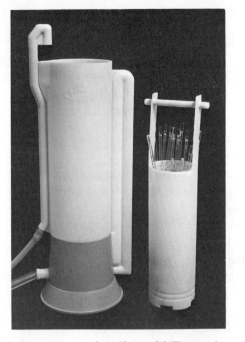

Fig. 8.4. *Siphon pipette washer. The model illustrated is made of polypropylene. Versions are available made of stainless steel, allowing the pipette carrier to be transferred directly to a drier.*

5.
Turn over valve to deionized water (see Fig. 4.3), or wait until last tap water finally runs out, turn off tap water, and empty and fill three times with deionized water (automatic deionized rinse cycle in machine).
6.
Transfer to pipette drier or drying oven and dry with tips uppermost.
7.
Plug with cotton (Fig. 8.5). Alternatively, pipette plugs may be dispensed with and a short sterile filter tube placed on the pipette bulb during use (see Fig. 5.2). To avoid cross-contamination, this must be replaced before starting work and every time that you change to a new cell strain.
8.
Sort pipettes by size and store dust free.

Sterilization
1.
Place pipettes in pipette cans (square aluminum or stainless steel with silicone cushions at either end; square cans do not roll on the bench) and label both ends of the cans. Put one pipette size per can with a few cans containing an assortment of 1-ml,

Fig. 8.5. *Semiautomatic pipette plugger (Bellco).*

2-ml, 10-ml, and 25-ml pipettes in the ratio 1:1:3:2.
2.
Seal with sterile-indicating tape and sterilize by dry heat for 1 hr at 160°C. This is different from the sterile indicating tape used in autoclaves, as the sterilizing temperature is higher. Most tend to char and release traces of volatiles from the adhesive, which can leave a deposit on the oven, and potentially on the pipettes. Use the smallest amount of tape possible or replace with temperature indicator tabs (Bennett), which are small and have less volatile material. The temperature should be measured in the center of the load, to ensure that this, the most difficult part to reach, attains the minimum sterilizing conditions. Leave spaces between cans when loading the oven, to allow for circulation of hot air.
3.
Remove from oven, allow to cool, and transfer cans to tissue culture laboratory. If you anticipate that pipettes will lie for more than 48 hr before use, seal cans around the cap with adhesive tape.

Screw Caps

There are two main types of caps for glass bottles in common use: (1) aluminum or phenolic plastic caps with synthetic rubber or silicone liners and (2) wadless polypropylene caps, reusable (Duran) or disposable (Sterilin). The reusable polypropylene Duran caps are the best choice, as they are deeply shrouded and have ring inserts to improve the seal and improve pouring (although pouring is not recommended in sterile work). Disposable polypropylene caps will only seal if screwed down very tightly on a bottle with no chips or imperfections on the lip of the opening.

Do not leave aluminum caps, or any other aluminum items in alkaline detergents for more than 30 min, as they will corrode. Do not have glassware in the same detergent bath or the aluminum may contaminate the glass. Avoid machine washing detergents as they are very caustic.

Reusable caps. Soak 30 min in detergent and rinse thoroughly for 2 hr (make sure all caps are submerged). Liners should be removed and replaced after rinsing. Rinsing may be done in a beaker (or pail) with running tap water led by a tube to the bottom. Stir the caps by hand every 15 min. Alternatively, place in a basket, or better in a pipette washing attachment, and rinse in an automatic washing machine, but do not use detergent in machine.

Disposable caps. These should not need to be washed unless reused. They may be washed and rinsed by hand as above (extending the detergent soak if necessary). Because these caps may float, they must be weighted down during soaking and rinsing. For automatic washers, use pipette washing attachment and normal cycle with machine detergent.

Stoppers. Use silicone or heavy metal-free white rubber in preference to natural rubber. Wash and sterilize as for disposable caps. (There will be no problem with floatation in washing and rinsing).

Sterilization. Place caps in a glass petri dish with the open side down. Wrap in cartridge paper or steam permeable nylon film (Portex), and seal with autoclave tape (Fig. 8.6). Autoclave for 20 min at 121°C, 100 kPa (1 bar, 15 lb/in^2) (see Fig. 8.8).

Keep organic matter out of the oven. Do not use paper tape or packaging unless you are sure that it will not release volatile products on heating. Such products will eventually build up on the inside of the oven, making it smell when hot, and some deposition may occur inside the glassware being sterilized.

Selection of Detergent

Solicit samples from local suppliers and test them: (1) for their ability to wash heavily soiled glassware; (2) for the quality of the growth surface afterwards; and (3) for the toxicity of the detergent (see also Chapters 11 and 19).

1. Washing efficiency

Materials

Standard 75-cm^2 or 120-cm^2 glass culture vessels

samples of detergents made up to working strength HBSS

Protocol

(i)
Autoclave flasks carrying cell monolayers, standing vertically, for 20 min at 120°C.

(ii)
Soak overnight in detergent, three flasks per detergent.

(iii)
Rinse out detergent with water, note flasks with residue and brush as necessary to get them clean.

(iv)
Rinse completely four times in water and three times in deionized water.

(v)
Dry and sterilize by dry heat as above.

(vi)
Check flasks again for apparent cleanliness and any sign of residue. Add a little BSS containing phenol red to each flask (approximately 1 ml/100 cm^2), rinse over the inside of the bottle, and look for any color change. If it becomes pink (alkaline), detergent has not been completely rinsed out of the bottle. When you are satisfied that the flasks are clean, sterilize them by dry heat (see above).

Fig. 8.6. *Packaging screw caps for sterilization.*

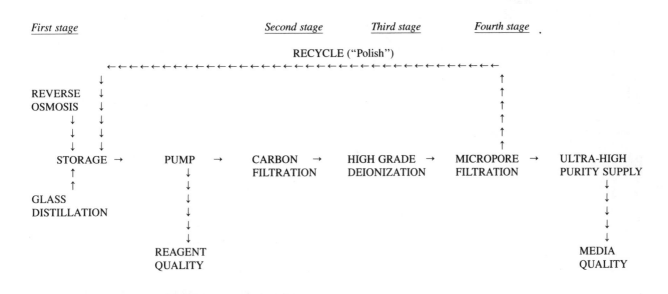

Fig. 8.7. *Suggested layout for high purity water supply. Tap water is fed through reverse osmosis or glass distillation to a storage container. This semi-purified water is then recycled via carbon filtration, deionization and micropore filtration back to the storage container. Reagent quality water is available at all times from* *storage; media quality water is available from the micropore filter supply (right of diagram). If the apparatus recycles continuously then the highest purity water will be collected first thing in the morning for the preparation of medium (Based on Elga system).*

2. Quality of growth surface

Materials

Monolayer cells with good cloning efficiency (10% or more)

medium

serum

(for materials for fixing and staining cells, see Chapter 13).

Protocol

(i)

Taking flasks from 1 (vi) (above), add cell suspension to each flask and to three control flasks (disposable tissue culture grade plastic of same surface area), using a cell concentration suitable for cloning (e.g., 20 cells/ml, 5 cells/cm^2, CHO-K1; 100 cells/ml 25 cells/cm^2, MRC-5; see Chapter 11). Use the minimum concentration of serum which

will allow the cells to clone (see under Testing Serum—Chapter 7).

(ii)

Incubate for 10–20 d, depending on the growth rate of the cells (see Chapters 11 and 19), fix stain, and count colonies (see Chapters 11 and 13).

(iii)

Determine relative plating efficiency (see Chapter 19) using disposable plastic flask as control, and record.

3. Cytotoxicity (See Chapter 19.)

Materials

Microtitration plate

medium

serum

detergent samples diluted to working strength and filter sterilized (see below; for fixing and staining cells, see Chapter 13)

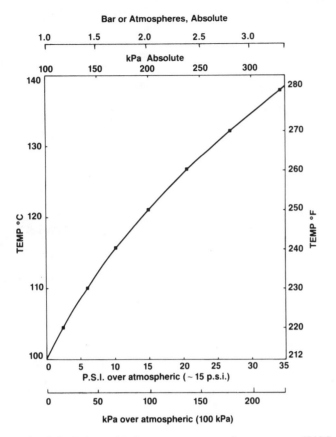

Fig. 8.8. *Relationship between pressure and temperature: 121°C and 100 kPa, or 1 Bar (250°F, 15 lb/in²) for 15–20 min are the conditions usually recommended.*

Protocol

(i)

Set up microtitration plate with suitable target cells 1,000/well, incubate to 20% confluence, and change to fresh medium, 100 μl/well.

(ii)

Add 100 μl of detergent diluted to usual working strength in medium and filter sterilized to the first well in each row and dilute serially across plate in that row, leaving the last two wells in each row as untouched controls.

(iii)

Continue growth for 3–5 d but *not beyond* the time that the control wells reach confluence.

Wash, fix, and stain plate.

(v)

Determine titration point (point at which cell number per well is reduced by approximately 50%) and record. A high detergent concentration at this point means low cytotoxicity.

Miscellaneous Equipment

Cleaning

All new apparatus and materials (silicone tubing, filter holders, instruments, etc.) should be soaked in detergent overnight, thoroughly rinsed, and dried. Anything which will corrode in the detergent—mild steel, aluminum, copper, or brass, etc.—should be washed directly by hand, without soaking, or soaking for 30 min only, using detergent if necessary, then rinsed and dried.

Used items should be rinsed in tap water and immersed in detergent immediately after use. Allow to soak overnight, rinse, and dry. Again, do not expose materials which might corrode to detergent for longer than 30 min. Aluminum centrifuge buckets and rotors must never be allowed to soak in detergent.

Particular care must be taken with items treated with silicone grease or silicone fluids. They must be treated separately and the silicone removed, if necessary, with carbon-tetrachloride. Silicones are very difficult to remove if allowed to spread to other apparatus, particularly glassware.

Packaging

Ideally, all apparatus for sterilization should be wrapped in a covering which will allow steam penetration but be impermeable to dust, microorganisms, and mites. Proprietary bags are available bearing sterile indicating marks which show up after sterilization. Semi-permeable transparent nylon film (Portex Plastics, Cedanco) is sold in rolls of flat tube of different diameters and can be made up into bags with sterile-indicating tape. Although expensive, it can be reused several times before becoming brittle.

Tubes and orifices should be covered with tape and paper or nylon film before packaging, and needles or other sharp points should be shrouded with a glass test tube or other appropriate guard.

Sterilization

The type of sterilization used will depend on the material (see Table 8.1). Metallic items are best sterilized by dry heat. Silicone rubber (which should be used in preference to natural rubber), Teflon, polycarbonate, cellulose acetate, and cellulose nitrate filters (see below for filters in holders), etc., should be autoclaved for 20 min at 121°C, 100 kPa (1 bar, 15 lb/in²) with pre-evacuation and postevacuation steps in the cycle. In small bench-top autoclaves and pressure-cookers, make sure that the autoclave boils vigorously for 10–15 min before pressurizing to displace all the air. (Take care that enough water is put in at the start to allow for this.) After sterilization, the steam is released and the items removed to dry off in an oven or rack.

◇ Take care releasing steam and handling hot items to avoid burns. Wear elbow length insulated gloves and keep face well clear of escaping steam.

Sterilizing filters

Reusable filter holders should be made up and sterilized as follows:

1.
After thorough washing in detergent (see above) rinse in water, then deionized water and dry.
2.
Insert support grid in filter and place filter membrane on grid. If polycarbonate, apply wet to counteract static electricity.
3.
Place prefilters (glass fiber and others as required, see below) on top of filter.
4.
Reassemble filter holder, but do not tighten up completely (leave about one half turn on collars, one whole turn on bolts).
5.
Cover inlet and outlet of filter with aluminum foil.
6.
Pack filter in sterilizing paper or steam-permeable nylon film and close with sterile-indicating tape.
7.
Autoclave at 121°C, 100 kPa (1 bar, 15 lb/in^2) with no preevacuation or postevacuation ("liquids cycle" in automatic autoclaves).
8.
Remove and allow to cool.
9.
Do not tighten filter holder completely until the filter is wetted at the beginning of filtration (see below).

Alternative Methods of Sterilization

Many plastics cannot be exposed to the temperature required for autoclaving or dry heat sterilization. To sterilize such items, immerse in 70% alcohol for 30 min and dry off under uv light in a laminar flow cabinet. Care must be taken with plexiglass (Perspex, Lucite) as it may crack in alcohol or uv treatment due to release of stresses built in during manufacture.

Ethylene oxide may be used to sterilize plastics, but 2–3 wk are required for the Et$_2$O to clear from the plastic surface.

γ-irradiation, 2,000–3,000 by (20–100 krad), is the best method for plastics. Items should be packaged and sealed. Polythene may be used and sealed by heat welding.

REAGENTS AND MEDIA

The ultimate objective in preparing reagents and media is to produce them in a pure form (1) to avoid the accidental inclusion of inhibitors and substances toxic to cell survival, growth, and expression of specialized functions, and (2) to enable the reagent to be totally defined and the functions of its constituents fully understood.

Most reagents or media can be sterilized either by autoclaving if they are heat stable (water, salt solutions, amino acid hydrolysates) or by membrane filtration if heat labile. During autoclaving the container should be kept sealed if borosilicate glass or polycarbonate. Soda glass bottles are better left with the caps slack to minimize breakage. Evolution of vapor will help to prevent ingress of steam from the autoclave, but the liquid level will need to be restored with sterile distilled water later.

Media and reagents supplied on line to large scale culture vessels and industrial or semi-industrial fermentors can be sterilized on line by ultra high temperature treatment for a short time (Alfa-Laval). Adaptation of this process to media production might allow increased automation and ultimately reduce costs.

Water (see also Chapter 4)

Water must be of a very high purity for use in tissue culture, particularly if serum-free media are required. As water supplies vary greatly, the degree of purification that is required may vary. Hard water will need a conventional water softener on the supply before entering the purification system, but this will not be necessary with soft water.

There are four main approaches to water purification: distillation, deionization, carbon filtration, and ultrafiltration. Simple systems depend on the first two alone or combined, while the more efficient systems employ multiple stages, each operating on a different principle.

Double glass distillation. This was the first system to be used widely and still be quite effective in some areas. The still should be electric, automatically controlled if possible, and feed the first distillate into the second boiler. The heating elements should be glass- or silica-sheathed.

Deionization and glass distillation. Deionization is used to replace the first stage distillation, and again, feeds directly, with a level controller, into the distillation boiler (see Fig.4.4a). The quality of the deionized water should be monitored by conductivity at regular intervals and the cartridge changed when an increase in conductivity is observed.

Multi-stage purification (Fig.8.7). The first stage is usually reverse osmosis but can be replaced by distillation. Distillation has the advantage that the water is heat sterilized, but is more expensive and the still needs to be cleaned out regularly. Reverse osmosis (RO) depends on the integrity of the filtration membrane, and hence the effluent must be monitored. If the costs of both are deducted directly from your budget, RO will probably work out cheaper, but if power is supplied free, or costed independently of usage, then distillation will be cheaper. The type of RO cartridge that is used is determined by the pH of the water supply (see manufacturer for details).

The second stage is carbon filtration which will remove both organic and inorganic material. The third stage is high grade deionization and the final stage micropore filtration to eliminate microbial contamination. The water is collected directly from the final stage micropore filter without storage, to minimize pollution during storage. Water is only stored after the first stage, and can be used from this point as rinsing water. In the Elga system, water is recycled continuously from the micropore filter to the store, and if the supply from the first stage is turned off (e.g., overnight), then the stored water gradually "polishes", i.e., increases in purity. Hence, in this system, water should be used first thing in the morning for preparation of media.

In the Millipore system, an ultrafiltration stage can be inserted between deionization and micropore filtration to produce pyrogen-free water. Millipore do not recommend storage, so a system should be selected to give the rate of supply that you require on line.

Water is sterilized by autoclaving at 121°C, 100 kPa (15 lb/in², 1 bar) for 10–15 min. It should be dispensed in aliquots suitable for media preparation from concentrates. The bottles should be sealed during sterilization. This will require borosilicate glass (Pyrex) or polycarbonate (Nalgene). If bottles are unsealed (e.g. if soda glass is used it may break if sealed), allow 10% extra volume per bottle to allow for evaporation.

Balanced Salt Solutions

A selection of formulations are given in Chapter 7. The formula for Hanks' BSS [after Paul, 1975] contains magnesium chloride in place of some of the sulphate originally recommended and should be autoclaved below pH 6.5 and neutralized before use. Similarly, Dulbecco's PBS is made up without calcium and magnesium (PBSA), which are made up separately (PBSB) and added just before use. These precautions are designed to minimize precipitation of calcium and magnesium phosphate, during autoclaving and storage.

Most balanced salt solutions contain glucose. Since glucose may caramelize on autoclaving, it is best omitted and added later. If prepared as a × 10 concentrate (20% W/V) caramelization during autoclaving is reduced.

Salt solutions may be sterilized by autoclaving for 10–15 min at 121°C, 100 kPa (1 bar, 15 lb/in²) in sealed bottles as for water (above) (Fig. 8.8).

Media

During the preparation of complex solutions care must be taken to ensure that all the constituents dissolve and do not get filtered out during sterilization, and that they remain in solution after autoclaving or storage. Concentrated media are often prepared at a low pH (between 3.5 and 5.0) to keep all the constituents in solution, but even then some precipitation may occur. If properly resuspended this will usually redissolve on dilution; but if the precipitate has been formed by degradation of some of the constituents of the medium, then the quality of the medium may be reduced. If a precipitate forms, the medium performance should be checked by cell growth, cloning, and assay of special functions (see below).

The preparation of media is rather complex. It is convenient to make up a number of concentrated stocks, essential amino acids at × 50 or × 100, vitamins at × 1,000, tyrosine and tryptophan at × 50 in 0.1 N HCl, glucose, 200 g/l, and single strength BSS. The requisite amount of each concentrate is then mixed and filtered through a 0.2 μm porosity cellulose acetate, cellulose nitrate, or polycarbonate (Nuclepore) filter and diluted with the BSS (see below and Fig. 8.9, Table 8.2), e.g.,

Amino acid concentrate, $\times$ 100 (in water)	100 ml
Tyrosine and tryptophan, $\times$ 50 (in 0.1 N HCl)	200 ml
Vitamins, $\times$ 1,000 (in water)	10 ml
Glucose, $\times$ 100 (in 200 g/1 BSS)	100 ml

Mix and sterilize by filtration. Store frozen. For use dilute 41-ml of concentrate mixture with 959 ml sterile $\times$ 1 BSS.

The advantage of this type of recipe is that it can be varied; extra nutrients (ketoacids, nucleosides, minerals, etc.) can be added or the major stock solutions altered to suit requirements, but, in practice, this procedure is so laborious and time-consuming that few laboratories make up their own media from basic constituents unless they wish to alter individual constituents regularly. Commercial media, which were quite unreliable in the 1950s and early 1960s, have now improved greatly, largely due to the introduction of appropriate quality control measures. There are now several reputable suppliers (see Trade Index) of standard formulations, and many of them will prepare media to your own formulation.

Commercial media are supplied as: (1) working strength solutions, complete with sodium bicarbonate and glutamine; (2) $\times$ 10 concentrates without NaHCO$_3$ and glutamine, which are available as separate concentrates; or (3) powdered media, complete or without glutamine. Powdered media are the cheapest, and not a great deal more expensive than making up your own, if you include time for preparation, sterilization, and quality control, cost of raw materials of high purity, and overheads such as power and wages. Powdered media are quality controlled by the manufacturer for their growth-promoting properties but not, of course, for sterility. Tenfold concentrates cost about twice as much per liter of working strength medium but are purchased sterile. Buying media at working strength is the most expensive (about five times the cost of a $\times$ 10 concentrate) but is the most convenient as no further preparation is required other than the addition of serum, if required.

Preparation of medium from $\times$ 10 concentrate.

Sterilize aliquots of deionized distilled water (see above) of such a size that one aliquot will last from 1 to 3 wk. Add concentrated medium and other constituents as follows:

1.

For sealed culture flask: gas phase air, low buffering capacity, and low CO$_2$/HCO$_3^-$ concentration

Water	884
$\times$ 10 concentrate	100
200 mM glutamine	10
5.6% NaHCO$_3$	6
	1,000 ml

1 N NaOH to pH 7.2 at 20°C, 7.4 at 36.5°C. If extra constituents are added. e.g. HEPES buffer, extra amino acids, this becomes as in (2) below.

2.

For sealed culture flask; high buffering capacity, gas phase air; may be vented to atmosphere by slackening cap for some cell lines at a high cell density if a lot of acid is produced; atmospheric CO$_2$/HCO$_3^-$

Water	854
$\times$ 10 concentrate	100
200 mM glutamine	10
$\times$ 100 nonessential amino acids	10
1 M HEPES	20
5.6% NaHCO$_3$	6
	1,000 ml

3.

For cultures in open vessels in a CO$_2$ incubator or under CO$_2$ in sealed flasks, two suggested formulations are as follows:

	(1) 2% CO$_2$	(2) 5% CO$_2$
Water	858	854
$\times$ 10 concentrate	100	100
200 mM glutamine	10	10
1 M HEPES	20	—
5.6% NaHCO$_3$	12	36
	1,000 ml	1,000 ml

1 N NaOH to pH 7.4 at 36.5°C.

(1) Good buffering capacity, moderate CO$_2$/HCO$_3$ concentration, (2) moderate buffering capacity, high CO$_2$/HCO$_3$ concentration. Always equilibrate at 36.5°C as the solubility of CO$_2$ decreases with increased temperature and the pKa of the HEPES will change.

The amount of alkali needed to neutralize a × 10 concentrated medium (which is made up in acid to maintain solubility of the constituents) may vary from batch to batch and from one medium to another, and, in practice, titrating medium to pH 7.4 at 36.5°C can sometimes be a little difficult. When making up a new medium for the first time, add the stipulated amount of NaHCO₃, and allow samples with varying amounts of alkali to equilibrate overnight at 36.5°C in the appropriate gas phase. Check the pH the following morning, select the correct amount of alkali, and prepare the rest of the medium accordingly.

Some media are designed for use with a high bicarbonate concentration and elevated CO_2 in the atmosphere, e.g., Ham's F12, while others have a low bicarbonate concentration for use with a gas phase of air, e.g., Eagle's MEM with Hanks' salts. If the use of a medium is changed, and the bicarbonate concentration altered, it is important to check that the osmolality is still within an acceptable range.

The bicarbonate concentration is important in establishing a stable equilibrium with atmospheric CO_2, but regardless of the amount of bicarbonate used, if the medium is at pH 7.4 and 36.5°C, the bicarbonate concentration at each concentration of CO_2 will be as in Table 7.2 (see Chapter 7).

The osmolality should be checked (see Chapter 7), where alterations are made to a medium that are not in the original formulation.

If your consumption of medium is great (> 200 l/yr) and you are buying medium ready-made, then it may be better to get extra constituents included in the formulation as this will work out ot be cheaper. HEPES particularly is very expensive to buy separately.

Glutamine is supplied separately as it is unstable and should be kept frozen. The half-life in medium at 4°C is about 3 wk, and at 36.5°C, about 1 wk.

Care should be taken with × 10 concentrates to ensure that all of the constituents are in solution, or at least evenly suspended before dilution. Some constituents, e.g., folic acid or tyrosine, can precipitate and be missed at dilution. Incubation at 36.5°C for several hours may overcome this.

The final step is the addition of serum which, since it is close to isotonic, is added to the final volume, e.g.,

Complete medium   1,000 ml
Serum   111 ml (10% final)
                   1,111 ml

Remember to allow space for all additions when choosing the container and the volume of H₂O to be sterilized.

Having once tested a batch of medium for growth promotion, etc., and found it to be satisfactory, then it need not be tested each time it is made up to working strength. The sterility should be checked, however, by incubating the medium at 36.5°C for 48 hr before adding glutamine, serum, or antibiotic.

Powdered media. Select a formulation lacking glutamine. If there are other unstable constituents, they also should be omitted and added later as a sterile concentrate, just before use. Dissolve the powder in the recommended amount of water (choose a pack size that you can make up all at once and use within 3 months), taking care that all the constituents are completely dissolved. (Follow the manufacturer's instructions). Do not store, but filter sterilize immediately. Precipitation may occur on storage, and microbiological contamination may also appear.

Alternatively, for people using smaller amounts (< 1/l week) or several different types of medium, smaller volumes may be prepared, complete with glutamine, and filtered directly into storage bottles using the Falcon sterilizer (Fig. 8.9c). With this, and other negative pressure filtration systems, some dissolved CO_2 may be lost during filtration and the pH may rise. Provided the correct amount of NaHCO₃ is in the medium to suit the gas phase (see Chapter 7), the medium will re-equilibrate in the incubator, but this should be confirmed the first time used to make sure.

Sterilization. Filter through 0.2 μm polycarbonate, cellulose acetate, or cellulose nitrate filter (see below). Add glutamine before using.

Autoclavable Media

Some commercial suppliers offer autoclavable versions of Eagle's MEM and other media. Autoclaving is much less labor intensive, less expensive, and has a much lower failure rate than filtration. The procedure to follow is supplied in the manufacturer's instructions and is similar to that described above for BSS. The medium is buffered to pH 4.25 with succinate to stabilize the B vitamins during autoclaving and subsequently neutralized. Glutamine is replaced by glutamate or added sterile after autoclaving. As with BSS, care should be taken with evaporation and any deficit made up with sterile deionized distilled water.

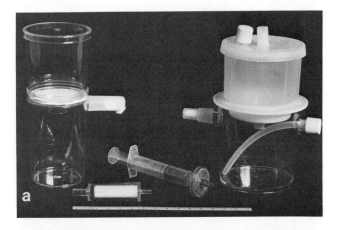

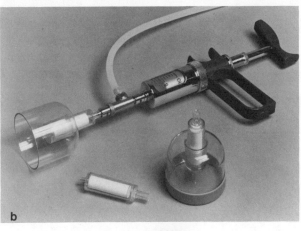

Fig. 8.9. *Disposable sterilizing filters. (a) Syringe type (Millex 25 mm disc, center, Sterivex cylindrical, Millipore), center left and 47 mm with reservoirs (Nalgene, left, Becton-Dickinson, right). (b) Sterivex in use with repeating syringe. (c) Bottle top fitting (Becton-Dickinson).*

Fig. 8.10. *Reusable filter holders (Millipore). (a) 47-mm in-line, polyproplene. (b) Millidisk range; stainless steel housings, high capacity cartridge-type filters. (Reproduced by permission of Millipore UK Ltd.)*

Filter Sterilization

This method is suitable for filtering heat-labile solutions (Figs. 8.9 to 8.12). Reusable filters are made up and sterilized by autoclaving, or presterilized disposable filters may be used. The latter are more expensive but less time-consuming to use and give fewer failures.

Preparation and sterilization of filter—see above.

Materials for filtration

Pressure vessel
pump
clamp to secure filter
sterile filter holder with filter
sterile container to receive filtrate with
    outlet at the base

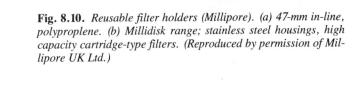

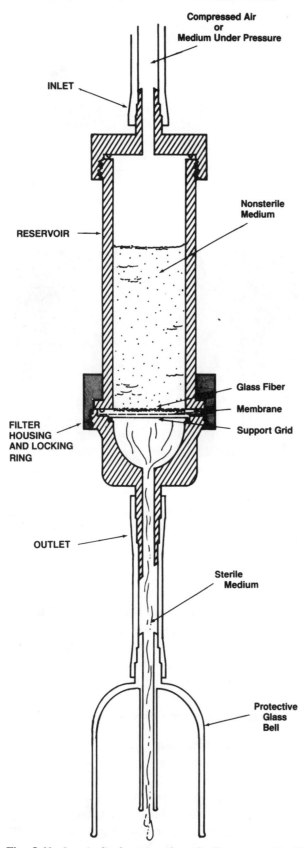

Fig. 8.11. *Longitudinal section through 47-mm reusable filter holder. The whole apparatus is sterilized. Fluid may be added directly to the reservoir or may be delivered from a pressurized tank (see Figs. 8.12 and 8.13).*

Fig. 8.12. *In-line filter assembly connected to receiver flask (a) and to pressurized reservoir (b). Only those items in (a) need be sterilized. Normally the glass bell would be covered in protective foil (left off here for purposes of illustration).*

TABLE 8.3. Filter Size and Fluid Volume

Filter size or designation	Disposable (D) or Reusable (R)	Approximate volume which may be filtered	
		Crystalloid	Colloid
25 mm	D	1–100 ml	1–20 ml
47 mm or Sterivex cartridge	R,D	0.1–1 l	100–250 ml
90 mm	R	1–10 l	0.2–2 l
Millipak-20	D	2–10 l	200 ml–2 l
Millipak-40	D	10–20 l	2–5 l
Millipak-60	D	20–30 l	5–7 l
Millipak-100	D	30–75 l	7–10 l
Millipak-200	D	75–150 l	10–30 l
Millidisk	D	30–300 l	5–50 l
142 mm	R	10–50 l	1–5 l
293 mm	R	50–500 l	5–20 l

tubing
spring clip and glass bell
(see Figs. 8.12, 8.13)
sterile bottles with caps and foil

Protocol
1.
Choose an appropriate filter size from the range available (Table 8.3), secure the filter holder in position, assemble the sterile and nonsterile components, and make the necessary connections (Figs. 8.12, 8.13).
2.
Decant the medium into the pressure vessel.
3.
Position the sterile receiver under the outlet of the filter.
4.
For reusable filters, turn on the pump just long enough to wet the filter. Stop the pump and tighten up the filter holder.
5.
Switch on the pump to deliver 100 kPa (15 lb/in^2). When the receiver starts to fill, draw off aliquots into medium stock bottles of the desired volume, cap, and store.

When sterilizing small volumes (100–200 ml), the solution may fit within the filter holder (Fig. 8.11) and a pressure vessel and receiver will not be required. Collect directly into storage bottle.

Positive pressure is recommended for optimum filter performance and to avoid removal of CO_2 which results from negative pressure filtration. However, the latter method is often used for filtering small volumes as the equipment is simple. It is important to incorporate a filter and trap on the outlet from the filter flask (Figs. 8.9a,c and 14).

Sterility Testing

Positive pressure may also be applied using a peristaltic pump (Fig. 8.15) in line between a nonsterile reservoir and a disposable in-line filter such as the Millipak (Millipore). No receiver is required and only the filter, which is bought sterile and is disposable, and the media bottles need be sterile. This is a simple, effective, and inexpensive method for batch filtration applicable to small, medium, and even large laboratories.

(1) When filtration is complete and all the liquid has passed through the filter, disconnect the outlet, and raise the pump pressure until bubbles form in the effluent from the filter. This is the "bubble point" and should occur at more than twice the pressure used for filtration (see manufacturer's instructions). If the filter bubbles at the sterilizing pressure (100 kPa) or lower, then it is perforated and should be discarded. Any filtrate which has been collected should then be regarded as nonsterile and refiltered.

(2) If the filter passes the "bubble point" test, withdraw aliquots from the beginning, middle, and end of the run and test for sterility by incubating at 36.5°C for 72 hr. If the samples become cloudy, discard them and resterilize the batch. If there are signs of contamination in the other stored bottles, the whole batch should be discarded.

For a more thorough test, take samples, as in (2), dilute one-third of each into nutrient broths, e.g., beef heart hydrolysate and thioglycollate. Divide each in two and incubate one at 36.5°C and one at 20°C for 10 days, with uninoculated controls. If there is any doubt after this incubation, mix and plate out aliquots on nutrient agar.

Alternatively, place a demountable sterile filter in the effluent line from the main sterilizing filter. Any contamination that passes due to failure in the first filter will be trapped in the second. At the end of the run, remove the second filter and place on nutrient agar. If colonies grow, discard or refilter. This method has the advantage that it monitors the entire filtrate and not just a small fraction, but it does not cover risks of contamination during bottling and capping.

Sterility testing of autoclaved stocks is much less essential, provided proper monitoring (temperature and

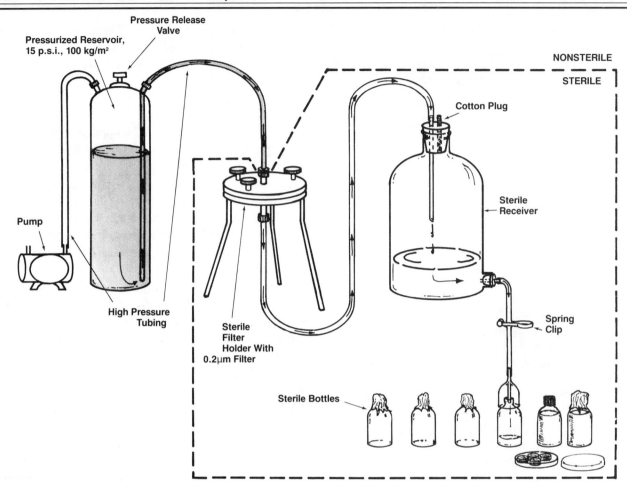

Fig. 8.13. *Diagram of complete system illustrated in Figure 8.12. The broken line encloses those items which are sterilized.*

time at the sterilizing temperature) of the autoclave is carried out.

Culture Testing

Media which have been produced commercially will have been tested for their ability to sustain growth of one or more cell lines. (If they have not, then you should change your supplier!) However, there are certain circumstances when you may wish to test your own media for quality, e.g.: (1) if it has been made up in the laboratory from basic constituents; (2) if any additions are made to the medium; (3) if the medium is for a special purpose that the commercial supplier is not able to test; and (4) if the medium is made up from powder and there is a risk of losing constituents during filtration.

Contamination of medium with toxic substances can arise during filtration. Some filters are treated with traces of detergent to facilitate wetting, and this may leach out into the medium during filtration. Such filters should be washed by passing PBS or BSS through before use, or by discarding the first aliquot of filtrate. Polycarbonate filters, e.g., Nuclepore, are wettable without detergents and are preferred by some workers, particularly where the serum concentration in the medium is low.

There are three main types of culture test: (1) plating efficiency; (2) growth curve at regular passage densities and up to saturation density; and (3) expression of a special function, e.g., differentiation in the presence of an inducer, virus propagation, or expression of a specific antigen. All of these should be performed on the new batch of medium with your regular medium as a control.

Plating efficiency (see Chapter 19). This is the most sensitive test as it will detect minor deficiencies and

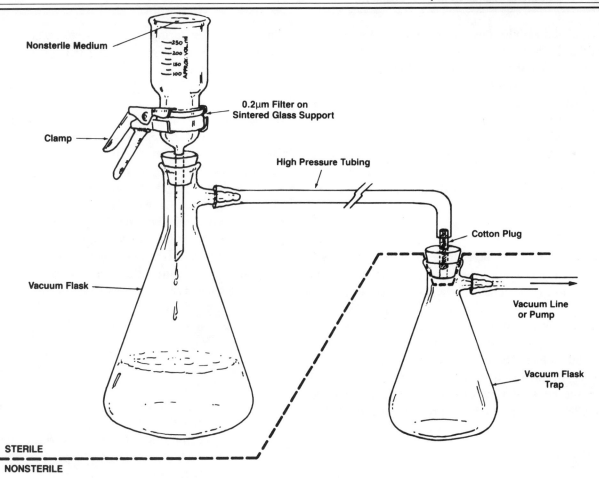

Nonsterile Medium

0.2μm Filter on
Sintered Glass Support

Clamp

High Pressure Tubing

Vacuum Flask

Cotton Plug

Vacuum Line
or Pump

Vacuum Flask
Trap

STERILE

NONSTERILE

Fig. 8.14. *Apparatus for vacuum filtration. Only those items above the broken line need be sterilized. Vacuum filtration is also possible with some disposable filters (Nalgene and Falcon, see Fig. 8.8).*

low concentrations of toxins not apparent at higher cell densities. Ideally it should be performed at a range of serum concentrations from 0% to 20% as serum may mask deficiencies in the medium.

Growth curve. Clonal assay will not always detect insufficiencies in the amount of particular constituents. For example, if the concentration of one or more amino acids is low, it may not affect clonal growth but could influence the maximum cell concentration attainable.

A growth curve (see Chapter 18) gives three parameters of measurement: (1) the lag phase before cell proliferation is initiated after subculture, indicating whether the cells are having to adapt to different conditions; (2) the doubling time in the middle of the exponential growth phase, indicating the growth-promoting capacity; and (3) the terminal cell density. In cell lines which are not sensitive to density limitation of growth, e.g., continuous cell lines (see Chapters 2 and 15), this indicates the total yield possible and usu-

ally reflects the total amino acid or glucose concentration. Remember that a medium which gives half the terminal cell density is costing twice as much per cell produced.

Special functions. In this case a standard test from the experimental system you are using, e.g., virus titer in medium after a set number of days, should be performed on the new medium alongside the old.

A major implication of these tests is that they should be initiated well in advance of the exhaustion of the current stock of medium (1) so that proper comparisons may be made and (2) so that there is time to have fresh medium prepared if the medium fails any of these tests.

Storage

Opinions differ as to the shelf-life of different media. As a rough guide, media made up without glutamine should last 2–3 months at 4°C. Once glutamine is added, storage time is reduced to 2–3 wk. Media which contain labile constituents should either be used within 2–3 wk of preparation or stored at −20°C.

Some forms of room fluorescent lighting will cause deterioration of riboflavin and tryptophan into toxic by-products [Wang, 1976]. Incandescent lighting should be used in cold rooms where media is stored and the light extinguished when the room is not occupied. Bottles should not be exposed to fluorescent lighting for longer than a few hours. A dark freezer is recommended for long-term storage but is not always practicable.

Serum

This is one of the more difficult preparative procedures in tissue culture because of variations in the quality and consistency of the raw material and because of the difficulties encountered in sterile filtration. However, it is also the highest constituent of the cost of doing tissue culture, accounting for 30–40% of the total budget if bought from a commercial supplier. Buying sterile serum is certainly the best approach from the point of view of consistency and simplicity (see Chapter 7) but it may be prepared as follows.

Outline

Collect blood, allow to clot, and separate the serum. Filter serum through gradually reducing porosity of filters, bottle and freeze.

Protocol

Large scale, 20–100 l per batch.

Collection

Arrangements may be made to collect whole blood from a slaughterhouse. It should be collected directly from the bleeding carcass and not allowed to lie around after collection. Alternatively, blood may be withdrawn from live animals under proper supervision. The second routine, if performed consistently on the same group of animals, gives a more reproducible serum but a lower volume for greater expenditure of effort. If done carefully, it may be collected aseptically.

Clotting

Allow the blood to clot by standing overnight in a covered container at 4°C. This so called "natural clot" serum is superior to serum physically separated from the blood cells by centrifugation and defibrination, as platelets release growth factor into the serum during clotting. Separate the serum from the clot and centrifuge at 2,000 g for 1 hr to remove sediment.

Sterilization. Serum is usually sterilized by filtration through a 0.2-μm-porosity sterilizing filter, but because of its viscosity and high particulate content, it should be passed through a graded series of cartridge-type glass-fiber prefilters before passing through the final sterilizing filter. Only the last filter, a 350-mm in-line disc filter or equivalent, need be sterilized.

The prefilter assemblies may be stainless steel with replaceable cartridges (Pall) or may be a single bonded unit (Gelman). The latter are easier to use but more difficult to clean and reuse. Reuse is possible, however. For smaller volumes, less than one liter, graded filters may be stacked in a single unit (Fig. 8.16).

Materials
        Pump
        pressure vessel
        clamp
        142 mm filter holder with series of nonsterile stacked filters, e.g., glass fiber, 5, 1.2, 0.45 μm
        millidisk disposable filter (Millipore) or equivalent
        sterile receiving vessel with outlet at base
        sterile bottles with caps and foil

For volumes of 5–20 l
1.
Connect a nonsterile reusable filter holder (142 mm) or cartridge-type filter (Pall, Millipore) in line with a sterile Millipak 200 or Millidisk 500 (disposable) or sterile 142-mm (reusable) filter holder (Fig. 8.17) with a 0.2 μm-porosity filter, and connected to a sterile receiver.
2.
Place a 0.45-μm-porosity filter on the support screen of the nonsterile holder, a 1.2-μm filter on top of it, and a 5-μm filter on top of that. Finally place a glass-fiber filter on the top of the 5-μm filter, wet the filters with sterile water, and close up the holder.
3.
Connect a pressure reservoir (see Figs. 8.12, 8.13) to the top of the nonsterile filter.

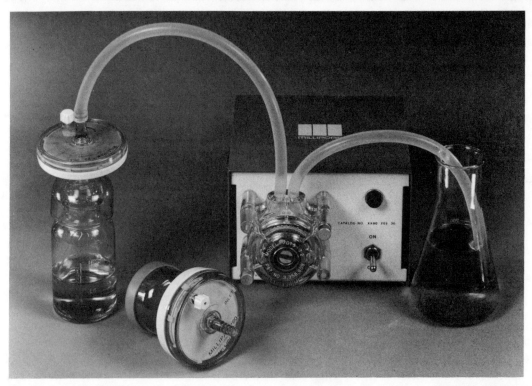

Fig. 8.15. *Sterile filtration with peristaltic pump between non-sterile reservoir and sterilizing filter Millipak, Millipore. (Reproduced by permission of Millipore UK Ltd.)*

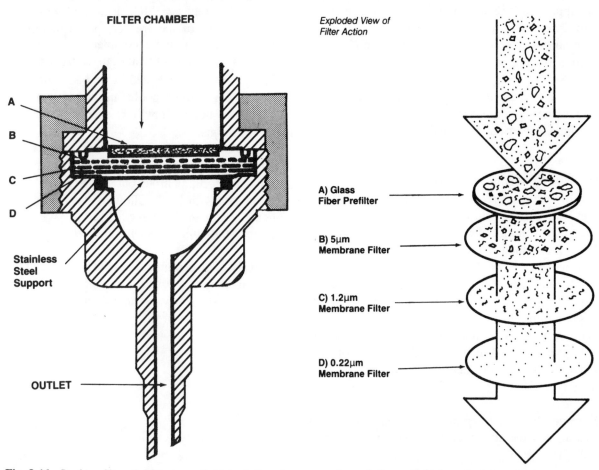

Fig. 8.16. *Stacking filters for filtering colloidal solutions (e.g., serum) or solutions with high particulate content.*

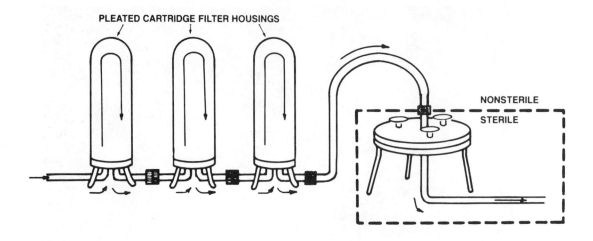

Fig. 8.17. *Series of pleated cartridge filters for large-scale filtration of colloidal solutions. Sterile filtration is completed by a conventional membrane disc, in-line filter.*

4.

Add serum to the pressure vessel, close, and apply pressure (15 kPa, 2 lb/in^2) until the first sign of liquid appears leaving sterile filter. Stop and tighten sterile filter.

5.

Reapply pressure and continue filtration, checking for leaks or blockages. Increasing the filtration pressure will increase the rate of filtration but may cause packing or clogging of the filters.

6.

When all serum is through, bubble-point test the filter (see above), then bottle and freeze the serum, taking samples out previously to test (see above).

Small-scale serum processing. If small amounts (< 1-l) of serum are required, then the process is similar but can be scaled down. After clot retraction (see above) small volumes of serum may be centrifuged (5–10,000 g), and then filtered through a reusable in-line filter assembly (47 or 90 mm) containing glass fiber, 5-μm, 1.2-μm, and 0.45-μm filters as described above and finally through a 47-mm, 0.2-μm-porosity sterile disposable filter.

With very small volumes (10–20 ml), centrifuge at 10,000 g and filter directly through one or more disposable 25 mm, 0.2-μm filters. A graded series of the syringe-type filters (e.g., Millex, Millipore) is now available.

Storage. Bottle the serum in sizes that will be used up within 2–3 wk after thawing. Freeze the serum as rapidly as possible and if thawed, do not refreeze unless further prolonged storage is required.

Serum is best used within 6–12 months of preparation if stored at −20°C, but more prolonged storage may be possible at −70°C. The bulk of serum stocks usually make this impractical. Polycarbonate or high-density polypropylene bottles will eliminate the risk of breakage if storage at −70°C is desired. Regardless of the temperature of the freezer, or the nature of the bottles, do not fill them completely. Allow for the anomalous expansion of water during freezing.

Human serum. Outdated blood bank human blood or plasma can be used, pooled instead of or in addition to bovine or horse. It will be sterile and free from major infections.

◇ Care must be taken with donor serum to ensure that it is screened for hepatitis, AIDS virus, tuberculosis, and so forth. Titrate out the heparin or citrate anticoagulant with Ca^{2+}, allow to clot overnight, separate the serum and freeze.

Quality control. Use same procedures as for medium.

Dialysis. For certain studies, the presence of small molecular weight constituents (amino acids, glucose, nucleosides, etc.) may be undesirable. These may be removed by dialysis through conventional dialysis tubing.

Materials

  Dialysis tubing
  beaker with distilled water

bunsen
tripod with wire gauze
serum to be dialyzed
HBSS at 4°C
magnetic bar and stirrer
measuring cylinder
sterile stacked filters: 0.22, 0.45, 1.2,
    5 μm, glass fiber
sterile bottles with caps

Protocol

1.
Boil five pieces, 30-mm × 500-mm dialysis tubing, in three changes of distilled water.
2.
Transfer to Hanks' Balanced Salt Solution (HBSS) and allow to cool.
3.
Tie double knots at one end of each tube.
4.
Half-fill each dialysis tube with serum (20 ml).
5.
Express air and knot other end leaving a space between the serum and the knot of about half the tube.

6.
Place in 5 l of HBSS and stir on a magnetic stirrer overnight at 4°C.
7.
Change HBSS and repeat twice.
8.
Collect serum into measuring cylinder and note volume. (If volume is reduced, add HBSS to make up to starting volume. If increased, make due allowance when adding to medium later).
9.
Sterilize through graded series of filters (see above).
10.
Bottle and freeze.

Preparation and Sterilization of Other Reagents

Individual recipes and procedures are given in the Appendix. On the whole most reagents are sterilized by filtration if heat-labile and by autoclaving if stable (see summary Table 8.2).

Filters with low binding properties (e.g. Millex-GV) are available for filter sterilization of proteins and peptides.

Chapter 9
Disaggregation of the Tissue and Primary Culture

A primary cell culture may be obtained either by allowing cells to migrate out of fragments of tissue adhering to a suitable substrate or by disaggregating the tissue mechanically or enzymatically to produce a suspension of cells, some of which will ultimately attach to the substrate. It appears to be essential for most normal untransformed cells (with the exception of hemopoietic cells) to attach to a flat surface in order to survive and proliferate with maximum efficiency. Tumor cells, on the other hand, particularly cells from transplantable animal tumors, are often able to proliferate in suspension.

The enzymes used most frequently are crude preparations of trypsin, collagenase, elastase, hyaluronidase, DNase, pronase, dispase, or various combinations. Trypsin and pronase give the most complete disaggregation but may damage the cells. Collagenase and dispase give incomplete disaggregation but are less harmful. Hyaluronidase can be used in conjunction with collagenase to digest intracellular matrix, and DNase is employed to disperse DNA released from lysed cells as it tends to impair proteolysis and promote reaggregation.

Although each tissue may require a different set of conditions, certain common requirements are shared by most primary cultures.

(1) Fat and necrotic tissue are best removed during dissection.

(2) The tissue should be chopped finely with minimum damage.

(3) Enzymes used for disaggregation should be removed subsequently by gentle centrifugation.

(4) The concentration of cells in the primary culture should be much higher than that normally used for subculture, since the proportion of cells from the tissue which survive primary culture may be quite low.

(5) A rich medium, such as Ham's F10 or F12, should be used in preference to a simple, basal medium, such as Eagle's BME; and if serum is required, fetal bovine often gives better survival than calf or horse. Isolation of specific cell types may require selective media (see Chapter 20).

(6) Embryonic tissue disaggregates more readily, yields more viable cells, and proliferates more rapidly in primary culture than adult.

ISOLATION OF THE TISSUE

Before attempting to work with human or animal tissue, make sure that your work fits within medical ethical rules or current animal experiment legislation. Recent changes in the law in the United Kingdom mean that the use of embryos or fetuses beyond 50% gestation or incubation comes under the Animal Experiments (Scientific Procedures) Act, 1986.

An attempt should be made to sterilize the site of the dissection with 70% alcohol if likely to be contaminated. Remove tissue aseptically and transfer to tissue culture laboratory in BSS or medium as soon as possible. If a delay is unavoidable, refrigerate the tissue. (Viable cells can be recovered from chilled tissue several days after explantation.)

Mouse Embryos

Outline
Remove uterus aseptically from timed pregnant mouse and dissect out embryos.

Materials
70% alcohol in wash bottle
bunsen burner
70% alcohol to sterilize instruments
sterile BSS in 50-ml sterile beaker to cool instruments after flaming
HBSS (with antibiotics if required) in 25–50 ml screw-capped vial or tube
timed pregnant mice

Protocol

1.

If males and females are housed separately, when they are put together for mating, estrus will be induced in the female 3 d later, when the maximum number of successful matings will occur. This enables the planned production of embryos at the appropriate time. The timing of successful matings may be determined by examining the vaginas each morning for a hard mucous plug. The day of detection of the plug (the "plug date") is noted as day zero, and the development of the embryos timed from this date. Full term is about 19–21 d. The optimal age for preparing cultures from whole disaggregated embryo is around 13 d, when the embryo is relatively large (Figs. 9.1, 9.2) but still contains a high proportion of undifferentiated mesenchyme. It is from this mesenchyme that most of the culture will be derived.

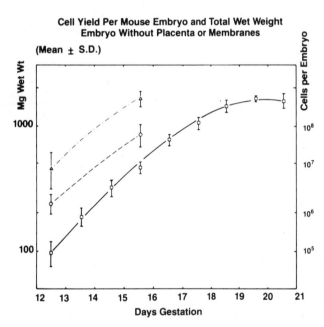

Fig. 9.1. *Total wet weight and yield of cells per mouse embryo. Total wet weight of embryo without placenta or membranes, mean ± standard deviation (squares) [From Paul et al., 1969]. Cell yield per embryo after incubation in 0.25% trypsin at 36.5°C for 4 hr (circles). Cell yield per embryo after soaking in 0.25% trypsin at 4°C for 5 hr and incubation at 36.5°C for 30 min (triangles; see text).*

Fig. 9.2. *Mouse embryos from the 12th, 13th, and 14th d of gestation. The 12-d embryo (bottom) came from a small litter (three) and is larger than would normally be found at this stage. Scale, 10 mm between marks.*

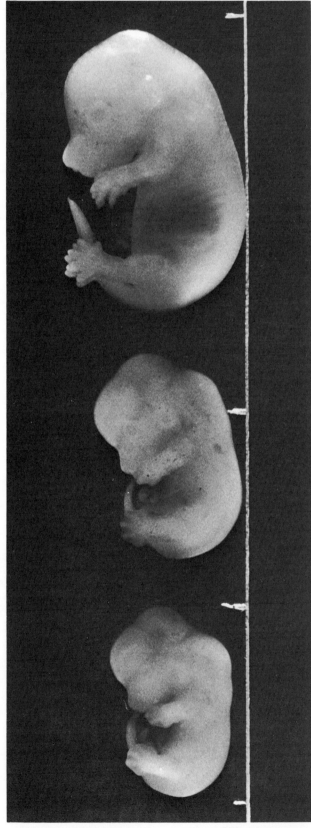

Most individual organs, with the exception of brain and heart, begin to form about the 9th d of gestation but are difficult to isolate until about the 11th d. Dissection is easier by 13–14 d, and most of the organs are completely formed by the 18th d.

2.

Kill the mouse by cervical dislocation and swab the ventral surface liberally with 70% alcohol. Tear the ventral skin transversely at the median line just over the diaphragm and, grasping the skin on both sides of the tear, pull in opposite directions* to expose the untouched ventral surface of the abdominal wall (Fig. 9.3a–c).

3.

Cut longitudinally along the median line with sterile scissors, revealing the viscera. At this stage, the uteri filled with embryos will be obvious posteriorly and may be dissected out into a 25-ml or 50-ml screw-capped tube containing 10 or 20 ml BSS (Fig. 9.3d–f). Antibiotics may be added to the BSS where there is a high risk of infection (see Appendix, Dissection BSS (DBSS).

All of the preceding steps should be done outside the tissue culture laboratory; a small laminar flow hood and rapid technique will help to maintain sterility. Do not take live animals into the tissue culture laboratory; they may carry contamination. If the carcass must be handled in the tissue culture area, make sure it is immersed in alcohol briefly, or thoroughly swabbed, and disposed of quickly after use.

4.

Take the intact uteri to the tissue culture laboratory and transfer to a fresh dish of sterile BSS.

5.

Dissect out the embryos (Fig. 9.3g,h). Tear the uterus with two pairs of sterile forceps, keeping their points close together to avoid distorting the uterus and bringing too much pressure to bear on the embryos. As the uterus is torn apart, the embryos may be freed from the membranes and pla-

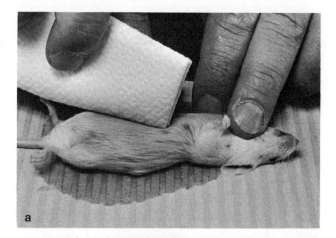

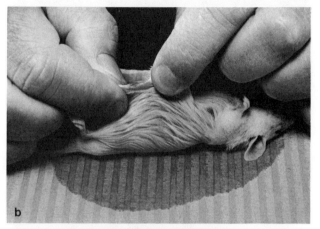

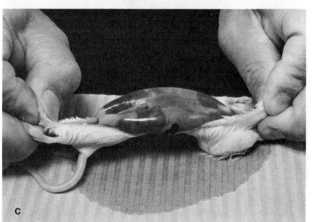

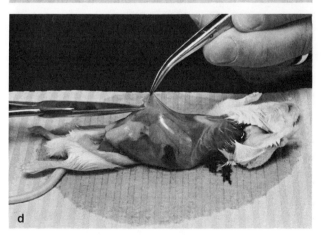

Fig. 9.3. *Stages in the aseptic removal of mouse embryos for primary culture (see text). a. Swabbing the abdomen. b,c. Tearing the skin to expose the abdominal wall. d. Opening the abdomen. e. Uterus in situ. f. Removing the uterus. g,h. Dissecting embryos from uterus. i. Removing membranes. j. Chopping embryos.*

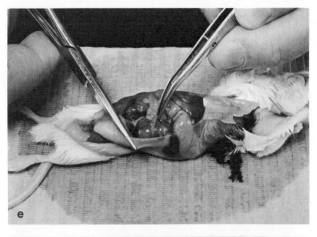

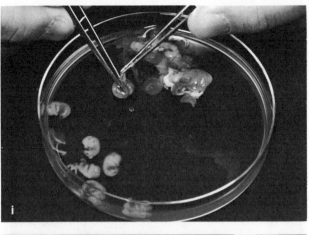

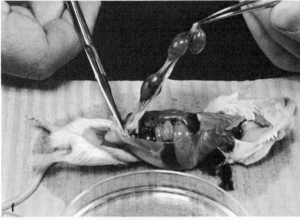

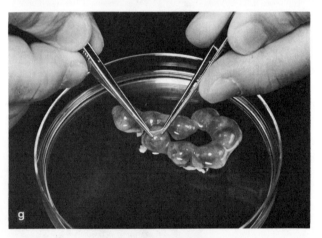

Fig. 9.3. *(continued; see legend and text on page 109)*

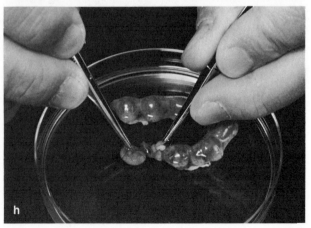

centa and placed to one side of the dish to bleed, then transferred to a fresh dish. If a large number of embryos is required (more than four or five litters), it may be helpful to place the last dish on ice (for subsequent dissection and culture, see below).

Hen's Egg

Outline
Remove embryo from egg and transfer to dish.

Materials
70% alcohol
swabs
small beaker 20–50 ml
forceps—straight and curved
9-cm petri dishes
BSS
11-day embryonated eggs
humid incubator (no CO_2 above atmospheric)

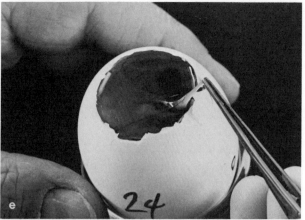

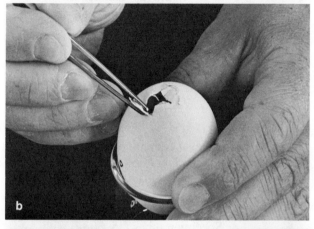

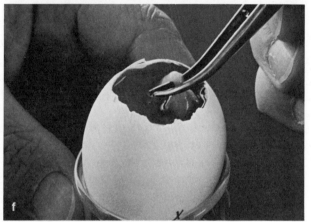

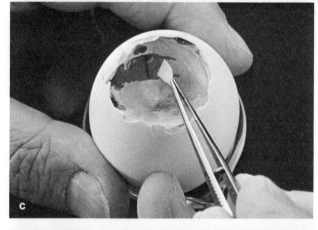

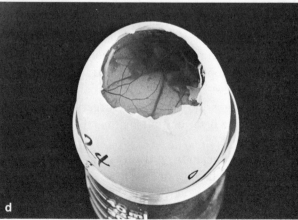

Fig. 9.4. *Stages in the aseptic removal of chick embryo from the egg (see text).*

Protocol

1.

Incubate the eggs at 38.5°C in a humid atmosphere and turn through 180° daily. Although hen's eggs hatch at around 20 to 21 d, the lengths of the developmental stages are different from the mouse. For culture of dispersed cells from the whole embryo, the egg should be taken at about 8 d, from isolated organ rudiments, at about 10 to 13 d.

2. Swab the egg with 70% alcohol and place with blunt end uppermost in a small beaker (Fig. 9.4a)

3.

Crack the top of the shell and peel off to the edge of the air sac with sterile forceps (Fig. 9.4b).

4.

Resterilize the forceps (dip in a beaker of alcohol, burn off alcohol, and cool in sterile BSS) and peel off the white shell membrane to reveal the chorioallantoic membrane (CAM) below, with its blood vessels (Fig. 9.4c,d).

5.

Pierce the CAM with sterile curved forceps and lift out the embryo by grasping gently under the head. Do not close the forceps completely or the neck will sever (Fig. 9.4e–g).

6.

Transfer embryo to a 9-cm petri dish containing 20 ml DBSS. (For subsequent dissection and culture, see below.)

Human Biopsy Material

Handling human biopsy material presents certain problems not encountered with animal tissue. It will usually be necessary to obtain consent: (1) from the hospital ethical committee; (2) from the attending physician or surgeon; and (3) from the patient or the patient's relatives. Furthermore, biopsy sampling is usually performed for diagnostic purposes and hence the needs of the pathologist must be met first. This is less of a problem if extensive surgical resection or non-pathological tissue (e.g., placenta or umbilical cord) is involved.

The operation will be performed by one of the resident staff at a time that is not always convenient to the tissue culture laboratory, so some formal collection or storage system must be initiated for times when you or someone on your staff cannot be there. If delivery to your lab is arranged, there must be a system for receipt of specimens, recording details, and alerting the person who will perform the culture, otherwise valuable material may be lost or spoiled.

◊ Biopsy material carries a risk of infection (See Chapter 6), so it should be handled in a Class II biohazard Cabinet, and all media and apparatus disinfected after use by autoclaving or immersion in a suitable disinfectant (see Chapter 6). If possible the tissue should be screened for infections such as hepatitis, AIDS, tuberculosis, and so forth, unless the patient has already been tested.

Outline

Provide labeled container(s) of medium, consult with hospital staff, and collect sample from operating room or pathologist.

Materials

Specimen tubes (15–30ml) with leakproof caps about one-half full with culture medium containing antibiotics (see Appendix, Collection Medium), and labeled with your name, address, and telephone number

Protocol

1.

Provide a container of collection medium clearly labeled, and either arrange to be there to collect it after surgery, or have your name and address on it and, preferably, your telephone number, so that it can be sent to you immediately and you can be informed easily when it has been dispatched.

2.

Transfer sample to tissue culture laboratory. Usually, if kept at 4°C, biopsy samples will survive for at least 24 hr and even up to 3 or 4 d, although the longer the time from surgery to culture, the more the deterioration that may be expected.

3.

Decontamination. A disinfectant wash is given before skin biopsy, and an oral antibiotic before gut surgery. Most surgical specimens, however, from the needs of surgery, are sterile when removed though problems may arise with subsequent hand-

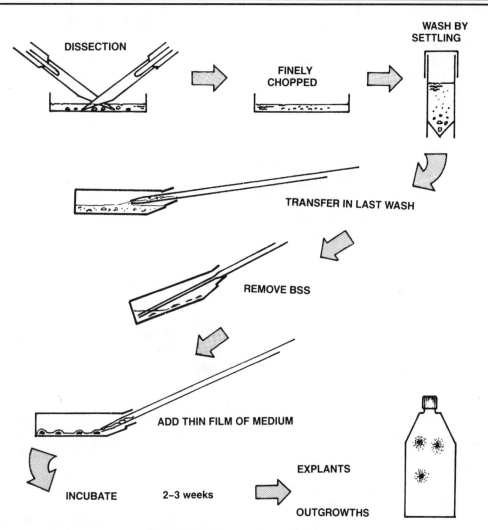

Fig. 9.5. *Primary explant technique.*

Fig. 9.6. *Primary explant from human mammary carcinoma. Dense area in center is an undisaggregated tissue fragment and cells are seen migrating radially from the explant. This is a good size for an explant, promoting maximal radial outgrowth, but routinely it is difficult to dissect below 0.5 mm. Scale bar, 100 μm.*

ling. Superficial (skin biopsies, melanomas, etc.) and gastrointestinal tract (colon and rectal samples) are particularly prone to infection. It may be advantageous to consult a medical microbiologist to determine what flora to expect in a given tissue and choose your antibiotics accordingly. If the surgical sample is large enough, (200 mg or more), a brief dip (30 s – 1 min) in 70% alcohol will help to reduce superficial contamination without causing much harm to the center of the tissue sample.

PRIMARY CULTURE

Primary Explant Technique

This was the original method developed by Harrison [1907], Carrel [1912], and others for initiating a tissue culture. A fragment of tissue was imbedded in blood plasma or lymph, mixed with embryo extract and serum, and placed on a slide or coverslip. The plasma clotted and held the tissue in place. The embryo extract plus serum both supplied nutrients and stimulated migration out of the explant across the solid substrate. Heterologous serum was used to promote clotting of the plasma. This technique is still used but has largely been replaced by the simplified method below.

Outline

The tissue is chopped finely, rinsed, and the pieces seeded onto the culture surface in a small volume of medium with a high concentration (40–50%) of serum, such that surface tension holds the pieces in place until they adhere spontaneously to the surface (Fig. 9.5). Once this is achieved, outgrowth of cells usually follows (Fig. 9.6).

Materials

        Petri dishes
        100 ml BSS
        forceps
        scalpels
        10-ml pipettes
        15- or 20-ml centrifuge tubes or
            universal containers
        culture flasks
        growth medium

The size of flasks and volume of growth medium will depend on the amount of tissue—roughly five 25-cm^2 flasks per 100 mg tissue, and initially 1 ml medium per flask, building up to 5 ml per flask over the first 3–5 d.

Protocol

1.
Transfer tissue to fresh sterile BSS and rinse.
2.
Transfer to a second dish, dissect off unwanted tissue such as fat or necrotic material, and chop finely with crossed scalpels (Fig. 9.5) to about 1-mm cubes.
3.
Transfer by pipette to a 15- or 50-ml sterile centrifuge tube or universal container (wet the inside of the pipette first or the pieces will stick). Allow the pieces to settle.
4.
Wash by resuspending the pieces in BSS, allowing the pieces to settle, and removing the supernatant fluid, two or three times.
5.
Transfer the pieces (remember to wet the pipette)

to a culture flask, about 20–30 pieces per 25-cm^2 flask.
6.
Remove most of the fluid and add about 1 ml growth medium per 25-cm^2 growth surface.
7.
Tilt the flask gently to spread the pieces evenly over the growth surface.
8.
Cap the flask and place in incubator or hot room at 36.5°C for 18–24 hr.
9.
If the pieces have adhered, the medium volume may be made up gradually over the next 3–5 d to 5 ml per 25 cm^2 and then changed weekly until a substantial outgrowth of cells is observed (see Fig. 9.6).
10.
The explants may then be picked off from the center of the outgrowth with a scalpel and transferred by prewetted pipette to a fresh culture vessel. (Return to step 7 above.)
11.
Replace medium in the first flask until the outgrowth has spread to cover at least 50% of the growth surface, at which point the cells may be passaged (see Chapter 10).

This technique is particularly useful for small amounts of tissue such as skin biopsies where there is a risk of losing cells during mechanical or enzymatic disaggregation. Its disadvantages lie in the poor adhesiveness of some tissues and the selection of cells in the outgrowth. In practice, however, most cells, fibroblasts, myoblasts, glia, epithelium, particularly from the embryo, will migrate out successfully.

Both adherence and migration may be stimulated by a plasma clot:

1.
Place the tissue pieces in position in the culture flask as in step 7 above.
2.
Mix 2 parts of chicken plasma with 1 part chicken embryo extract and 1 part fetal bovine serum. Immediately pipette gently over the tissue pieces, spacing the pieces evenly on the surface of the dish as you do so.

3.

Allow to clot and place at 37°C [see Paul, 1975, for further description].

Alternatively a glass coverslip may be placed on top of the explant, with the explant near the edge of the coverslip, or the plastic dish may be scratched through the explant to attach the tissue to the flask.

Enzymatic Disaggregation

Mechanical and enzymatic disaggregation of the tissue avoids problems of selection by migration, but more important perhaps, yields a higher number of cells, more representative of the whole tissue, in a shorter time. However, just as the primary explant technique selects on the basis of cell migration, dissociation techniques select cells resistant to the method of disaggregation and still capable of attachment.

Embryonic tissue disperses more readily and gives a higher yield of proliferating cells than newborn or adult. The increasing difficulty in obtaining viable proliferating cells with increasing age is due to several factors including the onset of differentiation, an increase in fibrous connective tissue and extra-cellular matrix, and a reduction of the undifferentiated proliferating cell pool. Where procedures of greater severity are required to disaggregate the tissue, e.g., longer trypsinization or increased agitation, the more fragile components of the tissue may be destroyed. In fibrous tumors, for example, it is very difficult to obtain complete dissociation with trypsin while still retaining viable carcinoma cells.

Crude trypsin is by far the most common enzyme used in tissue disaggregation [Waymouth, 1974] as it is tolerated quite well by many cells, is effective for many tissues, and any residual activity left after washing is neutralized by the serum of the culture medium. A trypsin inhibitor (e.g., soya bean trypsin inhibitor, Sigma) must be included when serum-free medium is used.

It is important to minimize the exposure of cells to active trypsin to preserve maximum viability. Hence when trypsinizing whole tissue at 36.5°C, dissociated cells should be collected every half hour, the trypsin removed by centrifugation and neutralized with serum in medium. Soaking the tissue for 6–18 hr in trypsin at 4°C (see below) allows penetration with minimal tryptic activity, and digestion may then proceed for a much shorter time (20–30 min) at 37°C [Cole and

Paul, 1966]. Although the cold trypsin method gives a higher yield of viable cells and is less effort, the warm trypsin method is still used extensively and is presented here for comparison.

Disaggregation in Warm Trypsin

Outline

The tissue is chopped and stirred in trypsin for a few hours, collecting dissociated cells every half hour, and the dissociated cells are then centrifuged and pooled in medium containing serum (Fig. 9.7).

Materials

> 250-ml Erlenmeyer flask
> magnetic bar
> magnetic stirrer
> 100 ml 0.25% trypsin (crude: Difco
>   1:250, Flow or Gibco) in CMF
>   or PBSA (see Appendix)
> two 50-ml centrifuge tubes
> hemocytometer or cell counter
> growth medium with serum (e.g., Ham's
>   F12 with 10% fetal bovine serum)
> culture flasks, 5–10 g tissue (will vary
>   depending on cellularity of tissue)

Protocol

1.

Proceed as for primary explant, although the pieces need only to be chopped to about 3-mm diameter. As described here, this method requires about 20 times as much tissue as the primary explant technique, although it can be scaled down if desired.

2.

After washing as in step 4 for primary explantation, transfer the chopped pieces to the trypsinization flask (Bellco makes one specially designed for the purpose, Fig. 9.8, but a 250-ml conical Erlenmeyer flask will do), and add 100 ml trypsin.

3.

Stir at about 200 rpm for 30 min at 36.5°C.

4.

Allow pieces to settle, collect supernatant, centrifuge at approximately 500 g for 5 min, resuspend pellet in 10 ml medium with serum, and store cells on ice.

5.

Add fresh trypsin to pieces and continue to stir and incubate for a further 30 min. Repeat steps 3

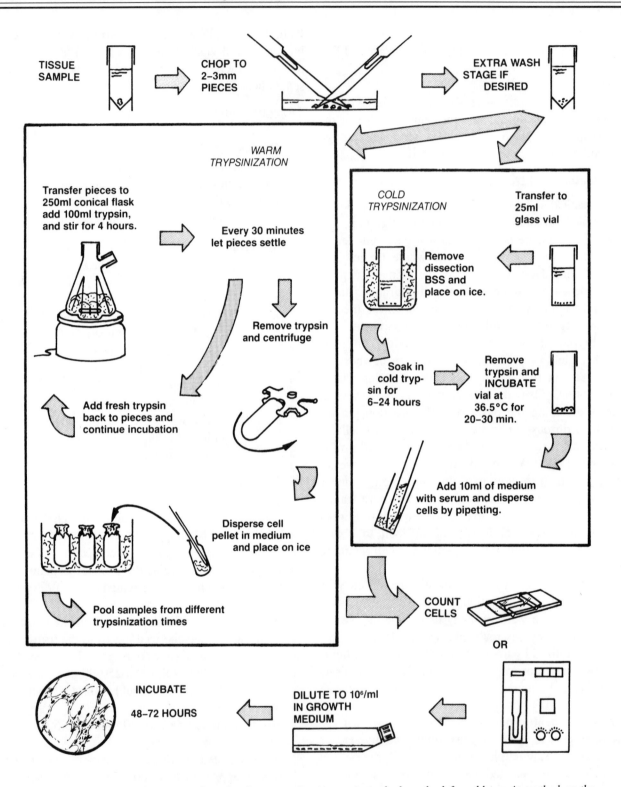

Fig. 9.7. *Preparation of primary culture by disaggregation in trypsin method on the left, cold trypsin method on the right.*

TABLE 9.1. Relative Cell Yield From 12-d Mouse Embryos by Warm or Cold Trypsin Methods

Trypsin temperature and duration of exposure (hr)		Total cell number recovered per embryo ($\times 10^{-7}$)	Viable cells per embryo		Percentage of total recovered after 24 hr in culture	Percentage of viable cells recovered
4°C	36.5°C		No. ($\times 10^{-7}$)	% Viability by dye exclusion (Trypan blue)		
—	4*	1.69	1.45	86	47.2	54.9
5.5	0.5†	3.32	1.99	60	74.5	124
24	0.5†	3.40	2.55	75	60.3	80.2

*Stirred continuously at 36.5°C; fractions were *not* collected at 30-min intervals.
†Incubated without agitation.

Fig. 9.8. *Trypsinization flask (Bellco). The indentations in the side of the flask improve mixing, and the rim around the neck, below the side arm, allows the cell suspension to be poured off while leaving the stirrer bar and any larger fragments behind.*

to 5 until complete disaggregation occurs or until no further disaggregation is apparent.

6.
Collect and pool chilled cell suspensions, count by hemocytometer or electronic cell counter (see Chapter 18).
Remember you are dealing with a very heterogeneous population of cells; electronic cell counting will initially require confirmation with a hemocytometer, as a "plateau" (see Chapter 18) is rather difficult to obtain.

7.
Dilute to 10^6 per ml in growth medium and seed as many flasks as are required with approximately

2×10^5 cells per cm². Where the survival rate is unknown or unpredictable, a cell count is of little value (e.g., tumor biopsies where the proportion of necrotic cells may be quite high). In this case, set up a range of concentrations from about 5 to 25 mg tissue per ml. Change the medium at regular intervals (2 to 4 d as dictated by depression of pH).

This technique is useful for the disaggregation of large amounts of tissue in a relatively short time, particularly whole mouse embryo or chick embryo. It does not work as well with adult tissue where there is a lot of fibrous connective tissue, and mechanical agitation can be damaging to some of the more sensitive cell types such as epithelium.

Trypsin at 4°C

One of the disadvantages of using trypsin to disaggregate tissue is the damage that may result from prolonged exposure to trypsin at 36.5°C—hence the need to harvest cells after 30-min incubations in the warm trypsin method, rather than have them exposed for the full time (3 to 4 hr) required to disaggregate the whole tissue. A simple method of minimizing damage to the cells during disaggregation is to soak the tissue in trypsin at 4°C to allow penetration of the enzyme with little tryptic activity (Table 9.1). Following this, the tissue will require much shorter incubation at 36.5°C for disaggregation [Cole and Paul, 1966].

Outline

Chop tissue and place in trypsin at 4°C for 6 to 18 hr. Incubate after removing the trypsin, and disperse the cells in warm medium (Fig. 9.7).

Materials

Petri dish
BSS
forceps—straight and curved
0.25% crude trypsin
scalpels
25-ml screw-capped Erlenmeyer
    flask(s),
    25-cm^2 or 75-cm^2 culture flasks
pipettes
culture medium

Protocol

1.

Follow steps 1—4 as for primary explants, but collect tissue in glass tube or vial to facilitate chilling (see below). Tissue need only be chopped to 3–4-mm pieces. Embryonic organs, if they do not exceed this size, are better left whole.

2.

After washing, place the container on ice, remove the last BSS wash, and replace with 0.25% trypsin in PBSA at 4°C (approximately 1 ml for every 100 mg of tissue).

3.

Place at 4°C for 6–18 hr.

4.

Remove and discard the trypsin carefully, leaving the tissue with only the residual trypsin.

5.

Place tube at 36.5°C for 20–30 min.

6.

Add warm medium, approximately 1 ml for every 100 mg, and gently pipette up and down until the tissue is completely dispersed.

7.

If some tissue is left undispersed, the cell suspension may be filtered through sterile muslin or stainless steel mesh (100–200 μm), or the larger pieces may simply be allowed to settle. Where there is a lot of tissue, increasing the volume of suspending medium to 20 ml for each gram of tissue will facilitate settling and subsequent collection of supernatant fluid. Two to 3 min should be sufficient to get rid of most of the larger pieces.

8.

Determine the concentration of the cell suspension and seed the vessels at 10^6 cells per ml (2 × 10^5 cells per cm^2).

The cold trypsin method usually gives a higher yield of cells (see Fig. 9.1 and Table 9.1) and preserves more different cell types than the warm method. Cultures from mouse embryos contain more epithelial cells when prepared by the cold method, and erythroid cultures from 13-d fetal mouse liver respond to erythropoietin after this treatment but not after warm trypsin or mechanical disaggregation [Cole and Paul, 1966; Conkie, personal communication]. The method is also convenient as no stirring or centrifugation is required, and the incubation at 4°C may be done overnight. This method does take longer, however, and is not as convenient where large amounts of tissue (greater than 10 g) are being handled. A particular advantage in the cold trypsin method is the handling of small amounts of tissue, such as embryonic organs. Taking 10–13-day chick embryo as a starting point, the following procedure gives good reproducible cultures with evidence of several different cell types characteristic of the tissue of origin.

Chick Embryo Organ Rudiments

Outline

Dissect out individual organs or tissues, and place, preferably whole, in cold trypsin overnight. Remove the trypsin, incubate briefly, and disperse in culture medium. Dilute and seed cultures.

Materials

Petri dish
BSS
scalpels (No. 11 blade for most steps)
iridectomy knives (Beaver, blade 21)
curved and straight fine forceps
pipettes (Pasteur, 2 ml, 10 ml)
0.25% crude trypsin in CMF or
    PBS on ice
10–15 ml test tubes with screw caps
25-cm^2 culture flasks
culture medium (e.g., Ham's F12 +
    10% fetal bovine serum)

Protocol

1.

Remove the embryo from the egg as described above and place in sterile BSS.

2.

Remove the head (Fig. 9.9a,b).

3.

Remove an eye and open carefully, releasing the

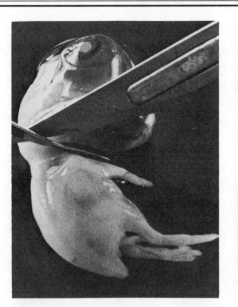

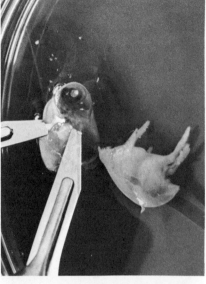

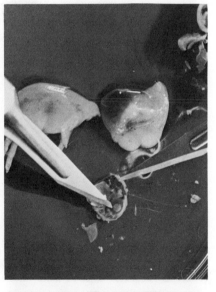

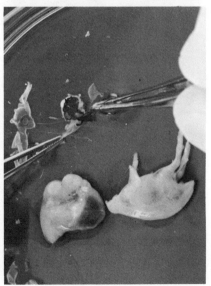

Fig. 9.9. *Dissection of chick embryo. a,b. Removing the head. c. Removing the eye. d. Dissecting out the lens. e. Peeling off the retina. f. Scooping out the brain. g. Halving the trunk. h. Teasing out the heart and lungs from the anterior half. i. Teasing out the liver and gut from the posterior half. j. Inserting the tip of the scalpel between the left kidney and the dorsal body wall. k. Squeezing out the spinal cord. l. Peeling skin off the back of the trunk and hind leg. m. Slicing muscle from the thigh. n. Organ rudiments arranged around periphery of dish; from the right, clockwise: brain, heart, lungs, liver, gizzard, kidneys, spinal cord, skin, and muscle.*

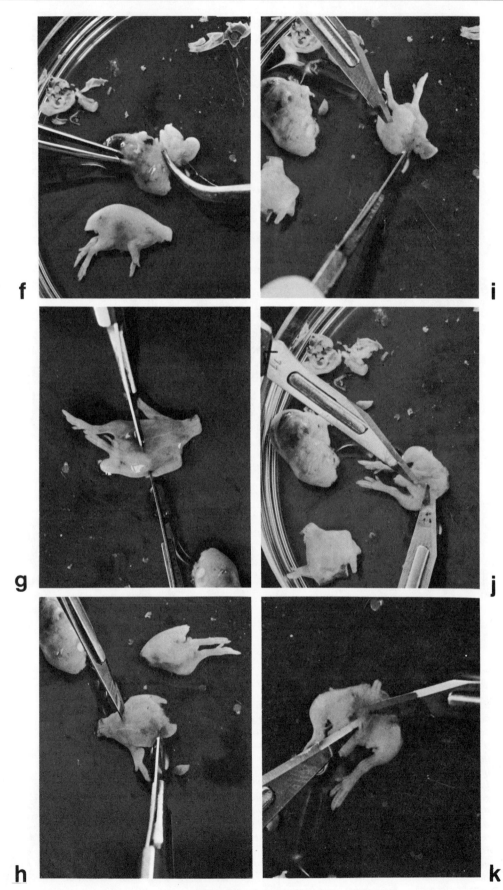

Fig. 9.9. *(continued)*

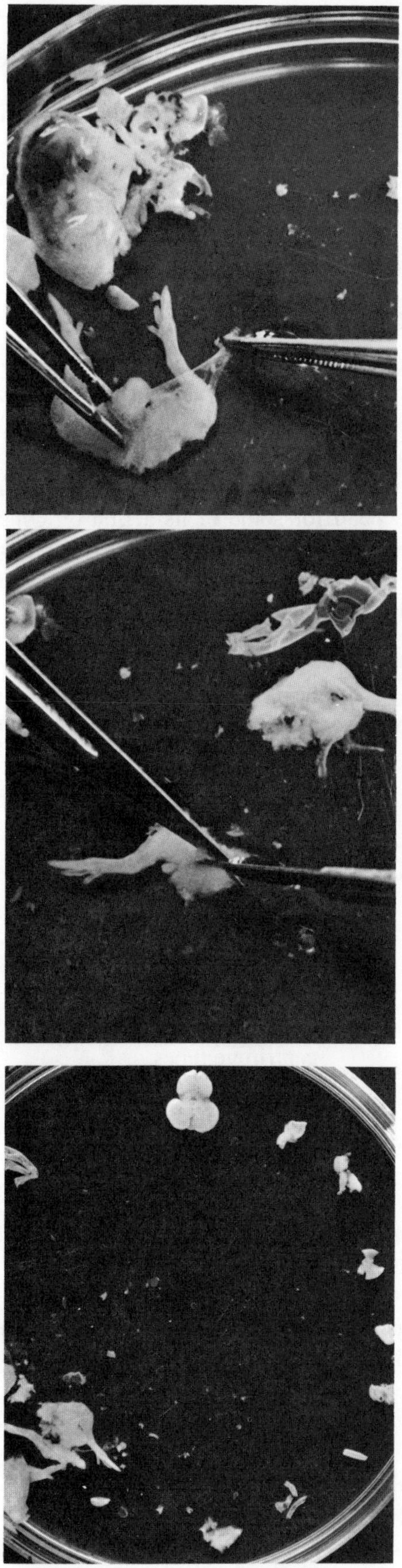

l

m

n

Fig. 9.9. *(continued)*

lens and aqueous and vitreous humors (Fig. 9.9 c,d).

4.
Grasp the retina in two pairs of fine forceps and gently peel the pigmented retina off the neural retina and connective tissue (Fig. 9.9e). (A brief exposure to 0.25% trypsin in 1 mM EDTA will separate the two tissues more easily.) Put tissue to one side.

5.
Pierce the top of the head with curved forceps and scoop out the brain (Fig. 9.9f).

6.
Halve the trunk transversely where the pink color of the liver shows through the ventral skin (Fig. 9.g). If the incision is made on the line of the diaphragm, it will pass between the heart and the liver; but sometimes the liver will go to the anterior instead of the posterior half.

7.
Gently probe into the cut surface of the anterior half and draw out the heart and lungs (Fig. 9.9h; tease the organs out and do not cut until you have identified them). Separate.

8.
Probe the posterior half, and draw out the liver with the folds of the gut enclosed in between the lobes (Fig. 9.9i). Separate.

9.
Fold back the body wall to expose the inside of the dorsal surface of the body cavity in the posterior half. The elongated lobulated kidneys should be visibly parallel to and on either side of the midline.

10.
Gently slide the tip of the scalpel under each kidney and tease away from the dorsal body wall (Fig. 9.9j). Carefully cut free and place on one side.

11.
Place the tips of the scalpels together on the midline at the posterior end and, advancing the tips forward, one over the other, express the spinal cord like toothpaste from a tube (Fig. 9.9k).

12.
Turn the posterior trunk of the embryo over and strip the skin off the back and upper part of the legs (Fig. 9.9l). Collect and place on one side.

13.
Dissect off muscle from each thigh and collect together. (Fig. 9.9m).

14.

Transfer all of these tissues, and any others you may want, to separate test tubes containing 1 ml of 0.25% trypsin in PBSA and place on ice. Make sure the tissue slides right down the tube into the trypsin.

15.

Leave 6–18 hr at 4°C.

16.

Remove trypsin carefully; tilting and rolling the tube slowly will help.

17.

Incubate the tissue in the residual trypsin for 15–20 min at 36.5°C.

18.

Add 4 ml medium to each of two 25-cm^2 flasks for each tissue to be cultured.

19.

Add 2 ml medium to tubes after step 18, and pipette up and down gently to disperse the tissue.

20.

Allow any large pieces to settle, pipette off supernatant fluid into the first flask, mix, and transfer 1 ml of diluted suspension to second flask. This gives two flasks at different cell concentrations and avoids the need to count the cells. Experience will determine the appropriate cell concentration to use in subsequent attempts.

21.

Change the medium as required (e.g., with brain it may need to be changed after 24 hr, but pigmented retina will probably last 5–7 d), and check for characteristic morphology and function. After 3 to 5 d, contracting cells may be seen in the heart cultures, colonies of pigmented cells in the pigmented retina culture, and the beginning of myotubes in skeletal muscle cultures.

Other Enzymatic Procedures

Disaggregation in trypsin can be damaging (e.g., to some epithelial cells) or ineffective (e.g., for very fibrous tissue such as fibrous connective tissue), so attempts have been made to utilize other enzymes. Since the intracellular matrix often contains collagen, particularly in connective tissue and muscle, collagenase has been the obvious choice [Coon and Cahn, 1966; Lasfargues, 1973; Freshney, 1972]. Other bacterial proteases such as pronase [Wiepjes and Prop, 1970; Prop and Wiepjes, 1973; Gwatkin, 1973] and dispase [Matumura, et al., 1975]; (Boehringer-Mann-

heim Biochemicals) have also been used with varying degrees of success. The participation of carbohydrate in intracellular adhesion has led to the use of hyaluronidase [Berry and Friend, 1969] and neuraminidase in conjunction with collagenase. It is not possible here to describe all the primary disaggregation techniques that have been used, but the following method has been found to be effective in several normal and malignant tissues.

Collagenase

This technique is very simple and effective for embryonic and normal and malignant adult tissue. It is of greatest benefit where the tissue is either too fibrous or too sensitive to allow the successful use of trypsin [Freshney, 1972]. Crude collagenase is often used and may depend for some of its action on contamination with other nonspecific proteases. More highly purified grades are available if nonspecific proteolytic activity is undesirable but may not be as effective.

Outline

Place finely chopped tissue in complete medium containing collagenase and incubate. When tissue is disaggregated, remove collagenase by centrifugation, seed cells at a high concentration, and culture (Fig. 9.10).

Materials

    Pipettes
    25-cm^2 culture flasks
    growth medium
    collagenase (2,000 units/ml)
        Worthington CLS or Sigma 1A
    centrifuge tubes
    centrifuge

Protocol

1.

Proceed as for primary explant up to step five, but transfer 20–30 pieces to one 25-cm^2 flask and 100 to 200 pieces to a second.

2.

Drain off BSS and add 4.5 ml growth medium with serum to each flask.

3.

Add 0.5 ml crude collagenase, 2,000 units/ml, 1 to give a final concentration of 200 units/ml collagenase.

4.

Incubate at 36.5°C for 4–48 hr without agitation.

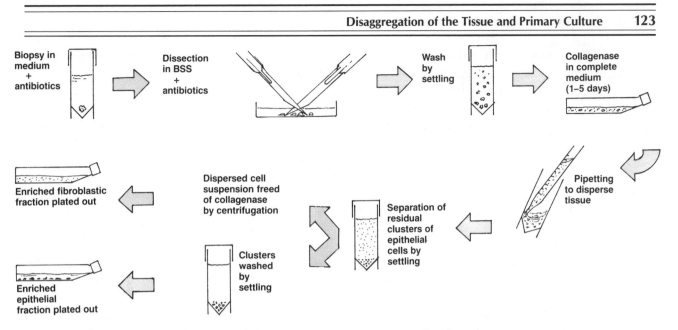

Fig. 9.10. *Stages in disaggregation of tissue for primary culture by collagenase.*

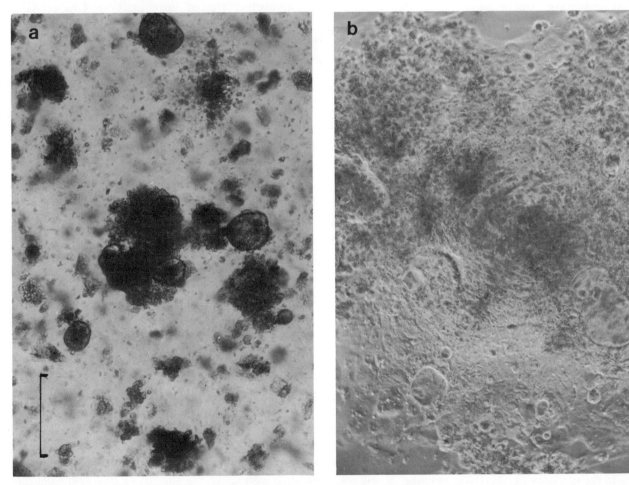

Fig. 9.11. *Cells and cell clusters from human colonic carcinoma after 48-hr dissociation in crude collagenase (Worthington CLS grade). a. Before removal of collagenase. b. After removal of collagenase, further disaggregation by pipetting and culture for 48 hr. The clearly defined clusters in (a) form epithelial-like sheets in (b) and the more irregularly shaped clusters form fibroblastic areas.*

Tumor tissue may be left up to 5 d or more if disaggregation is slow, although it may be necessary to centrifuge the tissue and resuspend in fresh medium and collagenase before then if an excessive drop in pH is observed (to less than pH 6.5).

5.

Check for effective disaggregation by gently moving the flask; the pieces of tissue will "smear" on the bottom of the flask and, with moderate agitation, will break up into single cells and small clusters (Fig. 9.11). With some tissues (e.g., lung, kidney, and colon or breast carcinoma) small clusters of epithelial cells can be seen to resist the collagenase and may be separated from the rest by allowing them to settle for about 2 min. If these clusters are further washed with BSS by resuspension and settling and the sediment seeded in medium, they will form healthy islands of epithelial cells. Epithelial cells generally survive better if not completely dissociated.

6.

Where complete disaggregation has occurred, or when the supernatant cells are collected after allowing clusters to settle, centrifuge at 50–100 *g* for 3 min. Discard supernatant DBSS, resuspend, and combine pellets in 5 ml medium, and seed in a 25-cm^2 flask. If the pH fell during collagenase treatment (to pH 6.5 or less by 48 hr), dilute twofold to threefold in medium after removing the collagenase.

7.

Replace medium after 48 hr.

Some cells, particularly macrophages, may adhere to the first flask during the collagenase incubation. Transferring the cells to a fresh flask after collagenase treatment (and removal) removes many of the macrophages from the culture. The first flask may be cultured as well if required. Trypsinization will remove any adherent cells other than macrophages.

Disaggregation in collagenase has proved particularly suitable for the culture of human tumors, mouse kidney, human adult and fetal brain, lung, and many other tissues, particularly epithelium. It is gentle and requires no mechanical agitation or special equipment. With more than 1 g of tissue, however, it becomes tedious at the dissection stage and can be expensive due to the amount of collagenase required. It will also release most of the connective tissue cells, accentuating the problem of fibroblastic outgrowth; so it may require to be followed by selective culture or cell separation (see Chapters 7, 12, and 20).

Many epithelial tissues (e.g., kidney tubules and glomeruli, clusters of carcinoma cells of breast and gastrointestinal tract and lung alveoli) are not disaggregated by collagenase and may be separated from connective tissue cells by allowing the epithelial clusters or tubules to sediment for 5–10 min. Connective tissue cells are completely dissociated by the collagenase and remain in suspension. If the sediment is resuspended in BSS and allowed to settle twice more, the final sediment is enriched for epithelial cells. The survival of this epithelium is probably enhanced by culturing the cells as undissociated clusters and tubule fragments.

The discrete clusters of epithelial cells produced by disaggregation in collagenase can be selected under a dissection microscope and transferred to individual wells in a microtitration plate, alone or with irradiated or mitomycin-C-treated feeder cells (see Chapter 11).

The addition of hyaluronidase aids disaggregation by attacking terminal carbohydrate residues on the surface of the cells. This combination has been found to be particularly effective for dissociating rat or rabbit liver, by perfusing the whole organ *in situ* [Berry and Friend, 1969; Seglen, 1975] and completing the disaggregation by stirring the partially digested tissue in the same enzyme solution for a further 10–15 min, if necessary.

This technique gives a good yield of viable hepatocytes and is a good starting point for further culture (see Chapter 20).

Collagenase may also be used in conjunction with trypsin. Cahn and others [Cahn et al., 1967] developed a formulation including chick serum, collagenase, and trypsin. Chick serum has a moderating effect on the activity of the trypsin but does not inhibit it to the extent expected of other sera.

Mechanical Disaggregation

The outgrowth of cells from primary explants is a relatively slow process and can be highly selective. Enzymatic digestion (see above) is rather more labor intensive, though potentially it gives a culture which is more representative of the tissue. As there is a risk of damaging cells during enzymatic digestion, many people have chosen the alternative of mechanical disaggregation, e.g., collecting the cells which spill out when the tissue is carefully sliced [Lasfargues, 1973],

DISAGGREGATION OF TISSUE BY SIEVING

(1) Press tissue
gently through
100μm sieve.

SYRINGE PISTON

SIEVE

MEDIUM OR SERUM
WITH BSS HEPES-BUFFERED

PETRI DISH

(2) Lift sieve with
strong forceps or
artery clamp and
wash dispersed cells
and clumps through sieve
leaving debris behind.

(3) Transfer suspension of
cells and clumps to second
finer mesh sieve (~50μm),
and repeat steps (1) and (2).

Repeat with 20μm mesh
sieve if desired
(see text).

(4) Count on hemocytometer
with dye exclusion
viability stain.

(5) Dilute in medium
with serum.
Seed at 10^6 viable
cells per ml,
and incubate.

b

Fig. 9.12. *a. Stainless steel sieves suitable for disaggregating tissue. b. Disaggregation of tissue by sieving.*

pressing the cells through sieves of gradually reduced mesh (Fig. 9.12), or forcing cells through a syringe and needle [Zaroff et al., 1961], or simply repeatedly pipetting. This gives a cell suspension more quickly than with enzymatic digestion but may cause mechanical damage. The following is one method of mechanical disaggregation found to be moderately successful with soft tissues such as brain.

Outline

The tissue in culture medium is forced through sieves of gradually reduced mesh until a reasonable suspension of single cells and small aggregates is obtained. The suspension is then diluted and cultured directly.

Materials

Forceps
sieves (1 mm, 100 μm, 20 μm; Fig. 9.12a)
9-cm petri dishes
scalpels
medium
disposable plastic syringes (2 ml or 5 ml)
culture flasks

Protocol

1.
After washing and preliminary dissection (see above) chop tissue into pieces about 5–10 mm across, and place a few at a time into a stainless steel sieve of 1-mm mesh in a 9-cm petri dish (Fig. 9.12b).
2.
Force the tissue through the mesh into medium by applying gentle pressure with the piston of a disposable plastic syringe. Pipette more medium into the sieve to wash the cells through.
3.
Pipette the partially disaggregated tissue from the petri dish into a sieve of finer porosity, perhaps 100-μm mesh and repeat step 2.
4.
The suspension may be diluted and cultured at this stage or sieved further through 20-μm mesh if it is important to produce a single cell suspension. In general, the more highly dispersed, the higher the sheer stress required, and the lower the resulting viability.
5.
Seed culture flasks at 10^6 cells/ml and 2×10^6 cells/ml by dilution of cell suspension in medium.

Only such soft tissues as spleen, embryonic liver, embryonic and adult brain, and some human and animal soft tumors respond at all well to this technique. Even with brain, where fairly complete disaggregation can be obtained easily, the viability of the resulting suspension is lower than that achieved with enzymatic digestion, although the time taken may be very much less. Where the availability of tissue is no limitation and the efficiency of the yield unimportant, it may be possible to produce as many viable cells as with enzymatic digestion in a shorter time but at the expense of very much more tissue.

Separation of Viable and Nonviable Cells

When an adherent primary culture is prepared from dissociated cells, nonviable cells will be removed at the first medium change. With primary cultures maintained in suspension, nonviable cells are gradually diluted out when cell proliferation starts. If necessary, however, nonviable cells may be removed from the primary disaggregate by centrifuging the cells on a mixture of Ficoll and sodium metrizoate (e.g., Hypaque or Triosil) [Vreis et al., 1973]. This technique is similar to the preparation of lymphocytes from peripheral blood described in Chapter 23. Up to 2×10^7 cells in 9 ml medium may be layered on top of 6 ml Ficoll/Hypaque ("Lymphoprep," Flow Laboratories) in a 25-ml screw-capped centrifuge bottle, centrifuged, and viable cells collected from the interface.

The disaggregation of tissue and preparation of the primary culture is the first, and perhaps most vital, stage in the culture of cells with specific functions. If the required cells are lost at this stage, then the loss is irrevocable. Many different cell types may be cultured by choosing the correct techniques (see Chapters 7, 12 and 20). In general, trypsin is more severe than collagenase but sometimes more effective in creating a single cell suspension. Collagenase does not dissociate epithelial cells readily, but this can be an advantage in separating them from stromal cells. Mechanical disaggregation is much quicker but will damage more cells. The best approach is to try out the techniques described above and select the method which works best in your system. If none of these is successful, try additional enzymes such as pronase, dispase, and DNase and consult the literature.

The first subculture represents an important transition for the culture. The need to subculture implies that the primary culture has increased to occupy all of the available substrate. Hence cell proliferation has become an important feature. While the primary culture may have a variable growth fraction (see Chapter 18) depending on the type of cells present in culture, after the first subculture the growth fraction is usually high (80% or more).

From a very heterogeneous primary culture, containing many of the cell types present in the original tissue, a more homogeneous cell line emerges. In addition to its biological significance, this has also considerable practical importance as the culture can now be propagated, characterized, and stored, and the potential increase in cell number and uniformity of the cells opens up a much wider range of experimental possibilities (Table 10.1).

Once a primary culture is subcultured (or "passaged," or "transferred"), it becomes known as a "cell line." This term implies the presence of several cell lineages either of similar or distinct phenotypes. If one cell lineage is selected, by cloning (see Chapter 11), by physical cell separation (see Chapter 12), or by any other selection technique, to have certain specific properties which have been identified in the bulk of the cells in the culture, this becomes known as "cell

TABLE 10.1. Subculture

Advantages	Disadvantages
Propagation	Selection, overgrowth
More cells	Loss of differentiated
Cloning	properties (may be
	indicable)
Homogeneity	Genetic instability
Characterisation of	Trauma of disaggregation,
replicate samples	enzymatic and mechanical
Frozen storage	damage
eventually	

strain." Some commonly used cell lines and cell strains are listed in Table 10.2. If a cell line transforms *in vitro* this gives rise to a continuous cell line (see Chapters 2 and 15) and if selected or cloned and characterised it is known as a continuous cell strain. The relative advantage and disadvantages of finite cell lines or continuous cell lines are listed in Table 10.3.

NOMENCLATURE

The first subculture gives rise to a "secondary" culture, the secondary to a "tertiary" and so on, although in practice this nomenclature is seldom used beyond tertiary. Since the importance of culture lifetime was highlighted by Hayflick and others with diploid fibroblasts [Hayflick and Moorhead, 1961], where each subculture divided the culture in half ("split ratio"—1:2), passage number has often been confused with "generation number." Cell lines with limited culture life-spans ("finite" cell lines) behave in a fairly reproducible fashion (see Chapter 2). They will grow through a limited number of cell generations, usually between 20 and 80 population doublings, before extinction. The actual number depends on strain differences and culture conditions but is consistent for one cell line grown under the same conditions. It is, therefore, important that reference to a cell line should express the approximate generation number or number of doublings since explantation, "approximate" because the number of generations which have elapsed in the primary culture is difficult to assess.

The cell line should also be given a code or designation (e.g., NHB, normal human brain), a cell strain or cell line number (if several cell lines were derived from the same source), NHB1, NHB2, etc., and if cloned, a clone number, NHB2-1, NHB2-2, etc. Together with the number of population doublings, this becomes, for example, NHB2/2 and will increase by one for a split ratio of 1:2 (NHB2/2, NHB2/3, etc.), by two for a split ratio of 1:4 (NHB2/2, NHB2/4, etc.), and so on. For publication, each cell line should be prefixed with a code designating the laboratory in which it was derived, e.g., WI, Wistar Institute [Federoff, 1975].

ROUTINE MAINTENANCE

Once a culture is initiated, whether it be a primary culture or a subculture of a cell line, it will need a periodic medium change ("feeding") followed eventually by subculture if the cells are proliferating. In nonproliferating cultures, the medium will still need to

TABLE 10.2. Cell Lines and Cell Strains in Regular Use

Name	Morphology*	Origin	Age†	Tissue‡	Ploidy	Characteristic	Reference
Finite cell lines							
MRC5	Fibroblast	Human lung	Emb	Nor	Diploid	Susceptible to human viral infection	[Jacobs, 1970]
MRC9	Fibroblast	Human lung	Emb	Nor	Diploid	Susceptible to human viral infection	[Jacobs, 1979]
WI38	Fibroblast	Human lung	Emb	Nor	Diploid	Susceptible to human viral infection	[Hayflick, 1971]
IMR90	Fibroblast	Human lung	Emb	Nor	Diploid	Susceptible to human viral infection	[Nichols, 1977]
Continuous cell lines							
A9	Fibroblast	Mouse subcutaneous	Ad	Neo	Aneuploid	HGPRT-ve; deriv. L929	[Littlefield, 1964]
BHK21 C13	Fibroblast	Syrian hamster kidney	NB	Nor	Aneuploid	Transformable by polyoma virus	[Macpherson and Stoker, 1962]
BRL3A	Epithelial	Rat liver	NB	Nor		Produce MSA	[Coon, 1968]
CHOK1	Fibroblast	Chinese hamster ovary	Ad	Nor	Diploid	Simple karyotype	[Puck, 1958]
EB-3	Lymphocytic	Human	Ju	Neo	Diploid	EB virus +ve	[Epstein and Barr, 1964]
GH1, GH3	Epithelial	Rat	Ad	Neo	Aneuploid	Produce growth hormone	[Yasamura et al., 1968]
HeLa	Epithelial	Human	Ad	Neo	Aneuploid	G6PD Type A	[Gey et al., 1952]
L1210	Lymphocytic	Mouse	Ad	Neo	Aneuploid	Rapidly growing; suspension	[Law et al., 1949]
L5178Y	Lymphocytic	Mouse	Ad	Neo	Aneuploid	Rapidly growing suspension	
L929	Fibroblast	Mouse	Ad	Nor	Aneuploid	Clone of L-cell	[Sanford et al., 1948]
LS	Fibroblast	Mouse	Ad	Neo	Aneuploid	Grow in suspension; deriv. L929	[Paul and Struthers, unpubl.]
MCF7	Epithelial	Human breast pleural effusion	Ad	Neo	Aneuploid	Estrogen recep +ve	[Soule et al., 1973]
P388D₁	Lymphocytic	Mouse	Ad	Neo	Aneuploid	Grow in suspension	[Dawe and Potter, 1957; Koren et al., 1975]
S180	Fibroblast	Mouse	Ad	Neo	Aneuploid	Cancer chemotherapy screening	[Foley et al., 1959]
3T3-L1	Fibroblast	Mouse Swiss	Emb	Nor	Aneuploid	Adipose diff.	[Green et al., 1974]
3T3-A31	Fibroblast	Mouse BALB/c	Emb	Nor	Aneuploid	Contact inhibited; readily transformed	[Aaronson and Todaro, 1968]
NRK49F	Fibroblast	Rat kidney	Ad	Nor	Aneuploid	Induction of suspension growth by transforming growth factors	[DeLarco, 1978]
Vero	Fibroblast	Monkey kidney	Ad	Nor	Aneuploid	Viral substrate and assay	[Yasumura, 1963]

*Fibroblast, epithelial, lymphocytic refer to appearance and not lineage.
†Emb=embryonic; Ad=adult; NB=newborn; Ju=juvenile.
‡Nor, from normal tissue; neo, from neoplastic tissue. See also Table 2.2 and American Type Culture Collection catalogue.

TABLE 10.3. Advantages and Disadvantages of Finite and Continuous Cell Lines

	Finite	Continuous
Ploidy	Diploid	Heteroploid
	Euploid	Aneuploid
Transformation	Normal	Transformed
Tomorigenicity	Non-tumorigenic	Tumorigenic
Anchorage dependence	Yes	No
Contact inhibition	Yes	No
Density limitation of growth	Yes	No (or less so)
Mode of growth	Monolayer	Monolayer or suspension
Maintenance	Cyclic	Steady state possible
Serum requirement (in simple media)	High	Low
Cloning efficiency	Low	High
Markers	May be tissue specific	Chromosomal, enzymic
Special functions (e.g., virus susceptibility, differentiation)	May be retained	Often lost
Growth rate	Slow (24–96 hr doubling time)	Rapid (12–24 hr doubling time)
Yield	Low ($< 10^6$ cells/ml, $< 10^5$ cells/cm^2)	High ($> 10^6$ cells/ml, $> 10^5$ cells/cm^2)
Control features	Generation number *In vivo* markers	Strain characteristics

TABLE 10.4. Cell Dissociation for Transfer or Counting—Procedures of Gradually Increasing Severity

1.	Shake-off	Mitotic or other loosely adherent cells
2.	Trypsin* in PBS (0.01–0.5% as required, usually 0.25%, 5–15 min)	Most continuous cell lines
3.	Prewash with PBS or CMF, then 0.25% trypsin* in PBS or saline-citrate	Some strongly adherent continuous cell lines and many cell lines at early passage stages
4.	Prewash with m*M* EDTA in PBS or CMF then 0.25% trypsin* in citrate	Some strongly adherent early passage cell lines
5.	Prewash with 1 m*M* EDTA, then EDTA 2nd rinse, and leave on, 1 ml/5 cm	Epithelial cells, although some may be sensitive to EDTA
6.	EDTA prewash, then 0.25% trypsin* with 1 m*M* EDTA	Strongly adherent cells, particularly epithelial and some tumor cells (note: EDTA can be toxic to some cells)
7.	1 m*M* EDTA prewash, then 0.25% trypsin* and collagenase*, 200 units/ml PBS or saline-citrate or EDTA/PBS	Thick cultures, multilayers, particularly collagen-producing dense cultures
8.	Scraping	All cultures, but may cause mechanical damage and usually will not give a single cell suspension
9.	Add dispase (0.1–1.0 mg/ml) or pronase (0.1–1.0 mg/ml) to medium and incubate till cells detach	Will dislodge most cells, but requires centrifugation step to remove enzyme not inactivated by serum. May be harmful to some cells

*Digestive enzymes are available (Difco, Worthington, Boehringer Mannheim, Sigma) in varying degrees of purity. Crude preparations, e.g., Difco trypsin 1:250 or Worthington CLS grade collagenase, contain other proteases which may be helpful in dissociating some cells but may be toxic to others. Start with a crude preparation and progress to purer grades if necessary. Purer grades are often used at a lower concentration (mg/ml) as their specific activities (enzyme units/ g) are higher. Purified trypsin at 4°C has been recommended for cells grown in low serum concentrations or in the absence of serum [McKeehan, 1977], and will generally be found to be more consistent. Batch testing and reservation, as for serum, may be necessary for some applications.

be changed periodically as the cells will still metabolize, and some constituents of the medium will become exhausted or will degrade spontaneously. Intervals between medium changes and between subcultures vary from one cell line to another depending on the rate of growth and metabolism; rapidly growing cell lines such as HeLa are usually subcultured once per week and the medium changed 4 d later. More slowly growing cell lines may only require to be subcultured every 2, 3, or even 4 wk, and the medium changed weekly between subcultures. (For a more detailed discussion of the growth cycle, see below).

Replacement of Medium

Four factors indicate the need for the replacement of culture medium (see also Chapter 7).

A drop in pH. The rate of fall and absolute level should be considered. Most cells will stop growing as the pH falls from pH 7.0 to pH 6.5 and will start to lose viability between pH 6.5 and pH 6.0, so if the medium goes from red through orange to yellow, the medium should be changed. Try to estimate the rate of fall; a culture which falls at 0.1 pH units in 1 d will not come to harm if left a day or two longer before feeding, but a culture that falls 0.4 pH units in 1 d will need to be fed 24–48 hr later and cannot be left over a weekend.

Cell concentration. Cultures at a high cell concentration will use up the medium faster than at a low concentration. This is usually evident in the pH change but not always.

Cell type. Normal cells (e.g., diploid fibroblasts) will usually stop dividing at a high cell density (density limitation of growth; see Chapter 15) due to growth factor depletion and other factors. The cells block in the G1 phase of the cell cycle and deteriorate very little even if left for 2–3 wk. Transformed cells, continuous cell lines, and some embryonic cells, however, will deteriorate rapidly at high cell densities unless the medium is changed daily or they are subcultured.

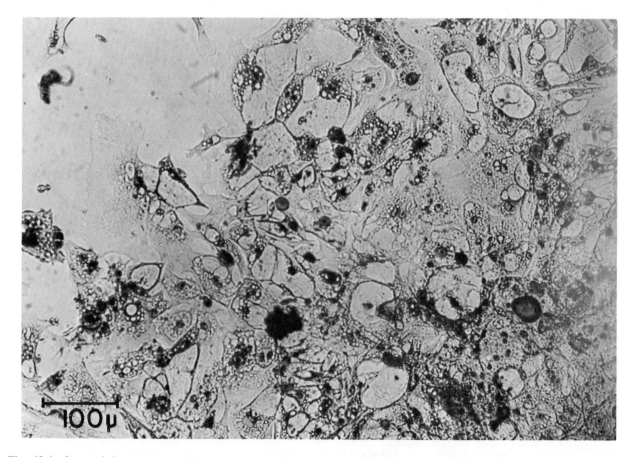

Fig. 10.1. *Signs of deterioration of the culture. Cytoplasm of cells becomes granular, particularly around the nucleus, and vacuolation occurs. Cells may become more refractile at the edge if cell spreading is impaired.*

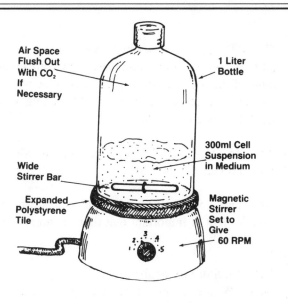

Fig. 10.2. *Simple stirrer culture for cells growing in suspension. An expanded polystyrene mat (dark shaded area below bottle) should be interposed between the bottle and the magnetic stirrer to avoid heat transfer from the stirrer motor.*

Cell morphology. When checking a culture for routine maintenance, be alert to signs of morphological deterioration: granularity around the nucleus, cytoplasmic vacuolation, and rounding up of the cells with detachment from the substrate (Fig. 10.1). This may imply that the culture requires a medium change, or may indicate a more serious problem, e.g., inadequate or toxic medium or serum, microbiological contamination, or senescence of the cell line. During routine maintenance, the medium change or subculture frequency should prevent such deterioration.

Volume, Depth, and Surface Area

The usual ratio of medium volume to surface area is 0.2–0.5 ml/cm^2. The upper limit is set by gaseous diffusion through the liquid layer and the optimum will depend on the oxygen requirement of the cells. Cells with a high O_2 requirement will be better in shallow medium (2 mm) and those with a low requirement may do better in deep medium (5 mm). If the depth is

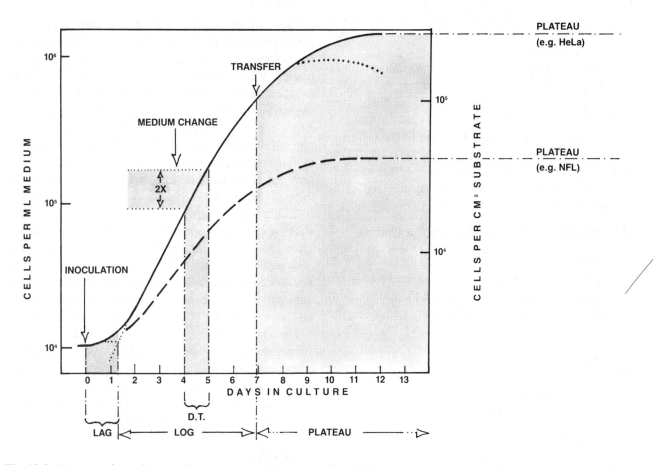

Fig. 10.3. *Diagram of growth curve of a continuous cell line such as HeLa and a finite cell line, NFL (normal fetal lung fibroblasts).* *The solid line represents growth of HeLa; the dashed; line illustrates the lower plateau obtained with NFL (see text).*

greater than 5 mm, then gaseous diffusion may become limiting. With monolayer cultures this can be overcome by rolling the bottle or perfusing the culture with medium and arranging for gas exchange in an intermediate reservoir (see Chapter 23). When the depth of suspension culture is increased, it should be stirred with a bar magnet (see Chapter 23). To prevent frothing, the depth of stirrer cultures must be a minimum of 5 cm. For intermediate depths of medium between 5 mm and 5 cm, use a roller bottle (see Table 7.1).

"Holding Medium"

A holding medium may be used where stimulation of mitosis, which usually accompanies a medium change, even at high cell densities, is undesirable. Holding media are usually regular media with the serum concentration reduced to 1 or 2% or eliminated completely. This will not stimulate mitosis in most untransformed cells unless a special serum-free formulation is used (see Chapter 7). Transformed cell lines are unsuitable for this procedure as they may either continue to divide successfully or the culture may deteriorate as transformed cells do not block in a regulated fashion in G_1.

Holding medium is also used to maintain cell lines with a finite life-span without using up the limited number of cell generations available to them (see Chapter 2). Reduction of serum and cessation of cell proliferation also promotes expression of the differentiated phenotype in some cells [Schousboe et al., 1979; Maltese and Volpe, 1979]. Medium used for the collection of biopsy samples can also be referred to as "holding medium."

Changing the Medium or "Feeding" a Culture

Outline

Examine the culture by eye and on an inverted microscope. If indicated, remove the old medium and add fresh medium. Return the culture to the incubator.

Materials

Pipettes
medium
(both sterile)

Protocol

1.
Examine culture carefully for signs of contamination or deterioration (see Figs. 16.1 and 10.1).

2.
Check the criteria described above—pH, cell density, or concentration, and, based on your knowledge of the behavior of the culture, decide whether or not to replace the medium. If feeding is required, proceed as follows.
3.
Take to sterile work area, remove and discard medium (see Chapter 5).
4.
Add same volume of fresh medium, prewarmed to 36.5°C if it is important that there is no check in cell growth.
5.
Return culture to incubator.
Note. Where a culture is at a low density and growing slowly, it may be preferable to "half-feed." In this case, remove only half the medium at step 3 and replace it in step 4 with the same volume as was removed.

Subculture

The growth of cells in culture usually follows the pattern depicted in Figure 10.3. A lag following seeding is followed by a period of exponential growth (log phase). When all the available substrate is occupied, or when the cell concentration exceeds the capacity of the medium, growth ceases or is greatly reduced. Then either the frequency of medium changing must increase or the culture must be divided. The usual practice in subculturing an adherent cell line involves removal of the medium and dissociation of the cells in the monolayer with trypsin, although some loosely adherent cells (e.g., HeLa-S_3) may be subcultured by shaking the bottle and collecting the cells in the medium, and diluting as appropriate in fresh medium in new bottles. Exceptionally, some cell monolayers cannot be dissociated in trypsin and require the action of alternative proteases such as pronase, dispase, or collagenase (Table 10.1) [Foley and Aftonomos, 1973].

The attachment of cells to each other and to the culture substrate is mediated by cell surface glycoproteins and Ca^{2+} and Mg^{2+} ions. Other proteins, derived from the cells and from the serum, become associated with the cell surface and the surface of the substrate and facilitate cell adhesion.

Outline

Remove medium, expose cells briefly to trypsin, incubate and disperse cells in medium.

Materials

  Pipettes (sterile)
  medium (sterile)
  PBSA (sterile)
  0.25% trypsin in PBSA, saline citrate,
    or EDTA (sterile) (see Table 10.3)
    culture flasks
  hemocytometer or cell counter

Protocol

1.

Withdraw medium and discard.

2.

Add PBSA prewash (5 ml/25 cm^2) to the side of the flask opposite the cells, so as to avoid dislodging cells, rinse the cells, and discard rinse. This step is designed to remove traces of serum which would inhibit the action of the trypsin.

3.

Add trypsin (3 ml/25 cm^2) to the side of the flask opposite the cells. Turn the flask over to cover the monolayer completely. Leave 15–30s and withdraw the trypsin, making sure beforehand that the monolayer has not detached. Using trypsin at 4°C helps to prevent this.

4.

Incubate until cells round up; when the bottle is tilted, the monolayer should slide down the surface (this usually occurs after 5–15 min). Do not leave longer than necessary, but do not force the cells to detach before they are ready to do so, or clumping may result.

  Note. In each case the main dissociating agent, be it trypsin or EDTA, is present only briefly and the incubation is performed in the residue after most of the dissociating agent has been removed. If difficulty is encountered in getting cells to detach, and, subsequently, in preparing a single cell suspension, alternative procedures, as described in Table 10.4, may be employed.

5.

Add medium (0.1–0.2 ml/cm^2) and disperse cells by repeated pipetting over the surface bearing the monolayer. Finally, pipette the cell suspension up and down a few times, with the tip of the pipette resting on the bottom corner of bottle. The degree of pipetting required will vary from one cell line to another; some disperse easily, others require vigorous pipetting. Almost all will incur mechanical damage from shearing forces if pipetted too vigorously; primary suspensions and early passage

cultures are particularly prone to damage due partly to their greater fragility and partly to their larger size. Pipette up and down sufficiently to disperse the cells into a single cell suspension. If this is difficult, apply a more aggressive dissociating agent (see Table 10.4) [Toshiharu et al., 1975].

  A single cell suspension is desirable at subculture to ensure an accurate cell count and uniform growth on reseeding. It is essential where quantitative estimates of cell proliferation or of plating efficiency are being made and where cells are to be isolated as clones.

6.

Count cells by hemocytometer or electronic particle counter (see Chapter 18).

7.

Dilute to the appropriate seeding concentration (a) by adding the appropriate volume of cells to a premeasured volume of medium in a culture flask, or (b) by diluting the cells to the total volume required and distributing that among several flasks. Procedure (a) is useful for routine subculture when only a few flasks are used and precise cell counts and reproducibility are not critical, but (b) is preferable when setting up several experimental replicate samples as the total number of manipulations is reduced and the concentration of cells in each flask will be identical.

8.

Cap the flask(s) and return to the incubator. Check after about 1 hr for pH change. If the pH rises in a medium with a gas phase of air, return to aseptic area and gas culture briefly (1–2s) 5% CO_2. Since each culture will behave predictably in the same medium, you will know eventually which cells to gas when they are passaged, without having to incubate them first. If the medium already has a 5% CO_2 gas phase, either increase to 7% or 10% or add sterile 0.1 N HCl.

  Expansion of air inside plastic flasks causes larger plastic flasks to swell and prevents them from lying flat. Release the pressure by slackening the cap briefly, or, alternatively, this may be prevented by compressing the top and bottom of large flasks before sealing them. Incubation then restores the correct shape. Care must be taken not to crack the flasks.

For finite cell lines, it is convenient to reduce the cell concentration at subculture by two-, four-, eight-, or 16-fold, making the calculation of the number of population doublings easier ($2 \equiv 1$, $4 \equiv 2$, $8 \equiv 3$, $16 \equiv 4$), e.g., a culture divided eightfold will require three doublings to achieve the same cell density. With continuous cell lines, where generation number is not usually recorded, the cell concentration is more conveniently reduced to a round figure, e.g., 5×10^4 cells/ml. In both cases, the cell number should be recorded so that growth rate can be estimated at each subculture and consistency monitored (see below under Growth Cycle).

Propagation in Suspension

The preceding instructions refer to subculture of monolayers, as most primary cultures or continuous lines grow in this way. Cells which grow continuously in suspension, either because they are nonadhesive (e.g., many leukemias and murine ascites tumors) or because they have been kept in suspension mechanically, or selected (see also Chapter 23), may be subcultured like micro-organisms. Trypsin treatment is not required and the whole process is quicker and less traumatic for the cells. Medium replacement is not usually carried out with suspension cultures as this would require centrifugation of the cells. Routine maintenance is, therefore, reduced to one of two alternative procedures, i.e., subculture by dilution, or increase of the volume without subculture.

Outline

Count cells, withdraw cell suspension, and add fresh medium to restore cell concentration to starting level.

Materials

Culture flasks (sterile)
medium (sterile)
pipettes (sterile)
bar magnet (sterile)
magnetic stirrer
hemocytometer or cell counter

Protocol

1.
Mix cell suspension and disperse any clumps by pipetting.
2.
Remove sample and count.
3.
Add medium to fresh flask.
Note. Any culture flask with a reasonably flat surface may be used for cells which grow sponta-

neously in suspension. Where stirring is required, e.g., for larger cultures, or cells which would normally attach, use standard round reagent bottles, or aspirators, siliconized, if necessary (see Appendix), and insert a magnetic stirrer bar, Teflon coated, and with a ridge around the middle (Fig. 10.2) or suspended from the top of the bottle. Select the appropriate size of bottle to give between 5 and 8 cm depth with the volume of medium that you require (See also Chapter 23).
4.
Add sufficient cells to give a final concentration of 10^5 cells/ml for slow-growing cells (24–48 hr doubling time) or 2×10^4/ml for rapidly growing cells (12–18-hr doubling time).
5.
Cap and return culture to incubator.
6.
Culture flasks should be laid flat as for monolayer culture. Stirrer bottles should be placed on a magnetic stirrer and stirred at 60–100 rpm. Take care that the stirrer motor does not overheat the culture. Insert a polystyrene foam mat under the bottle if necessary. Induction-drive stirrers generate less heat and have no moving parts.

Suspension cultures have a number of advantages (see Table 10.5). The production and harvesting of large quantities of cells may be achieved without increasing the surface area of the substrate (see Chapter 23). Furthermore, if dilution of the culture is continuous and the cell concentration kept constant, a steady state can be achieved; this is not readily achieved in monolayer culture. Maintenance of monolayer cultures is essentially cyclic with the result that growth rate and metabolism varies depending on the phase of the growth cycle.

Monolayers are convenient for cytological and immunological observations, cloning, mitotic "shake off" (for cell synchronization of chromosome preparation) and *in situ* extractions without centrifugation.

SLOW CELL GROWTH

Even in the best-run laboratories, problems may arise in routine cell maintenance. Some may be attributed to microbiological contamination (see Chapter 16), but often the cause lies in one or more alterations in culture conditions. The following check list may help to track these down:

1. Any change in procedure or equipment?

TABLE 10.5. Properties of Monolayer and Suspension Cultures

	Monolayer	Suspension
Maintenance	Cyclic pattern of propagation (see text)	Can be maintained at "steady state"
	Require dissociation	Simple dilution at passage
	Dependent on availability of substrate	Dependent on medium volume only (with adequate gas exchange)
Results of differences in geometry	Cell Interaction: metabolic cooperation, junctional communication contact inhibition of movement and membrane activity, density limitation of growth	Homogeneous suspension
		Cell density limited by nutrient and hormonal concentration of the medium only
	Diffusion boundary of effects (see text)	Shearing effects in stirred cultures may damage some cells
	Establishment of polarity, differentiation	
	Cell shape and cytoskeleton— spreading, motility, overlapping, underlapping	
Sampling and analysis	Good cytological preparation, chromosomes, immunofluorescence, histochemistry	Bulk production of cells
		Ease of harvesting (no trypsinization required)
	Enrichment of mitoses by "shake-off" (see Chapter 23)	
	Serial extractions *in situ* possible without centrifugation	
Which cells?	Most cell types except some hemopoietic cells and ascites tumors	Transformed cells and lymphoblastoid cell lines

2. Medium:
   a. Medium adequate?—check against other media (see Chapter 8).
   b. Frequency of changing correct?
   c. pH: check that it is within 7.0–7.4 during culture.
   d. Osmolality: check on osmometer.
   e. Component missed out: make up fresh batch.
   f. New batch of stock medium which is faulty?
   g. If BSS-based, is BSS satisfactory? (Check with other users.)
   h. If water-based, is water satisfactory (check with other users, or against fresh 1X medium, bought in).
   i. Check still—deionizer—conductivity, contamination—glass boiler—residue.
   j. Storage vessel, for algal or fungal contamination; chemical traces in plastic.
   k. HCO_3^-.
   l. Antibiotics.
3. Serum:
   a. New batch? Check supplier's quality control.
   b. Check concentration. Too low or too high?
   c. Reconfirm lack of toxicity, growth promotion and plating efficiency.
4. Glassware or plastics:
   a. If new, check against previous stock.
   b. Wash-up—other cells showing symptoms? Other users have trouble?
   c. Trace contamination of glass? Check growth on plastic.
5. Cells (if other people's cells are all right):
   a. Contamination (see also Chapter 16).
      i) Bacterial, fungal—grow up without antibiotics.
      ii) Mycoplasma:
         (A) Stain Culture with Hoechst 33258 (Chapter 16).
         (B) Check for cytoplasmic DNA (incorporation of radioactive thymidine) by autoradiography.
         (C) Get commercial test done (e.g., Flow Laboratories or Microbiological Associates).

    iii) Viral—difficult to detect—try E M or fluorescent antibody.

  b. Seeding density too low at transfer.

  c. Transferred too frequently.

  d. Allowed to remain for too long in plateau before transfer.

6. Subculture routine:

  a. Change in batch of trypsin or other dissociation agent.

  b. Severity of dissociation—too long, agent too concentrated or too high specific activity.

  c. Pipetting during dissociation too vigorous.

  d. Sensitive to EDTA (if EDTA used).

7. Hot room and incubators: check temperature and stability.

  a. Faulty thermostats.

  b. Access too frequent.

  c. Humidity of humid CO2 incubators.

  d. CO_2 concentration (check pH *in situ*).

At subculture a fragile or slowly growing line should be split 1:2; and a vigorous, rapidly growing line, 1:8 or 1:16. Once a cell line becomes continuous (usually taken as beyond 150 or 200 generations) the generation number is disregarded and the culture should simply be cut back to between 10^4 and 10^5 cells/ml. The split ratio or dilution is also chosen to establish a convenient subculture interval (perhaps 1 or 2 wk), and to ensure that the cells (1) are not diluted below that concentration which permits them to reenter the growth cycle with a lag period of 24 hr or less, and (2) do not enter plateau before the next subculture.

Even when a standard split ratio is employed, cell counts should still be performed to ensure a consistent growth rate is maintained. Otherwise minor alterations are not detected for several passages.

Routine passage leads to the repetition of a standard growth cycle. It is essential to become familiar with this cycle for each cell line that is handled as this controls the seeding concentration, the duration of growth before subculture, the duration of experiments, and the appropriate times for sampling to give greatest consistency. Cells at different phases of the growth cycle behave differently with respect to proliferation, enzyme activity, glycolysis and respiration, synthesis of specialized products, and many other properties.

Chapter 11
Cloning and Selection of Specific Cell Types

It can be seen from the preceding two chapters that a major recurrent problem in tissue culture is the preservation of a specific cell type and its specialized properties. While environmental conditions undoubtedly play a significant role in maintaining the differentiated properties of a culture (see Chapter 14), the selective overgrowth of unspecialized cells is still a major problem.

CLONING

The traditional microbiological approach to the problem of culture heterogeneity is to isolate pure cell strains by cloning, but the success of this technique in animal cell culture is limited by the poor cloning efficiencies of most primary cultures.

A further problem of cultures derived from normal tissue is that they may only survive for a limited number of generations (see Chapter 2), and by the time a clone has produced a usable number of cells, it may already be near to senescence (Fig. 11.1). Cloning is most successful in isolating variants from continuous cell lines, but even then considerable heterogeneity may arise within the clone as it is grown up for use (see Chapter 17).

Coon and Cahn [1966] were able to clone cartilage- and pigment-producing cell strains. Under the correct conditions, these cultures were able to retain their specialized functions over many generations. Similarly, Clark and Pateman [1978] isolated a Kupffer cell line from Chinese hamster liver by cloning the primary culture.

Cloning has also been used to isolate specific biochemical mutants and cell strains with marker chromosomes and may help to reduce the heterogeneity of a culture (see below).

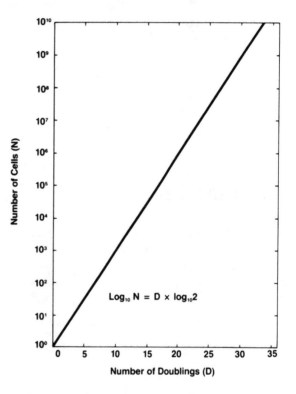

Fig. 11.1. *Relationship of cell yield in a clone to the number of population doublings e.g., 20 doublings are required to produce 10^6 cells.*

Dilution Cloning [Puck and Marcus, 1955]

Outline
Seed cells, at low density, incubate until colonies form, isolate, and propagate into cell strain (Fig. 11.2).

Materials
Pipettes
medium
trypsin

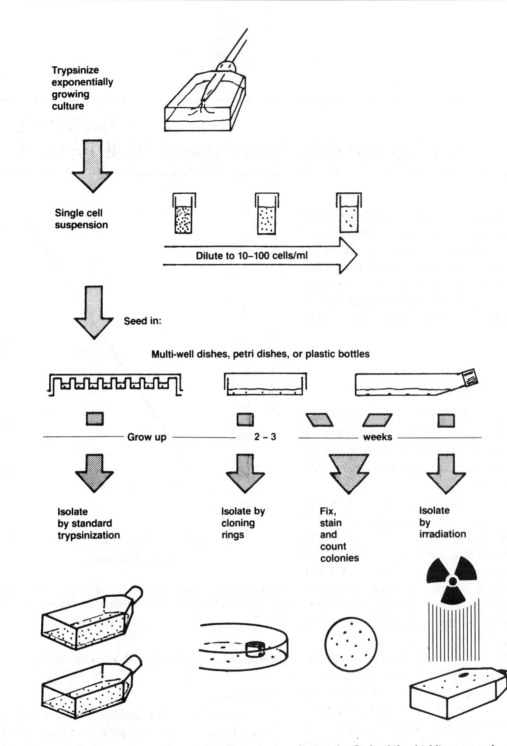

Fig. 11.2. *Cloning cells in monolayer culture. When clones form, they may be (1) isolated (a) directly, from multi-well dishes, (b) by the cloning ring technique (center left of figure), (c) by irra- diating the flask while shielding one colony (bottom right-hand side of figure), or (2) fixed, stained, and counted (center right of figure) for analysis.*

culture flasks or dishes
tubes for dilution
hemocytometer or cell counter

Protocol

1.

Trypsinize cells (see Chapter 10) to produce a single cell suspension. Undertrypsinizing will produce clumps; overtrypsinizing will reduce viability.

2.

While cells are trypsinizing, number flasks or dishes and measure out medium for dilution steps. (Four dilution steps may be necessary to reduce a regular monolayer to a concentration suitable for cloning.)

3.

When cells round up and start to detach, disperse the monolayer in medium containing serum or trypsin inhibitor, count, and dilute to desired seeding concentration. If cloning the cells for the first time, choose a range of 10, 50, 100, and 200 cells/ml (Table 11.1).

4.

Seed petri dishes or flasks with the requisite amount of medium (see Chapter 10), place petri dishes in a humid CO_2, incubator or gassed sealed container (2–10% CO_2, see Chapter 7), or gas flasks with CO_2, seal with cap, and place in dry incubator.

5.

Leave untouched for 1 wk. If colonies have formed, isolate (see below); if not, replace medium and continue to culture for a further week, or feed again and culture for 3 wk if necessary. If no colonies have appeared by 3 wk, it is unlikely that they will do so.

TABLE 11.1. Relationship of Seeding Density to Plating Efficiency

Expected plating efficiency (%)	Optimal cell number to be seeded	
	Per ml	Per cm^2
0.1	10^4	2×10^3
1.0	10^3	200
10	100	20
50	20	4
100	10	2

Stimulation of Plating Efficiency

When cells are plated at low densities, the survival falls in all but a few cell lines. This does not usually present a severe problem with continuous cell lines where the plating efficiency seldom drops below 10%, but with primary cultures and finite cell lines, the plating efficiency may be quite low—0.5%–5% or even zero. Numerous attempts have been made to improve plating efficiencies, based on the assumption either that cells require a greater range of nutrients at low densities or that cell-derived diffusible signals or conditioning factors are present in high-density cultures and absent or too dilute at low densities. The intracellular metabolic pool of a leaky cell in a dense population will soon reach equilibrium with the surrounding medium, while that of an isolated cell never will. This was the basis of the capillary technique of Sandford et al. [1948], when the L929 clone of L-cells was first produced. The confines of the capillary tube allowed the cell to create a locally enriched environment mimicking the higher cell density state. In microdrop techniques developed later, the cells were seeded as a microdrop under liquid paraffin. Keeping one colony separate from another, as in the capillary techniques, colonies could be isolated subsequently. As media improved, however, plating efficiencies increased, and Puck and Marcus [1955] were able to show that cloning cells by simple dilution (as described above) in association with a feeder layer of irradiated mouse embryo fibroblasts (see below) gave acceptable cloning efficiencies, although subsequent isolation required trypsinization from within a collar placed over each colony.

Some modifications which may improve clonal growth are listed below.

Medium.   Choose a rich medium such as Ham's F12 or one which has been optimized for the cell type in use, e.g., MCDB 110 [Ham, 1984] for human fibroblasts, Ham's F12 or MCDB 301 for CHO [Ham 1963; Hamilton and Ham, 1977] (see Chapters 7 and 20).

Serum.   Where serum is required, fetal bovine is generally better than calf or horse. Select a batch for cloning experiments which gives a high plating efficiency during tests.

Conditioning.   (1) Grow homologous cells, embryo fibroblasts, or another cell line to 50% of confluence, change to fresh medium, incubate for a further 48 hr, and collect the medium. (2) Filter through a 0.2-μm sterilizing filter (the medium may need to be clarified first by centrifugation 10,000 g, 20 min, or filtration through 5-μm and 1.2-μm filters) (see Chapter 9, section on sterilization of serum). (3) Add to cloning medium 1 part conditioned medium to 2 parts cloning medium.

Feeder layers (Fig. 11.3, regular feeder layer). (1) Trypsinize embryo fibroblasts from primary culture (see Chapters 9 and 10) and reseed at 10^5 cells/ml. (2) At 50% confluence, add mitomycin-C, 2μg/10^6 cells, 0.25 μg/ml, overnight [MacPherson and Bryden, 1971], or irradiate culture with 30 Gy (3,000 rad). (3) Change the medium after treatment, and after a further 24 hr, trypsinize the cells and reseed in fresh medium at 5×10^4 cells/ml (10^4 cells/cm^2). (4) Incubate for a further 24–48 hr and then seed cells for cloning. The feeder cells will remain viable for up to 3 wk but will eventually die out and are not carried over if the colonies are isolated.

Other cell lines or homologous cells may be used to improve the plating efficiency but heterologous cells have the advantage that if clones are to be isolated later, chromosome analysis will rule out accidental contamination from the feeder layer.

Hormones. Insulin, 1–10 IU/ml has been found to increase the plating efficiency of several cell types [Hamilton and Ham, 1977]. Dexamethasone, 2.5×10^{-5} M, ~10 μg/ml (a soluble synthetic hydrocortisone analogue) improves the plating efficiency of human normal glia, glioma, fibroblasts, and melanoma, and chick myoblasts, and will give increased clonal growth (colony size) if removed 5 d after plating [Freshney et al., 1980a,b]. Lower concentrations (10^{-7} M) have been found preferable for epithelial cells (see Chapter 20).

Intermediary metabolites. Keto acids, e.g., pyruvate or α-ketoglutarate, [Griffiths and Pirt, 1967; McKeehan and McKeehan, 1979] and nucleosides [α-medium, Stanners et al., 1971], have been used to supplement media and are already included in the formulation of a rich medium like Ham's F12. Pyruvate is also added to Dulbecco's modification of Eagle's MEM [Dulbecco and Freeman, 1959; Morton, 1970].

Carbon dioxide. CO_2 is essential to obtain maximum cloning efficiency for most cells. While 5% is most usual, 2% is sufficient for many cells, and may even be slightly better for human glia and fibroblasts. HEPES (20 mM) may be used with 2% CO_2, protecting the cells against pH fluctuations during feeding and in the event of failure of the CO_2 supply. (Using 2% CO_2 also cuts down in the consumption of CO_2.) At

the other extreme, Dulbecco's modification of Eagle's MEM is normally equilibrated with 10% CO_2 and is frequently used for cloning myeloma hybrids for monoclonal antibody production. The concentration of bicarbonate must be adjusted if the CO_2 tension is altered so that equilibrium is reached at pH 7.4 (see Table 7.2).

Treatment of substrate. Polylysine improves the plating efficiency of human fibroblasts in low serum concentrations [McKeehan and Ham, 1976] (see Chapter 7). (1) Add 1 mg/ml poly-D-lysine in water to plates (~5 ml/25 cm^2). (2) Remove and wash plates with 5 ml PBSA per 25 cm^2. The plates may be used immediately or stored for several weeks before used.

Fibronectin also improves the plating of many cells [Barnes and Sato, 1980]. The plates should be pretreated with 5 μg/ml fibronectin incorporated in the medium.

Trypsin. Pure, twice recrystallized, trypsin used at 0.05 μg/ml may be preferable to crude trypsin, but there are conflicting reports on this. McKeehan [1977] noted a marked improvement in plating efficiency when trypsinization (pure trypsin) was carried out at 4°C.

Multiwell Dishes

If clones are to be isolated, cloning by dilution directly into microwells (microtitration dishes or 24-well plates, see Fig. 7.4) makes subsequent harvesting easier. The plates must be checked regularly after seeding, however, to confirm that either only one cell is present per well at the start or, if there is more than one cell per well, they are not clumped and that only one cell gives rise to a colony, i.e., that the colonies which form are truly clonal in origin, and only one colony forms in the well.

Semisolid Media

Some cells, particularly hemopoietic stem cells and virally transformed fibroblasts, will clone readily in suspension. To hold the colony together and prevent mixing, the cells are suspended in agar or methocel and plated out over an agar underlay or into nontissue culture grade dishes.

Cloning in agar. See Figure 11.4 and Chapters 15, 20, and 21.

Outline

Agar is liquid at high temperatures but is a gel at 36.5°C. Cells are suspended in warm agar, and,

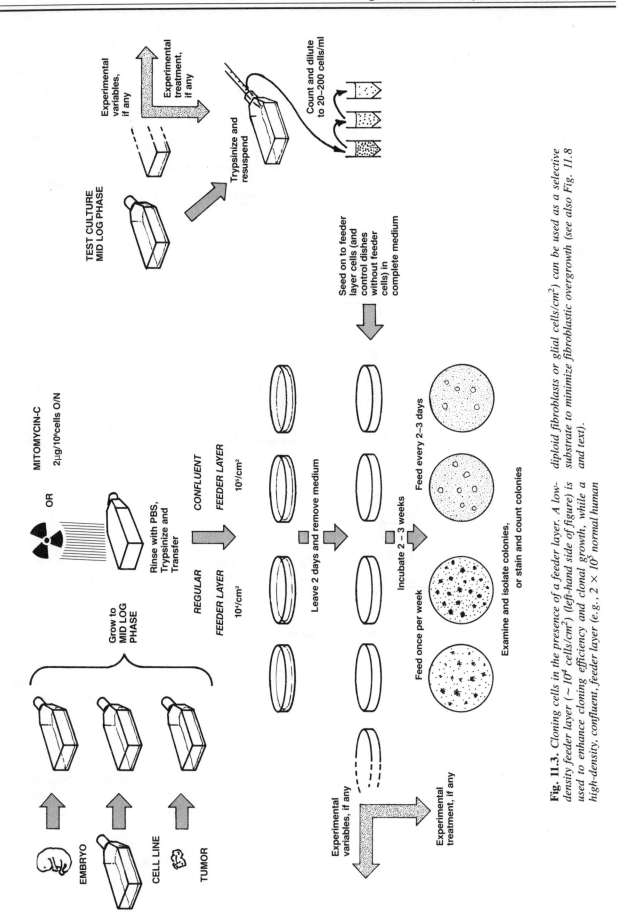

Fig. 11.3. *Cloning cells in the presence of a feeder layer. A low-density feeder layer (~ 10⁴ cells/cm²) (left-hand side of figure) is used to enhance cloning efficiency and clonal growth, while a high-density, confluent, feeder layer (e.g., 2 × 10⁵ normal human diploid fibroblasts or glial cells/cm²) can be used as a selective substrate to minimize fibroblastic overgrowth (see also Fig. 11.8 and text).*

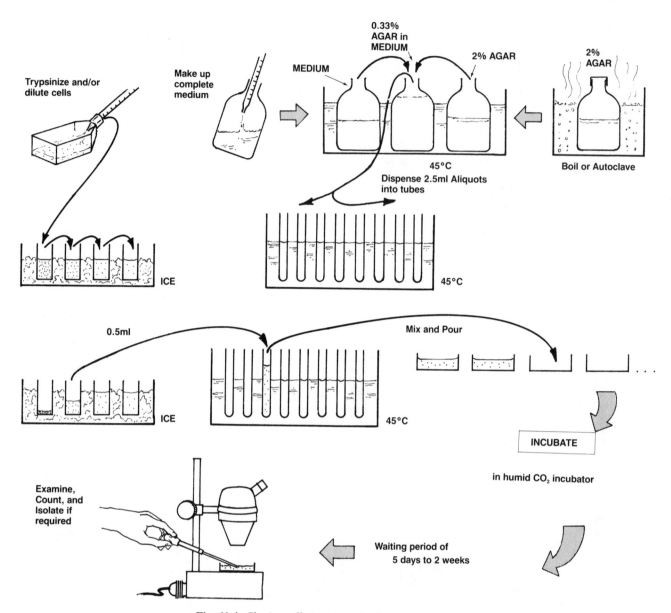

Fig. 11.4. *Cloning cells in suspension in agar.*

when incubated after the agar gels, will form discrete colonies which may be isolated easily.

Materials

2% agar

medium, e.g., Ham's F12, RPMI 1640, or CMRL 1066

fetal bovine serum

pipettes

35-mm petri dishes

50-mm × 8-mm test tubes

boiling water bath

water bath at 45°C

ice tray

Note. Before preparing medium and cells, work out cell dilutions and label petri dishes or multi-well plates. Convenient cell numbers per 35-mm dish are 1,000, 333, 111, 37.

Protocol

1.

Dissolve agar by placing bottle in a boiling water bath or autoclave. Cool to 80°C and transfer to 45°C water bath. (If you are delayed, keep agar at 60°C).

2.

Make up medium A: 70 ml medium and 30 ml FBS, keep at 45°C.

3.

Make up medium B: 60 ml medium A and 12 ml 2% agar, keep at 45°C.

4.

Count the cell suspension and dilute to give 2,000/ml, 667/ml (1:3), 222/ml (1:3), 74/ml (1:3). Place on ice.

5.

Place 5-ml test tubes in rack and keep at 45°C.

6.

Add 2.5 ml medium B to test tubes.

7.

Add 0.5-ml cell suspension to one tube, mix, and pour into dish immediately.

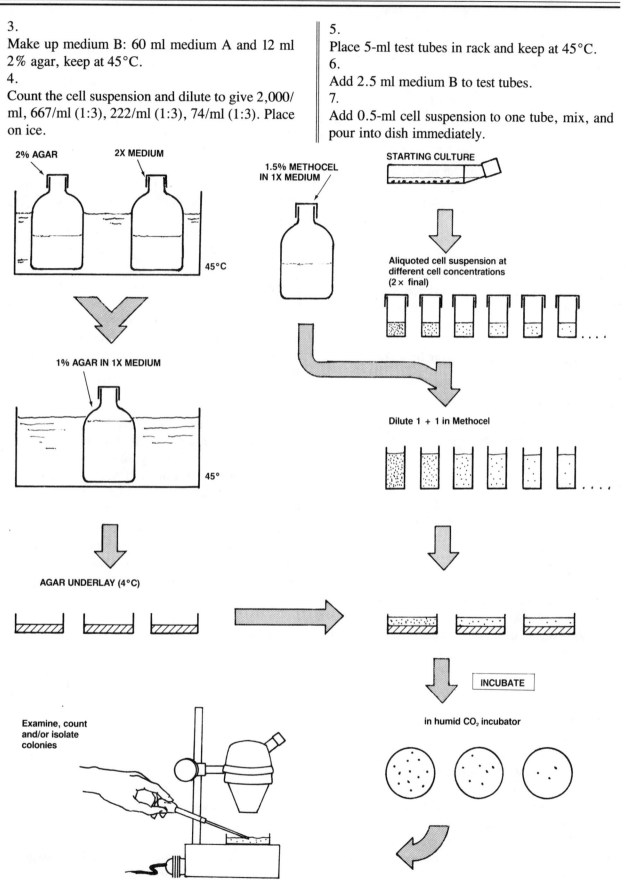

2% AGAR   2X MEDIUM
45°C

1.5% METHOCEL IN 1X MEDIUM

STARTING CULTURE

Aliquoted cell suspension at different cell concentrations (2 × final)

1% AGAR IN 1X MEDIUM
45°

Dilute 1 + 1 in Methocel

AGAR UNDERLAY (4°C)

INCUBATE

in humid CO_2 incubator

Examine, count and/or isolate colonies

Fig. 11.5. *Cloning cells in suspension in Methocel over an agar underlay.*

8.

After all dishes have been poured, place them at 4°C for 10 min.

9.

Before incubating the dishes in a humid CO_2 incubator, it is advisable to put them into another container to try to avoid contamination of the cultures from the moist atmosphere of the incubator. (a) Place petri dishes or multiwell plates in a plastic box with a lid and containing a dish of water. (The box should be washed first with 70% alcohol and allowed to dry.) (b) When using 35-mm petri dishes, two can be put into a 10-cm petri dish (nontissue culture grade) with a third 35-mm petri dish containing 3 ml sterile water.

Note. Rinse all pipettes used for agar with hot water before discarding, or use plastic pipettes.

Cloning in Methocel over agar base [Buick et al., 1979].

Outline

Suspend cells in medium containing Methocel and seed into dishes containing gelled agar medium (Fig. 11.5)

Materials

As for agar cloning; 1.36% Methocel 4 Nsm^{-2} (4,000 cps) in deionized distilled water.

Protocol

1.

Prepare agar underlay by heating sterile 1.0–2.0% agar to 100°C, bring to 45°C, and dilute with an equal volume of double-strength medium at 45°C (prepare from × 10 concentrate to half the recommended final volume and add twice the normal concentration of serum). Plate out 1 ml immediately into 35-mm dishes or 6 × 35 mm multiwell plates, and allow to gel at 4°C for 10 min.

2.

Trypsinize or collect cells from suspension and dilute to double the required final concentration.

3.

Dilute the cells with an equal volume of methocel and plate out 1 ml over the agar underlay (10–1,000 cells per dish for continuous cell strains but up to 5 × 10^5 per dish may be needed for primary cultures).

4.

Incubate until colonies form. Since the colonies form at the interface between the agar and the Methocel, fresh medium may be added, 1 ml per dish or well, after 1 wk and removed and replaced with more fresh medium after 2 wk without disturbing the colonies.

Many of the recommendations applying to medium supplementation for monolayer cloning also apply to suspension cloning. In addition, sulphydryl compounds such as mercaptoethanol (5 × 10^{-5} M), glutathione (1 mM), or α-thioglycerol (7.5 × 10^{-5} M) [Iscove et al., 1980] are sometimes used. Macpherson [1973] found the inclusion of DEAE dextran was beneficial for cloning.

Most cell types clone in suspension with a lower efficiency than in monolayer, some cells by two or three orders of magnitude. Isolation of colonies is, however, much easier.

Isolation of Clones

Monolayer clones—multiwell plates. If cells are cloned directly into multiwell plates (see above), colonies may be isolated by trypsinizing individual wells. It is necessary to confirm the clonal origin of the colony during its formation by regular microscopic observation.

Cloning rings. If cloning is performed in petri dishes, there is no physical separation between colonies. This must be created by removing the medium and placing a ring around the colony to be isolated (Fig. 11.6).

Outline

The colony is trypsinized from within a porcelain, Teflon, or stainless steel ring and transferred to one of the wells of a 24- or 12-well plate, or directly to a 25-cm^2 flask (see step 3 above) (Fig. 11.7).

Materials

Cloning rings
silicone grease
Pasteur pipettes with bent end
0.25% trypsin
medium
24-well plate and/or 25 cm^2 flasks
sterile forceps

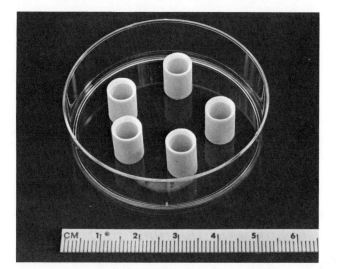

Fig. 11.6. *Cloning rings. Porcelain rings (Fisher) are illustrated, but thick-walled stainless steel rings (e.g., roller bearings) or plastic (e.g., cut from nylon or Teflon thick-walled tubing) can be used. Whatever the material, the base must be smooth, to seal with silicone grease onto the base of the petri dish, and the internal diameter just wide enough to enclose one whole clone, without overlapping adjacent clones.*

Protocol

1.

Sterilize cloning rings and silicone grease separately in glass petri dishes, by dry heat.

2.

Prepare about 20 bent Pasteur pipettes by heating briefly in a bunsen flame and allowing about 12 mm of the tip to drop under gravity. If the pipette is held at 30° above horizontal, the bend will be 120°. Place pipettes in sterile test tubes and allow to cool before use.

3.

Examine clones and mark those that you wish to isolate with a felt tip marker on the underside of the dish.

4.

Remove medium from dish.

5.

Using sterile forceps, take one cloning ring, dip in silicone grease, and press down on dish alongside silicone grease to spread the grease round the base of the ring.

6.

Place ring around desired colony.

7.

Repeat steps 5 and 6 for two or three other colonies in same dish.

8.

Add sufficient 0.25% trypsin to fill the hole in ring (~0.4 ml), leave 20 s, and remove.

9.

Close dish and incubate for 15 min.

10.

Add 0.4 ml medium to each ring.

11.

Taking each clone in turn, pipette medium up and down to disperse cells, and transfer to a well of a 24-well plate, or to a 25-cm² flask standing on end. Use a separate pipette for each clone.

12.

Wash out ring with a second 0.4 ml medium, and transfer to same well.

13.

Close plate and incubate, or if using flasks, add 1 ml medium and incubate standing on end.

14.

When clone grows to fill well, transfer up to 25-cm² flask, incubated conventionally with 5 ml medium. If using up-ended flask technique, remove medium when end of flask confluent, trypsinize cells, resuspend in 5 ml medium, and lay flask down flat. Continue incubation.

The cloning ring technique may be applied when cells are cloned in a plastic flask by swabbing the flask with alcohol and slicing the top off with a heated sterile scalpel or hot wire. Thereafter proceed as for petri dishes.

Irradiation. Alternatively, where an irradiation source is available, clones may be isolated by shielding one and irradiating the rest of the monolayer (30 Gy, 3,000 rads).

Outline

Invert the flask under an x-ray machine or ^{60}Co source, screening the desired colony with lead.

Materials

   X-ray or cobalt source
   piece of lead 2 mm thick
   PBSA
   0.25% trypsin
   medium

Protocol

1.
Select desired colony.
2.
Invert flask under x-ray or cobalt source.
3.
Cover colony with a piece of lead 2 mm thick.
4.
Irradiate with 30 Gy (3,000 rads).

5.
Return to sterile area and remove medium, trypsinize, and allow cells to reestablish in the same bottle, using the irradiated cells as a feeder layer.

If irradiation and trypsinization is carried out when the colony is about 100 cells in size, then the trypsinized cells will reclone. Three serial clonings may be performed within 6 wk by this method.

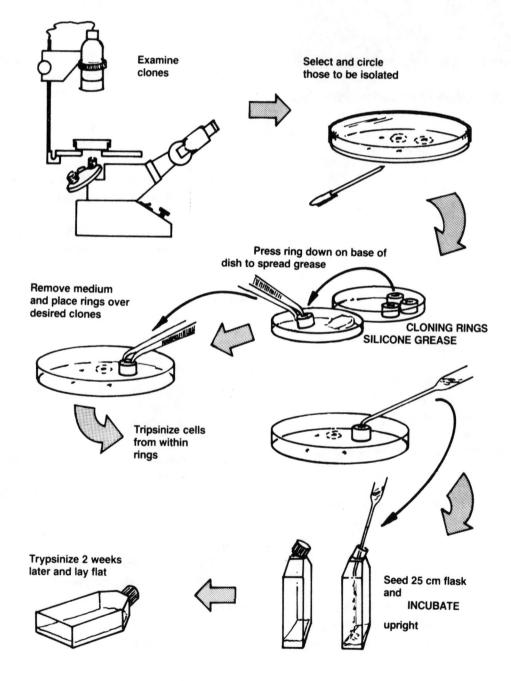

Examine clones

Select and circle those to be isolated

Press ring down on base of dish to spread grease

CLONING RINGS
SILICONE GREASE

Remove medium and place rings over desired clones

Tripsinize cells from within rings

Seed 25 cm flask and

INCUBATE

upright

Trypsinize 2 weeks later and lay flat

Fig. 11.7. *Isolation of clones with cloning rings.*

Other isolation techniques include: (1) distributing small coverslips or broken fragments of coverslips on the bottom of a petri dish. When plated out at the correct density, some colonies are found singly distributed on a piece of glass and may be transferred to a fresh dish or multiwell plate; (2) capillary technique of Sanford et al. [1948]. A dilute cell suspension is drawn into a glass capillary tube (e.g., 50-μl Drummond Microcap) allowing colonies to form inside the tube. The tube is then carefully broken on either side of a colony and transferred to a fresh plate, and (3) Petri-perm dish. This is a petri dish with a thin gas-permeable base (see Chapter 7), which may be cut with scissors or a scalpel to isolate colonies. Since this means keeping the outside of the dish sterile, it needs to be handled aseptically and kept inside a larger sterile petri dish.

Suspension Clones

Outline
Draw colony into micropipette and transfer to a flask or the well of a multiwell plate.

Materials
24-well plates
medium
microcapillary pipettes
dissecting microscope
25-cm^2 culture flask

Protocol
Picking colonies is best done on a dissecting microscope.
1.
Pipette 1 ml of medium into each well of a 24-well plate.
2.
Using a separate 50-μl microcapillary pipette for each clone, place the tip of the pipette against the colony to be isolated and gently draw in the colony.
3.
Transfer to a 24-well dish and flush out colony with medium. If from Methocel, the colony will settle, adhere, and grow out. If from agar, you may need to pipette the colony up and down a few times in the well to remove the agar.
4.
Clones may also be seeded directly into a 25-cm^2 plastic flask standing on end (see above).

SELECTIVE MEDIA

Manipulating the culture conditions by using a selective medium is a standard method for selecting microorganisms. Its application to animal cells in culture is limited, however, by the basic metabolic similarities of most cells isolated from one animal in terms of their nutritional requirements. The problem is accentuated by the effect of serum, which tends to mask the selective properties of different media. Peehl and Ham [1980] were able to demonstrate that by using two different media, MCDB 105 and MCDB 151, with minimal amounts of dialyzed serum, either fibroblasts or epithelial cells could be grown preferentially from human foreskin (see also Chapter 7, Table 7.7).

Gilbert and Migeon [1975, 1977] replaced the L-valine in the culture medium with D-valine and demonstrated that cells possessing D-amino acid oxidase would grow preferentially. Kidney tubular epithelium and epithelial cells from fetal lung and umbilical cord may be selected this way.

Much of the effort in developing selective conditions has been aimed at suppressing fibroblastic overgrowth. Whei-Yang Kao [1977] used cis-OH-proline for this purpose, although this substance can prove toxic to other cells. Fry and Bridges [1979] found phenobarbitone inhibited fibroblastic overgrowth in cultures of hepatocytes and Braaten et al. [1974] were able to reduce the fibroblastic contamination of neonatal pancreas by treating the culture with sodium ethylmercurithiosalicylate. One of the more successful approaches was the development of a monoclonal antibody to the stromal cells of a human breast carcinoma [Edwards et al., 1980]. Used with complement, this antibody proved cytotoxic to fibroblasts from several-tumors and helped to purify a number of malignant cell lines (see also Chapter 21).

Selective media are also commonly used to isolate hybrid clones from somatic hybridization experiments. HAT medium, a combination of hypoxanthine, aminopterin, and thymidine, selects hybrids with both hypoxanthine guanine phosphoribosyltransferase and thymidine kinase from parental cells deficient in one or the other enzyme (see Chapter 23) [Littlefield, 1964].

ISOLATION OF GENETIC VARIANTS

The following protocol for the development of mutant cell lines which amplify the dihydrofolate reduc-

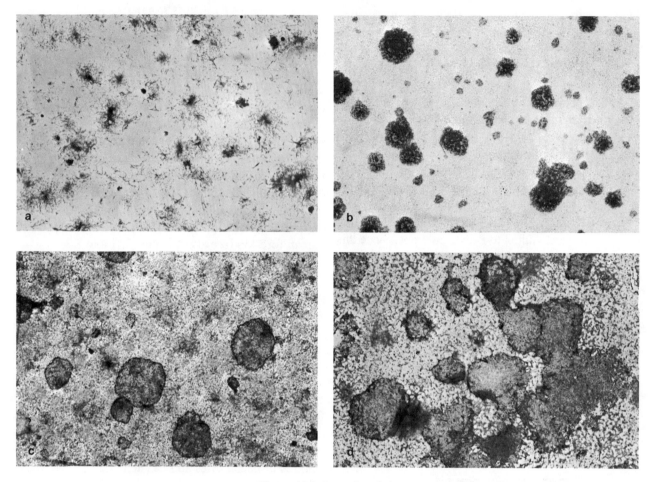

Figure 11.8. *Legend on facing page.*

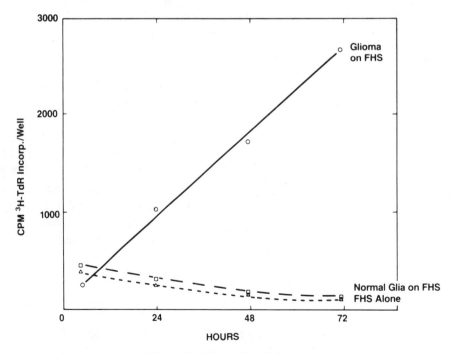

Figure 11.9. *Legend on facing page.*

tase (DHFR) gene has been contributed by June Biedler, Memorial Sloan-Kettering Cancer Center, New York, New York.

Principle

Cells exposed to gradually increasing concentrations of folic acid antagonists such as methotrexate (MTX) over a prolonged period of time will develop resistance to the toxic effects of the drug [Biedler et al., 1972]. Resistance resulting from amplification of the DHFR gene generally develops the most rapidly, although other mechanisms, e.g., alteration in anti-folate transport and/or mutation affecting enzyme structure or affinity, may confer part or all of the resistant phenotype.

Outline

Expose cells to a graded series of concentrations of MTX for 1–4 wk, periodically replacing the medium with fresh medium containing the same drug concentration. Select for subculturing those flasks in which a small percentage of cells survive and form colonies; repeatedly subculture such cells in the same and in two- to tenfold higher MTX concentrations until cells acquire the desired degree of resistance.

Materials

Chinese hamster cells or rapidly growing mouse cell lines

Fig. 11.8. *Selective cloning of breast epithelium on a confluent feeder layer. a. Colonies forming on plastic alone after seeding 4,000 cells from a breast carcinoma culture/cm^2 (2 × 10^4 cells/ml). Small dense colonies are epithelial cells, larger stellate colonies are fibroblasts. b. Colonies of cells from the same culture, seeded at 400 cells/cm^2 (2,000) cells/ml) on a confluent feeder layer of FHS 74 Int cells [Owens et al., 1974]. The epithelial colonies are much larger than in a, the plating efficiency is higher, and there are no fibroblastic colonies. c. Colonies from a different breast carcinoma culture plated onto the same feeder layer. Note different colony morphology with lighter stained center and ring at point of interaction with feeder layer. d. Colonies from normal breast culture seeded onto FHI cells (fetal human intestine similar to FHS 74 Int). There are a few small fibroblastic colonies present in c and d (after a technique described by Dr. A.J. Hackett, personal communication).*

Fig. 11.9. *Selective growth of glioma on confluent feeder layer. Cells were seeded at 2 × 10^4/ml (4 × 10^3/cm^2) onto confluent, mitomycin-C treated feeder layers (see text) of FHS 74 Int cells [Owens et al., 1974] and labeled at intervals thereafter with ^{3}H-thymidine (see text), extracted, and counted.*

Methotrexate Sodium Parenteral (Lederle Laboratories)
0.15 M NaCl
tissue culture flasks, pipettes
culture medium (e.g., Eagle's MEM with 10% fetal bovine serum)
inverted microscope
ultralow temperature cabinet or liquid nitrogen freezer

Protocol

1.

Clone parental cell line to obtain a rapidly growing, genotypically uniform population to be used for selection.

2.

Dilute MTX with sterile 0.15 M (0.85%) NaCl. Drug packaged for use in the clinic is in solution at 2.5 mg/ml and is stable indefinitely at −20°C.

3.

Inoculate 2.5 × 10^5 cells into replicate 25-cm^2 flasks containing no drug or 0.01, 0.02, 0.05, and 0.1 μg/ml of MTX in complete tissue culture medium. Adjust pH and incubate at 37°C for 5–7 d.

4.

Observe cultures with an inverted microscope. Replace medium with fresh medium containing the same amount of MTX in cultures showing clonal growth of a small proportion of cells amid a background of enlarged, substrate-adherent and probably dying cells and re-incubate.

5.

Allow cells to grow for another 5–7 d, changing growth medium as necessary but continuously exposing the cells to methotrexate. When cell density has attained 2–10 × 10^6 cells/flask, subculture cells at 2.5 × 10^5 cells/flask into new flasks containing the same and two- to tenfold higher drug concentrations.

6.

After another 5–7 d, observe new passage flasks as well as cultures from the previous passage which had been exposed to higher drug concentrations; change medium and select for viable cells as before.

7.

Continue selection with progressively higher drug concentrations at each subcultivation step until the desired level of resistance is obtained: 2–3 months for Chinese hamster cells with low to moderate levels of resistance, increase in DHFR activity, and/or transport alteration; 6 months or more for high levels of resistance and enzyme overproduction, when Chinese hamster, mouse, or fast growing human cells are used.

8.

Periodically freeze samples of developing lines at −70°C or in liquid nitrogen (see Chapter 17).

Analysis

Characterize resistant cells for levels of resistance to drug in a clonal or cell growth assay (see Chapter 19), for increase in activity or amount of DHFR by biochemical or gel electrophoresis techniques [Albrecht et al., 1972; Melera et al., 1980], and/or for increase in copy number of the reductase gene by Southern or dot blots [Scotto et al., 1986] with DHFR-specific probes to determine the mechanism(s) of resistance.

Variations

Cell culture media other than Eagle's MEM can be used; medium composition, e.g., folic acid content, can be expected to influence rate and type of MTX resistance development. Cells can be treated with chemical mutagens prior to selection [Thompson and Baker, 1973]; this treatment may also alter rate and type of mutant selection.

Selection can also be done using cells plated in drug at low density in 100-mm tissue culture dishes (with isolation of individual colonies using cloning cylinders; see above), as single cells in 96-well cluster dishes, or in soft agar, to enable isolation of one or multiple clonal populations at each or any step during resistance development.

Cells can be made resistant to a number of other agents, such as antibiotics, other antimetabolites, toxic metals, etc. by similar techniques; differences in the mechanism of action or degree of toxicity of the agents, however, may require that treatment with the agent be intermittent rather than continuous and may increase the time necessary for selection.

Cell lines of different species or with slower growth rates, such as some human tumor cell lines, may require different (usually lower) initial drug concentrations, longer exposure times at each concentration, and smaller increases in concentration between selection steps.

Solubilization of MTX other than the Lederle product will require addition of equimolar amounts of NaOH and sterilization through a 0.2-μm filter.

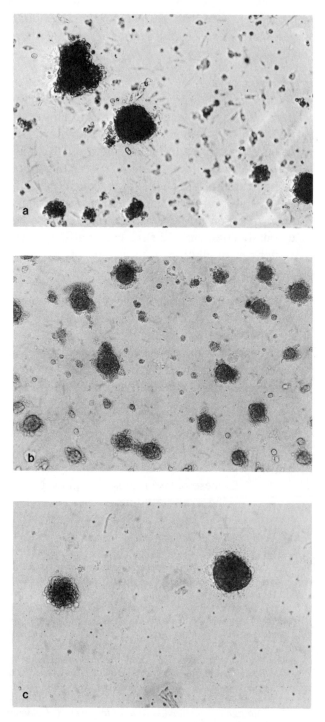

Fig. 11.10. *Growth of melanoma, fibroblasts, and glia in suspension. Cells were plated out at 5 × 10⁵ per 35-mm dish (2.5 × 10⁵ cells/ml) in 1.5% methocel over a 1.25% agar underlay. Colonies were photographed after 3 wk. a. Melanoma. b. Human normal embryonic skin fibroblasts. c. Human normal adult glia. d. Colony-forming efficiency of normal and malignant glial cells in suspension. Unshaded bars, colonies of over eight cells (approximately), and stippled bars, colonies of over 32 cells (approximately). Colony counts were done on an Artek Colony Counter at different threshold settings.*

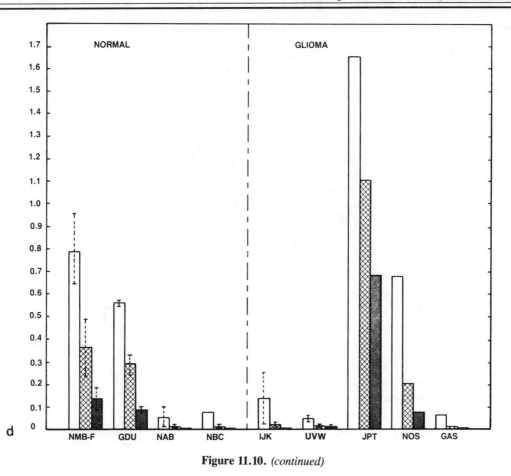

Figure 11.10. *(continued)*

INTERACTION WITH SUBSTRATE

Selective Adhesion

Different cell types have different affinities for the culture substrate and will attach at different rates. If a primary cell suspension is seeded into one flask and transferred to a second after 30 min, a third after 1 hr, and so on, the most adhesive cells will be found in the first flask and the least adhesive in the last. Macrophages will tend to remain in the first flask, fibroblasts in the next few flasks, then epithelial, and finally hemopoietic cells in the last flask. Polinger [1970] used a similar procedure for the separation of embryonic heart muscle cells from fibroblasts.

If collagenase in complete medium is used for primary disaggregation of the tissue (see Chapter 9), most of the cells released will not attach within 48 hr unless the collagenase is removed. However, macrophages migrate out of the fragments of tissue and attach during this period and can be removed from other cells by transferring the disaggregate to a fresh flask after 48–72-hr treatment with collagenase. This technique works well during disaggregation of biopsy specimens from human tumors.

Selective Detachment

Treatment of a heterogeneous monolayer with trypsin or collagenase will remove some cells more rapidly than others. Periodic brief exposure to trypsin removed fibroblasts from cultures of fetal human intestine [Owens et al., 1974] and skin [Milo et al., 1980; Lasfargues, 1973] found exposure of cultures of breast tissue to collagenase for a few days at a time removed fibroblasts and left the epithelial cells. EDTA, on the other hand, may release epithelial cells more readily than fibroblasts [Paul, 1975].

Dispase II (Boehringer, Mannheim) selectively dislodges sheets of epithelium from human cervical cultures grown on feeder layers of 3T3 cells (see below) without dislodging the 3T3 cells (see Chapter 20). This technique may be effective in subculturing epithelial cells from other sources, excluding stromal fibroblasts.

Nature of Substrate

The hydrophilic nature of most culture substrates (see also Chapter 7) appears to be necessary for cell attachment, but little is known about variations in charge distribution on the cell surface and how different mosaic patterns may interact with different substrates. Since cell sorting in the embryo is a highly selective process and probably relates to differences in the distribution of charged molecules and specific receptor sites on the cell surface, qualitative and quantitative variations in substrate affinity should be anticipated in cultured cells. The relative infrequency with which this is actually found probably illustrates our ignorance of the subtlety of cell-cell and cell-substrate interactions (see also Chapter 14).

The selective effect of substrates on growth may depend on both differential rates of attachment and growth, although in practice the two are indistinguishable. Polyacrylamide layers allow the cloning of tumor cells but not normal fibroblasts [Jones and Haskill, 1973, 1976]. Transformed cells proliferate on Teflon while most other cells will not [Parenjpe et al., 1975]. Macrophages will also attach to Teflon but do not proliferate. The dermal surface of freeze-dried pig skin was shown to allow growth of epidermal cells but not fibroblasts [Freeman et al., 1976]. Collagen, presumably the basis of the selection, has also been used in gel form to favor epithelial cell growth [Lillie et al., 1980] and in its denatured form to support endothelial outgrowth from aorta into a fibrin clot [Nicosia and Leighton, 1981].

Becton Dickinson has introduced a range of plastics in the Falcon series called Primaria. These have a different charge on the plastic surface from conventional tissue culture plastics and are designed to enhance epithelial growth relative to fibroblasts. A number of companies are also supplying plastics coated in natural or synthetic matrices which may facilitate growth of more fastidious cell types but are probably not selective.

Feeder layers. The conditioning of the substrate by feeder layers will be discussed in Chapter 14. Feeder layers can also be used for the selective growth of epidermal cells [Rheinwald and Green, 1975] and for repressing stromal overgrowth in cultures of breast (Fig. 11.8) and colon carcinoma (see Chapter 21). [Freshney et al., 1981]. The author has also been able to demonstrate that human glioma will grow on confluent feeder layers of normal glia while cells derived

from normal brain will not [MacDonald et al., 1985] (Fig. 11.9; see Chapter 15).

Semisolid supports. Transformation of many fibroblast cultures reduces anchorage dependence of cell proliferation (see Chapter 15) [Macpherson and Montagnier, 1964]. By culturing the cells in agar (see above) after viral transformation, it is possible to isolate colonies of transformed cells and exclude most of the normal cells. Most normal cells will not form colonies in suspension with the same high efficiency as virally transformed cells, although they will often do so with low plating efficiencies. Colonies of hemopoietic cells

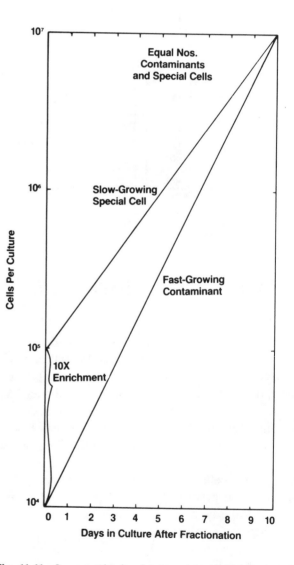

Fig. 11.11. *Overgrowth of a slow-growing cell line by a rapidly growing contaminant. This is a hypothetical example, but it demonstrates that a 10% contamination with a cell population which doubles every 24 hr will reach equal proportions with a cell population which doubles every 36 hr after only 10 d growth.*

will form in semisolid media [Metcalf, 1970], but these usually mature to nondividing differentiated cells and cannot be subcultured (see also Chapter 20). The difference between virally transformed fibroblasts and untransformed cells is not seen as clearly in attempts at selective culture of spontaneously arising tumors. Experiments in the author's laboratory have shown that normal glia and fetal skin fibroblasts will form colonies in suspension just as readily as glioma and melanoma (Fig. 11.10).

Cell cloning and the use of selective conditions have a significant advantage over physical cell separation techniques (next chapter) in that contaminating cells are either eliminated entirely by clonal selection or repressed by constant or repeated application of selective conditions. Even the best physical cell separation techniques will still allow some overlap between cell populations such that overgrowth will recur. As long as this situation exists, a steady state cannot be achieved and the constitution of the culture is altering continuously. From Figure 11.11 it can be seen that a 90% pure culture of line A will be 50% overgrown by a 10% contamination with line B in 10 days, given that B grows 50% faster than A. For continued culture, therefore, selective conditions are required in addition to, or in place of, physical separation techniques.

While cloning or selective culture conditions are the preferred methods for purifying a culture (see Chapter 11), there are occasions when cells do not grow with a high enough plating efficiency to make cloning possible or when appropriate selection conditions are not available. It may then be necessary to resort to a physical separation technique such as rate or density sedimentation. Physical separation techniques have the advantage that they give a high yield more quickly than cloning although not with the same purity.

The more successful separation techniques (Fig. 12.1) depend on differences in: (1) cell size; (2) cell density (specific gravity); (3) cell surface charge; (4) cell surface chemistry (affinity for lectins, antibodies, or chromatographic media); (5) total light scatter per cell; and (6) fluorescence emission of one or more cellular constituents or adsorbed antibody. The apparatus required ranges from about $10-worth of glassware to $200,000-worth of complex laser and computer technology; the choice depends on the parameter that you are obliged to use (see Table 12.1) and on your budget.

CELL SIZE AND SEDIMENTATION VELOCITY

The relationship between particle size and sedimentation rate at 1 g, though complex for submicron particles is fairly simple for cells and can be expressed approximately as

$$v = \frac{r^2}{4} \qquad \text{(eq. 12.1)}$$

[Miller and Phillips, 1969], where v = sedimentation rate in mm/hr and r = radius of the cell in μm (see Table 18.1).

TABLE 12.1. Cell Separation Methods

Method	Basis for separation	Comments	Reference
Sedimentation velocity at 1 g	Cell size	Simple technique	[Miller and Phillips, 1969]
Sedimentation by centrifugation "isokinetic gradient"	Cell size, density, and surface configuration	Computer-designed gradient in zonal rotor	[Pretlow, 1971]
Centrifugal elutriation	Cell size, density, and surface configuration	Rapid; high cell yield	[Meistrich et al., 1977a,b]
Isopycnic sedimentation	Cell density	Simple and rapid	[Pertoft and Laurent, 1977]
Flow cytophotometry (fluorescence-activated cell sorting)	Cell surface area fluorescent markers, fluorogenic enzyme substrates multiparameter	Complex technology and expensive; very effective; high resolution but low yield	[Kreth and Herzenberg, 1974]
Affinity chromatography	Cell surface antigens, cell surface carbohydrate	Elution of cells from columns difficult, better in free suspension	[Edelman, 1973]
Counter current distribution	Affinity of cell surface constituents for solvent phase	Poor viability after separation	[Walter, 1977]
Electrophoresis in gradient or curtain	Surface charge		[Platsoucas et al, 1979] [Kreisberg et al, 1977]

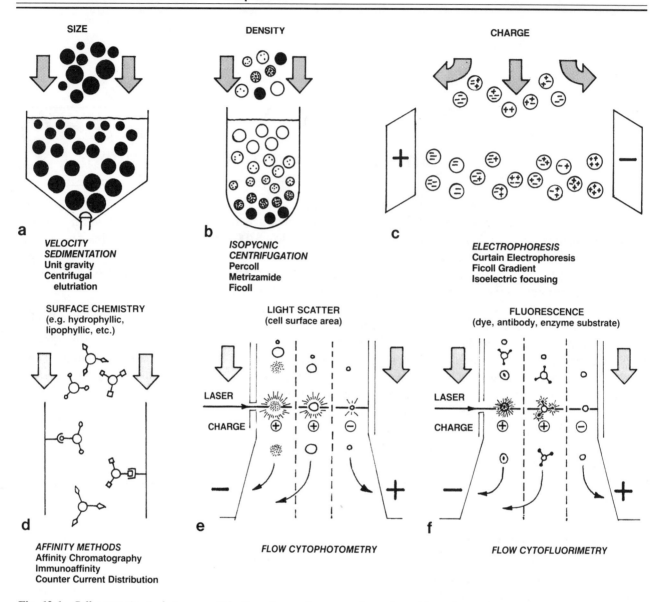

Fig. 12.1. *Cell separation techniques. a. Velocity sedimentation, influenced by cell size mainly, but also by cell density and surface area at elevated g. b. Isopycnic sedimentation. Cells sediment to a point in a density gradient equivalent to their own density. c. Electrophoresis. Cells migrate to either polarized plate according to their net surface charge. d. Affinity methods. Cells are separated by their differential affinities for: (1) chromatographic media; (2) antibodies or lectins bound to chromatographic media; or (3) two-phase aqueous polymer systems. e. Flow cytophotometry. Cells are diverted to either of two charged plates according to their light-scattering potential (proportional to surface area). f. Flow cytofluorimetry. Specific fluorochromes are used to label cells and electrophoretic separation is based on fluorescence emission. Both e and f are carried out on a single cell stream passing through the flow chamber of an instrument such as the Fluorescence-Activated Cell Sorter (FACS) (Becton Dickinson), the Cytofluorograph (Ortho), or the Coulter Cell Sorter.*

Unit Gravity Sedimentation

The apparatus required is illustrated in Figure 12.2 and can be assembled from routine laboratory glassware. To ensure stability of the column of liquid supporting the cells in the sedimentation chamber, it is formed from a serum, Ficoll, or bovine serum albu-men gradient, and run into the chamber through a baffle to prevent turbulence.

The height of the separation chamber determines how long the sedimentation may run. Since this is usually 2–4 hr, in a low-viscosity medium, 10 cm is approximately correct. A longer sedimentation time

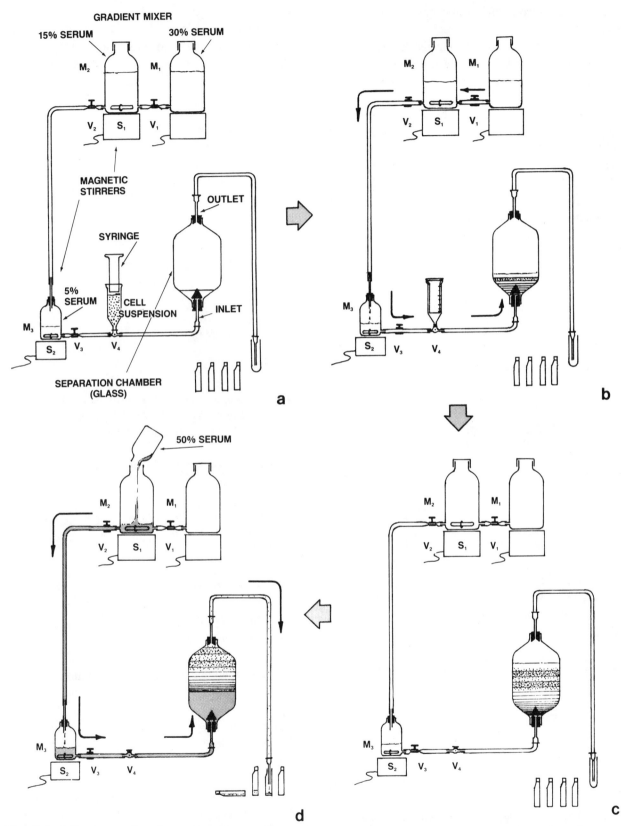

Fig. 12.2 *Cell separation by velocity sedimentation at unit gravity. Apparatus and position of valves at different stages of the procedure. a. At start, loading cells. b. Running in gradient. c.* *After cell sedimentation. d. Harvesting [Miller and Phillips, 1969]. V, valve; M, mixer vessel; S, stirrer.*

may give better resolution but may cause deterioration of the cells.

The width of the chamber controls the number of cells that may be loaded onto the gradient as the cell layer should be kept thin ($\sim$ 5 mm) and the cell concentration low ($\sim 10^6$/ml).

The height and width (i.e., volume) also affect the filling rate. The chamber must not be filled too rapidly or turbulence will result, and it cannot be filled too slowly or the cells will sediment faster than the liquid level rises. The dimensions given in Figure 12.3 are optimal for separating about 2×10^7 cells of 15–18 μm diameter or up to 10^8 cells of 10–12 μm diameter.

The procedure for separating a typical cell suspension of average cell diameter 15 μm is as follows.

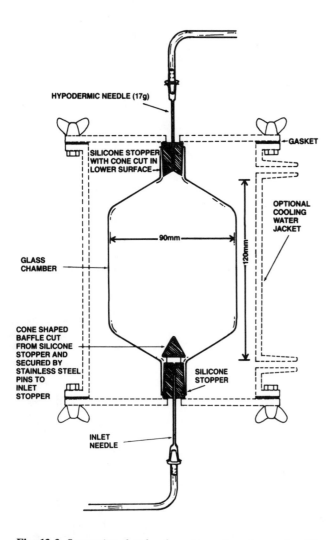

Fig. 12.3 *Separation chamber for unit gravity sedimentation. The dotted outline is a cooling jacket for carrying out sedimentations at 4°C [modified from Miller and Phillips, 1969].*

Outline

Float cells on top of a gradient of serum in medium, allow cells to sediment through the gradient for about 3 hr, and run off gradient into culture vessels (see Fig. 12.2).

Materials

300 ml Eagle's MEMS (suspension salts, see Table 7.4 + 30% serum
300 ml MEMS + 15% serum
20 ml MEMS + 5% serum
10 ml 0.25% trypsin-citrate (see Appendix)
30 ml PBS
20 ml MEMS + 3% serum
hemocytometer or cell counter
flotation medium (1M sucrose or 20% Ficoll)
25-cm^2 flasks
growth medium

Protocol

1.
Prepare apparatus as in Figure 12.2. Incorporate Luer connections to allow for disassembly for sterilization. Package and autoclave.
2.
Assemble and check that valves V_1, V_2, V_3, and V_4, are closed.
3.
Add 300 ml 30% serum in medium to mixer vessel M_1 and 300 ml 15% serum in medium to mixer vessel M_2.
4.
Check that stirrer S_1 is functioning.
5.
Add 20 ml 5% serum in medium to mixer M_3 and check that stirrer S_2 is functioning.
6.
Open V_4 to connect syringe to separation chamber and insert 20 ml PBS into separation chamber.
7.
Open V_4 to M_3 line, open V_3, and draw a little 5% serum into the syringe (just enough to fill line). Close V_3 and V_4.
8.
Prepare cell suspension, e.g., by trypsinizing primary culture for 15 min in 0.25% trypsin-citrate. Disperse cells carefully in 3% serum in medium and check that a single cell suspension is formed.

9.

Take up 20 ml at 10^6/ml (maximum) into syringe and connect to V_4 inlet.

10.

With syringe held vertically, open V_4 to M_3 line, open V_3, and draw a little 5% serum into syringe to clear any bubbles from M_3 line and V_4.

11.

Turn valve V_4 to separation chamber line and draw a little PBS into syringe to clear any bubbles from this line.

12.

Insert cell suspension slowly into chamber; avoid mixing cell suspension with the overlaying PBS layer. Take care to stop while a little fluid is left in the syringe to avoid injecting any air bubbles back into the line. If difficulty is encountered injecting cells smoothly, without turbulence, remove piston from syringe and allow cells to run in under gravity alone by raising V_4.

13.

Start stirrers S_1 and S_2.

14.

Open V_4 to connect M_3 line to separation chamber.

15.

Open V_1 and adjust flow rate by opening V_2 to give 15 ml/min (~ 5 drops/s) at M_3. Cell suspension will now float up into separation chamber on gradient of serum. Check for turbulence at baffle as suspension and gradient run in. If there is any, reduce flow rate at M_3 by closing V_2.

16.

When gradient mixers M_1 and M_2 are empty, but before M_3 empties, close V_3 and V_2.

17.

It should be possible to see the cell layer in the sedimentation chamber and to follow the cells as they sediment. As they do, the cell band will become wider and more diffuse. Check for signs of "streaming" in the early stages of sedimentation (tails of cells which sediment ahead of the main band). This occurs when the cell concentration is too high or the step between the cells and the gradient is too steep.

18.

After about 20 min, close V_1 and add 90 ml 50% serum to M_2. Open V_3 and adjust flow rate at M_3 to 15 ml/min. Stop when M_2 is empty but before M_3 empties, by closing V_3 and V_2.

19.

Add 500 ml flotation medium (1 M sucrose or 20% Ficoll) to M_1 and M_2, open V_1 and V_2, and let some of the flotation medium run into M_3. Close V_2.

20.

When sedimentation is complete, i.e., cell band midway down separation chamber, open V_3, and adjust V_2 to give a flow rate of 15 ml/min in M_3.

21.

Collect eluate from top of chamber via elution line and run into graduated culture vessels, e.g., 25-cm^2 flasks, 10 ml per flask. Mix the contents of each flask and take sample for cell counting.

22.

Seal and incubate flasks for 24 hr, replace medium with fresh medium at standard serum concentration and volume.

Variations

Gradient medium. If serum and regular culture medium are used, then it is possible to culture cells directly from the eluate. Fetal bovine serum causes less reaggregation than calf or horse serum. If serum is found to be unsuitable, gradients can be formed from bovine serum albumen [Catsimpoulas et al., 1978] or Ficoll (Pharmacia).

Aggregation. Aggregation can be reduced by enclosing the separation chamber in a cooling jacket and running the whole process at 4°C. Mixers M_1, M_2, and M_3 must all be kept cold also. Water-driven magnetic stirrers (Calbiochem) may be used at S_1 and S_2 to minimize overheating of gradient.

Pump. Gravity is used in this example as the cheapest and simplest method for generating the flow of gradient medium, but if desired, a peristalic pump (pulse free) may be inserted at V_3.

Sedimentation of cells at unit gravity is a simple low-technology method of separating cells. It works well for many cell types, e.g., brain [Cohen et al., 1978], hemopoietic cells [McCool et al., 1970; Petersen and Evans, 1967], and HeLa/fibroblast mixtures (Fig. 12.4) and can be performed in regular physiological media. The cells must be singly suspended, however, and there is a practical limit of about 10_8 cells that may be separated.

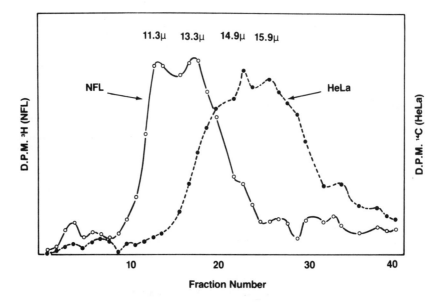

Fig. 12.4 *Elution profiles of artificial mixtures of HeLa and NFL (normal human fetal lung fibroblasts) after sedimentation at 1 g for 3 hr. Cultures were prelabeled with ^{3}H-leucine or ^{14}C-thymidine and the distribution of cells in the eluate was determined by dual-isotope scintillation counting. Open circles and solid line, NFL, solid circles and broken line, HeLa. The numbers above each peak are the calculated cell diameters derived from the sedimentation velocity by equation 12.1. The values obtained for HeLa were confirmed by micrometry, but the values for NFL did not agree with the micrometer readings which were around 16–18 μm [after Freshney et al., 1982a, reproduced with permission from the publisher].*

Pretlow and others [Pretlow, 1971; Pretlow et al., 1978; Hemstreet, et al., 1980] have used specially formed gradients of Ficoll (isokinetic gradients) to separate cells by sedimentation velocity at higher *g* forces on a zonal rotor. The gradients are shallow and of relatively low density to minimize the effect of cell density on sedimentation rate. Cells of many different types have been separated by this method, and it appears to have a wide application.

Centrifugal Elutriation

The centrifugal elutriator (Beckman) is a device for increasing the sedimentation rate and improving the yield and resolution by performing the separation in a specially designed centrifuge and rotor (Fig. 12.5). Cells in the suspending medium are pumped into the separation chamber in the rotor while the rotor is turning. While the cells are in the chamber, centrifugal force will tend to force the cells to the outer edge of the rotor (Fig. 12.6). Meanwhile the suspending medium is pumped through the chamber such that the centripetal flow rate approximates to the sedimentation rate of the cells. If the cells were uniform, they would remain stationary, but since they vary in size, density, and cell surface configuration, they tend to sediment at different rates.

As the sedimentation chamber is tapered, the flow rate increases toward the edge of the rotor and a continuous range of flow rates is generated. Cells of differing sedimentation rates will, therefore, reach equilibrium at different positions in the chamber. The sedimentation chamber is illuminated by a stroboscopic light and can be observed through a viewing port. When the cells are seen to reach equilibrium, the flow rate is increased and the cells are pumped out into receiving vessels. The separation can be performed in complete medium and the cells cultured directly afterward.

Equilibrium is reached in a few minutes and the whole run may take 30 min. On each run 10^8 cells may be separated and the run may be repeated as often as necessary. The apparatus is, however, fairly expensive and a considerable amount of experience is required before effective separations may be made. A number of cell types have been separated by this method [Greenleaf et al., 1979; Schengrund and Repman, 1979; Meistrich et al., 1977a], as have cells of different phases of the cell cycle [Meistrich et al., 1977b].

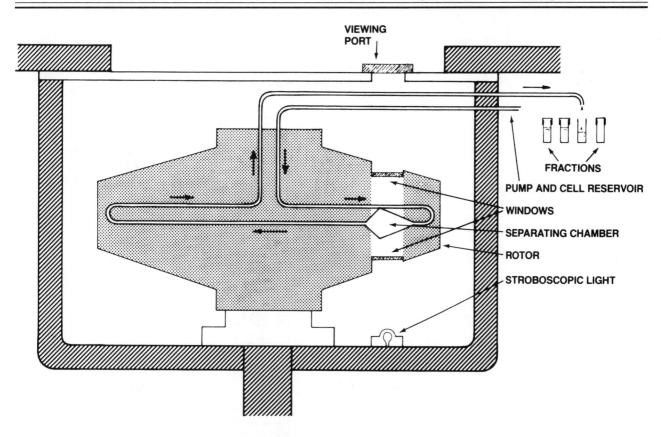

Fig. 12.5. *Centrifugal elutriator rotor (Beckman). Cell suspension and carrier liquid enter at the center of the rotor and are pumped to the periphery to enter the outer end of the separation chamber. The return loop is via the opposite side of the rotor to maintain balance.*

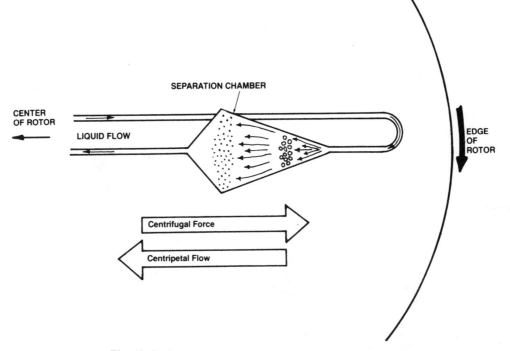

Fig. 12.6. *Separation chamber of elutriator rotor.*

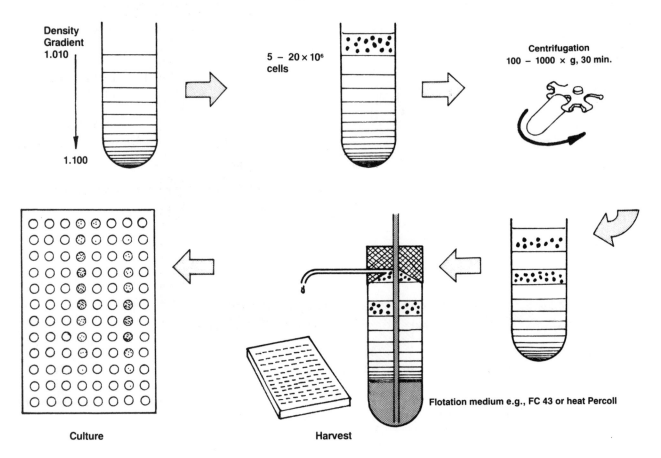

Fig. 12.7. *Cell separation by isopycnic centrifugation.*

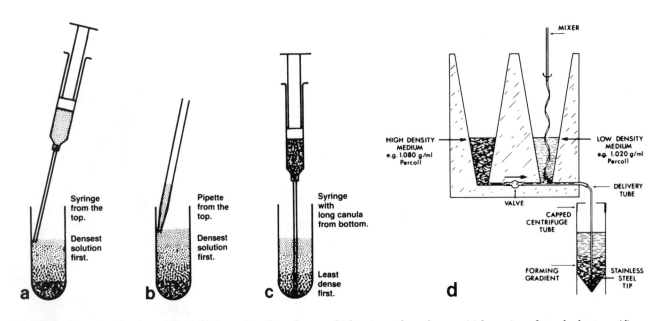

Fig. 12.8. *Layering density gradients (a) by syringe from the top, (b) by pipette from the top, (c) by syringe from the bottom, (d) by gradient mixing device (Buchler).*

CELL DENSITY AND ISOPYCNIC SEDIMENTATION

Separation of cells by density can be performed at low or high g using conventional equipment. The cells sediment in a density gradient to an equilibrium position equivalent to their own density (isopycnic sedimentation). Physiological media must be used and the osmotic strength carefully monitored. The density medium should be nontoxic, nonviscous at high densities (1.10 g/ml), and exert little osmotic pressure in solution. Serum albumen [Turner et al., 1967], dextran [Schulman, 1968], Ficoll (Pharmacia) [Sykes et al., 1970], metrizamide (Nygaard) [Munthe Kaas and Seglen. 1974], and Percoll (Pharmacia) [Pertoft, 1968; Pertoft and Laurent, 1977; Wolff and Pertoft, 1972] have all been used successfully; Percoll (colloidal silica) is one of the more effective media currently available. For isolation of lymphocytes on Ficoll/Metrizoate, see Chapter 23.

Outline

Form gradient: (1) by layering different densities of Percoll; (2) by high-speed spin; or (3) with special gradient former. Centrifuge cells through Percoll gradient (or allow to sediment at unit gravity), collect fractions, and culture directly (Fig. 12.7).

Materials

    Culture medium (sterile)
    medium + 20% Percoll (sterile)
    25-ml centrifuge tubes (sterile)
    PBSA (sterile)
    0.25% trypsin (sterile)
    syringe or gradient harvester (sterile)
    24-well plates or microtitration plates
        (sterile)
    refractometer or density meter
    hemocytometer
    cell counter

Protocol

1.
Prepare gradient: (1) Prepare two media, one regular culture medium and one with 20% Percoll; (2) adjust the density of the Percoll solution to 1.10 g/ml and its osmotic strength to 290 mOsm/kg: (3) mix the two media in varying proportions to give the desired density range (e.g., 1.020–1.100 g/ml) in ten or 20 steps; and (4) layering one step over another, build up a stepwise density gradient in a 25-ml centrifuge tube (see Fig. 12.8a). Gradients may be used immediately or left overnight.

Alternatively, place medium containing Percoll of density 1.085 g/ml in a tube and centrifuge at 20,000 g for 1 hr. This generates a sigmoid gradient (Fig. 12.9), the shape of which is determined

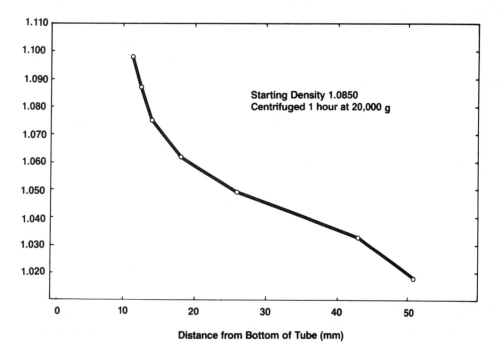

Starting Density 1.0850
Centrifuged 1 hour at 20,000 g

Distance from Bottom of Tube (mm)

Fig. 12.9. *Gradient generated by spinning Percoll at 20,000* g *for 1 hr.*

by the starting concentration of Percoll, the duration and centrifugal force of the centrifugation, the shape of the tube, and the type of rotor.

A continuous linear gradient may be produced by mixing, for example, 1.020 g/ml with 1.08 g/ml Percoll in a gradient-forming device (Fig. 12.8b)(MSE/Fisons, Pharmacia, Buchler).

2.

Trypsinize cells and resuspend in medium plus serum. Check that they are singly suspended.

3.

Layer up to 2×10^7 cells in 2-ml medium on top of the gradient.

4.

The tube may be allowed to stand on the bench for 4 hr or centrifuged for 20 min at between 100 and 1,000 g.

5.

Collect fractions using a syringe, or a gradient harvester (Fig. 12.10)(MSE/Fisons). Fractions of 1 ml may be collected into a 24-well plate or 0.1 ml into microtitration plates. Samples should be taken at intervals for cell counting and determination of the density (ρ) of the gradient medium. Density may be measured on a refractometer (Hilger) or density meter (Paar).

6.

Add equal volume of medium to each well and mix (to ensure cells settle to bottom of well). Change the medium to remove the Percoll after 24–48 hr incubation.

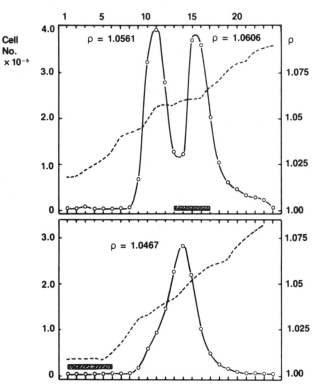

Fig. 12.11 *Incorporation of metrizamide (Nygaard) into cells during isopynic centrifugation. MRC-5 cells (human diploid embryonic lung fibroblasts) were layered in metrizamide-containing medium of the appropriate density in the center of the gradient or in medium alone at the top of the gradient. After centrifugation, the cells were eluted and counted, and samples were taken from each fraction to determine the density [reproduced from Freshney, 1976a, with permission of the publisher]. Slashed bars, position of cells at start.*

Variations. Cells may be incorporated into the gradient during formation by centrifugation. Only one spin is required although spinning the cells at such a high g force may damage them.

Other media. Ficoll is one of the most popular media as it, like Percoll, can be autoclaved. It is a little more viscous at high densities and may cause agglutination of some cells. Metrizamide (Nygaard), a nonionic derivative of metrizoate, which is a radio-opaque-iodinated substance used in radiography (Isopaque, Hypaque, Renografin) and in lymphocyte purification (e.g., Lymphoprep) (see Chapter 23), is less viscous at high densities [Rickwood and Birnie, 1975] but may be incorporated into some cells (Fig. 12.11) as is Isopaque

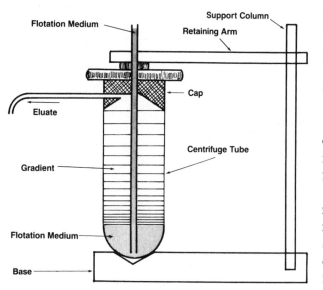

Fig. 12.10. *Gradient harvester (Fisons/M.S.E.). Fluorochemical FC43 is pumped down the inlet tube to the bottom of the gradient and displaces the gradient and cells upward and out through the delivery tube. (After an original design by Dr. G.D. Birnie.)*

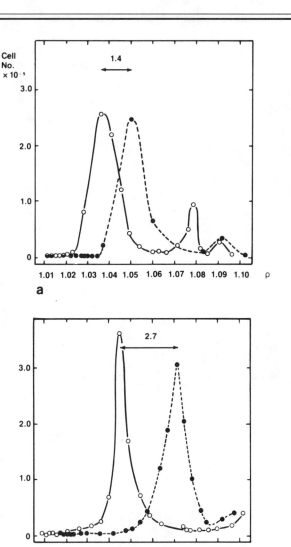

Fig. 12.12. *Sedimentation profiles of HeLa and MRC-5 cells, centrifuged to equilibrium in gradients of metrizamide in culture medium. a,b. Cells taken from log phase of growth. c,d. Cells taken from plateau phase. Gradients in a and c contained no serum; those in b and d, 10% fetal bovine serum. Numbers over the arrows are the differences in density between the peaks, multiplied by 100. Solid circles, HeLa; open circles, MRC-5. [From Freshney, 1976a, with permission of the publisher.]*

[Splinter et al., 1978]. Where such media are used, cells should always be layered on top of the gradient and not mixed in during formation.

Marker beads. Pharmacia manufacture colored marker beads of standard densities which may be used to determine the density of regions of the gradient.

Isopycnic sedimentation is quicker than velocity sedimentation at unit gravity and gives a higher yield of cells for a given gradient volume. It is ideal where clear differences in density exist between cells. Cell density may be affected by the gradient medium (e.g., metrizamide, (Fig. 12.11), by the position of the cells in the growth cycle, and by serum (Fig. 12.12).

This type of separation can be done on any centrifuge, as high g forces are not required, and can even be performed at 1 g.

FLUORESCENCE-ACTIVATED CELL SORTING

This technique [Herzenberg et al., 1976; Kreth and Herzenberg, 1974] operates by projecting a single cell stream through a laser beam in such a way that the light scattered from the cells is detected by a photomultiplier and recorded (Figs. 12.13, 12.14). If the cells are pretreated with a fluorescent stain (e.g., propidium iodide or Chromomycin A_3 for DNA) or flu-

Fig. 12.13. *Fluorescence-Activated Cell Sorter (FACS). This is the Becton Dickinson version of the flow cytophotometer. a. Flow chamber panel on left and computer/readout on right. b. Close-up of flow chamber compartment (see also Fig. 12.14).*

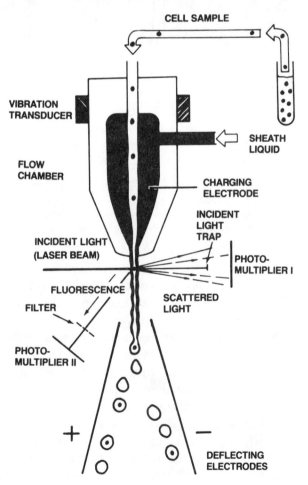

Fig. 12.14. *Principle of operation of flow cytophotometer (see text). [Reproduced from Freshney et al., 1982a, with permission of the publisher.]*

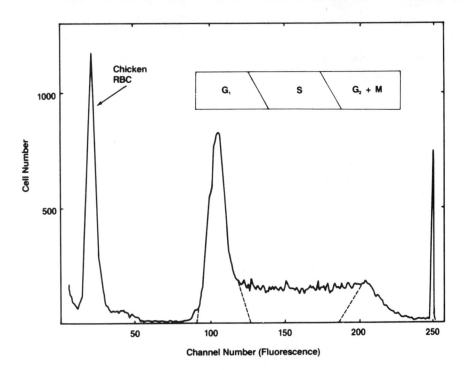

Fig. 12.15. *Printout from FACS II. Friend (murine erythroleukemia) cells were fixed in methanol as a single cell suspension and stained with Chromomycin A₃. The cell suspension was then mixed with chicken erythrocytes, fixed, and stained in the same way, and run through the FACS II. The printout plots cell number on the vertical axis and channel number (fluorescence) on the horizontal axis. Since fluorescence is directly proportional to the DNA per cell, the trace gives a distribution analysis of the cell population by DNA content. The lowest DNA content is found in the chicken erythrocytes, included as a standard. The major peak around channel 100 represents those cells as the GI phase of the cell cycle. Cells around channel 200 are, therefore, G2 and metaphase cells (double the amount of DNA per cell) and cells with intermediate values are in S(DNA synthetic phase) (see also Chapters 13 and 23. Cells accumulated in channel 250 are those where DNA value is off scale or cells which have formed aggregates (by courtesy of Dr. B.D. Young).*

orescent antibody, the fluorescence emission excited by the laser is detected by a second photomultiplier tube. This information is processed and displayed as a two (Fig. 12.15) or three-dimensional graph on an oscilloscope. If specific coordinates are then set to delineate sections of the display, the cell sorter will divert cells with the properties that would place them within these coordinates (e.g., high or low light scatter, high or low fluorescence) into a receiver tube placed below the cell stream. The cell stream is deflected by applying a charge to it as it passes between two oppositely charged plates. The charge is applied briefly and at a set time after the cell has cut the laser beam such that only one cell is deflected into the receiver. A low cell concentration in the cell stream is required such that the gap between cells is sufficient to prevent two cells being deflected together. [(Morasca and Erba, 1986)].

All cells having similar properties will be collected into the same tube. A second set of coordinates may be set and a second group of cells collected simulta-

neously into a second tube by changing the polarity of the cell stream and deflecting the cells in the opposite direction. All remaining cells will be collected in a central reservoir.

This method may be used to separate cells with any differences that may be detected by light scatter (e.g., cell size) or fluorescence (e.g., DNA, RNA, protein, enzyme activity, specific antigens). It is an extremely powerful tool but limited by cell yield (about 10^7 cells is a reasonable maximum) and the very high cost of the instrument (approximately \$100,000). It also requires a full-time, skilled operator.

OTHER TECHNIQUES

The many other techniques which have been used successfully to separate cells are too numerous to describe in detail. They are summarized below and listed in Table 12.1.

Electrophoresis either in a Ficoll gradient [Platsoucas et al., 1979] or by curtain electrophoresis; the second technique is probably more effective and has been used to separate kidney tubular epithelium [Kreisberg et al., 1977].

Affinity chromatography on antibody [Varon and Manthorpe, 1980; Au and Varon, 1979] or plant lectins [Pereira and Kabat, 1979] bound to nylon fiber [Edelman, 1973] or Sephadex (Pharmacia). These techniques appear to be useful for fresh blood cells but less so for cultured cells.

Counter current distribution [Walter, 1975, 1977] has been used to purify murine ascites tumor cells, but the viability may be too low for subsequent culture.

As so many techniques exist, it is difficult for the novice to know where to start. Like so many other areas of investigation, it is best to start with a simple technique such as velocity sedimentation at unit gravity or isopycnic centrifugation. Density gradient analysis may also be used in conjunction with velocity sedimentation in a two-stage fractionation. If there are still problems of resolution or yield, then it may be necessary to employ high-technology methods such as centrifugal elutriation, curtain electrophoresis, or fluorescence-activated cell sorting.

INTRODUCTION

There are four main reasons for characterizing a cell line:

(1) Correlation with the tissue of origin: (a) identification of the lineage to which the cell belongs, (b) position of the cells within that lineage, i.e., the precursor or differentiated status, and (c) whether transformed or not.

(2) Monitoring for instability and variation (see Chapter 17).

(3) Checking for cross contamination (see Chapter 16) and confirmation of species of origin.

(4) Identifying selected sublines or hybrid cell lines requiring demonstration of unique features.

Species Identification

Chromosomal analysis (see below) is the best method of distinguishing between species. Isoenzyme electrophoresis is also a good diagnostic test and is quicker than chromosomal analysis, but requires the appropriate apparatus and reagents (see below). In practice a combination of the two is often used and will give unambiguous results [Hay, 1986].

Lineage or Tissue Markers

Cell surface antigens. These markers are particularly useful in sorting subspecies of lymphocytes, and have also been effective in discriminating epithelium from stroma with antibodies such as anti-EMA [Heyderman et al., 1979] and anti-HMFG 1 and 2 [Burchell et al., 1983] and neuroectodermally derived cells (anti-2AB5) [Dickson et al., 1983] from other germ layer derived cells.

Intermediate filament proteins. These are among the most widely used lineage or tissue markers [Ramackers et al., 1982]. Glial fibrillary acidic protein (GFAP) for astrocytes and desmin for muscle are the most specific, while cytokeratin will mark epithelial cells and mesothelium. Vimentin, though usually re-stricted to mesodermally derived cells *in vivo*, can appear in other cell types *in vitro*.

Differentiated products. Hemoglobin for erythrocytic cells, myosin or tropomyosin for muscle, melanin for melanocytes, and serum albumin for hepatocytes are among the best examples of specific cell type markers, but, like all differentiation markers, depend on the complete expression of the differentiated phenotype.

Enzymes. Three parameters are available in enzymic characterization: (1) the constitutive level (i.e., in the absence of inducers or repressors); (2) the response to inducers and repressors; and (3) isozymic differences (see below). Creatine kinase BB isozyme is characteristic of neuronal and neuroendocrine cells as is neurone specific enolase; lactic dehydrogenase is present in most tissues but in different isozymic forms, and a high level of tyrosine aminotransferase, inducible by dexamethasone, is generally regarded as specific to hepatocytes.

Special functions. Transport of inorganic ions and water is characteristic of some epithelia such that, grown as monolayers, they will produce "domes," blisters in the monolayer caused by transport of water from the medium to the underside of the monolayer. This is found in kidney epithelium and some secretory epithelia. Other specific functions which can be expressed *in vitro* include muscle contraction and depolarization of nerve cell membrane.

Regulation. Although differentiation is usually regarded as an irreversible process, the level of expression of many differentiated products is under the regulatory control of environmental influences such as hormones, matrix, and adjacent cells (see Chapter 14). Hence the measurement of specific lineage markers may require preincubation of the cells in, for example, a hormone such as hydrocortisone, or growth of the cells on collagen of the correct type. Maximum expression of both tyrosine aminotransferase in liver cells and glutamine synthetase in glia require prior induction with dexamethasone. Glutamine synthetase is also repressed by glutamine, so glutamate should be substituted in the medium 48 hr before assay [DeMars, 1957].

Lineage fidelity. Although many of the markers described above have been claimed as lineage markers, they are more properly regarded as tissue or cell type markers, as they are often more characteristic of the function of the cell than its embryologic origin. Cytokeratins occur in mesothelium and kidney epithelium although both of these derive from the mesoderm. Neurone specific enolase and creatine kinase BB are expressed in neuroendocrine cells of the lung although these are now recognized to derive from the endoderm and not from neurectoderm as one might expect of neuroendocrine type cells.

Unique Markers

These include specific chromosomal aberrations, enzymic deficiencies, isozymes, and drug resistance.

Transformation

This will be dealt with in Chapter 15.

MORPHOLOGY

This is the simplest and most direct technique used to identify cells. It has, however, certain shortcomings which should be recognized. Most of these are related to the plasticity of cellular morphology in response to different culture conditions, e.g., epithelial cells growing in the center of a confluent sheet are usually regular, polygonal, and with a clearly defined edge, while the same cells growing at the edge of a patch may be more irregular, distended, and if transformed, may break away from the patch and become fibroblastoid in shape. Subconfluent fibroblasts from hamster kidney or human lung or skin assume multipolar or bipolar shapes and are well spread on the culture surface, but at confluence they are bipolar and less well spread. They also form characteristic parallel arrays and whorls which are visible to the naked eye. Mouse 3T3 cells and human glial cells grow like multipolar fibroblasts at low cell density but become epithelial-like at confluence (Fig. 13.1). Alterations in the substrate [Gospodarowicz, 1978; Freshney, 1980], and the constitution of the medium [Coon and Cahn, 1966] can also affect cellular morphology. Hence, comparative observations should always be made at the same stage of growth and cell density in the same medium, and growing on the same substrate (see Fig. 13.1).

The terms "fibroblastic" and "epithelial" are used rather loosely in tissue culture and often describe the appearance rather than the origin of the cells. Thus a bipolar or multipolar migratory cell, the length of which is usually more than twice its width, would be called "fibroblastic," while a monolayer cell which is polygonal, with more regular dimensions, and which grows in a discrete patch along with other cells, is usually regarded as "epithelial." However, where the identity of the cells has not been confirmed, the terms "fibroblast-like" or "fibroblastoid" and "epithelial-like" or "epithelioid" should be used.

Frequent brief observations of living cultures, preferably with phase-contrast optics, are more valuable than infrequent stained preparations studied at length. They will give a more general impression of the cell's morphology and its plasticity and will also reveal differences in granularity and vacuolation which bear on the health of the culture. Unhealthy cells often become granular and then display vacuolation around the nucleus (see Fig. 10.1).

It is useful to keep a set of photographs for each cell line as a record in case a morphological change is suspected. This record can be supplemented with photographs of stained preparations.

Staining

A polychromatic blood stain, such as Giemsa, provides a convenient method of preparing a stained culture. The recommended procedure is as follows.

Outline
Fix the culture in methanol and stain directly with Giemsa. Wash and examine wet.

Materials
  BSS
  undiluted Giemsa stain
  methanol
  deionized water

Protocol
1.
Remove medium and discard.
2.
Rinse monolayer with BSS, and discard rinse.
3.
Add BSS:methanol, 1:1, 5 ml per 25 cm^2.
4.
Discard 50% methanol/BSS mixture and replace with fresh methanol. Leave for 10 min.
5.
Discard methanol and replace with fresh anhy-

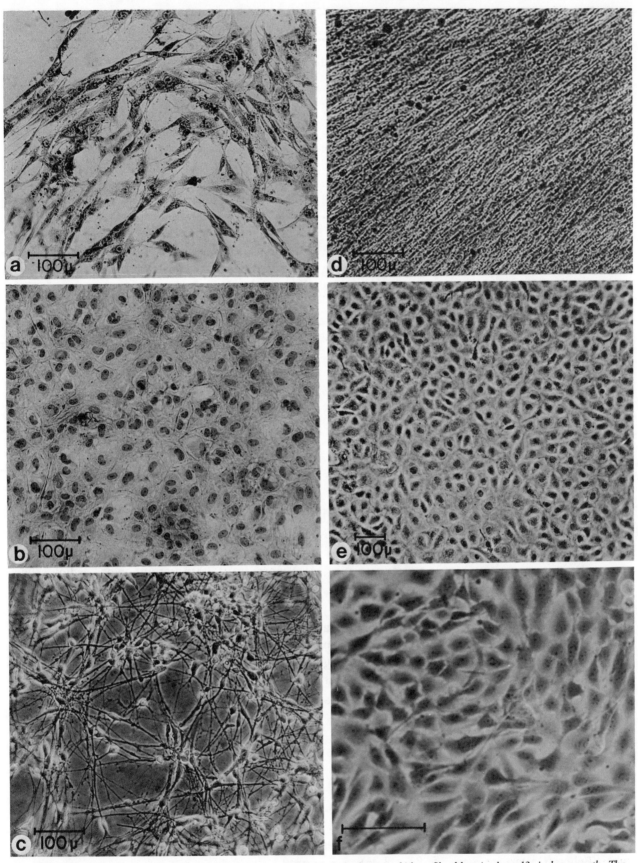

Fig. 13.1. *Examples of variations in cell morphology. a. BHK-21 (baby hamster kidney fibroblasts), clone 13, in log growth. The culture is not confluent, and the cells are well spread and randomly oriented (although some orientation is beginning to appear). b. Cells of an epithelial-like morphology from fetal human intestine (FHI). c. Astrocytes from human astrocytoma. This pattern is quite characteristic but is lost as the cells are passaged, and a morphology not unlike b, e, or f develops. d. Plateau phase BHK-21 C13 cells. The cells are smaller, more highly condensed, and have assumed a parallel orientation with each other. e. Bovine aortic endothelium. Similar regular appearance to b, though cells more closely packed. f. Again a similar pavementlike appearance as found in b and c but now produced by 3T3 cells, mouse fibroblasts. With experience, these cell types may be distinguished, but their similarity underlines the need for criteria for identification other than morphology. a, b, and d, Giemsa stained. c, e, and f, phase contrast. e, by courtesy of Dr. P. Del Vecchio.*

drous methanol, rinse monolayer, and discard methanol.

6.

At this point, the flask may be dried and stored or stained directly. It is important that staining should be done directly from fresh anhydrous methanol even with a dry flask. If the methanol is poured off and the flask is left for some time, water will be absorbed by the residual methanol and will inhibit subsequent staining. Even "dry" monolayers can absorb moisture from the air.

7.

Add neat Giemsa stain, 2 ml per 25 cm^2, making sure the entire monolayer is covered and remains covered.

8.

After 2 min, dilute stain with 8 ml water and agitate gently for a further 2 min.

9.

Displace stain with water so that scum which forms will be floated off and not left behind to coat the cells. Wash vigorously in running tap water until any pink cloudy background stain (precipitate) is removed but stain is not leached out of cells.

10.

Pour off water, rinse in deionized water, and examine on microscope while monolayer is still wet. Store dry and rewet to examine.

Note. Giemsa staining is a simple procedure giving a good high-contrast stain but precipitated stain may have a spotted appearance to the cells. This occurs (1) due to oxidation at the surface of the stain forming a scum, and (2) throughout the solution, particularly on the surface of the slide when water is added. Washing off stain by replacement rather than pouring off or removing slides is designed to prevent slides coming in contact with scum. Vigorous washing at the end is designed to remove precipitate left on the slide.

Culture Vessels for Cytology — Monolayer Cultures

(1) Regular 25 cm^2 flasks or 50 mm petri dishes.

(2) Coverslips (glass or Thermanox (Lux)) in multiwell dishes (see Fig. 7.4), petri dishes, or Leighton tubes (Bellco, Costar; see Figs. 7.5, 7.6, 13.2a).

(3) Microscope slides in 90-mm petri dishes or with attached multiwell chambers (Lab-Tek, Bellco; Fig. 13.2b).

(4) Petriperm dishes (Heraeus), cellulose acetate or polycarbonate filters, Melinex, Thermanox, and Teflon-coated coverslips have all been used for E.M. cytology studies. Some pretreatment of filters or Teflon may be required (e.g., gelatin, collagen, fibronectin, or serum coating; see Chapter 7).

(5) The Gabridge chamber/dish (Bionique, Northumbria Biologicals) can be used with a variety of different plastic membranes or standard glass or plastic coverslips, allowing culture directly on a thin, optically clear, surface for high-resolution microscopy (Fig. 13.2).

Suspension Culture

The following are four ways of preparing cytological specimens from suspension cultures.

Smear (as used in the preparation of blood films).

Materials

Concentrated cell suspension
serum
microscope slides
methanol

Protocol

1.

Place a drop of cell suspension 10^6 cells/ml or more, in 50–100% serum, on one end of a slide.

2.

Dip the end of a second slide into the drop and move it up the first slide, distributing a thin film of cells on the slide (Fig. 13.3).

3.

Dry off quickly and fix in methanol.

Centrifugation. The Cytospin (Shandon) (Fig. 13.4) is a centrifuge with sample compartments specially designed to spin cells down onto a microscope slide.

Materials

Same as for smear preparation.

Protocol

1.

Place approximately 100,000 cells in 250 μl of

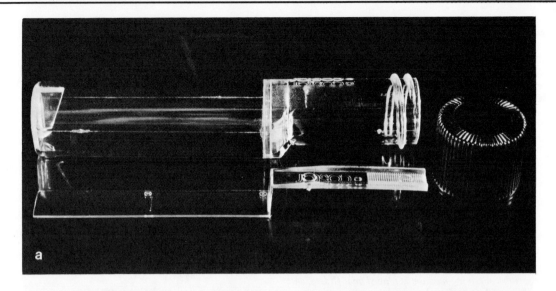

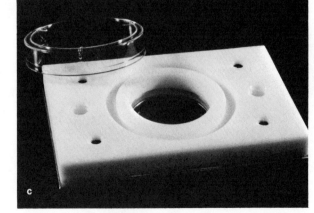

Fig. 13.2. *Culture vessels designed for cytological observation. a. Costar disposable Leighton tube with coverslip with handle for easy retrieval. b. Lab-Tek plastic chambers on regular microscope slide. One, two, four, and eight chambers per slide are available. c. Reusable chamber to take a regular glass coverslip (Bionique) (courtesy of Dr. M. Gabridge).*

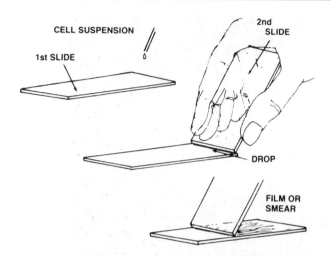

CELL SUSPENSION

1st SLIDE

2nd SLIDE

DROP

FILM OR SMEAR

Fig. 13.3. *Preparing a cell smear. Top: A drop of cell suspension in serum or serum-containing medium is placed on a slide. Center and bottom: A second slide is used to spread the drop.*

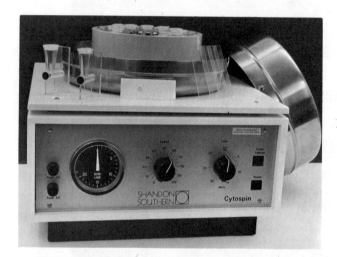

Fig. 13.4. *Shandon Cytospin. Centrifuge for making slide preparations from cell monolayers (photograph by courtesy of Shandon Scientific).*

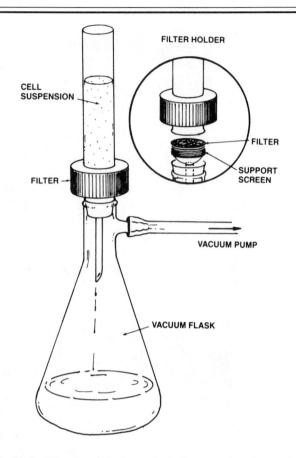

FILTER HOLDER

CELL SUSPENSION

FILTER

FILTER

SUPPORT SCREEN

VACUUM PUMP

VACUUM FLASK

Fig. 13.6. *Filter assembly for cytological preparations (see text).*

medium in the sample block.

2.

Switch on and spin the cells down onto the slide at 100 *g* for 5 min.

3.

Dry off slide quickly and fix in MeOH.

Note. Some centrifuges (e.g., Damon-IEC) have centrifuge buckets designed for preparing cytological preparations (Fig. 13.5). Fixation in this case is performed *in situ*.

Drop Technique. Same as for chromosome preparation (see below, this chapter) but omit Colcemid and hypotonic treatment.

Filtration. This technique is used in exfoliative cytology (see manufacturers' instructions for further details: Gelman, Millipore, Nuclepore).

Fig. 13.5. *Centrifuge carriers for spinning cells onto a slide (Damon-IEC).*

Materials

Filters (e.g., 25-mm Nuclepore, 0.5 μm porosity)

filter holder (Gelman, Nuclepore, Millipore) stand

cell suspension ($\sim 10^6$ cells in 5–10 ml medium with 20% serum)

20 ml BSS

50 ml methanol

vacuum pump or tap siphon

Giemsa stain

mountant (DePex or Permount)

Protocol

1.

Set up filter assembly (Fig.13.6) with 25 mm diameter, 0.5 μm porosity polycarbonate filter.

2.

Draw cell suspension on to filter using a vacuum pump. Do not let all the medium run through.

3.

Add 10 ml BSS gently when cell suspension is down to 2 ml.

4.

Repeat when BSS is down to 2 ml.

5.

Add 10 ml methanol to BSS and repeat until pure methanol is being drawn through filter.

6.

Switch off vacuum before all the methanol runs through.

7.

Lift out filter and air dry.

8.

Stain filter in Giemsa and dry.

9.

Mount on a slide in DePex or Permount by pressing the coverslip down to flatten the filter.

Photography

There are two major frame sizes you may wish to consider: 35mm, best for routine color transparencies and high-volume black and white, and 3½ × 4½ in (12 cm × 9 cm) Polaroid with positive/negative film (type 665) for low-volume black and white. With Polaroid, you obtain an instant result and know that you have the record without having to develop and print a film. The unit cost is high, but there is a considerable saving in time. Polaroid film also has a much shorter shelf-life.

Specific instructions are supplied with microscope cameras, but a few general guidelines may be useful.

1.

Choose the film and set the exposure meter before bringing out the culture.

2.

Make sure the culture is free of debris, e.g., change the medium on a primary culture before photography.

3.

Choose the appropriate field quickly, avoiding imperfections or marks on the flask (always label on the side of the flask or dish and not on the top or bottom). Rinse medium over the upper inside surface if condensation has formed and let it drain down before attempting to photograph.

4.

Focus carefully first the microscope and then the camera eyepiece if there is one.

5.

Turn up the light, check the focus and exposure, and expose; then turn down the light immediately to avoid overheating the culture.

Note. An infrared filter may be incorporated to minimize overheating.

6.

If Polaroid, check exposure on finished print and repeat if necessary at a different setting. If 35 mm, bracket the exposure by rephotographing at half and double the exposure.

7.

Return the culture to the incubator.

8.

Label Polaroid prints immediately. If 35 mm, keep a record against the frame number; otherwise the prints will be difficult to identify.

9.

File photographs in a readily accessible way, e.g., albums or filing cabinet sheets with transparent pockets.

CHROMOSOME CONTENT

Chromosome content is one of the most characteristic and well-defined criteria for identifying cell lines and relating them to the species and sex from which they were derived. See the Committee on Standardized Genetic Nomenclature for Mice [1972] for mouse karyotype; Committee for a Standardized Karyotype of Rattus norvegicus [1973] for rat, and Paris Conference [1975], or An International System for Human Cytogenetic Nomenclature [1978] for human. There is also an Atlas of Mammalian Chromosomes [Hsu and Benirscnke, 1967], although it predates chromosome banding (see below). Chromosome analysis can also distinguish between normal and malignant cells as the chromosome number is more stable in normal cells (except in mice where the chromosome complement of normal cells can change quite rapidly after explantation into culture).

Chromosome Preparations [Rothvels and Siminovitch, 1958; Worton and Duff, 1979]

Outline

Cells arrested in metaphase and swollen in hypotonic medium are fixed and dropped on a slide, stained, and examined (Fig.13.7).

Materials

Cultures of cells in log phase
10^{-5}M colcemid in BSS
PBSA
0.25% crude trypsin
centrifuge tubes
centrifuge
hypotonic solution:
    0.04M KCl
    0.025 M sodium citrate
fixative:
    1 part glacial acetic acid plus 3 parts
    anhydrous methanol or ethanol,
    made up fresh and kept on ice
vortex mixer
Pasteur pipettes
ice
slides
Giemsa stain
slide dishes
00 coverslips
DPX or Permount

Protocol

1.
Set up 75-cm^2 flask culture at between 2 and 5 × 10^4 cells/ml (4 × 10^3–10^4 cells/cm^2) in 20 ml.

2.
Approximately 5 d later when cells are in the log phase of growth, add colcemid (10^{-7} M, final) to the medium already in the flask.

3.
After 4–6 hr, remove medium gently, add 5 ml

0.25% trypsin, and incubate 10 min.

4.
Centrifuge cells in trypsin and discard supernatant trypsin.

5.
Resuspend the cells in 5 ml of hypotonic solution and leave for 20 min at 36.5°C.

6.
Add equal volume of freshly prepared ice-cold acetic methanol, mixing constantly, and then centrifuge at 100 g for 2 min.

7.
Discard the supernatant mixture, "buzz" the pellet on a vortex mixer, and slowly add fresh acetic methanol with constant mixing.

8.
Leave 10 min on ice.

9.
Centrifuge for 2 min at 100 g.

10.
Discard supernatant acetic methanol and resuspend pellet with "buzzing" in 0.2 ml acetic methanol, to give a finely dispersed cell suspension.

11.
Draw one drop into the tip of a Pasteur pipette and drop on a cold slide. Let the drop run down the slide as it spreads.

12.
Dry off rapidly over a beaker of boiling water and examine on microscope. If cells are evenly spread and not touching, prepare more slides at same cell concentration. If piled up and overlapping, dilute two- to fourfold and make a further drop preparation. If satisfactory, prepare more slides. If not, dilute further and repeat.

13.
Stain with Giemsa: (a) Immerse slides in neat stain for 2 min; (b) place dish in sink and add approximately 10 V water, allowing surplus stain to overflow from top of slide dish; (c) leave for a further 2 min; (d) displace remaining stain with running water and finish by running slides individually under tap to remove precipitated stain (pink, cloudy appearance on slide); and (e) check staining under microscope. If satisfactory, dry slide thoroughly and mount 00 coverslip in DePex or Permount.

Chromosome Banding

This group of techniques [see Yunis, 1974] was devised to enable individual chromosome pairs to be

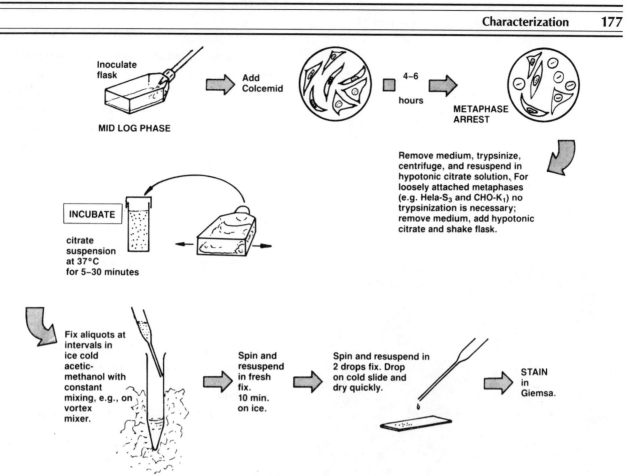

Fig. 13.7. *Chromosome preparation.*

identified where there is little morphological difference between them [Wang and Fedoroff, 1972, 1973]. For Giemsa banding, the chromosomal proteins are partially digested by crude trypsin, producing a banded appearance on subsequent staining. Trypsinisation is not required for quinacrine banding. The banding pattern is characteristic for each chromosome pair (Fig. 13.8).

The following protocols for chromosome banding have been contributed by Marie Ferguson-Smith, Department of Medical Genetics, Yorkhill Hospital, Glasgow.

Trypsin/Giemsa Banding

Materials

    Sterile saline
    SSC (17.538 g NaCl+8.823 g Na Citrate in 1 l Dist. H_2O)
    1 vial Difco Bacto trypsin
    50%, 95% and 100% ethanol
    Giemsa stain
    pH 6.8, 0.025 M phosphate buffer

    xylene
    DPX

Protocol

1.
Place fixed chromosome spreads in SSC overnight.
2.
Transfer to saline for 10 min.
3.
Place in 50% ethanol for 1 min, 95% ethanol for 1 min, and 100% ethanol for 1 min.
4.
Leave to air dry.
5.
Trypsin is rehydrated with 10 ml sterile saline and diluted to make 1% trypsin working solution and then cooled to 15°C in the refrigerator.
6.
Add trypsin to one slide or coverslip preparation for 4½ min (15°C).
7.
Change to saline for 1 min.
8.
Giemsa (60 ml Giemsa + 240 ml buffer, pH 6.8) for 5½ min.

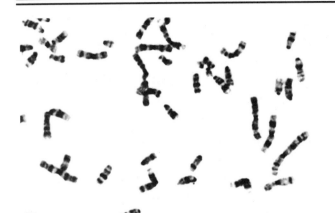

a

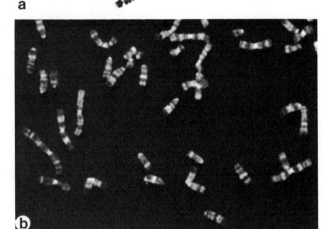

b

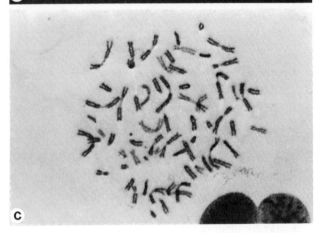

c

Fig. 13.8. *Chromosome staining. a. Human chromosomes banded by standard trypsin-Giemsa technique. b. Same preparation as a stained with Hoechst 33258. c. Human/mouse hybrid stained with Giemsa at pH 11. Human chromosomes are less intensely stained than mouse. Several human/mouse chromosomal translocations can be seen (By courtesy of Dr. R.L. Church).*

9.
Wash twice in buffer (pH 6.8) for 1 min each.
10.
Allow preparation to air dry.
11.
Examine unmounted on microscope: (a) If understained, return to stain for a further 1–5 min. (b)

Select trypsin treatment giving clearest banding. If no banding is apparent, increase trypsin treatment (duration, temperature, or concentration). If chromosomes are fuzzy or all of the protein is digested out leaving "ghosts," reduce trypsin treatment. (c) Trypsinize and stain remaining slides. (Preparations may be destained by soaking in methanol for 10 min, dried, and re-exposed to trypsin if necessary.)
12.
Immerse in xylene before mounting cell side down on glass microscope slides in DPX. Two coverslips can be mounted on the same slide.

Q-banding

Quinacrine mustard or dihydrochloride stains the interband regions unstained in the Giemsa technique.

Materials

Q-banding buffer: 0.1 M citrate, 0.2 M $Na_2 HPO_4$, pH 5.6
stain: 0.5% quinacrine dihydrochloride in Q-banding buffer

Protocol
1.
Hydrate slides in 100%, 70%, 50%, and 20% ethanol for 2 min.
2.
Place in buffer for 5 min.
3.
Place in stain for 20 min.
4.
Place in three changes of buffer for 1, 1, and 5 min.
5.
Drain off buffer and mount in distilled water. Seal edges of coverslip.
6.
Observe under uv light at BP330–500 nm excitation and (BG12) and LP520 emission, with a dichroic mirror, DS510.

Analysis. 1. Count chromosome number per spread for between 50 and 100 spreads (need not be banded). Closed-circuit television or a camera lucida attachment may help. You should attempt to count all the mitoses that you see and classify them (a) by chromosome number or (b), if counting is impossible, as "near diploid uncountable" or "polyploid uncountable." Plot the results as a histogram (see Fig. 12.2).

2. Prepare karyotype. Photograph about 10 or 20 good spreads of banded chromosomes and print on 20 × 25 cm high-contrast paper. Cut out the chromosomes, sort into sequence, and stick down on paper (see Fig. 13.9, 13.10).

Variations. *Metaphase block.* (1) Vinblastine, 10^{-6}M, may be used instead of colcemid. (2) Duration of the metaphase block may be increased to give more metaphases for chromosome counting, but chromosome condensation will increase, making banding very difficult.

Collection of mitosis by "shake-off" technique. Some cells, e.g. CHO and HeLa, detach readily when in metaphase. This allows trypsinization for collection of metaphases to be eliminated. (1) Add colcemid. (2) Remove carefully and replace with hypotonic citrate/ KCl. (3) Shake the flask to dislodge cells in metaphase either before or after incubation in hypotonic medium. (4) Fix as before.

Hypotonic treatment. Substitute 0.075 M KCl alone or HBSS diluted to 50% with distilled water. Duration of hypotonic treatment may be varied from 5 min to 30 min to reduce lysis or increase spreading.

Spreading. There are perhaps more variations at this stage than any other, designed to improve the degree and flatness of the spread. They include: (1) dropping cells onto slide from a greater height. Clamp the pipette and mark the position for the slide using a trial run with fixative alone. (2) Flame drying. Dry slide after dropping cells by heating over a flame or actually burn off the fixative by igniting the drop on the slide as it spreads (this may make banding more difficult later). (3) Ultracold slide. Chill slide on solid CO_2 before dropping on cells. (4) Refrigerate fixed cell suspension overnight before dropping. (5) Drop cells on a chilled slide (e.g., steep in cold alcohol and dry off), then place over a beaker of boiling water. (6) Tilt slide or blow drop across slide as it spreads.

Banding. (1) Giemsa-banding: use trypsin + EDTA rather than trypsin alone. (2) Q-banding [Caspersson et al., 1968]: stain in 5% (w/v) quinacrine dihydrochloride in 45% acetic acid, rinse, and mount in deionized water at pH 4.5 [Lin and Uchida, 1973; Uchida and Lin, 1974]. (3) C-banding: this technique emphasizes the centromeric regions. The fixed preparations are pretreated for 15 min with 0.2 N HCl, 2 min with 0.07 N NaOH, and then treated overnight with SSC (either 0.03 M sodium citrate, 0.3 M NaCl, or 0.09 M sodium

citrate, 0.9 M NaCl) before staining with Giemsa stain [Arrighi and Hsu, 1974].

Techniques have been developed for discriminating between human and mouse chromosomes, principally to aid the karyotypic analysis of human and mouse hybrids. These include fluorescent staining with Hoechst 33258, which causes mouse centromeres to fluoresce more brightly than human [Lin et al., 1974; Hilwig and Gropp, 1972] and alkaline staining with Giemsa ("Giemsa-11") [Bobrow et al., 1972; Friend et al., 1976].

Chromosome counting and karyotyping will allow species identification of the cells and, when banding is used, will distinguish cell line variation and marker chromosomes. Banding and karyotyping is time consuming and chromosome counting with a quick check on gross chromosome morphology may be sufficient to confirm or exclude a suspected cross contamination.

DNA CONTENT

The amount of DNA per cell is relatively stable in normal cell lines such as human fibroblasts and glia, and chick and hamster fibroblasts, but varies in cell lines from the mouse and from many neoplasms. DNA can be measured by microdensitometry of Feulgen-stained cells [Pearse, 1968] or by ethidium bromide fluorescence and microphotometry. The advent of flow cytophotometry (flow cytofluorimetry, fluorescence-activated cells sorting; see Chapter 12) has made the assay of DNA per cell much more quantitative and reproducible (see Fig. 12.15; see also Chapter 18).

RNA AND PROTEIN

Histochemical reactions and flow cytophotometry also enable measurement of RNA and protein per cell [Morasca and Erba, 1986]. These are prone to considerable fluctuations, but in some cases, the ratio of RNA: DNA or protein: DNA may be found characteristic of the cell type if measured under standard culture and assay conditions.

Qualitative analysis of total cell protein (see Chapter 18) will reveal differences between cells when whole cell, or cell membrane extracts, are run on two-dimensional gels [O'Farrell, 1975]. This produces a characteristic "fingerprint" similar to polypeptide maps of protein hydrolysates, but contains so much information that interpretation can be difficult. Labeling the cells with [^{32}P], [^{35}S] methionine, or a combination of ^{14}C-labeled amino acids, followed by autoradiography, may make analysis easier, but it is not a technique

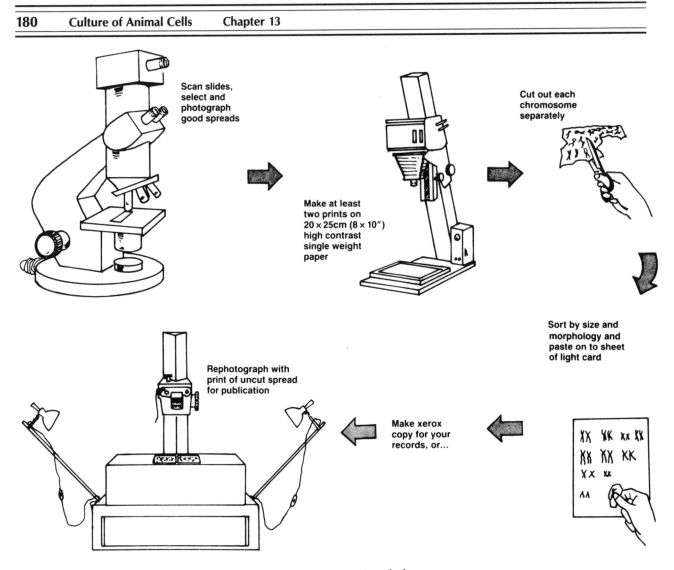

Scan slides, select and photograph good spreads

Make at least two prints on 20 × 25cm (8 × 10″) high contrast single weight paper

Cut out each chromosome separately

Sort by size and morphology and paste on to sheet of light card

Rephotograph with print of uncut spread for publication

Make xerox copy for your records, or...

Fig. 13.9. *Steps in the preparation of a karyotype.*

suitable for routine use unless the technology for preparing 2D gels is currently in use in the laboratory.

ENZYME ACTIVITY

Specialized functions *in vivo* are often expressed in the activity of specific enzymes, e.g., urea cycle enzymes in liver, alkaline phosphatase in endothelium. Unfortunately, many enzyme activities are lost *in vitro* for the reasons discussed in Chapter 2 and are no longer available as markers of tissue specificity. Liver parenchyma loses arginase activity within a few days and cell lines from endothelium lack high alkaline phosphatase activity. However, some cell lines do express specific enzymes such as tyrosine aminotransferase in the rat hepatoma HTC cell lines [Granner et al., 1968]. When looking for specific marker enzymes, the constitutive (uninduced) level and the induced level should be measured and compared with a

number of control cell lines. Glutamyl synthetase activity, for instance, characteristic of astroglia in brain is increased severalfold when the cells are cultured in the presence of glutamate instead of glutamine [DeMars, 1957].

Induction of enzyme activity will require specialized conditions for each enzyme and these may be obtained from the literature. Common inducers are glucocorticoid hormones such as dexamethasone, polypeptide hormones such as insulin and glucagon, or alteration in substrate or product concentrations in the medium, as in the example above with glutamyl synthetase.

Isoenzymes

Enzyme activities can also be compared qualitatively between cell strains, as many enzyme activities are expressed by different, though related, species of

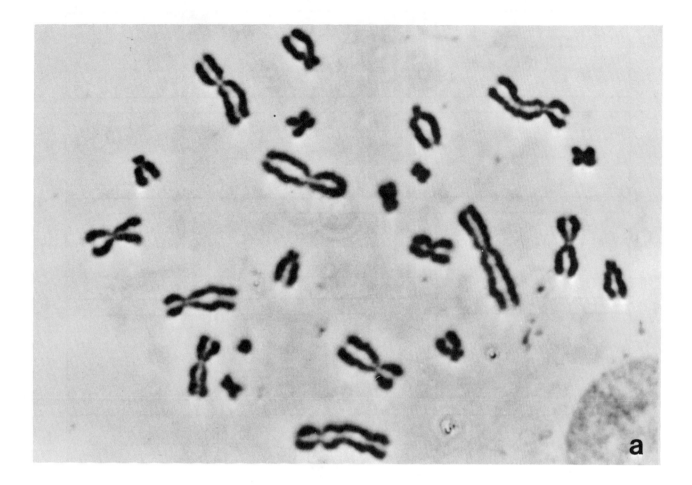

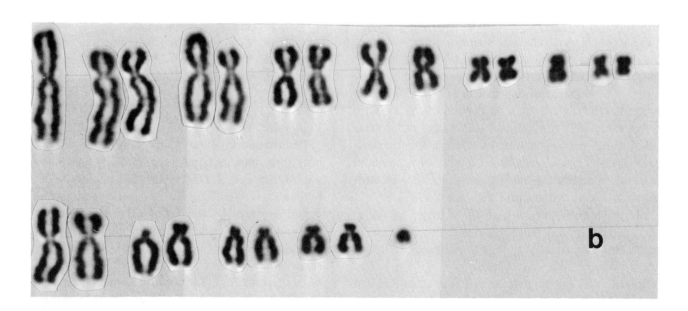

Fig. 13.10. *Example of karyotype. Chinese hamster cells recloned from the Y-5 strain of Yerganian and Leonard [1961] (Acetic-orcein stained).*

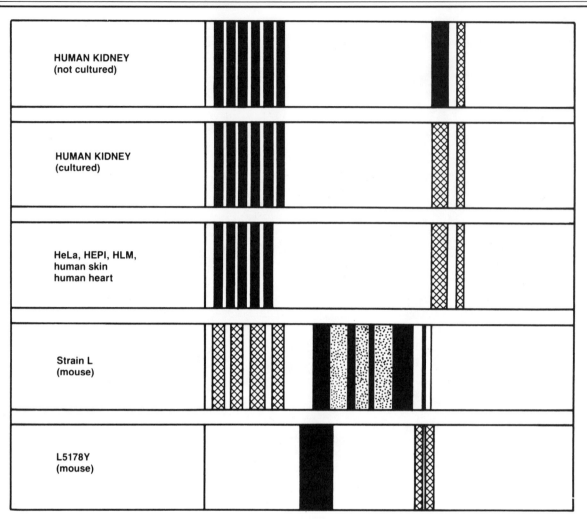

Fig. 13.11. *Isozyme patterns revealed by electrophoresis. Starch gel zymogram for esterase [from Paul and Fottrell, 1961].*

molecules. These so-called isoenzymes or isozymes may be separated chromatographically or electrophoretically and the distribution patterns (zymograms) found to be characteristic of species or tissue. Paul and Fottrell, [1961] demonstrated differences in esterase zymograms between normal and malignant mouse cells, and human cells (Fig.13.11), and O'Brien and Kleiner [1977] have described a number of very useful isozymic markers for human cell lines after electrophoresis of cell extracts [see also Macy, 1978].

Electrophoresis media include agarose, cellulose acetate, starch, and polyacrylamide. In each case, a crude enzyme extract is applied to one point in the gel and a potential difference applied across the gel. The different isozymes migrate at different rates and can be detected later by staining with chromogenic substrates. Stained gels may be read directly by eye and photographed, or scanned with a densitometer.

The following protocol for isoenzyme analysis has been contributed by Marvin L. Macy, American Type Culture Collection, 12301 Park Lawn Drive, Rockville, MD.

Principle

Determination of the electrophoretic mobility of only three enzymes, purine nucleoside phosphorylase (NP), E.C.2.4.2.1; glucose-6-phosphate dehydrogenase (G6PD), E.C.1.1.1.49; and lactate dehydrogenase (LDH), E.C.1.1.1.27, is adequate to provide an effective means for identification of species of origin of cultured cells in the majority of instances [Montes de Oca et al., 1969]. Cell lines representing 37 taxonomic groups are easily identifiable from each other by comparison of their NP, G6PD, and LDH mobility differences as determined by vertical starch gel electrophoresis.

Outline

Harvest cells, wash in PBS, resuspend at 5×10^7 cells/ml and prepare crude extract. Recover supernatant, aliquot, and store at $-70°C$. Prepare vertical starch gel apparatus, add cell extract, and electrophorese. Slice gel horizontally, apply enzyme stains, incubate, wash, and examine.

Materials

> Starch gel electrophoresis apparatus (Haake Buchler)
> electrostarch (Electrostarch)
> DC power supply
> gel slicing device (Haake Buchler)
> Eppendorf tubes, 1.5 ml
> microcentrifuge, Eppendorf (Brinkmann)
> stirring hot plate
> staining boxes (Haake Buchler)
> Hamilton syringe, 50 μl
> gloves, Zetex
> petrolatum (Petroleum jelly, Vaseline)
> TEB chamber buffer: 0.18 M tris, 4 mM Na$_2$ EDTA, 0.1 M boric acid, pH 8.6, 1 l. For LDH and G6PDH add 4 ml 5 mM NADP, mix, and place at 4°C

Protocol

1.

Harvest cells, resuspend in PBS, and count viable cells.

2.

Centrifuge at 300 g for 5–10 min to pellet cells and decant supernate. Cell pellet may be stored at $-70°C$ at this point.

3.

Resuspend at 5×10^7 cells/ml in a 1:15 mixture of Triton X-100 and 0.9% NaCl solution (pH 7.1) containing 6.6×10^{-4} M Na$_2$ EDTA. Aspirate the solution using a small bore pipette for several minutes until the cell membranes clump together.

4.

Transfer the homogenate to Eppendorf tubes (1.5 ml) and spin at top speed (8,733 g) for 2 min in a microfuge. Recover the supernate, dispense in desired volume aliquots, and store at $-70°C$.

5.

In a 1,000-ml filtering flask add 60 g electrostarch in 500 ml 0.1 $\times$ TEB buffer, heat over burner with shaking (wear gloves) until mixture thickens and begins to bubble (90°C). Apply a vacuum for about 1 min to completely degas, seen by lack of solution rise and removal of gas bubbles. Remove vacuum and allow gel to cool to 60°C with continuous swirling. For LDH and G6PD add 1 ml 5 mM NADP and mix.

6.

Pour into gel mold ($260 \times 160 \times 6$ mm) and allow to cool for 1–3 min. Place glass plate over top and seal edges with molten starch.

7.

Leave at room temperature for 30–60 min then place at 4°C. Gel is ready to use in 4 hr and is good for 24 hr.

8.

Thaw cell extracts and spin in microfuge for 1 min. Melt about 100 ml of petrolatum. Gently remove slot former and apply 40-μl sample in each slot. Cover sample chamber with melted petrolatum and allow to harden (5 min).

9.

Remove end pieces of gel mold and place the mold into the bottom electrode chamber and secure. Add 600 ml of cold chamber buffer to cathode, top chamber, and 400 ml to anode, bottom chamber. Connect the output terminal electrodes from the power supply to the upper (black) and lower (red) electrode chamber leads. Run the gel at 160 V for 16–18 hr at 4°C.

10.

After electrophoresis, remove gel from mold, cut off and discard 5 cm from both ends, slice the gel horizontally into three equal sections using the gel cutting device, separate, and place cut side up in staining boxes.

11.

Each stain solution is mixed with 25 ml of boiled 2% noble agar cooled to 45–50°C. For NP, combine 20 ml H$_2$O, 5 ml 0.1 M NaH$_2$PO$_4$.H$_2$O, 50 mg inosine, 0.1 ml xanthine oxidase (100 mg/ml), 1 ml MTT (5 mg/ml), and 2 ml PMS (1 mg/ml). For G6PD, combine 5 ml H$_2$O, 5 ml 0.5 M Tris pH 7.5, 5 ml 0.025 M glucose-6-phosphate, 5 ml 0.1 M MgCl$_2$, 5 ml 0.005 M NADP, 1 ml MTT, and 2 ml PMS. For LDH, combine 10 ml H$_2$O, 5 ml 0.5 M Tris pH 7.5, 5 ml 1.0 M Na lactate, 5 ml NAD (10 mg/ml), 1 ml MTT, and 2 ml PMS.

12.

Cover the entire gel surface with the individual noble agar-stain mixtures and incubate at 37°C in the dark. Most enzyme bands appear in 1–3 hr.

After development carefully wash off the agar-stain mixture with cold tap water.

Analysis

Record results obtained both in terms of number and intensity of enzyme bands with reference to the distance of band migration from the point of origin for each sample.

Variations

Cell extracts can be prepared by ultrasonication, freezing and thawing rapidly three times, or treatment with octyl alcohol [Macy, 1978].

Isoenzyme analysis of polymorphic gene-enzyme systems in cell cultures of human origin can also be performed for intraspecies identification [Povey et al., 1976; O'Brien et al., 1977]. The resolution of these allelic isoenzymes yields an allozyme genetic signature and provides a phenotype frequency product as a means of identifying individual human cell lines [O'Brien et al., 1980; Wright et al., 1981].

Polymorphic isoenzymes have also been reported for cell lines of murine origin as a means of identifying inbred strains of mice [Nichols and Ruddle, 1973; Krog, 1976].

Isoenzyme analysis for species identification can also be performed using an agarose gel system [Halton, 1983]. A complete kit for species identification via isoenzyme analysis using an agarose gel system is available from Innovative Chemistry.

Isozyme analysis has proved to be of great value in determining the chromosomal constitution of somatic cell hybrids (see Chapter 23) [Harris and Hopkinson, 1976; Nichols and Ruddle, 1973; Meera Khan, 1971; Van Someren et al., 1974].

ANTIGENIC MARKERS

This area has expanded more than any other in recent years due largely to the availability of affinity purified and monoclonal antibodies, many of which are now supplied by commercial companies. Regardless of the source of the antibody it is essential to be certain of the specificity of the antibody by using appropriate control material. This is true for monoclonal antibodies and polyclonal antisera alike; a monoclonal antibody is highly specific for a particular epitope but the generality or specificity of the expression of the epitope must still be demonstrated.

Immunological staining may be direct, i.e., the specific antibody is itself conjugated to a fluorochrome such as fluorescein or rhodamine or to horseradish peroxidase and used to stain the specimen directly, followed by direct observation on a fluorescence microscope or further development with a chromogenic peroxidase substrate and observation on a regular microscope. Alternatively, an indirect method may be used where the primary (specific) antibody is used in its native form to bind to the antigen in the specimen and this is followed by treatment with a second antibody, raised against the immunoglobulin of the first antibody. The second antibody may be conjugated to a fluorochrome [Coons and Kaplan, 1950; Kawamara, 1969] or peroxidase for subsequent visualisation [Avrameas, 1970; Taylor, 1978].

Various methods have been employed to enhance the sensitivity of detection of these methods, particularly the peroxidase linked methods. The commonest of these is the PAP or peroxidase-antiperoxidase technique, where a further tier is added by reacting with a peroxidase complex containing antibody from the same species as the primary antibody [Sternberber, 1974]. This is bound to the free valency of the second antibody. Even greater sensitivity has been obtained by using a biotin-conjugated second antibody with a streptavidin complex carrying peroxidase or alkaline phosphatase (Amersham) or gold-conjugated second antibody with subsequent silver intensification (Janssen).

Indirect Immunofluorescence

Outline

Fix cells and treat sequentially with first and second antibodies. Examine by uv light.

Materials

Culture grown on glass coverslip, slide, or polystyrene

freshly prepared fixative: 5% acetic acid in ethanol—place at -20°C

primary antibody diluted 1:100–1:1000 in culture medium with 10% FBS

second antibody raised against the species of the first, e.g., if first antibody was raised in rabbit then the second should be from a different species, e.g., goat anti-rabbit immunoglobin; the second antibody should be conjugated to fluorescein or rhodamine

HBSS or PBS without phenol red

mountant: 50% glycerol in PBS or HBSS without phenol red

Protocol

1.

Wash coverslip with cells in HBSS or PBS and place in suitable dish, e.g. 24-well plate for 13 mm coverslip.

2.

Place at -20°C for 10 min and add cold fixative for 20 min.

3.

Add 1 ml normal swine serum and leave at room temperature for 20 min.

4.

Rinse in PBS or HBSS, drain on tissue, and place inverted on a 50µl drop of diluted primary antibody. Place at 37°C for 30 min, room temperature for 1–3 h or overnight at 4°C. For the last, antibody may be diluted 1:1000.

5.

Rinse in PBS or HBSS and transfer to second antibody diluted 1:20 for 20 min at 37°C.

6.

Rinse in PBS or HBSS and mount in 50% glycerol in PBS with fluorescence bleaching retardant, e.g., Citifluor AFI.

7.

Examine on fluorescence microscope.

Variations. For cell surface or particularly fixation sensitive antigens, treat with antibodies first, then postfix as above. Where a glass substrate is used, cold acetone may be substituted for acid ethanol.

Indirect Peroxidase. Substitute peroxidase conjugated antibody at stage 5 and then transfer to peroxidase substrate (diaminobenzidine stained preparations can be dehydrated and mounted in DPX but ethyl carbazole must be mounted in glycerol as above).

PAP. Use unconjugated second antibody at stage 5. and then transfer to diluted PAP complex (1:100) in PBS or HBSS for 20 min. Rinse and add peroxidase substrate as above, incubate, wash and mount.

Specific cell surface antigens are usually stained in living cells (at 4°C in the presence of sodium azide to inhibit pinocytosis) while intracellular antigens are stained in fixed cells, sometimes requiring light trypsinization to permit access of the antibody to the antigen.

HLA and blood group antigens can be demonstated on many human cell lines, and serve as useful characterization tools, especially where the donor patient profile is known [Espmark et al., 1978; Pollack et al., 1981; Stoner et al., 1980]

DIFFERENTIATION

Many of the characteristics described under antigenic markers or enzyme activities may also be regarded as markers of differentiation and as such they can help to correlate cell lines with their tissue of origin. Other examples of differentiation, and as such highly specific markers of cell line identity, are given in Table 2.2, and the appropriate assays for these properties may be derived from the the references cited (see also Chapter 20).

While much of the interest in characterization is related to the study of specialized functions and their relationship to the cells behavior *in vivo*, these techniques are also important in confirming the identity of a cell line and excluding the possibility of cross contamination.

Chapter 14
Induction of Differentiation

As discussed in Chapter 2, when cells are cultured and propagated as a cell line, the resultant cell phenotype is often different from the characteristics predominating in the tissue from which it was derived. This is due to several factors, many of them as yet undefined, which regulate the geometry, growth, and function *in vivo* but which are absent from the tissue culture environment. Before considering these systematically. it is first necessary to state what is meant by *differentiation* in this context. The term will be used to define the process leading to the expression of phenotypic properties characteristic of the functionally mature cell *in vivo*. It does not imply that the process is complete or that it is irreversible. There will be processes such as the cessation of DNA synthesis in the erythroblast nucleus which are not normally reversible, but, for simplicity, no attempt will be made to distinguish these from reversible processes such as the induction of albumin synthesis in hepatocytes which is lost under certain circumstances but may be reinduced.

Differentiation will be used to describe the combination of constitutive and adaptive properties found in the mature cell. *Commitment*, on the other hand, will imply an irreversible progression from a stem cell to a particular defined lineage endowing the cell with the potential to express a limited repertoire of properties either constitutively or when induced to do so.

Terminal differentiation implies that a cell has progressed down a particular lineage to a point at which the mature phenotype is fully expressed and beyond which the cell cannot progress. In principle this need not exclude cells which can revert to a less differentiated phenotype and resume proliferation, such as a fibrocyte, but in practice the term tends to be reserved for cells like neurons, skeletal muscle, or keratinized squames where differentiation is irreversible.

Dedifferentiation has been used to describe the loss of the differentiated properties of a tissue when it becomes malignant or when it is grown in culture. As these are complex processes with several contributory factors including cell death, selective overgrowth, adaptive responses, and loss of certain phenotypic properties, the term should be used with great caution, or not at all. When used correctly dedifferentiation means the loss by a cell of the specific phenotypic properties associated with the mature cell. When it occurs, this is most probably an adaptive process implying that the differentiated phenotype may be regained given the right inducers (see also Chapter 2).

STAGES OF COMMITMENT AND DIFFERENTIATION

There are two main pathways to differentiation in the adult organism (Fig.14.1). Typically a small population of totipotent or pluripotent undifferentiated stem cells gives rise to committed precursor cells which will progress towards terminal differentiation, losing their capacity to divide as they reach the terminal stages. This gives rise to a fully mature, differentiated cell which will normally not divide. Alternatively, cells such as fibrocytes may respond to a local reduction in cell density and/or the presence of one or more growth factors by losing some of their differentiated properties, e.g., collagen synthesis, and re-entering the cell cycle. When the tissue has regained the appropriate cell density by division, cell proliferation stops and differentiation is reinduced.

The first process is used where continual renewal is required, such as in the hemopoietic system, the skin, and the gastric mucosa, while the second is used where regeneration is not continual and requires quick mounting in response to trauma, such as in wound repair or liver regeneration. In the first, amplification is possible by having several cell divisions at each precursor stage but takes longer than the rapid recruitment of a higher proportion of the total cell population into a limited number of divisions available in the second.

PROLIFERATION AND DIFFERENTIATION

As differentiation progresses, cell division is reduced and eventually lost. In most cell systems cell proliferation is incompatible with the expression of

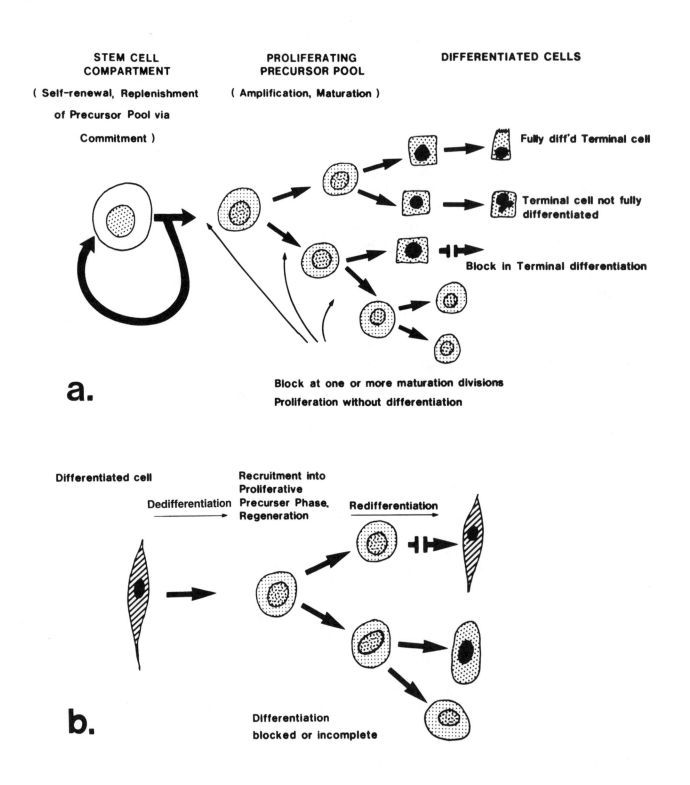

STEM CELL
COMPARTMENT

(Self-renewal, Replenishment

of Precursor Pool via

Commitment)

PROLIFERATING
PRECURSOR POOL

(Amplification, Maturation)

DIFFERENTIATED CELLS

Fully diff'd Terminal cell

Terminal cell not fully
differentiated

Block in Terminal differentiation

a.

Block at one or more maturation divisions
Proliferation without differentiation

Differentiated cell

Dedifferentiation

Recruitment into
Proliferative
Precurser Phase,
Regeneration

Redifferentiation

b.

Differentiation
blocked or incomplete

Fig. 14.1. *Alternative pathways for cell differentiation. a. Generation from stem cell b. Generation by recruitment from differentiated cells. The blocks refer to points at which expression of differentiation may be inhibited in vitro, but could apply equally to blocks found in neoplasia. (Reproduced from Freshney, 1985, by permission of the publisher).*

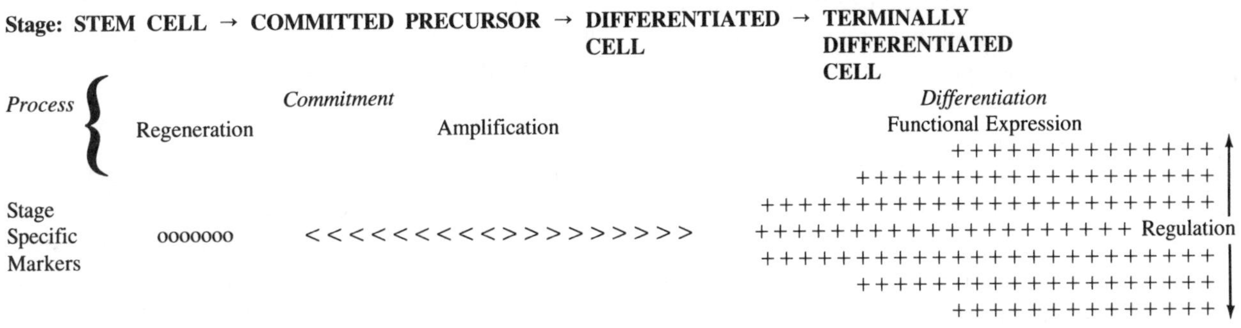

Stage: STEM CELL → COMMITTED PRECURSOR → DIFFERENTIATED → TERMINALLY
                                                      CELL   DIFFERENTIATED
                                                                              CELL

Process {

                    Commitment   *Differentiation*

Regeneration   Amplification   Functional Expression

+ + + + + + + + + + + + + ↑
    + + + + + + + + + + + + + + +
  + + + + + + + + + + + + + + + + + +

Stage
Specific   ooooooo   < < < < < < < < < > > > > > > > > >   + Regulation
Markers
  + + + + + + + + + + + + + + + + + +
    + + + + + + + + + + + + + + +
      + + + + + + + + + + + + + ↓

Fig. 14.2. *Stages in cell differentiation. Stem cells are capable of regeneration without commitment or commitment division leading to a precursor cell which is still capable of cell proliferation and hence amplification of the lineage. Successive divisions lead to further differentiation, in the presence of the appropriate inducer environment, until, finally, a stage may be reached where no further division is possible. This is terminal differentiation. Phenotypic expression may be regulated quantitatively at terminal differentiation from the constitutive to the induced state by the action of hormones, metabolites, etc.*

differentiated properties. Tumor cells can sometimes break this restriction, and in melanoma, for example, melanin continues to be synthesized while the cells are proliferating. Even in these cases, however, synthesis of the differentiated product increases when division stops.

There are severe implications for this relationship in culture, where the major objective for many years has been the propagation of cell lines and the production of large numbers of cells for biochemical or molecular analysis. Where the incompatibility of differentiation and proliferation is maintained, it is not surprising to find that the majority of cell lines do not express fully differentiated properties.

This fact was noted many years ago by the exponents of organ culture (see Chapter 22), who set out to retain three-dimensional, high cell density tissue architecture and prevent dissociation and selective overgrowth of undifferentiated cells. However, although of considerable value in elucidating cellular interactions regulating differentiation, organ culture has always suffered from the inability to propagate large numbers of identical cultures, particularly if large numbers of cells are required, and the heterogeneity of the sample, assumed to be essential for the maintenance of the tissue phenotye has, in itself, made the ultimate biochemical analysis of pure cell populations, and their responses, extremely difficult.

Hence, in recent years, there have been many attempts to reinduce the differentiated phenotype in pure populations of cells by recreating the correct environ-

ment, and, by doing so, defining individual influences exerted on the induction and maintenance of differentiation. This usually implies cessation of cell division and creation of an interactive high-density cell population as in histotypic or organotypic culture. This will be discussed in greater detail below.

COMMITMENT AND LINEAGE

Progression from a stem cell to a particular pathway of differentiation usually implies a rapid increase in commitment with advancing stages of progression (see Figs. 2.3 and 14.2). A hemopoietic stem cell after commitment to lymphocytic differentiation will not change lineage at a later stage and adopt myeloid or erythrocytic characteristics. Similarly, a primitive neurectodermal stem cell, once committed to become a neurone, will not change to a glial cell. This is not to say that if the inductive environment is altered early enough (before commitment) that a cell can alter its destiny or even adopt a mixed phenotype under artefactual or pathological circumstances.

Commitment may therefore be regarded as the point, between the stem cell and a particular precursor stage, where a cell or its progeny can no longer transfer to a separate lineage.

Many claims have been made in the past for cells transferring from one lineage to another. Perhaps the best substantiated of these is the regeneration of the amphibian lens by recruitment of cells from the iris [Clayton et al., 1980; Cioni et al., 1986]. Since the iris can be fully differentiated and still regenerate lens, this has been proposed as proper transdifferentiation. It is, however, one of the few examples, and most other claims have been from tumor cell systems where the origin of the tumor population may not be clear. Small cell carcinoma of the lung has been found to alter to squamous carcinoma following recurrence after relapse from chemotherapy. Whether this implies that one cell type, the Kulchitsky cell [de Leij et al., 1985], presumed to give rise to small cell lung carcinoma, changed its commitment or whether the tumor originally derived from a multipotent stem cell and on recurrence progresses down a different route is still not clear [Gazdar et al., 1983; Goodwin et al., 1983; Terasaki et al., 1984]. Similarly, the K562 cell line was isolated from a myeloid leukemia, but subsequently was shown capable of erythroid differentiation [Andersson et al., 1979b]. Rather than a committed myeloid precursor converting to erythroid, it seems more likely that the tumor arose in the common stem cell known to give rise to both erythroid and myeloid lineages. For some reason, as yet unknown, continued culture favoured erythroid differentiation rather than the myeloid features seen in the original tumor and early culture.

In some cases, again in cultures derived from tumors, a mixed phenotype may be generated. The C_6 glioma of rat expresses both astrocytic and oligodendrocytic features and these may be demonstrated simultaneously in the same cells.

In general, however, these cases are unusual and restricted to tumor cultures. Most cultures from normal tissues, although they may differentiate in different directions, once committed will not alter to a different lineage. This raises the question of the actual status of cell lines derived from normal tissues. This has been dealt with already in Chapter 2 and the conclusion reached that most cultures are derived from: (1) stem cells which may differentiate in one or more different directions, e.g., lung mucosa which can become squamous or mucin-secreting depending on the stimuli; (2) committed precursor cells which will stay true to lineage; or (3) differentiated cells such as fibrocytes which may dedifferentiate and proliferate, but still retain lineage fidelity. Some mouse embryo cultures, loosely called fibroblasts, probably more correctly belong to (1) as they can be induced to become adipocytes, muscle cells, and endothelium as well as fibrocytes.

Cell lines may have different degrees of commitment depending on the "stemness" or precursor status

of the cells from which they were derived; but unless the correct environmental conditions are re-established and proliferation is encouraged they will remain at the same position in the lineage.

MARKERS OF DIFFERENTIATION

Before studies of differentiation, its properties, and the regulation of its expression can be made, marker properties must be defined which will allow differentiation to be recognized. Markers expressed early and retained throughout subsequent maturation stages are generally regarded as lineage markers, e.g., intermediate filament proteins such as the cytokeratins (epithelium) [Moll et al., 1982] or glial fibrillary acidic protein (astrocytes) [Bignami et al., 1980; Eng and Bigbee, 1979]. Markers of the mature phenotype representing terminal differentiation are more usually specific cell products or enzymes involved in the synthesis of these products, e.g., hemoglobin in an erythrocyte, serum albumin in a hepatocyte, transglutaminase in a differentiating squame [Schmidt et al., 1985], or glycerol phosphate dehydrogenase in an oligodendrocyte [Breen and De Vellis, 1974] (see Table 2.2). These properties are often expressed well after commitment and are more likely to be reversible and under adaptive control by hormones, etc.

Differentiation should be regarded as the expression of one or preferably more than one of these marker properties. While lineage markers are helpful in confirming cell identity, the expression of the functional properties of the mature cells is the best criterion for terminal differentiation.

INDUCTION OF DIFFERENTIATION

There are four main parameters governing the control of differentiation and these are summarised in Figure 14.3.

Soluble Inducers

(Table 14.1) These include established endocrine hormones such as hydrocortisone, glucagon, and thyroxine (or triiodotyrosine), paracrine factors released by one cell and influencing adjacent cells, which as yet are poorly characterized (e.g., TGFβ from platelets, prostaglandins, NGF, glia maturation factor) [Lim and Mitsunobu, 1975], alveolar maturation factor [Post et al., 1984], vitamins such as vitamin D_3 and retinoic acid (see Table 14.1), and inorganic ions, particularly Ca^{2+}, where high Ca^{2+} promotes keratinocyte differentiation, for example (see Chapter 20).

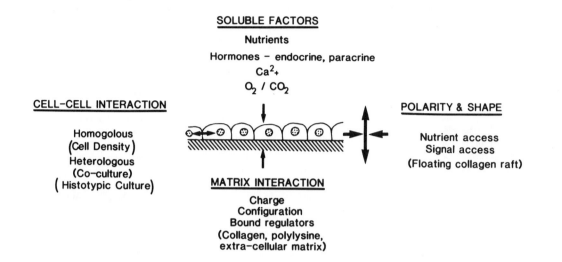

Fig. 14.3. *Parameters controlling expression of differentiation in vitro. (Reproduced from Freshney, 1985, by permission of the publisher).*

TABLE 14.1 Soluble Inducers of Differentiation—Physiological

Inducer	Cell type	Reference
Hydrocortisone	Glia glioma	[McLean et al., 1986]
	Hepatocytes	[Granner et al., 1968]
	Mammary epithelium	[Stockdale and Topper, 1966]
	Myeloid leukemia	[Sachs, 1978]
Melanotropin	Melanocytes	[Fuller and Meyskens, 1981]
Thyrotropin	Thyroid	[Chambard et al., 1983]
Erythropoietin	Erythroblasts	[Krantz and Goldwasser,]
Prolactin	Mammary epithelium	[Stockdale and Topper, 1966; Rudland et al., 1982]
Insulin	Mammary epithelium	[Stockdale and Topper, 1966; Rudland et al., 1982]
Prostaglandins	Melanocytes	[Fuller and Meyskens, 1981]
Nerve growth factor	Neurons	[Levi-Montalcini, 1979]
Glia maturation factor	Glial cells	[Lim and Mitsunobu, 1975]
Alveolar cell maturation factor	Type II pneumocytes	[Post et al., 1984]
TGF-β	Bronchial epithelium	[Lechner et al., 1981]
Retinoic acid	Bronchial epithelium	[Wu and Wu, 1986]
	Tracheal epithelium	[Klann and Marchok, 1982]
	Melanoma	[Lotan and Lotan, ; Meyskens and Fuller, 1980]
	Myeloid leukemia	[Breitman et al., 1980]
Vitamin E	Neuroblastoma	[Prasad et al., 1980]
Vitamin D_3	Myeloma	[Murao et al., 1983]
Ca^{2+}	Keratinocytes	[Boyce and Ham., 1983]

TABLE 14.2. Soluble Inducers of Differentiation—Nonphysiological

	Cell type	Reference
Planar-polar compounds DMSO	Murine erythroleukemia	[Rossi and Friend, 1967]
	Myeloma	[Tarella et al., 1982]
	Neuroblastoma	[Kimhi et al., 1976]
	Mammary epithelium	[Rudland et al., 1982]
Sodium butyrate	Erythroleukemia	[Anderson et al., 1979]
	Colon cancer	[Chung et al., 1985; Augeron and Laboisse, 1984]
N-methyl acetamide	Glioma	[McLean et al., 1986]
N-methyl formamide,		
Dimethyl formamide	Colon cancer	[Dexter et al., 1979]
Hexamethylenebis-		
acetamide	Erythroleukemia	[Osborne et al., 1982]
Benzodiazepines	Erythroleukemia	[Clarke and Ryan, 1980]
Cytotoxic drugs		
Cytosine arabinoside	Myeloid leukemia	[Takeda et al., 1982]
Mitomycin C		
Anthracyclines	Melanoma	[Raz, 1982]
Other compounds		
TPA	Bronchial epithelium	[Willey et al., 1984]
	Myeloma	[Rovera et al., 1972]
	Neuroblastoma	[Spinelli et al., 1982]
Cyclic AMP	Oat cell cancer	[Tsuji et al., 1976]

Non-physiological inducers (Table 14.2). Friend observed that mouse erythroleukemia cells treated with DMSO (to induce the production of Friend leukemia virus) turned red due to the production of hemoglobin [Rossi and Friend, 1967]. Subsequently it was demonstrated that many other cells, e.g., neuroblastoma, myeloma, and mammary carcinoma also responded to DMSO by differentiating. Many other compounds have now been added to this list of nonphysiological inducers—hexamethylene bisacetamide, N-methyl acetamide, sodium butyrate, benzodiazepines, whose action may be related to that of DMSO, and a range of cytotoxic drugs such as hydroxyurea, cytosine arabinoside, and mitomycin-C (see Table 14.2).

The action of these compounds is as yet unclear but may be mediated by changes in membrane fluidity (particularly the polar solvents like DMSO and the anesthetics and tranquilizers), or by alterations in DNA methylation. Induction of differentiation by polar solvents such as DMSO may be phenotypically normal but the induction by cytotoxic drugs may also induce gene expression unrelated to differentiation [McLean et al., 1986].

Tumor promoters such as TPA have been shown to induce squamous differentiation in bronchial mucosa although not in bronchial carcinoma [Willey et al., 1984]. Although these are not naturally occurring *in vivo*, they bind to specific receptors which are presumably for analogous *in vitro* derived compounds.

Cell Interaction

Homologous cell interaction occurs optimally at high cell density. It may involve gap junctional communication [Finbow and Pitts, 1981], where metabolites, second messengers such as cyclic AMP, or electrical charge, may be communicated between cells. This interaction probably harmonizes the expression of differentiation within a population of similar cells rather than initiating its expression.

Heterologous cell interaction, e.g., between meso-dermally- and endodermally-derived cells, is responsible for initiating and promoting differentiation. During and immediately following gastrulation in the embryo, and later during organogenesis, mutual interaction between cells originating in different germ layers promotes differentiation [Grobstein, 1953; Cooper, 1965]. For example, when endodermal cells form a diverticulum from the gut and proliferate within adjacent mesoderm, the mesoderm induces the formation of alveoli and bronchiolar ducts and is itself induced to become elastic tissue [Taderera, 1967].

The extent to which this process is continued in the adult is not clear, but evidence from epidermal maturation suggests that the underlying dermis is required for the formation of keratinized squames with fully cross-linked keratin [Bohnert et al., 1986].

Dissimilar cells do not form gap junctions readily, so the exchange of information may occur at the cell surface by an effector/receptor type of interaction, by the transmission of paracrine factors (see below), or by modification of the intercellular matrix. In practice, all of these may occur.

An interesting aspect of the dependence of differentiation on cellular interaction which has emerged in recent years is the response of certain epithelial tissues to hormones. It has been demonstrated that type II alveolar cells in the lung produce surfactant in response to dexamethasone *in vivo*. *In vitro* experiments have shown that this is dependent on the steroid binding to receptors in the stroma which then releases a peptide to activate the alveolar cells [Post et al., 1984]. Similarly, the response of epithelial cells in the mouse prostate to androgens is mediated by stroma, as receptor deficient epithelium (testicular feminization mutant, *tfm*) will respond in coculture with normal stroma, while normal epithelium will not respond in coculture with *tfm* stroma [Cunha 1984; Lasnitzki and Mizuno, 1979].

This implies a triple role, at least, for heterologous cell interaction: (1) direct inductive interaction by cell surface contact or diffusible inducers; (2) matrix interaction specified by contributions from both interactive partners; and (3) an inductive interaction with a diffusible inducer produced by hormonal interaction with the interacting cells.

Cell-Matrix Interactions

Surrounding the surface of most cells there is a complex mixture of glycoproteins and proteoglycans which is almost certainly highly specific for each tissue and even parts of a tissue. Reid has shown that the construction of artificial matrices from different constituents can regulate gene expression. Addition of liver derived matrix material will induce expression of the albumin gene in hepatocytes. Furthermore, collagen has been found to be essential for the functional expression of many epithelial cells [Yang et al., 1979] and for endothelium to mature into capillaries [Folkman and Haudenschild, 1980].

Attempts to mimic matrix effects by use of synthetic macromolecules have been partially successful using poly-D-lysine to promote neurite extension in neuronal cultures (see Chapter 20).

Polarity and Cell Shape

Studies with hepatocytes [Sattler et al., 1978] showed that full maturation required the growth of the cells on collagen gel and the subsequent release of the gel from the bottom of the dish using a spatula or bent Pasteur pipette. This allowed shrinkage of the gel and an alteration in cell shape from flattened to cuboidal or even columnar. Accompanying or following shape change, and also possibly due to access to medium through the gel as well as via the top cell surface, the cells developed polarity visible by electron microscopy; when the nucleus became asymmetrically distributed, nearer the bottom of the cell, an active Golgi complex formed and secretion towards the, now, apical surface was observed.

A similar establishment of polarity has been demonstrated in thyroid epithelium by Mauchamp [Chambard et al., 1983], using a filter well assembly. In this case the lower (basal) surface generated receptors for TSH and secreted triiodotyrosine and the upper (apical) surface released thyroglobulin. More recent studies by Guguen-Guillouzo [1986](see Chapter 20) with hepatocytes and Jetten and Smets [1985] with bronchial epithelium have suggested that floating collagen may not be essential but this point is not yet fully resolved.

DIFFERENTIATION AND MALIGNANCY

It is frequently observed that, with increasing progression of cancer, histology of the tumor indicates poorer differentiation, and from a prognostic standpoint patients with poorly differentiated tumors will generally have a lower survival rate than those with differentiated tumors. It has also been stated that cancer is principally a failure of cells to differentiate normally. It is therefore surprising to find that many tumors grown in tissue culture can be induced to differentiate (Table 14.2). Indeed, much of the fundamental data on cellular differentiation has been derived from the Friend murine leukemia, mouse and human myeloma, hepatoma, and neuroblastoma (see Table 14.2). Nevertheless, there appears to be an inverse relationship between the expression of differentiated and malignancy associated properties, even to the extent that the induction of differentiation has often been proposed as a mode of therapy [Spremulli and Dexter, 1984; Freshney, 1986].

Apparently tumor cells may often retain the ability to respond to inducers of differentiation, though not always those active on normal cells. Friend erythroleukemia responds to DMSO but not to erythropoietin, while normal bone marrow [Gross and Goldwasser, 1971] and foetal mouse liver [Cole and Paul, 1966] respond to erythropoietin, but as yet there are no reports of induction by DMSO.

If tumor cells are capable of differentiation, it is somewhat anomalous that they do not do so *in vivo* unless either the differentiation process is not complete and proliferation is still possible or there is a stem cell compartment which is insensitive to feedback inhibition or other forms of regulation and its uncontrolled proliferation is the key event regardless of whether complete or partial differentiation results.

Whatever the clinical outcome tumor cells remain useful models for the study of differentiation and the production of specialized products.

PRACTICAL ASPECTS

It is clear that, given the correct environmental conditions, and assuming that the appropriate cells are present, partial or even complete differentiation is achievable in cell culture. The conditions required for individual cell types are not all elaborated and would be difficult to review here, but some indications are given in Chapter 20. As a general approach to promoting differentiation as distinct from cell proliferation and propagation the following may be suggested.

(1) Select the correct cell type by use of appropriate isolation conditions and medium (see Chapter 20).

(2) Grow to high cell density ($> 10^5/cm^2$) on the appropriate matrix. This may be collagen of a type appropriate to the site of origin of the cells with or without fibronectin or laminin, or more complex tissue derived [Reid and Rojkind, 1979], cell derived [Gospodarowicz et al., 1980], or synthetic (e.g., poly-D-lysine for neurones [Yavin and Yavin, 1980]) matrix.

(3) Change to a differentiating medium rather than propagation medium, e.g., for epidermis increase Ca^{2+} to around 3 mM, for bronchial mucosa increase the serum concentration (see Chapter 20).

(4) Add differentiation inducing agents such as glucocorticoids, retinoids, vitamin D_3, DMSO, HMBA, prostaglandins, or peptide differentiating factors such as glia maturation factor, lung maturation factors, NGF, or melanocyte stimulating hormone (MSH) as appropriate for the type of cell (see Tables 14.1, 2).

(5) Add interacting cell type during growth phase [(2) above], or induction phase [(3) and (4) above], or both. Selection of the correct cell type is not always

obvious, but lung fibroblasts for lung epithelial maturation [Post et al., 1984], glial cells for neuronal maturation [Lindsay, 1979], and bone marrow adipocytes for hemopoietic cells (see Chapter 20) are some of the better characterized examples.

(6) Floating the culture on detached collagen rafts [Sattler et al., 1978] or in a filter well [Chambard et al., 1983] may be advantageous, particularly for certain epithelia.

Preparation of Collagen Gel

The following was adapted from a protocol by Ted Ebendal, Department of Zoology, University of Uppsala, Sweden, based in turn on the method of Elsdale and Bard [1972].

Principle

Collagen is soluble in liquid solution at low pH and salt concentration. Increasing the pH to 7.4 and increasing the ionic strength causes the collagen to gel.

Outline

Dissolve rat tail tendons in 0.5 M acetic acid and dialyze against one-tenth strength culture medium, pH 4.0. Dilute dialyzed collagen with culture medium to form gel.

Materials

> 10 X BME (Eagle's Basal Medium)
> 7.5% solution of $NaHCO_3$
> L-glutamine (200 mM)
> fetal calf serum
> distilled water
> 0.142 M NaOH
> 0.5 M acetic acid

Protocol A. To prepare collagen

1.

Cut off rat tail and soak in 70% EtOH for 1 min.

2.

Fracture progressively from the tip and pull out the tendons. Handle with sterile gloves and keep tendons sterile.

3.

Dissolve tendons from 20 tails in 200 ml of 0.5 M sterile acetic acid and filter through sterile muslin gauze.

4.

Dialyze against 4 1 sterile 1:10 Eagle's BME for 24 hr.

5.

Repeat stage 4 adjusting the pH of the medium to 4.0 beforehand.

6.

Centrifuge to clarify, 17,000 g for 24 hr or 50,000 g for 2 hr.

Protocol B. To prepare gel

1.

Dilution medium—mix the following components in a glass tube on an ice bath to prepare approximately 5 ml of gel:

455 μl 10 × BME

112 μl 7.5% $NaHCO_3$

50 μl L-glutamine

55 μl fetal bovine serum (if 1% serum in the gel is desired; for 10% serum take 555 μl)

383 μl distilled water (minus the volume of water in the NaOH solution). Immediately before use add 50–150 μl of a 0.142 M NaOH solution, the exact amount needed to raise the pH to 7.4, as indicated by the change in color of the medium. This must be tested in advance for each batch of collagen solution that will be used.

2.

Transfer 0.80 ml of the collagen solution to a 10-ml glass tube on ice (one tube for each dish).

3.

To prepare the gel take 0.21 ml of the dilution medium (for a gel intended to contain 10% serum take 0.31 ml) and mix thoroughly with 0.80 ml collagen solution (avoid blowing air bubbles which might be trapped in the gel) using a wide-bore pipette.

4.

Transfer this final mixture to the culture dish and allow the gel to set.

Variations

Collagen gel may be derivatized by carboimide to increase its adherence to the plastic substrate (see below). Collagen is available commercially through Flow Laboratories and the Collagen Corporation.

The following protocol for coating surfaces with cross-linked collagen has been contributed by Jeffrey D. Macklis, Department of Neuroscience, Children's Hospital, and Department of Neuropathology, Harvard Medical School, Boston, Massachusetts 02115.

Principle

A new type of collagen surface for culture of nervous system cells was described by Macklis et al. [1985], which allowed extended culture survival, improved microscopy, and dry storage of coated culture dishes. Collagen was derivatized to plastic culture dishes by a cross-linking reagent, 1-cyclohexyl-3-(2-morpholinoethyl)-carbidiimide-metho-p-toluenesulfonate (carbodiimide); comparison to conventional ammonia-polymerized or adsorbed surfaces showed superior culture viability and improved optical characteristics. Simple covalent bonding of collagen fibrils to active groups on tissue culture plastic is described below. A large supply of coated dishes can be prepared in a single 5-hour session and stored for later use.

Outline

Prepare stock collagen solution, then dilute into an aqueous solution of carbodiimide. Coat dishes, incubate, wash, air dry, sterilize under UV, and use or store dry.

Materials

collagen solution in dilute acetic acid at protein concentration of approximately 500 μg/ml, which can be purchased commercially or prepared by extraction from rat tails by the method of Bornstein and Murray [1958]

carbodiimide (Aldrich Chemical Co.)

tissue culture plates (Falcon Plastics, Becton Dickinson & Co.)

double distilled water, sterilized by autoclaving

Protocol

1.

Place approximately 2 μg of carbodiimide in each of several 15-ml sterile medium tubes. Seal and store at 4°C until used.

2.

Add 14 ml of sterile, double distilled water at room temperature to each tube containing carbodiimide to be used (each prepared tube will coat 15 35-mm culture dishes.

3.

Vortex each tube for approximately 10 s and set aside.

4.

Add 1 ml of stock collagen solution to one tube

containing carbodiimide solution (approximately 130 μg/ml) and rapidly vortex until uniform.

5.

Rapidly transfer collagen-carbodiimide solution to dishes, generously covering the bottoms (approximately 1 ml in a 35-mm dish). The rapidity of transfer minimizes derivitization to the solution tube and maximizes early contact with the dish.

6.

Incubate dishes at 25°C for 3 hr.

7.

Wash three times with sterile, double distilled water.

8.

Air dry at room temperature for 1 hr.

9.

Sterilize under ultraviolet irradiation for 1 hr.

10.

Use dishes immediately or store dry for later use.

All of these factors may not be required and the sequence they are presented in is meant to imply some degree of priority. Scheduling may also be important, e.g., matrix generally turns over slowly so prolonged exposure may be important while some hormones may be effective in relatively short exposures. Furthermore, the response to hormones may depend on the presence of the appropriate extracellular matrix, cell density, or heterologous cell interaction, so these components may be required to be stabilized prior to drug exposure.

Not all types of differentiation may be reproducible in culture but it seems likely that given the isolation and survival of the correct cell types and the elaboration of the correct inducers, many of which may be tissue derived peptides analogous to growth factors, production of functionally mature cells of many more types may be feasible in the near future.

Chapter

The Transformed Phenotype

WHAT IS TRANSFORMATION?

In microbiology, where the term was first employed in this context, transformation implies a change in phenotype dependent on the uptake of new genetic material. Although this is now possible in mammalian cells (see Chapter 23), it has been called "transfection" to distinguish it from transformation which, in tissue culture, implies a phenotypic modification not necessarily involving the uptake of new genetic material. Although transformation can arise from infection with a transforming virus such as polyoma, and incorporation of new genomic DNA, it can also arise spontaneously or following exposure to a chemical carcinogen. The primary alteration is usually considered to be genetic and irreversible, although recent studies have shown that some of the phenotypic properties of transformed cells can be restored to normal by chemical or hormonal inducers (see Chapter 14).

There are several properties associated with transformation *in vitro* (Table 15.1), the most important one being immortalization. Most normal cells have a finite life-span of 20–100 generations (see Chapter 2) but some cells, notably those from rodents and from most tumors, can produce continuous cell lines with an infinite life-span. The rodent cells are karyotypically normal at isolation and appear to go through a crisis after about 12 generations, in which most of the cells die out, but a few survive with an enhanced growth rate and give rise to a continuous cell line.

If continuous cell lines from mouse embryos (e.g., the various 3T3 cell lines) are maintained at a low cell density and not allowed to remain at confluence for any length of time, they remain sensitive to contact

TABLE 15.1. Properties of Transformed Cells*

Growth characteristics	Immortal
	Anchorage independent
	Loss of contact inhibition
	Growth on confluent monolayers of homologous cells "focus" formation (see Fig. 15.1)
	Reduced density limitation of growth, high saturation density
	Low serum requirement
	Growth factor independent
	High plating efficiency
	Shorter population doubling time
Genetic properties	High spontaneous mutation rate
	Aneuploid
	Heteroploid
Neoplastic properties	Tumorigenic
	Angiogenic
	Enhanced protease secretion, e.g., plasminogen activator
	Invasive

*Note: It is not implied in this table that all of these properties are expressed in transformed cell lines, but there is a higher probability of them occurring than in normal, finite cell lines. Furthermore, the degree of expression of individual properties will vary greatly among different lines.

inhibition and density limitation of growth [Todaro and Green, 1963] (see below). If, however, they are allowed to remain at confluence for extended periods, then foci of cells appear with reduced contact inhibition, begin to pile up, and will ultimately overgrow (Fig. 1A.).

The fact that these cells are not apparent at low densities, or when confluence is first reached, suggests that they arise *de novo*, by a further transformation event. They appear to have a growth advantage and subsequent subcultures will rapidly be overgrown by the randomly growing cell. This cell type is often found to be tumorigenic.

Studies on cell lines from normal human urothelium also indicate that *in vitro* transformation is a progressive, multi-step event [Christensen et al., 1984]. Normal urothelial cell cultures are made up of predominantly diploid cells which are nonangiogenic, noninvasive, and nontumorigenic. Over a period of up to 2 years these lines become first angiogenic, then aneuploid, and ultimately tumorigenic and invasive.

Hence transformation is apparently a multistep process often culminating in the production of neoplastic cells [Quinfanilla et al., 1986]. It is therefore strange to find that cell lines from malignant tumors, presumably already "transformed," can undergo further "transformation" with increased growth rate, reduced anchorage dependence, more pronounced aneuploidy, and immortalization. Such a transformation has been observed in a squamous lung carcinoma cell line in the author's laboratory, and the resultant continuous cell line actually lost tumorigenicity in Nude mice.

This suggests that a series of steps, not necessarily coordinated or interdependent and not necessarily individually tumorigenic, are required for malignant transformation. The same set of properties need not be expressed in every tumor and "progression" may imply expression of new properties or deletion of old ones which may induce metastasis or even spontaneous remission.

The process may be likened to a game of cards such as gin rummy, where the cards need not be obtained in any specific sequence and several different sets will allow the player to "go out." There are, therefore, several steps in transformation including increased immortalization and tumorigenicity (see Table 15.1); the sequence may be determined by environmental selective pressure. *In vitro*, where there is little restriction on growth imposed, the events need not necessarily follow in the same sequence. Hence transformation is not as easily defined in cell culture as it is in microbiology and should be used with caution and preferably qualified, e.g., "transformation into a continuous cell line," or "ploidy transformation," or "malignant" or "neoplastic transformation."

Since malignancy, as such, cannot be demonstrated *in vitro*, we are obliged to use a number of properties associated with cells from malignant tumors grown *in vitro*. In the present discussion, I shall describe markers which might be used in cell identification and do not imply a causal relationship between these properties and the expression of malignancy *in vivo*, but clearly many of the properties discussed have an obvious functional relationship to malignancy.

Two approaches have been used to explore malignancy-associated properties. (1) Cells have been cultured from malignant tumors and characterized. (2) Transformation *in vitro* with a virus or a chemical carcinogen has produced cells which were tumorigenic and which could be compared with the untransformed cells. The second system provides transformed clones of the same lineage, which can be shown to be malignant and they can be compared with untransformed clones which are not. Unfortunately, many of the characteristics of cells transformed *in vitro* have not been found in cells derived from spontaneous tumors. Ideally, tumor cells and equivalent normal cells should be isolated and characterized. Unfortunately, there have been relatively few instances where this has been possible, and even then, although the cells may belong to the same lineage, their position in that lineage is not always clear (see Fig. 2.3), and comparison not strictly justified.

At best, there are a number of generally accepted properties which can be recognized in many tumor cells *in vitro*. Many occur in normal cells, confirming the conclusion that malignancy is not the expression of abnormal characteristics *de novo* but rather the inappropriate and uncontrolled expression of normal properties, and none is common to all neoplastic cell lines.

ANCHORAGE INDEPENDENCE

Many of the properties associated with neoplastic transformation *in vitro* are the result of cell surface modifications [Hynes, 1974; Nicolson, 1976], e.g., changes in the binding of plant lectins [Ambrose et al., 1961; Aub et al., 1963; Willingham and Pastan, 1975; Reddy et al., 1979] and in cell surface glycoproteins [Hynes, 1976; Lloyd et al., 1979; Van Beek, 1979; Warren et al., 1978], which may be correlated with the development of invasion and metastasis *in vivo*. Fibronectin (LETS protein, large extracellular transformation sensitive) is lost from the surface of transformed fibroblasts [Hynes, 1973; Vaheri et al., 1976]. This may contribute to a decrease in cell-cell and cell-substrate adhesion [Easty et al., 1960] and to a decreased requirement for attachment and spreading for the cells to proliferate [MacPherson and Montagnier, 1964]. In addition, loss of cell-cell recognition, a product of reduced adhesion, leads to a disorganized growth pattern and loss of density limitation of growth (see below). This results in the ability of cells to grow detached from the substrate, either in stirred suspension culture or suspended in semisolid media such as agar or Methocel. There is an obvious analogy with detachment from the tissue in which a tumor arises and the formation of metastases in foreign sites, but how valid this analogy is, is not clear.

Suspension Cloning

MacPherson and Montagnier [1964] were able to demonstrate that polyoma-transformed BHK21 cells could be grown preferentially in soft agar while untransformed cells cloned very poorly. Subsequently, it has been shown that colony formation in suspension is frequently enhanced following viral transformation. The position regarding spontaneous tumors is less clear, however, in spite of the fact that Freedman and Shin [1974; Kahn and Shin, 1979] demonstrated a close correlation between tumorigenicity and suspension cloning in Methocel. Although Hamburger and Salmon [1977] have shown that many human tumors contain a small percentage of cells (<1.0%) which are clonogenic in agar, we [Freshney and Hart, 1982] and others [Laug et al., 1980] have shown that a number of normal cells will also clone in suspension (see Fig. 11.11) with equivalent efficiency. Since normal fibroblasts are among these cells which will clone in suspension, the value of this technique for assaying for the presence of tumor cells in short-term cultures from human tumors is in some doubt. It remains a valuable technique for assaying neoplastic transformation *in vitro* by tumor viruses and has been used extensively by Styles [1977] to assay for carcinogenesis.

The technique of cloning in suspension is described in Chapter 11. Variations with particular relevance to the assay of neoplastic cells are in the choice of suspending medium. It has been suggested [Neugut and Weinstein, 1979] that agar may only allow the most

highly transformed cells to clone while agarose (lacking sulphated polysaccharides) is less selective. Montagnier [1968] was able to show that untransformed BHK21 cells, which would grow in agarose but not in agar, could be prevented from growing in agarose by the addition of dextran sulphate.

Contact Inhibition and Density Limitation of Growth

The loss of contact inhibition (see Chapter 10) may be detected morphologically by the formation of a disoriented monolayer of cells or rounded cells in foci within the regular pattern of normal surrounding cells. This is illustrated in Figure 15.1 where 3T3 cells transformed by bovine papilloma virus DNA are compared with spontaneous transformants. Cultures of human glioma show a disorganized growth pattern and exhibit reduced density limitation of growth by growing to a higher saturation density than normal glial cell lines [Freshney et al., 1980a,b]. As variations in cell size will influence the saturation density, the increase in the labeling index with [^{3}H]-thymidine at saturation density is a better measurement of reduced density limitation of growth. Human glioma, labeled for 24 hr at saturation density with [^{3}H]-thymidine, gave a labeling index of 8% while normal glial cells gave 2%.

Outline

The culture is grown to saturation density in nonlimiting medium conditions and the percentage of cells labeling with [^{3}H]-thymidine determined autoradiographically.

Materials

Culture of cells ready for subculture
PBSA
0.25% trypsin
24-well plates containing 13-mm coverslips
medium
9-cm petri dishes (one per coverslip)
medium containing 0.1 μCi/ml [^{3}H]-thymidine (2 Ci/mmol)

Protocol

1.

Trypsinize cells and seed 10^5 cells/ml into 24-well plate, 1 ml/well, each well containing a 13-mm-diameter coverslip.

2.

Incubate in humidified CO_2 incubator for 1–3 d.

Fig. 15.1. *Transformation foci in a monolayer of normal, contact-inhibited NIH3T3 mouse fibroblasts. a. NIH3T3 mouse fibroblasts transformed by transfection with bovine papilloma virus DNA cloned in bacterial plasmid pAT-153, coprecipitated with* $Ca_3(PO_4)_2$. *b. Spontaneous transformant arising when cells reach a high density. By courtesy of D. Spandidos, photographs by M. Freshney.*

3.

Transfer the coverslips to 9-cm petri dishes containing 20 ml medium and return to CO_2 incubator.

4.

Continue culturing, changing medium every 2 d after cells become confluent on the coverslips. Trypsinize and count cells from two coverslips every 3–4 d. As cells become denser on the coverslip, it may be necessary to add 200–500 units/ml crude collagenase to the trypsin to achieve complete dissociation of the cells for counting.

5.

When cell growth ceases, i.e., two sequential counts show no significant increase, add 1.0 μCi/ml [³H]-thymidine (2 Ci/mmol), and incubate for 24 hr.

◊Note. Handle [³H]-thymidine with care. Al-

though a low energy β-emitter, it localizes to DNA and can induce radiolytic damage. Wear gloves, do not handle in horizontal laminar flow, but in a biohazard or cytotoxic drug handling cabinet (see Chapter 6), and discard waste liquids and solids by the appropriate route specified in the local rules governing the handling of radioisotopes.

6.

Transfer coverslips back to a 24-well plate and trypsinize the cells for autoradiography (see Chapter 23). They may be fixed in suspension and dropped on a slide as for chromosome preparations (without the hypotonic treatment), centrifuged onto a slide using the Shandon Cytospin, or trapped on Millipore or Nuclepore filters by vacuum filtration (see Chapter 13).

Note. It is necessary to trypsinize high-density

cultures for autoradiography because of their thickness and the weak penetration of β-emission from ^{3}H (mean path length in water is approximately 1 μm). Labeled cells in the underlying layers would not be detected by the radiosensitive emulsion due to the absorption of the β-particles by the overlying cells. If the cells remain as a monolayer at saturation density, this step may be omitted, and autoradiographs prepared by mounting the coverslips, cells uppermost, on a microscope slide.

Analysis. Count the number of labeled cells as a percentage of the total. Scan the autoradiographs under the microscope and count the total number of cells, and the proportion labeled in representative parts of the slide. A suggested scanning pattern for a circular array of cells (such as would be produced by the drop technique or the Cytospin) is given in Figure 18.6.

Growth on Confluent Monolayers

Aaronson et al. [1970] showed that transformed mouse 3T3 cells and human fibrosarcoma cells were able to form colonies on confluent monolayers of normal cells, while normal fibroblasts were not. In the author's laboratory, cells cultured from both normal and malignant breast tissue formed colonies on contact-inhibited monolayers of human fetal intestinal cells (FHI) and on normal mouse embryo fibroblasts (STO) [Freshney, Hart, and Russell, 1982] (Fig. 11.8) as do normal epidermal cells (Fig. 15.2). Rheinwald and Green [1975] had previously shown that normal epidermal cells grew on 3T3 monolayers. In some cases, therefore, where an appropriate control cell can be tested, this technique may discriminate between malignant and normal cells, but this may require that the normal cells are of the same lineage as the malignant cells. We have found that human glioma will grow on confluent glial feeder layers, but cells derived from normal glia will not (see Fig. 11.9), while some normal glial cell lines will grow on other feeder layers, e.g., fetal lung fibroblasts and FHI.

Outline

Prepare cells as for cloning and seed onto confluent monolayers of contact-inhibited cells (see Fig. 13.3).

Materials

    14 × 25-cm^2 flasks
    medium

    PBSA
    0.25% trypsin
    cultures of tumor cells
    tubes for dilutions
    hemocytometer or cell counter

Protocol

1.
Seed 14 25-cm^2 flasks with feeder cell (3T3 or other cell line which will become contact inhibited after reaching confluence).
2.
Feed cultures until all available substrate is covered with cells, and mitosis is greatly reduced. If a cell line is used which does not arrest at confluence, treat cells during exponential growth with mitomycin-C or irradiate (see Chapter 11) and reseed sufficient cells to produce confluent monolayers directly with no spaces between cells.
3.
Trypsinize putative tumor cells and suitable control cells (e.g., equivalent normal cells) and dilute to 10^4, 10^3, and 10^2 cells/ml in 20 ml and replace medium in each pair of feeder layer flasks with 5 ml of cell suspension, and with medium alone in two flasks.
4.
Seed two flasks at each concentration with each cell type without feeder layers.
5.
Incubate for 2–3 wk with medium changes every 2d.
6.
Wash, fix, and stain (see Chapter 13).

Analysis. Count foci of tumor cells (usually morphologically distinguishable) and express as percentage of number of cells seeded. Compare with flasks with no feeder layers.

Variations. It is possible to have proliferation of cells seeded on confluent monolayers without the formation of discrete colonies. In our experience, human fibrosarcoma and glioma seeded on confluent monolayers of fetal human intestine do not form colonies, but still proliferate, infiltrating the monolayer as they do so. In order to quantify this type of growth, DNA synthesis is first inhibited in the monolayer (see Chapter 11) by treatment with mitomycin-C or by irradiation. Growth of cells seeded onto this layer can then be monitored by measuring the incorporation of [^{3}H]-

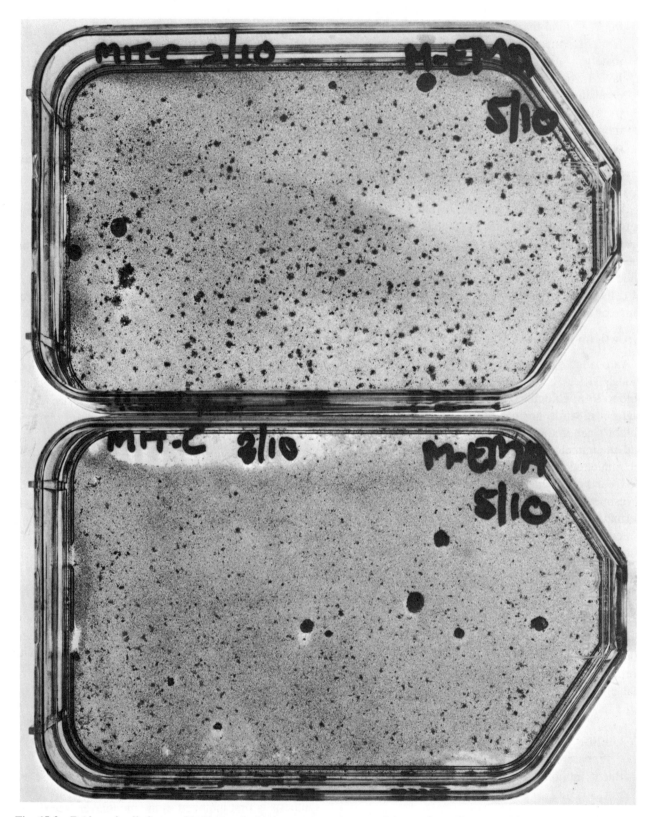

Fig. 15.2. *Epidermal cells from a skin biopsy of a benign naevus growing on a confluent monolayer of fetal human intestinal epithelial cells (FHS74Iht) [Owens et al., 1974]. The large, dark-staining, circular colonies resemble keratinocytes. The smaller colonies may have been melanocytes (fibroblasts do not normally form colonies readily in this system), but no further characteriza-tion was done on them. The lower flask was treated with 1 mM dibutyryl cyclic AMP inhibiting the small colony type but not the large (putative keratinocytes). Cyclic AMP stimulates growth of keratinocytes (see Chapter 20) and may have done so here (top flask, four colonies; bottom flask, eight colonies).*

thymidine by scintillation counting or autoradiography (see Chapters 10 and 24).

GENETIC ABNORMALITIES

Whether mutation is the prime cause of neoplastic transformation or not, chromosomal abnormalities and variations in DNA content per cell are found frequently in cells derived from malignant tumors.

Chromosomal Aberrations

Both changes in ploidy and increases in the frequency of individual chromosomal aberrations can be found [Biedler, 1976]. Figure 15.3 demonstrates variations in chromosome number found in human glioma and melanoma in culture. Chromosome analysis is described in Chapter 13 [see also Sandberg, 1980]. The frequency of chromosomal rearrangement can be determined by the sister chromatid exchange assay [Venitt, 1985] (see also Chapter 9).

DNA content. Microdensitometry of Feulgen-stained preparations [Wright and Dendy, 1976; Stolwijk et al., 1986] and flow cytofluorimetry [Traganos et al., 1977] show that the DNA content of tumor cells may vary from the normal. DNA analysis does not substitute for chromosome analysis, however, as cells with an apparently normal DNA content can yet have an aneuploid karyotype. Deletions and polysomy may

cancel out, or translocations may occur without net loss of DNA.

Some specific aberrations are associated with particular types of malignancy. The first of these to be documented was the Philadelphia chromosome in chronic myeloid leukemia (trisomy 13). Subsequently translocations of the long arms of Chromosomes 8 and 14 were found in Burkitt's lymphoma [Lebeau and Rowley, 1984; Yunis, 1983]. Several other leukemias also express similar and different translocations [Mark, 1971]. Meningiomas often have consistent aberrations and small cell lung cancer frequently has a 3p2 deletion [Wurster-Hill et al., 1984]. These aberrations constitute tumor specific markers which can be extremely valuable in cell line characterization and confirmation of neoplasia.

CELL PRODUCTS AND SERUM DEPENDENCE

Transformed cells have a lower serum dependence than their normal counterparts [Temin, 1966; Eagle et al., 1970]. Lindgren et al. [1975] showed that a short exposure to serum would trigger glioma cells into cycle while normal glial cells required serum to be present throughout G1.

A possible explanation for the low serum dependence of tumor cells lies in the demonstration of Todaro and de Larco [1978], and others, that tumor cells may secrete their own growth factors. While the production of these polypeptides is not assayed as easily as some of the foregoing properties, the increasing availability of specific antibodies against them could bring this approach within the reach of any laboratory with the appropriate immunological expertise. Westall et al. [1978] suggested some homology between a FGF-like peptide from brain and myelin basic protein, once claimed as a tumor marker.

These factors have been collectively described as autocrine growth factors. Implicit in this definition is: (1) the cell produces the factor; (2) the cell has receptors for it; and (3) the cell responds to it by entering mitosis.

Some of these factors may have an apparent transforming activity on normal cells (e.g., TGFα) which bind to the EGF receptor and induce mitosis [Richmond et al., 1985] although, unlike true transformation (see above), this is probably reversible. They also cause nontransformed cells to adopt a transformed phenotype and grow in suspension [Todaro and da Larco, 1978].

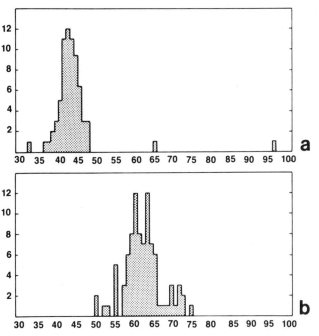

Fig. 15.3. *Chromosome counts for cells cultured from human anaplastic astrocytoma (a) and human metastatic melanoma (b). (2N = 46.)*

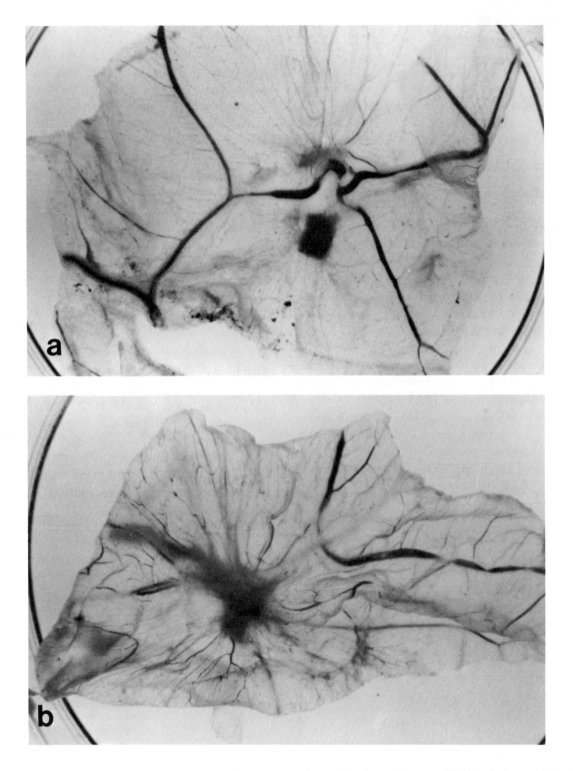

Fig. 15.4. *Induction of angiogenesis in chick chorioallantoic membrane by tumor cell extract. A crude extract of Walker 256 carcinoma cells absorbed into sterile filter paper was placed on the chorioallantoic membrane at 10 d incubation, and the membrane was removed 2 wk later. a. Control. b. Walker 256 extract. By courtesy of Margaret Frame.*

Tumor cells can also produce many hemopoietic growth factors such as the interleukins 1, 2, and 3, along with colony stimulating factor CSF [Metcalf, 1985a; Fontana et al., 1984].

It has been proposed [Cuttitta et al, 1984] that some factors such as bombesin and hCG, hitherto believed to be ectopic hormones produced by lung carcinomas, may in fact be autocrine growth factors. Production of such factors implies expression of the malignancy-associated phenotype and can also help in lineage classification.

Tumor Angiogenesis Factor (TAF)

Tumor cells also release a factor (or factors) capable of inducing blood vessel proliferation [Phillips et al., 1976; Folkman, 1985, 1986; Gullino, 1985]. Fragments of tumor, pellets of cultured cells, or cell extracts, implanted on the surface of the chorioallantoic membrane (CAM) of the hen's egg, promote an increase in vascularization which is apparent to the naked eye 6–8 days later (Fig. 15.4). Since this assay is not readily quantified, stimulation of cell migration [Gullino, 1985] or proliferation [Freshney et al., 1985] in monolayer cultures of vascular endothelium may provide the basis for a more quantitative assay.

Plasminogen Activator

Other cell products which can be recognized are proteolytic enzymes [Mahdavi and Hynes, 1979], long since associated with theories of invasive growth [Ossowski et al., 1979]. Since proteolytic activity may be associated with the cell surface of many normal cells and is absent on some tumor cells, an equivalent normal cell must be used as a control when using this criterion. Plasminogen activator is higher in some cultures from human glioma than in cultures from normal brain [Hince and Roscoe, 1978] (Fig. 15.5), and others have shown previously that plasminogen activator is associated with many different tumors [Rifkin et al., 1974; Nagy et al., 1977]. Plasminogen activator (PA) may be measured by clarification of a fibrin clot or release of free soluble ^{125}I from [^{125}I] fibrin [Strickland and Beers, 1976; Unkeless et al., 1974]. A simple chromogenic assay has also been developed by Whur et al. [1980].

It has been proposed that, for some carcinomas, soluble urokinase-like PA (uPA) is elevated more than

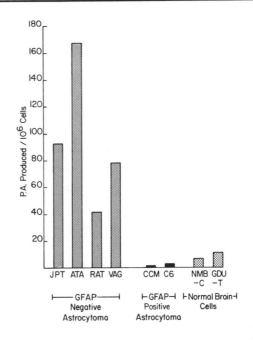

Fig. 15.5. *Plasminogen activator (PA) produced by tumor cells in vitro (arbitrary units). The four gliomas, JPT, ATA, RAT, and VAG were all higher than cells cultured from normal brain (NMB-C, GDU-T). It was also found that the only cells to produce the differentiated glial marker glial fibrillary acidic protein, CCM and C6, had the lowest PA of all.*

tissue-type PA (tPA) [Markus et al., 1980; Shyamala and Dickerman, 1982].

INVASIVENESS

An advantage of the CAM assay using tumor cells to induce angiogenesis is that the subsequent histology may reveal whether the tumor cells have penetrated the underlying basement membrane. Easty and Easty [1974] showed that invasion of the CAM could be demonstrated in organ culture, and others [Hart and Fidler, 1978] have attempted, with some limited success, to construct a chamber capable of quantitating the penetration of tumor cells across the CAM. Mareel et al. [1979] developed an *in vitro* model for invasion, using chick embryo heart fragments cocultured with reaggregated clusters of tumor cells. Invasion appears to be correlated with the malignant origin of the cells, is progressive, and causes destruction of the host tissue. The application of this technique to human tumor cells is now being explored.

TUMORIGENESIS

The only definition of malignancy that is generally accepted is the demonstration of the formation of invasive or metastasizing tumors *in vivo*. Transplantable tumor cells (10^6 or less) injected into isogeneic hosts will produce invasive tumors in a high proportion of cases while 10^6 normal cells of similar origin will not. To study the tumorgenicity of human tumors, a number of models have been developed using immune-suppressed or immune-deficient host animals. The genetically athymic "nude" mouse [Giovanella et al., 1974] and thymectomized irradiated mice [Selby et al., 1980; Bradley et al., 1978] have both been used extensively as hosts for xenografts. The take rate varies, however, and many clearly defined tumor cell lines and tumor biopsies have failed to produce tumors as xenografts, and frequently those which do, fail to metastatize, although they may be invasive locally.

TYPES OF MICROBIAL CONTAMINATION

Bacteria, yeasts, fungi, molds and mycoplasmas all appear as contaminants in tissue culture, and where protozoology is carried on in the same laboratory, some protozoa can infect cell lines. Usually the species or type of infection is not important unless it becomes a frequent occurrence. It is only necessary to note the type (bacterial rods or cocci, yeast, etc), how detected, the location where the culture was last handled, and the operator's name. If a particular type of infection recurs frequently, it may be beneficial to have it identified to help to find the origin. [For more detailed screening procedures for microbial contamination, see McGarrity, 1979; Fogh, 1973; Cour et al, 1979; Hay et al., 1979; Hay, 1986].

Characteristic features of microbial contamination are as follows: (1) Sudden change in pH, usually a decrease with most bacterial infections, very little change with yeast until contamination is heavy, and sometimes an increase in pH with fungal contamination. (2) Cloudiness in medium, sometimes with a slight film or scum on the surface or spots on the growth surface which dissipate when the flask is moved. (3) When examined on a low-power microscope ($\sim \times 100$), spaces between cells will appear granular and may shimmer with bacterial contamination. (Fig. 16.1c). Yeasts appear as separate round or ovoid particles which may bud off smaller particles (Fig. 16.1a). Fungi produce thin filamentous mycelia (Fig. 16.1b), and sometimes denser clumps of spores. With toxic infection, some deterioration of the cells will be apparent. (4) On high-power microscopy ($\sim \times 400$), it may be possible to resolve individual bacteria and distinguish between "rods" and cocci. At this magnification, the shimmering seen in some infections will be seen to be due to mobility of the bacteria. Some bacteria form clumps or associate with the cultured cells. (5) If a slide preparation is made, the morphology of bacteria can be resolved more clearly at $\times$ 1,000, but this is not usually necessary. Microbial infection may be confused with precipitates of media

constituents, particularly protein, or with cell debris, but can be distinguished by their regular particulate morphology. Precipitates may be crystalline or globular and irregular but are not usually as uniform in size. If in doubt, plate out a sample of medium on nutrient agar (see Chapter 8). (6) Mycoplasmal infections (Fig. 16.1d–f) cannot be detected by naked eye other than by signs of deterioration in the culture. The culture must be tested specially by fluorescent staining, orcein or Giemsa staining, autoradiography, or microbiological assay (see below). Fluorescent staining of DNA by Hoechst 33258 [Chen, 1977] is the easiest and most reliable method (see below) and reveals mycoplasmal infections as fine particulate or filamentous staining over the cytoplasm at $\times$ 500 magnification (Fig. 16.2). The nuclei of the cultured cells are also brightly stained by this method, as are any other microbial contaminations.

It is important to appreciate that mycoplasmas do not always reveal their presence with macroscopic alterations of the cells or media. Many mycoplasma contaminants, particularly in continuous cell lines, grow slowly and do not destroy host cells. However, they can alter the metabolism of the culture in subtle ways. As mycoplasmas take up thymidine from the medium, infected cultures show abnormal labeling with ^{3}H-thymidine. Immunological studies can also be totally frustrated by mycoplasmal infections as attempts to produce antibodies against the cell surface may raise antimycoplasma antibodies. Mycoplasmas can alter cell behavior and metabolism in many other ways [Barile, 1977; McGarrity, 1982], so there is an absolute requirement for routine, periodic assays for possible covert contamination of all cell cultures, particularly continuous or established cell lines.

Monitoring Cultures for Mycoplasmas

Superficial signs of chronic mycoplasmal infection include reduced rate of cell proliferation, reduced saturation density [Stanbridge and Doerson, 1978], and

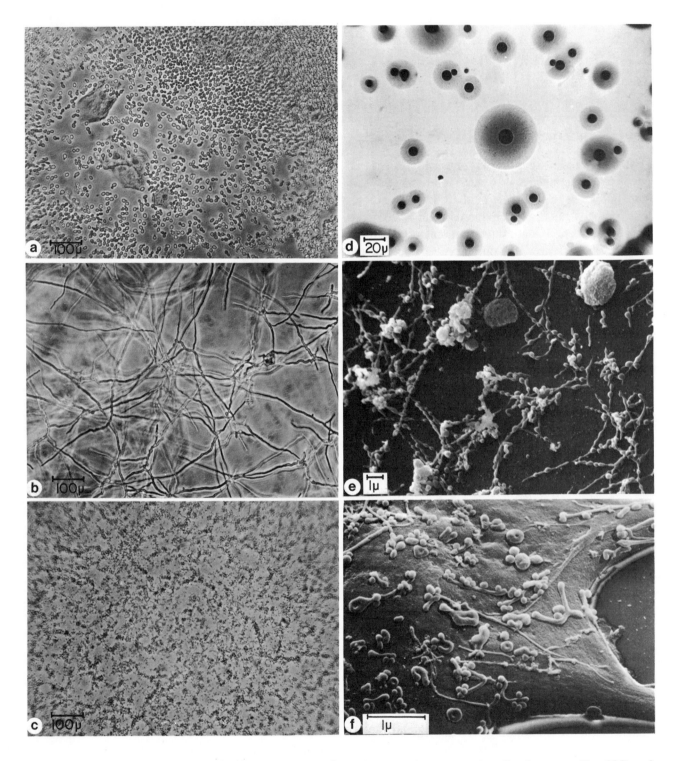

Fig. 16.1. *Examples of microorganisms found as contaminants of cell cultures (magnification approximately × 100). a. Yeast. b. Mold. c. Bacteria. d. Mycoplasma colonies growing on special nutrient agar (not as seen in cell culture—see Fig. 16.2). e,f. Scanning electron micrograph of mycoplasma growing on the surface of cultured cells. (d–f, by courtesy of Dr. M. Gabridge).*

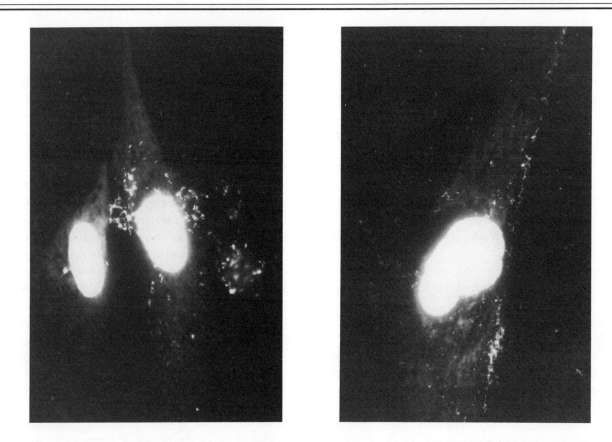

Fig. 16.2. *Human normal diploid lung fibroblasts, infected with mycoplasma, stained with Hoechst 33258. The nuclei fluoresce brightly from cellular DNA and extra-nuclear fluorescence due to mycoplasma is also apparent. Light cytoplasmic staining is present in these preparations, but its diffuse nature makes it easily distinguishable from the bright particulate or filamentous staining of the mycoplasma. Usually, there is no cytoplasmic background. The lack of nuclear detail is due to the overexposure of the photographs necessary to reveal the mycoplasma.*

agglutination during growth in suspension. Acute infection causes total deterioration with perhaps a few resistant colonies. "Resistant" colonies and resulting cell lines are not necessarily free of contamination and may carry a chronic infection.

Fluorescent Technique for Detecting Mycoplasmas

Principle

The cultures are stained with Hoechst 33258, a fluorescent dye, which binds specifically to DNA [Chen, 1977]. Since mycoplasmas contain DNA, they can be detected readily by their characteristic particulate or filamentous pattern of fluorescence on the cell surface and, if the contamination is heavy, in surrounding areas.

Outline

Fix and stain subconfluent cultures or smears and look for fluorescence other than in the nucleus.

Monolayer cultures

Materials

Hoechst 33258 stain
   50 ng/ml in BSS without phenol red
PBSA
deionized water
fresh acetic methanol (1:3, cold)
mountant: 50% glycerine in 0.044 M
   citrate, 0.111 M phosphate buffer,
   pH 5.5

Protocol

1.
Seed culture at regular passage density (2×10^4–10^5 cells/ml, 4×10^3–2.5×10^4 cells/cm^2) and incubate at 36.5°C until they reach 20–50% confluence. Allowing cultures to reach confluence will impair subsequent visualization of myco-

plasma. Cultures may be grown on coverslips in a multiwell plate, or in 35-mm petri dishes without coverslips (see Chapter 13).

2.

Remove medium and discard.

3.

Rinse monolayer with BSS without phenol red and discard rinse.

4.

Add fresh BSS diluted 50:50 with acetic methanol, rinse monolayer, and discard rinse.

5.

Add pure acetic methanol, rinse, and discard acetic methanol.

6.

Add fresh acetic methanol and leave for 10 min.

7.

Remove acetic methanol and discard.

8.

Dry monolayer completely if to be stored. (Samples may be accumulated at this stage and stained later.)

9.

If proceeding directly, wash off acetic methanol with deionized water and discard wash.

10.

Add Hoechst 33258 in BSS without phenol red and leave 10 min at room temperature.

11.

Remove stain and discard.

12.

Rinse monolayer with water and discard rinse.

13.

Mount a coverslip in a drop of mountant and blot off surplus from edges of coverslip.

14.

Examine by epifluorescence with 330/380 nm excitation filter, and LP 440 nm barrier filter.

Suspension cultures and infected media

Suspension cultures are a little more difficult to handle as the cells do not spread as well in cytological preparations and give less flat cytoplasm for critical examination, and, as preparations often involve centrifuging the cells onto a slide, they are often contaminated with particulate debris, some of which can contain DNA and stain positively with Hoechst 33258. To overcome this problem, to reveal low-level contaminations in resistant cell lines, and to screen potentially infected media,

many laboratories now use an indicator cell line such as 3T6 or Vero cells.

The indicator cell is grown to about 20% confluence, the medium removed, and the test medium or whole suspension culture added for 72 hr. If the test medium contains antibiotics it should be removed after 24 hr and culture of the monolayer continued for a further 48 hr in fresh antibiotic-free medium.

The monolayer is then fixed and stained as above. Control monolayer cells unexposed to test media should be stained and examined in parallel.

This method has proved useful in screening holding medium or primary culture medium where the culture cannot be sacrificed, as well as suspension cultures. It also has the advantage that a constant indicator cell is used so the operator becomes more familiar with its normal appearance. Furthermore, if a cell line, such as 3T6, is chosen for its known capacity to support mycoplasmas, then low-level or cryptic contaminations may become more apparent.

Analysis. Check for extranuclear fluorescence. Mycoplasmas give pin points or filaments of fluorescence over the cytoplasm and sometimes in intercellular spaces. The pin points are close to the limits of resolution with a $\times 50$ objective (0.1–1.0 μm) and are usually regular in size. Not all of the cells will necessarily be infected, so most of the preparation should be scanned before declaring the culture uninfected.

Fluorescence outside the nucleus can be observed in uninfected cultures where there is evidence of cell damage, e.g., primary cultures or cells recently recovered from frozen storage. Usually the fluorescent particles in this case are irregular in size and shape and disappear following subculture. Use of an indicator cell line helps to eliminate this problem (see above).

If there is any doubt regarding the interpretation of this test, it should be repeated, allowing about 1 wk for any low-level infection to increase. During this time quarantine the suspect cell line.

Alternative Methods

Several other methods have been reported for the detection of mycoplasmal infections such as the detection of mycoplasma-specific enzymes like arginine deiminase or nucleoside phosphorylase [see Levine and Becker, 1978; Schneider and Stanbridge, 1975], or toxicity with 6-methylpurine decoscyribase (mycotect BRL), but the DNA-fluorescence method is simpler and, though not specific for mycoplasmas, will detect any DNA-containing infection, which is, after all, the prime objective. Several other methods have been reported, however, which are of general importance. The first depends on microbiological culture of the organism although it is best not attempted unless you have the necessary expertise as these organisms are quite fastidious. The cultured cells are seeded into mycoplasma broth [Taylor-Robinson, 1978], grown for 6 days and plated out onto special nutrient agar [Hay, 1986]. Colonies form in about 8 days and can be recognized by their size ($\sim 200 \mu m$ diameter) and their characteristic "fried egg" morphology—dense center with lighter periphery (Fig. 16.1d).

It is necessary to grow known mycoplasmal cultures as a positive control to confirm that the culture conditions are adequate, an element of the microbiological method which is in itself a disadvantage to most tissue culture laboratories, which should be kept clear of mycoplasmas at all times.

While the use of selective culture conditions and examination of colony morphology enables the species of mycoplasma to be identified, the microbiological culture method is much slower and more difficult to perform than the fluorescence technique. Commercial screening for mycoplasma is available (e.g., Flow Laboratories, Microbiological Associates), using microbiological culture. Specific monoclonal antibodies are now available (BRL) which allow the characterization of mycoplasma infections.

Other methods use staining with aceto-orcein or Giemsa stain (see under Staining, Chapter 13). In both cases, particulate cytoplasmic staining is regarded as indicative of mycoplasmal infection but both are more difficult to interpret than the fluorescent method and can give false positives due to nonspecific precipitation of stain.

One other method which has been used quite successfully is autoradiography with [3H]thymidine [Nardone et al., 1965]. The culture is incubated overnight with 0.1 $\mu Ci/ml$ high specific activity [3H]thymidine and an autoradiograph prepared (see Chapter 23). Grains over the cytoplasm are indicative of infection (see Fig. 23.12) and this can be accompanied by a lack of nuclear labeling due to the thymidine being trapped at the cell surface by the mycoplasma.

DETECTION OF MICROBIAL CONTAMINATION

Potential sources of contamination are listed in Table 16.1 along with the precautions that should be taken to avoid them. Even in the best laboratories, however, contaminations do arise, so the following procedure is recommended.

1.
Check for contamination at each handling by eye and on microscope. Every month check for mycoplasmas.

2.
If a contamination is suspected but not obvious and cannot be confirmed *in situ,* clear the hood or bench of all but your suspected culture and one can of Pasteur pipettes. Because of the potential risk to other cultures, this is best done after all your other culture work is finished. Remove a sample from the culture and place on a microscope slide (Kovaslides are convenient for this as they do not require a coverslip). Check on microscope, preferably by phase contrast. If contamination is confirmed, discard pipettes, swab hood or bench with 70% alcohol containing a phenolic disinfectant, and do not use for at least 30 min.

3.
Note the nature of contamination, etc., on record sheet.

4.
If new contamination (not a repeat and not widespread), discard (a) culture, (b) medium bottle used to feed it, and (c) any other bottle, e.g., trypsin, which has been used in conjunction with this culture. Discard into disinfectant, preferably in a fume hood, and outside the tissue culture area.

5.
If new and widespread (i.e., in at least two different cultures), discard all media and stock solutions, trypsin, etc.

6.
If similar contamination is repeated, check stock solutions for contaminations by (a) incubation alone or in nutrient broth (see Chapter 8), or (b) plating out on nutrient agar (Oxoid, Difco) (see Chapter 8). (c) If (a) and (b) fail and contamination is still suspected, incubate 100 ml, filter through 0.2 μm filter, and plate out filter on nutrient agar.

7.

If contamination is widespread, multispecific, and repeated, check sterilization procedures, e.g., temperature of ovens and autoclaves particularly in the center of load, times of sterilization, packaging, storage (e.g., unsealed glassware (see Chapter 8) should be resterilized every 24 hr), and integrity of aseptic room and laminar flow hood filters.

8.

Do not attempt to decontaminate cultures unless they are irreplaceable. If necessary, decontaminate by (a) washing five times in BSS containing a higher than normal concentration of antibiotics (see DBSS in Appendix), and (b) adding antibiotics (as in DBSS) to medium for 3 subcultures. If possible, the infection should be tested for sensitivity to a range of individual antibiotics. (c) Remove antibiotics and culture without for a further 3 subcultures. (d) Recycle b and c twice. (e) Culture for 2 months without antibiotics to check that contamination has been eliminated. Check by phase-contrast microscopy and Hoechst staining (see above).

Tumor tissue can sometimes be decontaminated by animal passage [Van Diggelen et al., 1977]. Cytotoxic antibodies may be effective [Pollock and Kenny, 1963], particularly against mycoplasms.

Polyanethol sulphonate [Mardh, 1975], 5-bromouracil in combination with Hoechst 33258 and uv light [Marcus et al., 1980], and antibiotics such as Tylosin [Friend et al., 1966], Kannamycin, Gentamycin, and BM-Cycline (Boehringer) may be effective in removing mycoplasmas. Schimmelpfeng et al. [1968] were able to eliminate mycoplasmal infections by coculturing with macrophages.

The general rule should be, however, that contaminated cultures are discarded, and that decontamination is not attempted unless it is absolutely vital to retain the cell strain. Complete decontamination, especially with mycoplasmas, is difficult to achieve and attempts to do so may produce hardier, antibiotic-resistant strains of the contaminant.

CROSS CONTAMINATION

During the history of tissue culture, a number of cell strains have evolved with very short doubling times and high plating efficiencies. Although these properties make such cell lines valuable experimental material, they also make them potentially hazardous for cross-infecting other cell lines. The extensive cross contamination of many cell lines with HeLa cells [Gartler, 1967; Nelson-Rees and Flandermeyer, 1977] is now well known, but many operators are still unaware of the seriousness of the risk. To avoid cross contamination:

TABLE 16.1. Routes of Contamination

Source	Route or cause	Prevention*
Manipulations, pipetting, dispensing, etc.	Nonsterile surfaces and equipment. Spillage on necks and outside of bottles and on work surface. Touching or holding pipettes too low down, touching necks of bottles, screw caps. Splash-back from waste beaker. Sedimentary dust or particles of skin settling on culture or bottle. Hands or apparatus held over open dish or bottle.	Clear work area of items not in immediate use. Swab regularly with 70% alcohol. Do not pour if it can be avoided. Dispense or transfer by pipette, auto dispenser or transfer device (see Chapter 4). If you pour: (1) do so in one smooth movement, (2) discard the bottle that you pour from, and (3) wipe up any spillage with sterile swab moistened with 70% alcohol. Discard into beaker with funnel or, preferably, by drawing off into reservoir with vacuum pump (Figs. 4.5, 5.4). Do not work over (vertical laminar flow and open bench) or behind and over (horizontal laminar flow) an open bottle or dish.
Solutions	Nonsterile reagents and media. Dirty storage conditions. Inadequate sterilization procedures. Poor commercial supplier.	Filter or autoclave before use. Monitor performance of autoclave with recording thermometer or sterility indicator (see Trade Index). Check integrity of filters after use (bubble-point). Test all solutions after sterilization.
Glassware and screw caps	Dust and spores from storage. Ineffective sterilization, e.g., sealed bottles preventing ingress of steam.	Dry-heat sterilize or autoclave before use. Do not store unsealed for more than 24 hr. Check oven and autoclave regularly and monitor each load.

(continued)

TABLE 16.1. Routes of Contamination *(continued)*

Source	Route or cause	Prevention*
Tools, instruments, pipettes	Contact with nonsterile surface or material. Invasion by insects, mites or dust. Ineffective sterilization.	Sterilize by dry heat before use; monitor performance of oven. Resterilize instruments (70% alcohol, burn and cool off) during use. Do not grasp part of instrument or pipette which will later pass into culture vessel. Do not store for more than 24 hr, unless sealed with tape.
Culture flasks, media bottles in use	Dust and spores from incubator or refrigerator. Dirty storage or incubation conditions. Media under cap spreading to outside of bottle.	Use screw caps in preference to stoppers Wipe flasks and bottles with 70% alcohol before using. Flame necks and caps (without opening) before placing in laminar flow hood. Cover cap and neck with aluminum foil during storage or incubation.
Room air	Draughts, eddies, turbulence, dust, aerosols.	Clean filtered air. Reduce traffic and extraneous activity. Wipe floor and work surfaces regularly.
Work surface	Dust, spillage.	Swab before, after, and during work with 70% alcohol. Mop up spillage immediately.
Operating—hair, hands, breath, clothing	Dust from skin, hair, or clothing dropped or blown into culture. Aerosols from talking, coughing sneezing, etc.	Wash hands thoroughly. Keep talking to a minimum and face away from work if you do. Avoid working with a cold or throat infection or wear a mask. Tie back long hair, or wear a cap. Do not wear the same lab coat as in general lab area or animal house. Change to gown or apron.
Hoods	Perforated filter. Filter change needed. Spillages, particularly in crevices or below work surface.	Check filters regularly for holes and leaks. Check pressure drop across filter. Clear around and below work surface regularly.
Tissue samples	Infected at source or during dissection.	Do not bring animals into tissue culture lab. Incorporate antibiotics in dissection fluid (see Chapter 9). Dip large tissue samples in 70% alcohol for 30 sec.
Incoming cell lines	Contaminated at source or during transit.	Handle alone, after all other sterile work is finished, swab down bench or hood carefully after use, and do not use until next morning. Check for contamination by growing for 2 wk without antibiotics. (Keep duplicate in antibiotics at first subculture.) Check for contamination visually, by phase-contrast microscopy and Hoechst stain for mycoplasma. Using indicator cell allows screening before first subculture.
Mites, insects, and other infestation in wooden furniture, benches, incubators, and on mice, etc., taken from animal house.	Entry of mites, etc., into sterile packages.	Seal all sterile packs. Avoid wooden furniture if possible, use plastic laminate, one-piece, or stainless steel bench tops. If wooden furniture is used, seal with polyurethane varnish or wax polish and wash regularly with disinfectant. Keep animals out of tissue culture lab.
Anhydric incubators	Growth of molds and bacteria on spillages.	Wipe up any spillage with 70% alcohol on a swab. Clean out regularly.
CO_2, humidified incubators	Growth of molds and bacteria in humid atmosphere on walls and shelves. Spores, etc., carried on forced air circulation.	Clean out weekly with detergent and 70% alcohol. Enclose open dishes in plastic boxes with tight-fitting lids (but do not seal). Swab with 70% alcohol before opening. Fungicide or bacteriocide in humidifying water (but check first for toxicity).

*A one-to-one relationship between prevention and cause is not intended throughout this table. Preventative measures are interactive and may relate to more than one cause.

(1) Either obtain cell lines from a reputable cell bank where appropriate characterization has been performed (see Chapter 17 and Appendix) or perform the necessary characterization yourself as soon as possible (see Chapter 13).

(2) Do not have bottles or flasks of more than one cell line open simultaneously.

(3) Handle rapidly growing lines, such as HeLa, on their own and after other cultures.

(4) Never use the same pipette for different cell lines.

(5) Never use the same bottle of medium, trypsin, etc., for different cell lines.

(6) Whenever possible, do not put a pipette back into a bottle of medium, trypsin, etc., after it has been in a culture flask containing cells. Add medium and any other reagents to the flask first and then add the cells last.

(7) Do not use unplugged pipette, even micropipettes for routine maintenance.

(8) Check the characteristics of the culture regularly and suspect any sudden change in morphology, growth rate, etc. Confirmation of cross contamination may be obtained by chromosome [Nelson-Rees and Flandermeyer, 1977] or isoenzyme [O'Brien and Kleiner, 1977] analysis (see Chapter 13).

CONCLUSIONS

Check living cultures regularly for contaminations using normal and phase-contrast microscopy, and for mycoplasmas by fluorescent staining of fixed preparations.

Do not maintain all cultures routinely in antibiotics. Grow at least one set of cultures of each cell line in the absence of antibiotics for a minimum of 2 wk at a time to allow cryptic contaminations to become overt.

Do not attempt to decontaminate a culture unless it is irreplaceable and then do so under strict quarantine.

Quarantine all new lines that come into your laboratory until you are sure that they are uncontaminated.

Do not share media or other solutions among cell lines or among operators.

Check cell line characteristics (see Chapter 13) periodically to guard against cross contamination.

It cannot be overemphasized that cross contaminations can and do occur. It is essential that the above precautions be taken and regular checks of cell strain characteristics be made.

It is a fundamental property of living organisms that they diversify to provide sufficient variation for selection to be a useful mechanism in the adaptation of the species to its environment. It is, therefore, not surprising that cells in culture behave in a similar fashion. Spontaneous variation and selection occur as with any microorganism.

During the evolution of a cell line from a primary culture and during subsequent maintenance as a cell line or purified cell strain, there is evidence of both phenotypic and genotypic instability. This arises as a result of variations in culture conditions, selective overgrowth of constituents of the cell population, and genomic variations.

Since the constitution of a culture may vary from time to time it is important: (1) to standardize the culture conditions; (2) to select a period in the life history of the cell line where variation is at a minimum; (3) to select a pure cloned and characterized cell strain if possible; and (4) to preserve a seed stock to recall into culture at intervals, to maintain consistency.

ENVIRONMENT

Environment has been discussed already in Chapter 7. Once the appropriate conditions have been adopted, they should be adhered to, as alterations in media, serum, substrates, etc., will alter phenotypic expression. Test batches of serum, select one that has the required properties, and reserve enough to be sufficient for 6 months to 1 year. Repeat the process before changing to a new batch (see also Chapter 7 and below).

SELECTIVE OVERGROWTH, TRANSFORMATION, AND SENESCENCE

Following isolation of a primary culture, the predominant phenotype may change as cells with a higher proliferative capacity will tend to overgrow the more slowly dividing and nondividing cells (see Chapters 2,

9, 11, 20, and 21). As an example, consider a culture taken from carcinoma of the colon. Following explantation, these cultures can be predominantly epithelial; but after the first subculture, the epithelial cells steadily lose ground, and the fibroblasts take over. Preserving the epithelial component during this phase is a major problem and requires various selective culture and separation techniques (see Chapters 11 and 12). Occasionally, however, a transformed line may appear without using selective conditions. This may be a minority component of the original culture which has undergone transformation and now has increased growth capacity (shorter doubling time, infinite survival), and will ultimately outgrow the fibroblasts which grow more slowly and have a finite life-span in culture.

To counter cell line variation, finite cell lines should be used between certain generation limits; e.g., human diploid fibroblasts gain a fair degree of uniformity by the tenth generation and remain fairly stable up to about the 30th. Beyond 40 generations, senescence can be anticipated, although some lines may survive longer.

The culture should be grown through the first five or six generations until sufficient cells have been produced to freeze down a seed stock (see below).

GENETIC INSTABILITY

Evidence of genetic rearrangement can be seen in chromosome counts (see Fig. 2.2) and karyotype analysis. While the mouse karyotype is made up exclusively of small telocentric chromosomes, several metacentrics are apparent in many continuous murine cell lines (Robertsonian fusion). Furthermore, while virtually every cell in the animal has the normal diploid set, this is more variable in culture. In extreme cases, e.g., continuous cell strains such as HeLa-S3, less than half of the cells will have exactly the same karyotype, i.e., they are *heteroploid*.

Most continuous cell strains, even after cloning, contain a range of genotypes which are constantly changing. As transformation often involves chromosomal rearrangement, it is possible that it can only occur in cells with the capacity for chromosomal alterations. Alternatively, transformation may cause genetic instability to arise in a previously stable genotype. Hence, transformed continuous lines retain a capacity for genetic variation which is not apparent *in vivo* or in many finite cell lines.

There are two main causes of genetic variation: (1) the spontaneous mutation rate appears to be higher *in vitro*, associated, perhaps, with the high rate of cell proliferation, and (2) mutant cells are not eliminated unless their growth capacity is impaired. It is not surprising that phenotypic variation will arise as a result of this genetic variation. Minimal deviation rat hepatoma cells, grown in culture, express tyrosine aminotransferase activity constitutively and may be induced further by dexamethasone [Granner et al., 1968], but, as can be seen from Figure 17.1, subclones of a cloned strain of H4-II-E-C3 [Pitot et al., 1964] differed both in the constitutive level of the enzyme and in its capacity to be induced by dexamethasone.

PRESERVATION

In order to minimize genetic drift in continuous cell lines, to avoid senescence or transformation in finite cell lines, and to guard against accidental loss by contamination, it is now common practice to freeze aliquots of cells to be thawed out at intervals as required.

Selection of Cell Line

A cell line is selected with the required properties. If it is a finite cell line, it is grown to between the fifth and tenth population doubling to create sufficient bulk of cells for freezing. Continuous cell lines should be cloned (see Chapter 11) and an appropriate clone selected and grown up to sufficient bulk to freeze. Prior to freezing, the cells must be characterized (see Chapter 13) and checked for contamination (see Chapter 16), particularly cross contamination.

Continuous cell lines have the advantage that they will survive indefinitely, grow more rapidly, and can be cloned more easily; but they may be less stable genetically. Finite cell lines are usually diploid or close to it, stable between certain passage levels, but are harder to clone, grow more slowly, and will eventually die out or transform (Table 10.3).

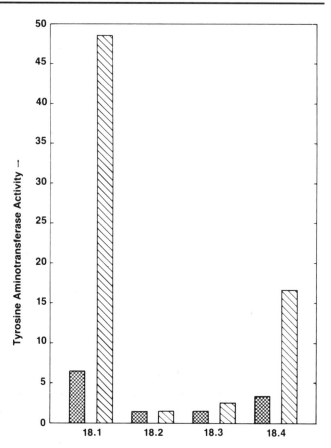

Fig. 17.1. *Variation in tyrosine aminotransferase activity between four subclones of clone 18 of a rat minimal deviation hepatoma cell strain. H-4-II-E-C3 cells were cloned; clone 18 was isolated, grown up, and recloned, and the second-generation clones were assayed for tyrosine aminotransferase activity with and without pretreatment of the culture with dexamethasone. Crosshatching, basal level; hatching, induced level. Data provided by J. Somerville.*

Standardization of Medium and Serum

The type of medium used will influence the selection of different cell types and regulate their phenotypic expression (see Chapters 2, 7, 14, and 20). Consequently, once a medium has been selected, standardize on that medium and, preferably, on one supplier if it is being purchased ready-made.

Variation in serum. Considerable variation [see Olmsted, 1967; Honn et al., 1975] may be anticipated between batches of serum resulting from differing methods of preparation and sterilization, different ages and storage conditions, and variations in animal stocks from which the serum was derived, including strain differences, pasture, climate, and so on. It is important to select a batch, use it for as long as possible, and replace it, eventually, with one as similar as possible (see Chapter 7). Better still, convert to serum-free medium.

Cell Freezing

When a cell line has been produced, or a cloned cell strain selected, with the desired characteristics and free of contamination, then a "seed stock" should be stored frozen. This will protect the cell line from change by genetic drift, from risk of contamination, and from technical problems such as incubator failure.

For a line in constant and general use it is a good plan to keep the "seed stock" inviolate and work from a "using stock." When the using stock is exhausted then seed stock may be used but only after making sure that is has not become depleted and replacing it if it has without allowing too many extra generations to pass.

Storage in liquid nitrogen (Fig. 17.2; see also Chapter 4) is currently the most satisfactory method of preserving cultured cells. The cell suspension, preferably at a high concentration, should be frozen slowly, at 1°C per min [Leibo and Mazur, 1971; Harris and Griffiths, 1977] in the presence of a preservative such as glycerol or dimethyl sulphoxide [Lovelock and Bishop, 1959]. The frozen cells are transferred rapidly to liquid nitrogen when they reach −50°C or below and are stored immersed in the liquid nitrogen or in the gas phase above the liquid. When required, the cells are thawed rapidly and reseeded at a relatively high concentration to optimize recovery.

If liquid nitrogen storage is not available, cells may be stored in a conventional freezer. The temperature should be as low as possible, but significant deterioration may yet occur even at −70°C. Little deterioration is found at −196°C [Greene et al., 1967].

Outline

The culture is grown to late log phase and a high cell density suspension is prepared and frozen slowly with a preservative (Fig. 17.3). When required, aliquots are thawed rapidly and reseeded at high cell density.

Materials

Cultures to be frozen
if monolayer: PBSA and 0.25% crude trypsin
medium
hemocytometer or cell counter
dimethylsulphoxide or glycerol

glass or plastic ampules—if glass, ampule sealer with gas/O$_2$ burner
canes or racks for storage
cotton wool
polystyrene box
forceps
protective gloves

To thaw:
Protective gloves and face mask
bucket of water at 37°C with lid
forceps
70% alcohol
swab
culture flask
centrifuge tube
culture medium

Protocol

1.
Check culture for (a) healthy growth, (b) freedom from contamination, and (c) presence of specific characteristics required of the line for subsequent use (viral propagation, differentiation, antigenic constitution, etc.).

2.
Grow up to late log phase and, if monolayer, trypsinize—if suspension, centrifuge.

3.
Resuspend at approximately 5×10^6–2×10^7 cells/ml in culture medium containing serum and a preservative such as dimethyl sulphoxide (DMSO) or glycerol at a final concentration of 5–10%. The preservative must be pure and free from contamination. DMSO should be colorless. Glycerol should be not more than 1-yr old as it may become toxic after prolonged storage.

It is *not* advisable to place ampules on ice in an attempt to minimize deterioration of the cells. A delay of up to 30 min at room temperature is not harmful when using DMSO and is beneficial when using glycerol

◊ DMSO is a powerful solvent. It will leach impurities out of rubber and some plastics and should, therefore, be kept in its original stock bottle or in glass tubes with glass stoppers. It can also penetrate many synthetic and natural membranes including *skin* and rubber gloves [Horita and Weber, 1964]. Consequently, any potentially harmful substances in regular use (e.g., carcinogens) may well be carried into the circulation

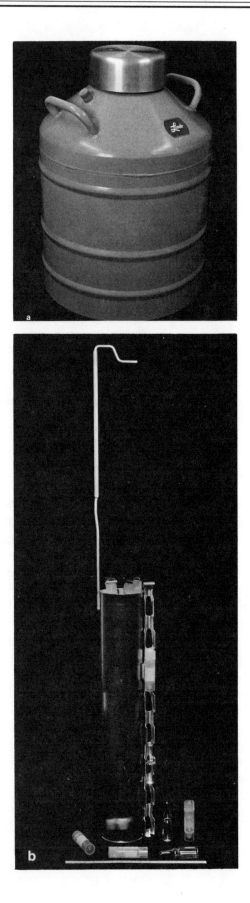

Fig. 17.2. *Liquid nitrogen freezers. a. Narrow-necked freezer with storage on canes in cannisters, b. c. Wide-necked freezer, storage in cannisters or in drawers, d; see also Fig. 17.6.*

through the skin and even through rubber gloves. DMSO should always be handled with caution because of its known hazardous potential, particularly in the presence of any toxic substances.
4.
Dispense cell suspensions into 1- or 2-ml prelabeled glass or plastic ampules and seal (Fig. 17.4, 17.5).
◇ Glass ampules are still widely used but must be perfectly and quickly sealed. If sealing takes too long, the cells will heat up and die, and the air in the ampule will expand and blow a hole in the top

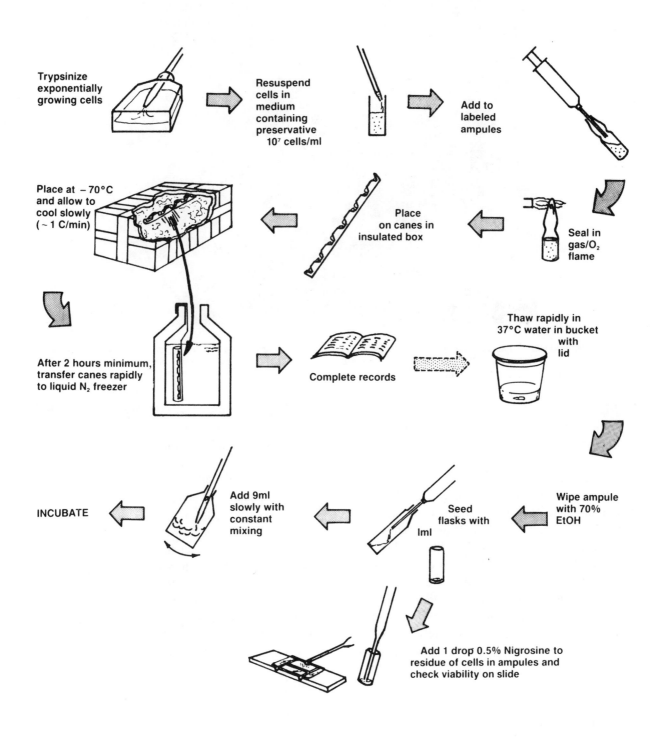

Fig. 17.3. *Preparation of cells for freezing and subsequent recovery after storage.*

Fig. 17.4. *Semiautomatic ampule sealer (Kahlenburg-Globe). Ampules are placed between the rollers and the flat plate by the operator's left hand. As the jig carrying the burner is moved* *forward, the ampule rotates in the flame and is finally ejected against the plate at the bottom of the picture.*

(Fig. 17.5d). If the ampule is not perfectly sealed it may inspire liquid nitrogen during freezing and storage in the liquid phase of the nitrogen freezer, and will subsequently explode (violently) on thawing, or it may become infected.

It is possible to check for leakage by placing ampules in a dish of 1% methylene blue in 70% alcohol at 4°C for 10 min. If the ampules are not properly sealed, the methylene blue will be drawn in and the ampule should be discarded. The ampules may need to be relabeled after this procedure. This step is inconvenient and time-consuming and may only be necessary when sealing glass ampules for the first few times. When experience is obtained, a well-sealed ampule may be recognized by the appearance of the tip (Fig. 17.5f).

Plastic ampules (e.g. Nunclon) are unbreakable but must be sealed with the correct torsion on the screw cap or they may leak. They are of a larger diameter and taller than equivalent glass ampules. Check that they will fit in the canes or racks used for storage. (Special canes for plastic ampules are available.)

Plastic ampules may be more convenient for the average experimental and teaching laboratory but repositories and cell banks generally prefer glass ampules as the long-term storage properties of glass are well characterized, and, correctly performed, sealing is absolute.

5.

Place ampules on canes for cannister storage or leave loose for drawer storage. Lay on cotton wool in a polystyrene foam box with a wall thickness of ~ 15 mm. This, plus the cotton wool, should provide sufficient insulation such that the ampules will cool at 1°C/min when the box is placed at −70°C or −90°C in a regular deep freeze or insulated container with solid CO_2. Tubular foam pipe insulation can also be used for ampules on canes. Most cultured cells survive best at this cooling rate; but if recovery is low, try a faster rate (i.e., less insulation) [Leibo and Mazur, 1971].

Programmable coolers are available to control the cooling rate (Fig. 17.6), usually by sensing the temperature of the ampule, and running in liquid N_2 at the correct rate, to achieve a preprogrammed cooling rate. They are, however, very expensive,

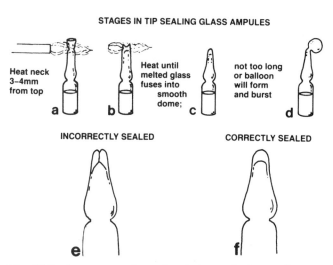

STAGES IN TIP SEALING GLASS AMPULES

Heat neck 3–4mm from top

Heat until melted glass fuses into smooth dome;

not too long or balloon will form and burst

a   b   c   d

INCORRECTLY SEALED   CORRECTLY SEALED

e   f

Fig. 17.5. *Appearance of ampules during and after sealing. a. Tip melting. b. Molten glass folds inward. c. Sides coalesce to form dome. d. Overheating can cause air in the ampule to blow out the glass forming a balloon, or bursting it completely. e. Incorrectly sealed ampule; fine capillary hole left. f. Correctly sealed ampule; glass fused evenly inside and outside.*

and have few advantages unless you wish to vary the cooling rate [e.g., Foreman and Pegg, 1979].

Union Carbide also markets a special neck plug which carries ampules and places them in the neck of the freezer at the desired level to obtain a set cooling rate (Fig. 17.7).

6.

When the ampules have reached $-70°C$ (1½–2 hr after placing at $-70°C$ if starting from 20°C ambient), transfer to liquid N_2 freezer. This must be done quickly as the cells will deteriorate rapidly if the temperature rises above $-50°C$.

◇ Protective gloves and face mask should be used when handling liquid nitrogen.

There are four main types of liquid nitrogen storage systems (Fig. 17.8), based on whether the storage vessel is wide-necked or narrow-necked, and whether storage is in the vapor or liquid phase. Wide-necked freezers are chosen for ease of access and maximum capacity, narrow for economy (slow evaporation rate). Storage in the vapor phase eliminates the explosion risk with sealed ampules while storage under the liquid means the container can be filled and the liquid nitrogen will therefore last longer. Storage capacities and static holding times (time to boil dry without being opened) are given in manufacturing literature.

◇ Biohazardous material *must* be stored in the gas phase and teaching and demonstrating are best done with gas-phase storage. Above all, if liquid-

phase storage is used, the user must be made aware of the explosion hazard of both glass and plastic, and wear a face shield or goggles.

7.

When ampules are safely located in freezer, make sure that the appropriate entries are made in the freezer index (Tables 17.1, 17.2). Records should contain (a) freezer index showing what is in each part of the freezer and (b) a cell strain index, describing the cell line, its designation, what its special characteristics are, and where it is located. This may be done on a conventional card index, but a computerized data file will give superior data storage and retrieval. Material stored on discs or tape must have back-up copies on disc, tape, or hard copy printout.

The ampules should also carry a label with the cell strain designation and preferably the date and user's initials, although this is not always feasible on the available space.

Remember, cell cultures stored in liquid nitrogen may well outlive you! They can easily outlive your stay in a particular laboratory. The record should be, therefore, readily interpreted by others and sufficiently comprehensive so that the cells may be of use to others.

Thawing

Protocol

1.

When you wish to recover cells from the freezer, check the index, label your culture flask, then retrieve the ampule from the freezer, check that it is the correct one, and place it in 10 cm of water at 37°C in a 5-1 bucket with a lid.

◇ A face shield or protective goggles, and gloves must be worn. If the ampule has been stored in the gas phase, a lid is not necessary; but ampules stored under liquid nitrogen may inspire the liquid, and on thawing, will explode violently. A plastic bucket with a lid is, therefore, essential in this case to contain any explosion.

2.

When thawed, swab the ampule with 70% EtOH and break open. (Prescored ampules are available; these are much easier to open.) Transfer the contents to a culture flask. The dregs in the ampule may be stained with nigrosine to determine viability (see Chapter 19).

TABLE 17.1. Record Cards for Cells in Frozen Storage—Freezer Card

Position:

Freezer no._____ Cannister/section no._____ Tube/drawer no._____

Cell strain/line_____ Freeze date_____ Frozen by_____

No. ampules frozen_____ No. cells/ampule_____ in_____ml

Growth medium_____ Serum_____ Conc._____ Freeze medium_____

Method of cooling_____ Cooling rate_____

Thaw date	No. of amp.	No. left	Seeding			Take*			Notes
			Conc.	Vol.	Medium	Dye exclusion	% Attached by 24 h	Cloning efficiency	

*Only one parameter need be used

TABLE 17.2. Record Cards for Cells in Frozen Storage—Cell Line/Cell Strain Card

Cell clone designation | Freezing date | Location

Origin:

Species_____Normal/Neoplastic Adult/fetal/NB

Tissue_____Site_____ Gen/Pass No.____

Author_____Lab/ref_____

Morphology_____ Mode of growth_____

Special characteristics e.g., karyotype, isoenzymes,

 differentiation markers, virus sensitivity, antigens, products:

_____

_____

_____

(*Further details on back of card*)

Mycoplasma:

Date of test____ Method_____ Result_____

Normal maintenance:

Subculture

Frequency_____ Min. cell conc_____ Agent_____

Medium change

Frequency____ Whole/part vol_____Medium_____Serum_____%_

Gas phase_____ Buffer_____ pH_____ Matrix?_____

Any other special conditions_____

_____

Freezing instructions:

Rate_____ Preservative_____%____

Thawing instructions:

Thaw rapidly (37°C), fresh water bath, dilute to 5/10/20/50 ml.

Centrifuge to remove preservative: Yes_____ No_____

Notes or special requirements:

_____

Biohazard precautions:

_____

Name of person completing card_____Date_____

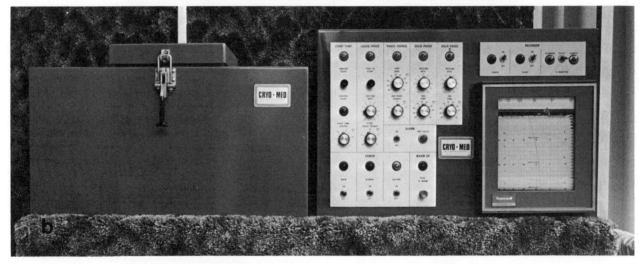

Fig. 17.6. *Programmable coolers capable of regulating the cooling rate before and after freezing. Ampules are placed in an insulated box and the cooling rate regulated by injecting liquid N$_2$ into the box at a rate determined by a sensor in the box and a preset program in the consol unit (Union Carbide, Cryo-med).*

Add medium slowly to the cell suspension: 10 ml over about 2 min added dropwise at the start and then a little faster, gradually diluting the cells and preservative. This is particularly important with DMSO where sudden dilution can cause severe osmotic damage and reduce survival by half.

In some cases, e.g., L5178Y lymphoma with DMSO, the preservative may be too toxic for the cell to survive on thawing, [M. Freshney, personal communication]. In these cases, dilute the cells slowly, as above, and centrifuge for 2 min at 100 *g*, discard the supernatant medium with preservative, and resuspend the cells in fresh medium for culture.

The number of cells frozen should be sufficient to allow for 1:10 or 1:20 dilution on thawing to dilute out the preservative but still keep the cell concentration higher than at normal passage, e.g., for cells subcul-

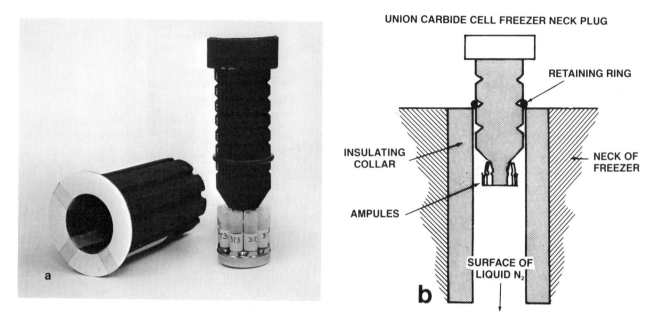

UNION CARBIDE CELL FREEZER NECK PLUG

RETAINING RING

INSULATING COLLAR

NECK OF FREEZER

AMPULES

SURFACE OF LIQUID N₂

Fig. 17.7. *Modified neck plug for narrow-necked freezers allowing controlled cooling at different rates. The "O" ring is used to set the height of the ampules within the neck of the freezer. The lower the height, the faster the cooling.*

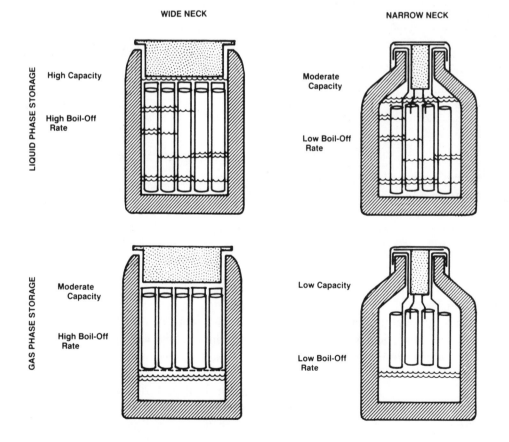

Fig. 17.8. *Variations in freezer design and usage.*

TABLE 17.3. Cell Banks and Indices*

U.S.A.	American Type Culture Collection (ATCC), Rockville, MD
	Catalog of Human and Other Animal Cell Cultures, Naval Biosciences Laboratory, Cell Culture Department Naval Supply Center, Oakland, CA
	Human Genetic Mutant Cell Culture and Ageing Cell Culture Repositories, Institute for Medical Research, Camden, NJ
Europe	Culture de Cellules Eucaryotes. Repertoire des Utilisateurs., M. Adolphe, D. Gourdji, A. Tixier-Vidal, R. Robineaux, Inserm Publications, 101 Rue de Tolbiac, 75654 Paris, Cedex 13
	Department of Medical Virology, Institute Pasteur, Paris
	European Collection for Animal Cell Cultures (ECACC), PHLS/CAMR, Porton, Salisbury, England
	European Human Genetic Mutant Cell Bank, Dept. of Clinical Genetics, Erasmus University, Rotterdam
Japan	Collection of Cancer Cell Lines, National Institute of Hygenic Sciences, Tokyo
	General Cell Bank, Institute of Physical and Chemical Research of RIKEN, Saitama

*Commercial companies such as Flow Laboratories and GIBCO will also supply cell cultures.

tured normally at 10^5 ml, 10^7 should be frozen in 1 ml, and the whole seeded, after thawing into 20 ml, giving 5×10^5/ml ($5 \times$ the normal seeding density) and diluting the preservative from 10% to 0.5%, at which concentration it is less likely to be toxic. Residual preservative may be diluted out as soon as the cells start to grow (for suspension cultures) or the medium changed as soon as the cells have attached (for monolayers).

The dye exclusion viability and the approximate take (e.g., proportion attached versus those still floating after 24 hours) should be recorded on the appropriate record card or file to assist in future thawing. One ampule should be thawed from each batch as it is frozen, to check that the operation was successful.

During a prolonged series of experiments, lasting more than 6 months, stock cultures should be replaced from the freezer at regular intervals, say every 3 months, thereby preserving the properties of the cell line.

A seed stock and using stock should be frozen down for valuable cell lines in regular use. The using stock should be sufficient for the forseeable future assuming culture stock replacement every 3 months. In addition, 10–12 ampules should be frozen at the lowest passage level possible (for primary cultures) or soon after receipt if an imported cell line. This seed stock must not be used unless the using stock runs out and if it is, the ampules must be replaced as soon as it falls below five ampules, with the least gain in passage level possible.

Although variability cannot be eradicated entirely from cultured cell lines, it may be minimized by adopting the procedures described above. In summary: (1) select finite lines around the fifth generation, or clone, isolate, and grow up a continuous cell strain; (2) confirm characteristics of the strain; (3) check for microbial and cross contamination; (4) freeze down a large number of ampules; (5) thaw one ampule to test survival; and (6) replace stock regularly from freezer.

CELL BANKS

To assist in the distribution of standardized, characterized cell lines and to provide secure repositories for valuable material, a number of cell banks have been set up in different parts of the world (Table 17.3). Since many cell lines may come under patent restrictions, particularly hybridomas and other genetically modified cell lines, it has also been necessary to provide patent repositories with limited access.

As a general rule it is generally preferable to obtain your initial seed stock from a reputable cell bank where the necessary characterization and quality control will have been done. Furthermore, it is a good plan to submit valuable cultures to a cell bank for preservation in addition to maintaining your own frozen stock, as this will protect you against loss of your own lines and allow distribution to others. If you feel that your cells should not be distributed, then they can be banked with that restriction placed on them.

In the design of cell culture experiments it is important to be aware of the growth state of the culture as well as the qualitative characteristics of the cell strain or cell line. Cultures will vary significantly in many of their properties between exponential growth and stationary phase. It is, therefore, important to take account both of the status of the culture at the time of sampling, and the effects of the duration of an experiment on the transition from one state to another. Adding a drug in the middle of the exponential phase and assaying later may give different results depending on whether the culture is still in exponential growth when harvested or whether it has entered plateau. Cells that have entered plateau have a greatly reduced growth fraction, a different morphology, may be more differentiated, and may become polarized. They generally tend to secrete more extracellular matrix and may be more difficult to disaggregate.

Quantitation of growth is also important in routine maintenance as an important element in monitoring the consistency of the culture and knowing the best time to subculture, the optimum dilution, and the estimated plating efficiency at different cell densities. Testing medium, serum, new substrates, etc., all require quantitative assessment.

The first part of this chapter will consider the various methods commonly used to quantify cells in culture, such as cell counting, biochemical determinations, and cytometry. The second part will deal with the kinetics of cell growth *in vitro* as determined by studying the growth cycle, the cell cycle, plating efficiency, and the growth fraction.

CELL COUNTING

While estimates can be made of the stage of growth of a culture from its appearance under the microscope, standardization of culture conditions and proper quantitative experiments are difficult unless the cells are counted before and after, and preferably during, each experiment.

Hemocytometer

The concentration of a cell suspension may be determined by placing the cells in an optically flat chamber under a microscope (Fig. 18.1). The cell number within a defined area is counted and the cell concentration derived from the count.

Materials

> Hemocytometer (Improved Neubauer)
> PBSA
> 0.25% crude trypsin
> medium
> tally counter
> Pasteur pipette
> microscope

Protocol

1.

Prepare the slide: (a) Clean the surface of the slide with 70% alcohol taking care not to scratch the semi-silvered surface. (b) Clean the coverslip and, wetting the edges very slightly, press it down over the grooves and semi-silvered counting area (see Fig. 18.1). The appearance of interference patterns ("Newton's rings"—rainbow colors between coverslip and slide like those formed by oil on water) indicates that the coverslip is properly attached, thereby determining the depth of the counting chamber.

2.

Trypsinize monolayer or collect sample from suspension culture. Approximately 10^5 cells/ml minimum are required for this method so the suspension may need to be concentrated by centrifuging (100 g for 2 min) and resuspending in a smaller volume.

3.

Mix the sample thoroughly and collect about 20 μl into the tip of a Pasteur pipette or micropipette.

227

Do not let the fluid rise in a Pasteur pipette or cells will be lost in the upper part of the stem.

4.

Transfer the cell suspension immediately to the edge of the hemocytometer chamber and let the suspension run out of the pipette and be drawn under the coverslip by capillarity. Do not overfill or underfill the chamber or its dimensions may change due to surface tension; the fluid should run to the edges of the grooves only. Reload the pipette and fill the second chamber if there is one.

5.

Blot off any surplus fluid (without drawing from under the coverslip) and transfer the slide to the microscope stage.

6.

Select × 10 objective and focus on grid lines in chamber (see Fig. 18.1). If focusing is difficult because of poor contrast, close down the field iris or make the lighting slightly oblique by tilting the mirror or offsetting the condenser.

7.

Move the slide so that the field that you see is the central area of the grid and is the largest area that you can see bounded by three parallel lines. This area is 1 mm^2. Using a standard 10 × objective, it will almost fill the field, or the corners will be slightly outside the field, depending on the field of view (see Fig. 18.1).

8.

Count the cells lying within this 1-mm^2 area, using the subdivisions (also bounded by three parallel lines) and single grid lines as an aid to counting. Count cells which lie on the top and left-hand lines of each square but not those on the bottom or right-hand lines to avoid counting the same cell twice. For routine subculture, attempt to count between 100 and 300 cells per mm^2; the more cells that are counted, the more accurate the count becomes. For more precise quantitative experiments, 500–1,000 cells should be counted.

9.

If there are very few cells (< 100/mm^2), count one or more additional squares (each 1 mm^2) surrounding the central square.

10.

If there are too many cells (> 1,000/mm^2), count only five small squares (each bounded by three parallel lines) across the diagonal of the larger (1-mm) square.

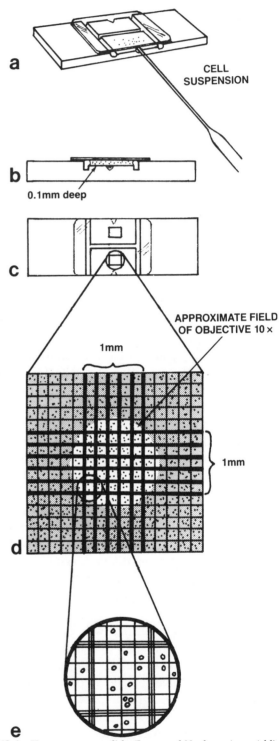

Fig. 18.1. *Hemocytometer slide (Improved Neubauer). a. Adding cell suspension to the assembled slide. b. Longitudinal section of slide showing position of cell sample in 0.1-mm-deep chamber. c. Top view of slide. d. Magnified view of total area of grid. Light central area is that area which would be covered by the average × 10 objective (depending on field of view of eye piece). This covers approximately the central 1 mm^2 of the grid. e. Magnified view of one of the 25 smaller squares, bounded by triple parallel lines, making up the 1-mm^2 central area. This is subdivided by single grid lines into 16 small squares to aid counting.*

11.

If the slide has two chambers, move to the second chamber and do a second count. If not, rinse the slide and repeat the count with a fresh sample.

Analysis. Calculate the average of the two counts, and derive the concentration of your sample as follows:

$$c = \frac{n}{v}$$

where c = cell concentration (cells/ml), n = number of cells counted, and v = volume counted (ml). For the Improved Neubauer slide, the depth of the chamber is 0.1 mm, and assuming only the central 1 mm^2 is used, v is 0.1 mm^3 or 10^{-4} ml. The formula becomes:

$$c = n \times 10^4.$$

Hemocytometer counting is cheap and gives you the opportunity to see what you are counting. If the cells are mixed previously with an equal volume of a viability stain (see below), a viability determination may be performed at the same time. The procedure is, however, rather slow and prone to error both in the method of sampling and the size of samples and requires a minimum of 10^5 cells/ml.

Most of the errors occur by incorrect sampling and transfer of cells to the chamber. Make sure the cell suspension is properly mixed before you take a sample, and do not allow the cells time to settle or adhere in the tip of the pipette before transferring to the chamber. Ensure also that you have a single cell suspension as aggregates make counting inaccurate. Larger aggregates may enter the chamber more slowly or not at all.

If aggregation cannot be eliminated during preparation of the cell suspension (see Table 10.1), lyse the cells in 0.1 M citric acid containing 0.1% crystal violet at 37°C for 1 hr and count the nuclei [Sanford et al., 1951].

Electronic Particle Counting

Although a number of different automatic methods have been developed for the counting of cells in suspension, the system devised originally by Coulter Electronics is the one most widely used. Briefly, cells drawn through a fine orifice change the current flow through the orifice, producing a series of pulses which are sorted and counted.

Coulter Counter

(Fig. 18.2). There are two main components of the system (Fig. 18.3): (1) an orifice tube connected to a pump and a mercury manometer by a two-way valve (Fig. 18.3a); and (2) an amplifier, pulse height analyzer, and scaler connected to two electrodes—one in the orifice tube and one in the sample beaker—the current to them controlled by switch points on the mercury manometer.

When the two-way valve is turned vertically, the mercury manometer and orifice tube are connected to the pump (Fig. 18.3a). While liquid is drawn through the orifice generating a signal on the cathode ray oscilloscope, the mercury is drawn up the manometer to a preset level determined by the negative pressure generated by the pump. When the valve is restored to the horizontal position (Fig. 18.3b), the pump is disconnected, and the mercury monometer is connected directly to the orifice tube. As the mercury returns to equilibrium, fluid carrying the cell suspension is drawn through the orifice.

As the mercury travels along the tube, it passes two switch points; the first starts the count cycle, the second stops it. The mercury displacement between the two switches is 0.5 ml, hence 0.5 ml of cell suspension is drawn in through the orifice during the count cycle.

As each cell passes through the orifice, it changes the resistance to the current flowing through the orifice by an amount proportional to the volume of the cell. This generates a pulse (amps^{-1}) which is amplified and counted. Since the size of the pulse is proportional to the volume of the cell or particle passing through the orifice, a series of signals of varying pulse height are generated. A threshold control is set on the front panel to eliminate electronic noise and fine particulate debris but to retain pulses derived from cells (see below). This setting controls a pulse height analyzer circuit between the amplifier and the scaler which only allows pulses to pass to the scaler above the preset threshold.

Operation of Coulter Cell Counter (Model DI)

Outline

A sample of cells is diluted in electrolyte (physiological saline or PBS) and placed under the orifice tube and counted by drawing 0.5 ml of the sample through the counter.

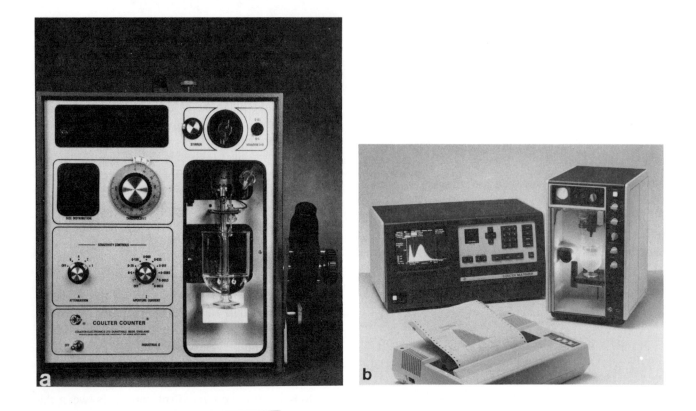

Fig. 18.2. *Coulter electronic cell counter and cell sizer. (a) D Industrial, suitable for routine laboratory use for cell counting. (b) Multisizer. Can be used for cell counting, but is primarily intended for cell sizing.*

Materials

    Culture
    PBSA
    0.25% crude trypsin
    medium
    counting cup
    counting fluid (see Appendix)

Protocol

1.

Trypsinize cell monolayer or collect sample from suspension culture. The cells must be well mixed and singly suspended.

2.

Dilute the sample of cell suspension 1:50 in 20 ml counting fluid in a 25-ml beaker or disposable sample cup. An automatic dispenser will speed up this dilution and improve reproducibility. Dispen-sing counting fluid rapidly can generate air bubbles which will be counted as they pass through the orifice. Consequently, the counting fluid should stand for a few moments before counting. If the fluid is dispensed first and cells added second this problem is minimized.

3.

Mix well and place under tip of orifice tube, en-suring that the orifice is covered and that the external electrode lies submerged in the counting fluid in the sample beaker.

4.

Set the two-way valve vertically until red light comes on and then extinguishes (mercury has passed both switch points).

5.

Clear the display.

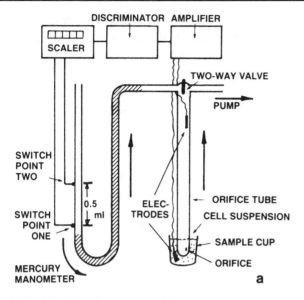

Fig. 18.3. *Principle of operation of electronic particle counter (based on Coulter counter). a. Manometer connected to pump, drawing mercury up to starting position. b. Mercury returning to* equilibrium, drawing sample though orifice, and activating count cycle. Inset:magnified view of orifice in section.

6.
Restore two-way valve to the horizontal position and allow count to proceed (red light comes on).
7.
When the red light extinguishes, the count is finished. Note the count and replace the sample with fresh counting fluid.

Analysis. The counter takes 0.5 ml of the 1:50 dilution, so multiply the final count on the readout by 100 to give the concentration of the cell suspension.
Problems. 1. Count stops, will not start, or counts slowly (i.e., takes longer than 25 s):
Orifice clogged. Free with the tip of finger or fine brush.
2. Count lower than expected, orifice blocks frequently. Cell suspension aggregated:
Disperse cells by pipetting original sample vigorously, redilute, and proceed.
3. High electrical activity on oscilloscope screen but will not count:
Electrode out of beaker or disconnected.
4. Red light comes on but will not go out, or will not come on at all:
(a) Blocked orifice as in 1, (b) insufficient negative pressure. Pump has failed or leakage in tubing. Check pump and connections.
5. Gurgling sound from pump:
Counting fluid from waste reservoir has been drawn into pump:

Switch off, disconnect, and dry out pump. Relubricate, reconnect, and start again. The level in the waste reservoir should be checked regularly so that it may be emptied before the pump becomes contaminated.
6. High background (a) line or radio interference, usually from electrical equipment (motors, fluorescent lights, incubators):
Check and eliminate by fitting suppressors to equipment. Line filters are available but are not always effective. Check grounding (earthing) of counter, particularly the case.
(b) Particulate matter in counting fluid:
Filter through disposable Millex or equivalent filter.
Calibration. To set the threshold in the correct position, perform a series of counts on the cell suspension, moving the threshold up in increments of 5 or 10 units from zero to 100. Plotting the counts against the threshold settings should produce a curve with a plateau (Fig. 18.4a). If the plateau is too short, decrease the attenuation and repeat. If the curve is too drawn out and does not reach a plateau, increase the attenuation.
Cell sizing. If the amplification is set such that the curve of cell counts versus threshold reaches a value at 100, which is <5% of the count at 5, then the bulk of the cell population has been analyzed and differentiating the curve (subtracting each count from the previous count at a lower threshold setting) will give a plot of relative size distribution (Fig. 18.4b). If standard latex particles are treated in the same way, a standard curve

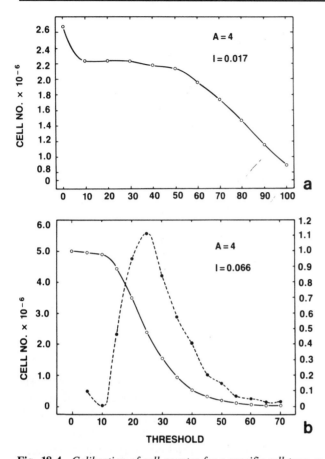

Fig. 18.4. *Calibration of cell counter for a specific cell type. a. A sample of LS cells was counted at a range of threshold values. The correct threshold setting is that equivalent to the center of the plateau, i.e., 20. b. Repeat counts (higher cell concentration) with aperture current reset to 0.066. Plateau is now between zero and ten (unsuitable for routine counting) and cell count falls to near zero by threshold setting 70. By differentiating the first curve (solid line) a cell size analysis is obtained (dotted line). (Counters with automatic simultaneous pulse height analysis will plot this curve automatically.) A = attenuation. I = aperture current.*

is obtained and the absolute cell size may be derived from the formula

$$v = KAIT$$

where v = cell volume, A = attenuation setting (inverse of amplification), I = current setting, T = threshold setting, and K = constant. Derive K from standard latex particles and substitute in above equation to derive v for your cell sample.

Counters are available with automatic cell sizing capabilities (Fig. 18.2b). They operate on the same principle but instead of controlling the threshold manually, the pulses from the amplifier are fed into a more sophisticated pulse height analyzer and sorter which gives an instant readout on an oscilloscope and will

print the size distribution histogram on an X-Y recorder. A counter with this facility will cost about three or four times as much as the simple version described above but makes cell sizing faster, easier, and more accurate.

Electronic cell counting is rapid and has a low inherent error due to the high number of cells counted. It is prone to misinterpretation, however, as cell aggregates, dead cells, and particles of debris of the correct size will all be counted indiscriminately. The cell suspension should be examined carefully before dilution and counting.

Electronic particle counters are expensive, but if used correctly, they are very convenient and give greater speed and accuracy to cell counting. There are now several such instruments available, but the Coulter D (Industrial) remains one of the cheapest.

Stained Monolayers

There are occasions when cells cannot be harvested for counting or are too few to count in suspension. This situation is encountered with some multiwell plates or Terasaki plates. In these cases, the cells may be fixed and stained *in situ* and counted by eye on a microscope. Since this procedure is tedious and subject to high operator error, isotopic labeling or estimation of total DNA or protein (see below) is preferable though these measurements may not correlate directly with cell number, e.g., if the ploidy of the cell varies. A rough estimate of cell number per well can also be obtained by staining the cells with crystal violet and measuring absorption on a densitometer. This method has also been used to calculate the number of cells per colony in clonal growth assays [McKeehan et al., 1977].

CELL WEIGHT

Wet weight is seldom used unless very large cell numbers are involved, as the amount of adherent intracellular liquid gives a high error. As a rough guide, however, there are about 2.5×10^8 HeLa cells (14–16 μm diameter) per g wet weight, about 8–10×10^8 cells/g for murine leukemias (e.g., L5178Y or Friend) (11–12 μm diameter) and about 1.8×10^8 for human diploid fibroblasts (16–18 μm diameter) (See Table 18.1).

Dry weight is, similarly, seldom used, as salt derived from the medium contributes to the weight of unfixed cells and fixed cells will lose some of their low molecular weight intracellular constituents and lipids.

TABLE 18.1. Relationship Between Cell Size, Volume, and Mass

Cell type	Diameter (μm)	Volume (μm^3)	Cells/gm $\times$ 10^{-6} Calculated	Measured
Murine leukemia				
e.g., L5178Y or Friend	11–12	800	1,250	1,000
HeLa	14–16	1,200	800	250
Human diploid fibroblasts	16–18	2,500	400	180

DNA CONTENT

In practice, apart from cell number, DNA and protein are the two most useful measurements for quantifying the amount of cellular material. DNA may be assayed by several fluorescence methods, including the reactions with DAPI [Brunk et al., 1979] or Hoechst 33258 [Labarca and Paigen, 1980]. DNA can also be measured by its absorbance at 260 nm where 50 μg/ml has an O.D. of 1.0. Because of interference from other cellular constituents, this method is only useful for purified DNA. The following is a relatively simple and straightforward assay for DNA.

Determination of DNA Content of Cells by Hoechst 33258 [Labarca and Paigen, 1980]

Principle

The fluorescence emission of Hoechst 33258 at 458 nm is increased by interaction of the dye with DNA at pH 7.4 and in high salt to dissociate the chromatin protein. This method gives a sensitivity of 10 ng/ml, but requires intact double stranded DNA.

Outline

Cells or tissue are homogenized in buffer and sonicated. Aliquots are mixed with H33258 and the fluorescence measured.

Materials

Buffer: 0.05 M NaPO$_4$, 2.0 M NaCl pH 7.4 containing 2 $\times$ 10^{-3} M EDTA

H33258: 1 μg ml in buffer for DNA above 100 ng/ml and 0.1 μg ml for 10–100 ng/ml

Protocol

1.
Homogenize cells in buffer, 10^5/ml for 1 min using a Potter homogenizer.

2.
Sonicate for 30 s.

3.
Dilute 1:10 in Hoechst buffer.

4.
Read fluorescence at 356 nm excitation and 492 emission using calf thymus DNA as a standard.

Determination of DNA Content of Cells by DAPI

The following was contributed by Edith Schwartz (based on the method of Bounk et al., 1979), Departments of Orthopedic Surgery and Physiology, Tufts University School of Medicine, 136 Harrison Avenue, Boston, MA 02111.

Principle

DAPI (4′,6-diamidino-2-phenylindole) bound to DNA fluoresces when excited by long wavelength uv.

Outline

Lysed cells are mixed with DAPI in the dark and fluorescence measured on a fluorimeter.

Materials

DNA buffer solution: 0.01 M NaCl + 0.005 M Hepes buffer, pH 7.0.

DNA standard: prepare a stock solution of DNA by dissolving at room temperature 80 mg of calf thymus DNA into a final volume of 10 ml of distilled water. Divide this solution into 1-ml aliquots into Pyrex test tubes and store them at $-20°$.

DAPI stock (4′,6-diamidino-2-phenylindole): dissolve 300 mg DAPI in 1 ml DNA buffer solution. This solution is 100 $\times$.

Working solution of DAPI: to prepare the working solution of DAPI, add 100 μl of DAPI stock solution to 900 μl of DNA buffer solution and mix well. This yields 1 ml of DAPI (10

×). Add 0.5 ml of DAPI (10 ×) to 4.5 ml of DNA buffer solution to give 5 ml of DAPI working solution.
Assay dilution buffer: 0.1 M sodium acetate, pH 6.2.

Protocol

1.
Prepare a set of blanks and standards (in triplicate) containing from 0 to 0.8 μg DNA.
2.
Prepare test samples by suspending 5×10^4–5×10^5 cells in 1 ml of 0.1 M sodium acetate buffer, pH 6.2. Sonicate for a total of 30 s.
3.
Add 150 μl sonicated cell suspension in triplicate to 850 μl of DNA buffer solution.
4.
Vortex standards and tests and blanks (no DNA).
5.
In the dark, add 50 μl of DAPI working solution to each sample and standard. Vortex each tube immediately after adding DAPI solution. Cover tubes with foil.
6.
Let the tubes stand in the dark for 30 min.
7.
Measure the fluorescence of standards and tests against mean of blanks at 372 nm excitation and 454 nm emission wavelengths.
8.
Plot fluorescence (%) vs. g DNA in standards. Use these values to determine DNA content of test samples.

PROTEIN

Protein content is widely used for estimating total cellular material and can be used in growth experiments or as a denominator in expression of specific activity of enzymes, receptor content, or intracellular metabolite concentrations. Proteins in solubilized cells can be estimated directly by measuring absorbance at 280 nm, and in this case, as the bulk of the cell's dry weight is protein interference from nucleic acids and other constituents is minimal. Absorbance at 280 nm can detect down to 100 μg or about 2×10^5 cells.

Colorimetric assays are more sensitive, and among these the Lowry method [Oyama and Eagle, 1956] and the Bradford reaction with Coomassie blue [Bradford, 1976] are the most widely used.

Solubilization of Sample

Both methods rely on a final colorimetric step and must be carried out on clear solutions. Cell monolayers and cell pellets may be dissolved in 0.5–1.0 N NaOH by heating to 100°C for 30 min or leaving overnight at room temperature. Alternatively, with 0.3N NaOH and 1% sodium lauryl sulphate, solution is complete after 30 min at room temperature.

Lowry Method

The cell sample is precipitated with ice-cold 10% trichloracetic acid to remove the amino acid pools and dissolved in alkali, and protein is estimated colorimetrically after sequential addition of alkaline copper sulphate and Folin Ciocalteau reagent [after Lowry et al., 1951].

Principle

In a two-step reaction, Folin's reagent reacts with the aromatic amino acids in protein, after treatment with alkaline copper, to give a blue color.

Outline

Solubilize sample in 0.1N NaOH, add alkaline copper solution and then diluted Folin's solution, and read in a spectrophotometer.

Materials

2% Sodium carbonate in 0.1N NaOH (A)
0.5% copper sulphate in 1% sodium citrate solution (B)
1 ml B:50 ml A—make up with constant mixing and use on same day (C)
1 ml Folin Ciocalteau:1.3 ml water (D)
BSA standard solution 50 μg/ml in 0.1N NaOH
1N NaOH

Protocol

1.
Dissolve cell pellet in N NaOH and dilute to 0.1N.
2.
Add 1.0 ml of reagent C to 200 μl of protein sample, mix, and leave 10 min.
3.
Add 100 μl of reagent D, with constant mixing, and leave 40 min.
4.
Read on a spectrophotometer at 700 nm against a reagent blank and with a BSA standard, 50 μg/ml.

Protein measurements by the Lowry method are dependent on the presence of tyrosine and phenylalanine residues and will give underestimates if the frequency of these amino acids is low, as in nuclear histone proteins. For this reason, the Bradford method, which is not dependent on specific amino acids, has become more popular.

Bradford Method

Samples may be solubilized as before although removal of amino acid pools in 10% TCA is not necessary. The Bradford assay is more sensitive than Lowry's and only 50–100,000 cells are required. Color is generated in one step with a short incubation and read within 30 min.

Principle

Coomassie Blue undergoes a spectral change on binding to protein in acidic solution. This gives a more sensitive assay than Lowry, independent of aromatic amino acid frequency in the protein.

Outline

Protein is dissolved and mixed with color reagent and the O.D. read after 10 min.

Materials

0.01% in Coomassie Brilliant Blue G-250 in 4.7% EtOH and 85% (w/v) phosphoric acid: dissolve 100 mg Coomassie Blue in 50 ml 95% EtOH, add 100 ml 85% phosphoric acid, and dilute to 1 l.

Protocol

1.
Solubilize protein (1–20 μg) or cells (around 10^6) in 0.1% sodium dodecyl sulphate in water or 0.3N NaOH.
2.
Add 1.0 ml Coomassie Blue to 100 μl protein solution, mix, and let stand for 10 min.
3.
Read on spectrophotometer at 595 nm against a reagent blank and with a BSA standard curve (1–50 μg/ml).

Variations. Reagents available in kit from BioRad.

Protein Synthesis

Colorimetric assays measure the total amount of protein present at any one time. Sequential observations over a period of time may be used to measure net protein accumulation or loss (protein synthesized–protein degraded), while the rate of protein synthesis may be determined by incubating with radio-isotopically labeled amino acid, such as ^{3}H-leucine or ^{35}S-methionine, and measuring (e.g., by scintillation counting) the amount of radioactivity incorporated per 10^6 cells or per mg protein over a set period.

◇Radioisotopes must be handled with care and according to local regulations governing permitted amounts, authorized areas and disposal, and so forth.

Materials

Sterile:
Cell culture, 10^4–10^6 cells in, e.g., 24-well plate
^{3}H-leucine 50 μCi/ml in culture medium without serum
(Specific activity unimportant as this will be determined by the leucine concentration in the medium)
Non-sterile:
0.3N NaOH with 1% sodium lauryl sulphate
scintillation vials
Eppendorf tubes
scintillation fluid with 10% water tolerance, e.g., Instagel or Ecoscint
monolayer cells in 24-well plate.

Protocol

1.
Incubate culture to appropriate cell density.
2.
Remove from incubator and add prewarmed isotopic solution, 1:10, e.g., 100 μl per 1 ml/well.
3.
Return to incubator as rapidly as possible.
4.
Incubate for 4–24 hr.
Note: Different proteins turn over at different rates. This protocol is not aimed at any specific subset but at total protein in rapidly proliferating cells. When assaying protein synthesis in a cell line for the first time, check that the rate of synthesis is linear over the incubation time chosen. A lag may be encountered if the amino acid pool is slow to saturate.
5.
Remove from incubator, withdraw medium carefully from wells into radioactive liquid waste.

6.

Wash cells gently with cold HBSS or PBS.

Note: Some monolayer cultures may detach during washing, particularly some loosely adherent continuous cell lines such as HeLa-S$_3$. In this case add methanol, after removing isotope, to fix monolayer. Leave for 10 min, remove carefully and dry monolayer (see also Chapter 19).

7.

Place plates on ice. Add 10% trichloracetic acid, at 4°C, to remove unincorporated precursor. Leave 10 min.

8.

Repeat step 7 twice, 5 min each.

9.

Wash with MeOH and dry.

10.

Add 0.5 ml 0.3N NaOH, 1% SLS, and leave 30 min at room temperature.

11.

Mix contents well and transfer to scintillation vials.

12.

Add 5 ml scintillant and count.

Note: New biodegradable scintillants, e.g., Ecoscint, are now available. They are less toxic to handle than toluene or xylene based scintillants and can be poured down the sink with excess water provided the levels of radioactivity fall within the legal limits.

For suspension cultures spin (1000 *g*, 10 min) at step 5 to remove medium and at 6 and after 7 and 8; omit 9.

DNA Synthesis

Measurements of DNA synthesis are often taken as representative of cell proliferation (see also Chapter 19). Incorporation of ^{3}H-thymidine (^{3}H-TdR) or ^{3}H-deoxycytidine are the usual precursors used. Exposure may be for short periods (1/2–1 hr) for rate estimations or for 24 hr or more to measure accumulated DNA synthesis where the basal rate is low, e.g., in high-density culture. ^{3}H-TdR should not be used for incubations longer than 24 hr or at high specific activities, as radiolysis of DNA will occur due to the short path length of β-emission ($\sim$ 1 μm) from decaying tritium releasing energy within the nucleus and causing DNA strand breaks. If prolonged incubations or high specific activities are required, ^{14}C-TdR or ^{32}P should be used.

Reagents

Sterile cells at suitable stage

^{3}H-TdR, 10 μCi/ml, 100 μl for each ml culture medium

non-sterile 10% TCA (on ice), 6 ml for each 1 ml culture

HBSS, ice cold, 2 ml per ml culture

2N PCA or 0.3N NaOH, 1% SLS, 0.5 ml per 1 ml culture

MeOH, 1 ml per 1 ml culture

scintillation vials

scintillant (10 $\times$ vol of PCA or NaOH/SLS)

Outline

Label cells with ^{3}H-TdR, extract DNA, and count.

Protocol

1.

Grow culture to desired density (usually mid log phase for maximum DNA synthesis or plateau for density-limited DNA synthesis (see Chapter 15).

2.

Add ^{3}H-TdR, 1.0 μCi/ml ($\sim$ 50 Ci/mol) in HBSS. Note: Some media, e.g., Ham's F10 and F12, contain thymidine which will ultimately determine the specific activity of added ^{3}H-TdR. Allowance will have to be made for this when judging the amount of isotope to add. While 1.0 Ci/ml may be sufficient for most media, 5 μCi/ml should be used with F10 or F12.

3.

Incubate for 1–24 hr as required.

4.

Remove radioactive medium carefully and discard into liquid radioactive waste.

5.

Wash carefully with 2 ml HBSS and add 2 ml ice-cold 10% TCA for 10 min. Fix in MeOH first if cells are loosely adherent (see Protein Synthesis above).

6.

Repeat TCA washes twice, 5 min each.

7.

Add 0.5 ml 2N PCA, place on hot plate at 60°C for 30 min, and allow to cool.

8.

Collect PCA, transfer to scintillant, and count.

9.

The residue may be dissolved in alkali for protein determination as above. If protein determinations are not required, the whole monolayer can be dissolved in 1% SLS in 0.3N NaOH and counted. In this case, and perhaps anyway, replicate cultures should be set up to provide cell counts to allow calculation of DNA synthesis by cell number, if DNA synthesis per mg protein is not suitable.

For suspension cultures spin (100 g, 10 min) at steps 4, 5, and 6, mix on vortex mixer to disperse pellet before each wash, 10% TCA and 2N PCA, at step 7. Spin after 7 (1,000 g, 10 min) to separate precipitate (for protein estimation if required) and supernatant (for scintillation counting).

◊^{3}H-thymidine represents a particular hazard because of the induced radiolytic damage mentioned above. Take care to avoid accidental ingestion, injection, or inhalation of aerosols. Work in a biohazard cabinet or on the open bench.

Incubation with isotopic precursor can provide several different types of data depending on the incubation conditions and subsequent processing. Incubation followed by a short wash in ice-cold BSS will give a measure of uptake and, if carried out over a few minutes duration, will give a fair measure of unidirectional flux. In uptake experiments, incorporation into acid insoluble precursors such as protein or DNA is assumed to be minimal, due to the short incubation time, and only the acid soluble pools are counted by extraction into cold 10% TCA. In longer incubations, 2-24 hr, it is assumed that the precursor pools become saturated. Equilibrium levels may be measured by cold TCA extraction and incorporation into polymers measured by extraction with hot 2N PCA (DNA), cold dilute alkali (RNA), or hot 1N NaOH (protein).

PREPARATION OF SAMPLES FOR ENZYME ASSAY AND IMMUNOASSAY

As the amount of cellular material available from cultures is often too small for efficient homogenization, other methods of lysis are required to release soluble products and enzymes for assay. It is convenient either to set up cultures of the necessary cell number in sample tubes (see below) or to trypsinize a bulk culture and aliquot cells into assay tubes. In either case the cells should be washed in HBSS, to remove serum, and lysis buffer added. The lysis buffer is chosen to suit the assay, but if unimportant, 0.15 M NaCl or PBS may be used. If the product to be measured is membrane bound, add 1% detergent (Na deoxycholate, Nonidet P40) to the lysis buffer. If the

cells are pelleted resuspend in the buffer by vortex mixing. Freeze and thaw the preparation three times by placing in EtOH containing solid CO_2 ($\sim$ $-90°C$) for 1 min and then in 37°C water for 2 min (longer for samples greater than 1 ml). Finally spin at 10,000 g for 1 min (e.g., in Eppendorf centrifuge) and collect supernatant for assay.

Alternatively, whole extract may be assayed for enzyme activity and the insoluble material removed by centrifugation later if necessary.

REPLICATE SAMPLING

As cultured cells can be prepared in a uniform suspension in most cases, the provision of large numbers of replicates for statistical analysis is often unnecessary. Usually three replicates are sufficient, and for many simple observations, e.g., cell counts, duplicates may be sufficient.

Many types of culture vessel are available for replicate monolayer cultures (see Chapter 4) and the choice is determined (1) by the number of cells required in each sample and (2) by the frequency or type of sampling. For example, if incubation time is not a variable, replicate sampling is most readily performed in multiwell plates such as microtitration plates or 24-well plates. If, however, samples are collected over a period of time, say 5 days, then constant removal of a plate for daily processing may impair growth in the rest of the wells. In this case, replicates are best prepared in individual tubes or 4-well plates. Plain glass or tissue culture treated plastic test tubes may be used, though Leighton tubes are superior as they provide a flat growth surface. Alternatively, if optical quality is not critical, flat bottomed glass specimen tubes and even glass scintillation vials may be good containers. If glass vials or tubes are used they must be washed as tissue culture glassware (Chapter 8) and not returned to tissue culture after use with scintillant.

Sealing large numbers of vials or tubes can become tedious, so many people seal tubes with vinyl tape rather than screw caps. Such tapes can also be color coded to identify different treatments.

Handling suspension cultures is generally easier as the shape of the container and its surface charge are less important. Multiple sampling can also be performed on one culture. This is done conveniently by sealing the bottle with a silicone rubber membrane closure (Pierce) and sampling via a syringe and needle (remembering to replace the volume removed with an equal volume of air).

GROWTH CYCLE

As described in Chapter 10, following subculture, cells will progress through a characteristic growth pattern of lag phase, exponential or "log" phase, and stationary or "plateau" phase (see Fig. 10.3). The log and plateau phases give vital information about the cell line, the population doubling time during log growth, and the maximum cell density achieved in plateau (saturation density). Measurement of the population doubling time is used to quantify the response of the cells to different inhibitory or stimulatory culture conditions such as variations in nutrient concentration, hormonal effects, or toxic drugs. It is also a good monitor of the culture during serial passage and enables the calculation of cell yields and the dilution factor required at subculture.

It must be emphasized that the population doubling time is an average figure and describes the net result of a wide range of division rates, including zero, within the culture.

Single time points are unsatisfactory for monitoring growth, without knowing the shape of the growth curve. A reduced cell count after, say, 5 days could be caused by a reduced growth rate of some or all the cells, a longer lag period implying adaptation or cell loss (difficult to distinguish), or a reduction in saturation density. This is not to say that growth curves are of no value. They can be useful for a rapid screen, and once the response being monitored is fully characterized and the type of response predictable, e.g., an increased doubling time, then single time point observations may be sufficient. Growth curves are particularly useful for the determination of saturation density, although growth at saturation density should be assessed by the labeling index with ^{3}H-thymidine (see below and Chapter 15).

The preferred method for analyzing growth and survival at lower cell densities is by clonal growth analysis (see Chapter 19). This technique will reveal differences in growth rate within a population and will distinguish between alterations in growth rate (colony size) and survival (colony number). It should be remembered, however, that cells may grow differently as isolated colonies at low cell densities. Fewer cells will survive even under ideal conditions, and all interaction is lost until the colony starts to form. Heterogeneity in clonal growth rates reflects differences in growth capacity between lineages within a population, but these need not necessarily be expressed in an interacting monolayer at higher densities where cell communication is possible.

The population doubling time (PDT) derived from a growth curve should not be confused with the cell cycle or generation time. The PDT is an average figure for the population and subject to the reservations stated above. The cell cycle time or generation time is measured from one point in the cell cycle until the same point is reached again (see below) and refers only to the growing cells in the population, while the PDT is influenced by nongrowing and dying cells. PDTs vary from 12–15 hr in rapidly growing mouse leukemias like the L1210, to 24–36 hr in many adherent continuous cell lines, and up to 60 or 72 hr in slow-growing finite cell lines.

A growth cycle is performed each time the culture is passaged and can be analyzed in more detail as described below.

Outline

Set up a series of cultures at three different cell concentrations and count the cells at daily intervals until they "plateau."

Materials

> Cell culture
> 24-well plates (sterile)
> 100 ml growth medium (sterile)
> 0.25% crude trypsin (for monolayer cultures only) (sterile)
> plastic box to hold plates
> CO_2 incubator or CO_2 supply to gas box

Protocol

1.
Trypsinize cells as for regular subculture (see above).

2.
Dilute cell suspension to 10^5 cells/ml, 3×10^4 cells/ml, and 10^4 cells/ml, in 25 ml medium for each concentration.

3.
Seed three 24-well plates, one at each cell concentration, with 1 ml per well. Add cell suspension slowly from the center of the well so that it does not swirl around the well. Similarly, do not shake the plate to mix the cells, as the circular movement of medium will concentrate the cells in the middle of the well.

4.
Place in a humid CO_2 incubator or sealed box gassed with CO_2.

5.
After 24 hr, remove plates from incubator and

count the cells in three wells of each plate: (a) remove medium completely from wells to be counted; (b) add 1 ml trypsin to each well; (3) incubate with trypsin; and (4) after 15 min, disperse cells in trypsin and transfer 0.4 ml to counting fluid and count on cell counter.

Note. Hemocytometer counting may be used but may be difficult at lower cell concentrations. Reduce trypsin volume to 0.1 ml and disperse cells carefully without frothing using a micropipette and transfer to hemocytometer.

6.

Return plate to incubator as soon as cell samples in trypsin are removed. The plate must be out of the incubator for the minimum length of time, to avoid disruption of normal growth.

7.

Repeat sampling at 48 and 72 hr as in steps 5 and 6.

8.

Change medium at 72 hr or sooner if indicated by pH drop (see above).

9.

Continue sampling at daily intervals for rapidly growing cells (doubling time 12–14 hr) but reduce frequency of sampling to every 2 days for slowly growing cells (doubling time > 24 hr), until plateau is reached.

10.

Keep changing medium every 1, 2, or 3 days as indicated by pH.

Analysis

1. Calculate cell number per well, per ml of culture medium (same figure), and per cm^2 of available growth surface in well. (Stain one or two wells (see Chapter 13) at each density to determine whether distribution of cells in wells is uniform and whether they grow up the sides of the well.)

2. Plot cell density (per cm^2) and cell concentration (per ml), on a log scale, against time on a linear scale (Fig. 10.3).

3. Determine the lag time, population doubling time, and plateau density (see below and Fig 10.3)

4. Establish which is the appropriate starting density for routine passage. Repeat growth curve at intermediate cell concentrations if necessary

Variations

1. Different culture vessels may be used, e.g., 25 cm^2 flasks, although more cells and medium will be required, or flat-bottomed glass sample tubes. Individual tubes have the advantage that the rest are not disturbed when samples are removed for counting.

2. Frequency of medium changing may be altered.

3. Different media or supplements may be tested.

Suspension cultures

1.

Add cell suspension in growth medium to wells at a range of concentrations as for monolayer.

2.

Sample 0.4 ml at intervals as per trypsin samples. Alternatively, seed two 75-cm^2 flasks with 20 ml for each cell concentration and sample 0.4 ml from each flask daily or as required. Mix well before sampling and keep flasks out of incubator for the minimum length of time. Do not feed cultures during growth curve.

The growth cycle (Fig. 10.3) is conventionally divided into three phases.

The Lag Phase

This is the time following subculture and reseeding during which there is little evidence of an increase in cell number. It is a period of adaptation during which the cell replaces elements of the glycocalyx lost during trypsinization, attaches to the substrate, and spreads out. During spreading the cytoskeleton reappears and its reappearance is probably an integral part of the spreading process. Enzymes, such as DNA polymerase, increase, followed by the synthesis of new DNA and structural proteins. Some specialized cell products may disappear and not reappear until cessation of cell proliferation at high cell density.

The Log Phase

This is the period of exponential increase in cell number following the lag period and terminating one or two doublings after confluence is reached. The length of the log phase depends on the seeding density, the growth rate of the cells, and the density at which cell proliferation is inhibited by density. In the log phase the growth fraction is high (usually 90–100%) and the culture is in its most reproducible form. It is the optimal time for sampling since the population is at its most uniform and viability is high. The cells are, however, randomly distributed in the cell cycle and, for some purposes, may need to be synchronized (see Chapter 23).

The Plateau Phase

Toward the end of the log phase, the culture becomes confluent—i.e., all the available growth surface is occupied and all the cells are in contact with surrounding cells. Following confluence the growth rate of the culture is reduced, and in some cases, cell proliferation ceases almost completely after one or two further population doublings. At this stage, the culture enters the plateau (or stationary) phase, and the growth fraction falls to between 0 and 10%. The cells may become less motile; some fibroblasts become oriented with respect to one another, forming a typical parallel array of cells. "Ruffling" of the plasma membrane is reduced, and the cell both occupies less surface area of substrate and presents less of its own surface to the medium. There may be a relative increase in the synthesis of specialized versus structural proteins and the constitution and charge of the cell surface may be changed.

The phenomenon of cessation of motility, membrane ruffling, and growth was described originally by Abercombie and Heaysman [1964] and designated "*contact inhibition.*" It has since been realized that the reduction of the growth of normal cells after confluence is reached is not due solely to contact but may also involve reduced cell spreading [Stoker et al., 1968; Folkman and Moscona, 1978], depletion of nutrients, and, particularly, growth factors [Stoker, 1973; Dulbecco and Elkington, 1973; Westermark and Wasteson, 1975] in the medium [Holley et al., 1978]. The term "*density limitation*" of growth has, therefore, been used to remove the implication that cell-cell contact is the major limiting factor [Stoker and Rubin, 1967], and "contact inhibition" is best reserved for those events directly contingent on cell contact, i.e., reduced cell motility and membrane activity, resulting in the formation of a strict monolayer and orientation of the cells with respect to each other.

Cultures of normal simple epithelial and endothelial cells will stop growing after reaching confluence and remain as a monolayer. Most cultures, however, with regular replenishment of medium, will continue to proliferate, although at a reduced rate, well beyond confluence, resulting in multilayers of cells. Human embryonic lung, or adult skin, fibroblasts, which express contact inhibition of movement, will continue to proliferate, laying down layers of collagen between the cell layers, until multilayers of six or more cells can be reached under optimal conditions [Kruse et al., 1970]. They still retain an ordered parallel array, how-

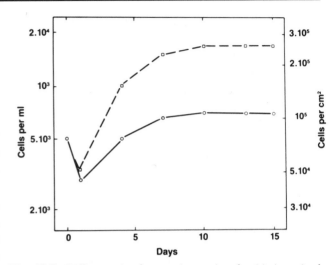

Fig. 18.5. *Difference in plateaus (saturation densities) attained by cultures from normal brain (circles, solid line) and a glioma (squares, broken line). Cells were seeded onto 13-mm coverslips and 48 hr later the coverslips were transferred to 9-cm petri dishes with 20 ml growth medium, to minimize exhaustion of the medium.*

ever. The terms "plateau" and "stationary" are not strictly accurate, therefore, and should be used with caution.

Cultures which have transformed spontaneously or have been transformed by virus or chemical carcinogens will usually reach a higher cell density in the plateau phase than their normal counterparts [Westermark, 1974] (Fig. 18.5). This is accompanied by a higher growth fraction and loss of density limitation. These cultures are often *anchorage independent* for growth—i.e., they can easily be made to grow in suspension (see Density Limitation of Growth in Chapter 15; also see Chapter 2).

The construction of a growth curve from cell counts performed at intervals after subculture enables the measurement of a number of parameters which should be found to be characteristic of the cell line under a given set of culture conditions. The first of these is the duration of the *lag period* or "lag time" obtained by extrapolating a line drawn through the points on the exponential phase until it intersects the seeding or inoculum concentration (see Fig 10.3), and reading off the elapsed time since seeding equivalent to that intercept. The second is the *doubling time*, i.e., the time taken for the culture to increase two-fold in the middle of the exponential, or "log," phase of growth. This should not be confused with the *generation time* or *cell cycle time* (see below), which are determined by measuring the transit of a population of cells through the cell cycle until they return to the same point in the cell cycle.

The last of the commonly derived measurements from the growth cycle is the "*plateau level*" or "*saturation density*." This is the cell concentration in the plateau phase and is dependent on cell type and frequency of medium replenishment. It is difficult to measure accurately as a steady state is not achieved as easily as in the log phase. Ideally the culture should be perfused; but a reasonable compromise may be achieved by growing the cells on a restricted area, say a small-diameter coverslip (15 mm) in a large-diameter petri dish (90 mm) with 20 ml of medium replaced daily (see Chapter 15). Under these conditions, medium limitation of growth is minimal, and cell density exerts the major effect. Counting the cells under these conditions gives a more accurate and reproducible measurement. "Plateau" does not imply complete cessation of cell proliferation but represents a steady state where cell division is balanced by cell loss.

With normal cells a "steady state" may be achievable by not replenishing the growth factors in the medium. In this case cells are seeded and grown and plateau reached without changing the medium. Clearly, the conditions used to attain "plateau" must be carefully defined.

The maximum cell concentration in suspension cultures, which are not limited by available substrate, is usually limited by available nutrients. By fortifying the medium with a higher concentration of amino acids, Pirt and others [Birch and Pirt, 1971; Blaker et al., 1971] were able to obtain a maximum cell concentration of 5×10^6 cells/ml for L"S" cells, far in excess of what can be achieved with attached cells.

PLATING EFFICIENCY

When cells are plated out as a single cell suspension at low cell densities (2–50 cells/cm^2), they will grow as discrete colonies (see Chapter 11). When these are counted the results are expressed as the plating efficiency:

$$\frac{\text{No. of colonies formed}}{\text{No. of cells seeded}} \times 100$$
$$= \text{plating efficiency.}$$

If it can be confirmed that each colony grew from a single cell, this term becomes the *cloning efficiency*.

Strictly according to the definition, plating efficiency measurements are derived from counting colonies over a certain size (usually 16–50 cells) growing

from a low inoculum of cells, and the term should not be used for the recovery of adherent cells after seeding at higher cell densities (e.g., 2×10^4 cells/cm^2). This is more properly called the *seeding efficiency*:

$$\frac{\text{No. of cells recovered}}{\text{No. of cells seeded}} \times 100$$
$$= \text{seeding efficiency.}$$

It should be measured at a time when the maximum number of cells has attached but before mitosis starts. This provides a crude measurement of recovery in, for example, routine cell freezing or primary culture.

CLONAL GROWTH ASSAY BY DILUTION CLONING

The protocol for dilution cloning [Puck and Marcus, 1955] is given in Chapter 11. When colonies have formed, remove medium, rinse carefully in BSS, and fix and stain colonies (see Chapter 13).

Analysis

1. Count colonies and calculate plating efficiency. Magnifying viewers (e.g., Fig. 18.6) help to make counting easier.

2. The size distribution of the colonies may also be determined (e.g., to assay the growth promoting ability of a test medium or serum, see Chapter 7) by counting the number of cells per colony or by densitometry. Fix in 1% glutaraldehyde, stain the colonies with crystal violet, and measure absorption on a densitometer [McKeehan et al., 1977].

Automatic Colony Counting

If the colonies are uniform in shape and quite discrete, they may be counted on an automatic *colony counter* (e.g., New Brunswick, Artek, Micromeasurements Ltd.) which scans the plate with a conventional TV camera and analyzes the image to give an instantaneous readout of colony number. A size discriminator gives size analysis based on colony diameter (not always proportional to cell number, as cells may pile up in the center of the colony).

Though expensive, these instruments can save a great deal of time and make colony counting more objective. They will not cope well with colonies which overlap, or have irregular outlines.

LABELING INDEX

If a culture is labeled with [^{3}H]-thymidine, cells that are synthesizing DNA will incorporate the isotope.

Fig. 18.6. a. *Simple magnifying colony counter. Versions are available with an electronically activated marking pen which records the count automatically.* b. *Bellco projection viewer. Magnifies the plate and allows more discrimination in scoring colonies.*

The percentage of labeled cells, determined by autoradiography [e.g., Westermark, 1974; Maciera-Coelho, 1973] (see Chapter 23 and Fig 23.12a), is known as the labeling index (L.I.). Measurement after a 30-min exposure to [³H]-thymidine shows a large difference between exponentially growing cells (L.I. = 10-20%) and plateau cells (L.I. ≤ 1%). Since the

L.I. is very low in plateau, exposure times may have to be increased to 24 hr. With this length of label normal cells can be shown to have a lower labeling index with thymidine than neoplastic cells (see Chapter 15).

Outline

Grow cells to appropriate density, label with [³H-thymidine for 30 min, wash, fix, remove unincorporated precursor, and prepare autoradiographs.

Materials

      Culture of cells
      multiwell plate(s) containing 13-mm
         Thermanox coverslips
      PBSA
      0.25% crude trypsin
      growth medium
      hemocytometer or cell counter
      [³H]-thymidine (2 Ci/mmol)
      HBSS
      acetic methanol (1 part glacial acetic acid
         to 3 parts methanol) icecold, freshly
         prepared
      microscope slides
      DPX
      10% trichloroacetic acid ice-cold
      deionized water
      methanol

Protocol

1.
Set up cultures at 2×10^4/ml–5×10^4/ml in 24-well plates containing coverslips.
2.
Allow cells to attach, start to proliferate (48–72 hr), and grow to desired cell density.
3.
Add [³H]-thymidine to medium, 5 μCi/ml (2 Ci/mmol) and incubate for 30 min.
Note. Some media, e.g., Ham's F10 and F12, contain thymidine. In these cases, the concentration of radioactive thymidine must be increased to give the same specific activity in the medium. In prolonged exposure to high specific activity, [³H]-thymidine causes radiolysis of the DNA. This can be reduced by using [¹⁴C]-thymidine or [³H]-thymidine of a lower specific activity.
4.
Remove labeled medium and discard (◇care, radioactive!). Wash coverslip three times with BSS.

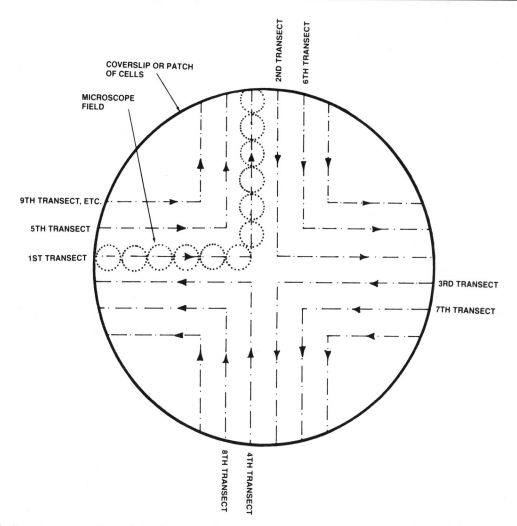

Fig. 18.7. *Scanning pattern for analysis of cytological preparations. Each dotted circle represents one microscope field and the whole, greater, circle, the extent of the specimen (e.g., a cover-* *slip, culture dish or well, or spot of cells on a slide). Guide lines can be drawn with a nylon tipped pen.*

Lift the coverslip off the bottom (but not right out of the well) at each wash to remove isotope from underneath.

5.

Add 1:1 BSS: acetic methanol, 1 ml per well and remove immediately.

6.

Add 1 ml acetic methanol at 4°C and leave 10 min.

7.

Remove coverslips and dry at fan.

8.

Mount coverslip cells uppermost on a microscope slide.

9.

When mountant is dry (overnight), place slides in 10% trichloroacetic acid at 4°C in a staining dish and leave 10 min. Replace trichloroacetic acid twice during this extraction. This removes unincorporated precursors.

10.

Rinse in deionized water, then in methanol, and then dry.

11.

Prepare autoradiograph (see Chapter 23).

Analysis. Count percentage labeled cells. To cover a representative area, follow the scanning pattern illustrated in Figure 18.7.

Growth Fraction

If cells are labeled with [³H]-thymidine for varying lengths of time up to 48 hr, the plot of labeling index against time increases rapidly over the first few hr and

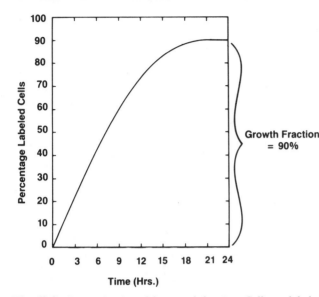

Fig. 18.8. *Determination of the growth fraction. Cells are labeled continuously with [³H]-thymidine and the percentage labeled cells determined at intervals afterward by autoradiography (see text).*

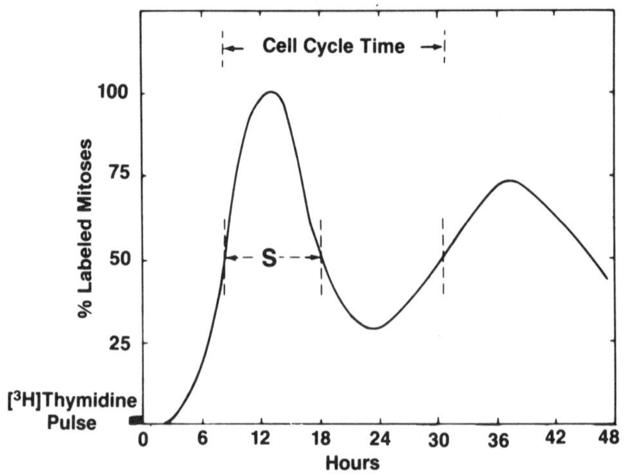

Fig. 18.9. *Determination of cell cycle time. The percentage of labeled mitosis relative to the total number of mitoses is plotted against time after removal of the thymidine. The time between the midpoint of the first ascending curve and the second rising curve is the cell cycle time [after Van't Hof, 1973; Maciera-Coelho, 1973; for further discussion, see Quastler, 1963; and Van't Hof, 1968].*

then flattens out to a very low gradient, almost a plateau (Fig. 18.8). The level of this plateau, read against the vertical axis, is the growth fraction of the culture, i.e., the proportion of cells in cycle at the time of labeling.

> **Outline**
>
> Label culture continuously for 48 hr, sampling at intervals for autoradiography.
>
> **Protocol**
>
> As for labeling index except that at step 3, incubation should be carried out for 15 min, 30 min, 1, 2, 4, 8, 24, and 48 hr.

Analysis. Count labeled cells as a percentage of the total, using scanning pattern from previous protocol. Plot labeling index against time (Fig. 18.8).

Note. Autoradiographs with ³H can only be prepared where the cells remain as a monolayer. If they form a multilayer, they must be trypsinized after labeling and slides prepared by the drop technique or by cytocentrifugation (see Chapter 13), as the energy of β-emission from ³H is too low to penetrate an overlying layer of cells.

MITOTIC INDEX

This is the fraction or percentage of cells in mitosis, determined by counting mitoses in stained cultures as a proportion of the whole population.

CELL CYCLE TIME (GENERATION TIME)

To determine the length of the cell cycle and its constituent phases, cells are labeled with [³H]-thymidine for 30 min, the label is removed, and the appearance of the label in mitotic cells is determined autoradiographically at 30-min intervals up to 48 hr. The plot of the percentage of labeled mitoses against time takes the form of Figure 18.9, for which the cell cycle time and the length of its constituent phases may be derived.

CYTOMETRY

Histochemical and other cytological techniques (see also Chapter 13) can measure the amounts of enzyme, DNA, RNA, protein, or other cellular constituents *in situ*. It is difficult, however, to interpret the results of such techniques in a quantitative fashion. The introduction of flow cytometry [Melamad et al., 1979; Crissman et al., 1975], while losing the relationship between cytochemistry and morphology, has added a new dimension to the measurement of cellular constituents and activities [Kurtz and Wells, 1979]. The potential amount of information that may be obtained about the constituent cells of a population is so vast that the problem becomes one of intellectual interpretation rather than data collection.

For a further description of flow cytometry, see Chapter 12.

Chapter 19
Measurement of Cytotoxicity and Viability

Because cultured cells are in an abnormal environment relative to their tissue of origin and are subject to many artefactual physical and chemical manipulations, those working *in vitro* will always be required to confirm the viability of their material. Furthermore, certain manipulations *in vitro* are specifically designed to assay viability or survival as the major parameter of response to a substance under study, thereby establishing a cheaper, more reproducible substitute for a test otherwise performed in animals.

Current legislation demands that new drugs, cosmetics, food additives, etc., go through extensive cytotoxicity testing before they are released [see also Berky and Sherrod, 1977]. This usually involves a large number of animal experiments which are very costly and raise considerable public concern. There is, therefore, much pressure, both emotional and economic, to perform at least part of cytotoxicity testing *in vitro*. The introduction of specialized cell lines, as well as the continued use of long-established cultures, may make this a reasonable proposition.

LIMITATIONS OF *IN VITRO* METHODS

It is important that any measurement can be interpreted in terms of the *in vivo* response, or at least with the understanding that clear differences exist between *in vitro* and *in vivo* measurements.

Measurement of toxicity *in vitro* is a purely cellular event as presently carried out. It would be very difficult to re-create the complex pharmacokinetics of drug exposure, for example, *in vitro*, and there will usually be significant differences in drug exposure time and concentration, rate of change of concentration, metabolism, tissue penetration, clearance, and excretion. Although it may be possible to stimulate these parameters, e.g., using multicellular tumor spheroids for drug penetration, most studies concentrate on a direct cellular response. They thereby gain their simplicity and reproducibility.

Many nontoxic substances become toxic after metabolism by the liver; and in addition, many substances, toxic *in vitro*, may be detoxified by liver enzymes. For testing *in vitro* to be accepted as an alternative to animal testing, it must be demonstrated that potential toxins reach the cells *in vitro* in the same form as they would *in vivo*. This may require additional processing by purified liver microsomal enzyme preparations [Sladek, 1973].

The nature of the response must also be considered carefully. A toxic response *in vitro* may be measured by changes in cell survival or metabolism (see below), while the major problem *in vivo* may be a tissue response, e.g., an inflammatory reaction or fibrosis. For *in vitro* testing to be more effective, construction of models of these responses will be required utilizing, perhaps, cultures reassembled from several different cell types and maintained in the appropriate hormonal milieu.

It should not be assumed that complex tissue and even systemic reactions cannot be simulated *in vitro*. Assays for the inflammatory response, teratogenic disorders, or neurological disfunctions may be feasible *in vitro*, given a proper understanding of cell-cell interaction and the interplay of endocrine hormones with local paracrine and autocrine factors.

NATURE OF THE ASSAY

The choice of assay will depend on the agent under study, the nature of the response, and the particular target cell. Assays can be divided into two major classes: (1) an immediate or short-term response such as an alteration in membrane permeability or a perturbation of a particular metabolic pathway, and (2) long-term survival, either absolute, usually measured by the retention of self-renewal capacity, or survival in altered state, e.g., expressing genetic mutation(s) or malignant transformation.

Short-Term Assays—Viability

Assays of this type are used to measure the proportion of viable cells following a potentially traumatic procedure such as primary disaggregation (Chapter 9), cell separation (Chapter 12), or freezing and thawing (Chapter 17).

Dye exclusion viability tends to overestimate viability, e.g., 90% of cells thawed from liquid nitrogen may exclude trypan blue but only 60% prove to be capable of attachment 24 h later.

Most viability tests rely on a breakdown in membrane integrity determined by the uptake of a dye to which the cell is normally impermeable (e.g., trypan blue, erythrosin, or nigrosin) or the release of a dye or isotope normally taken up and retained by viable cells (e.g., diacetyl fluorescein or ^{51}chromium).

Dye exclusion. Viable cells are impermeable to nigrosin, trypan blue, and a number of other dyes [Kaltenbach et al., 1958].

Outline

A cell suspension is mixed with stain and examined by low-power microscopy.

Materials

        cells
        PBSA
        0.25% trypsin
        growth medium
        hemocytometer
        viability stain (e.g., 0.4% trypan blue)
        Pasteur pipettes
        microscope
        tally counter

Protocol

1.
Prepared cell suspension at a high concentration ($\sim 10^6$ cells/ml) by trypsinization or centrifugation and resuspension.
2.
Take a clean hemocytometer slide and fix the coverslip in place (see above).
3.
Add one drop of cell suspension to one drop of stain on the open surface of the slide, mix, transfer to the edge of the coverslip, and allow to run into the counting chamber.
4.
Leave 1–2 min (do not leave too long or viable cells will deteriorate and take up stain).

5.
Place on microscope under a × 10 objective.
6.
Count the number of stained cells and the total number of cells.
7.
Wash hemocytometer and return to box.

Analysis. Calculate the percentage of unstained cells. This is the percentage viability by this method. If the volumes of cell suspension and stain are measured accurately at step 3, this method of viability determination can be incorporated into the hemocytometer cell counting protocol.

Dye uptake. Viable cells take up diacetyl fluorescein and hydrolyse it to fluorescein to which the cell membrane of live cells is impermeable [Rotman and Papermaster, 1966]. Live cells fluoresce green; dead cells do not. Nonviable cells may be stained with ethidium bromide or propidium iodide and will fluoresce red. Viability is expressed as the percentage of cells fluorescing green. This method may be applied to flow cytometry (see Chapters 12 and 18).

Outline

Stain cell suspension in a mixture of propidium iodide and diacetyl fluorescein and examine by fluorescence microscopy or flow cytometry.

Materials

        Single cell suspension
        fluorescein diacetate 10 μg/ml in HBSS
        fluorescence microscope
        propidium iodide 500 μg/ml
        filters:
            fluorescein: excitation 450/590 nm
            emission LP 515 nm
            propidium iodide:excitation 488 nm
            emission 615 nm

Protocol

1.
Prepare cell suspension as for dye exclusion above but in medium without phenol red.
2.
Add fluorescent dye mixture 1:10 to give final concentration of 1 μg/ml diacetylfluorescein and 50 μg/ml propidium iodide.
3.
Incubate at 36.5°C for 10 min.
4.
Place a drop of cells on a microscope slide, add a

coverslip, and examine by incident light 488 nm excitation.

Analysis. Cells fluorescing green are viable, those fluorescing red only, non-viable. Express viability as percentage fluorescing green as a proportion of the total.

Chromium release. Reduced $^{51}Cr^{3+}$ is taken up by viable cells and oxidized to $^{51}Cr^{2+}$ to which the membrane of viable cells is impermeable [Holden et al., 1973; Zawydiwski and Duncan, 1978]. Dead cells release the $^{51}Cr^{2+}$ into the medium. A reduction in viability is detected by γ-counting aliquots of medium from cultures labeled previously with $Na_2{}^{51}CrO_4$ for released ^{51}Cr. The test works well for a few hours but over longer periods spontaneous release of ^{51}Cr may be a problem.

Analysis. Express counts released as a percentage of total (medium + cells) and plot against time.

This method allows comparison of different toxic stimuli but does not give an absolute figure for percentage viable cells. It is often used to measure cytotoxic T-lymphocyte activity.

Metabolic tests. Alterations in glycolysis and respiration [Dickson and Suzangar, 1976] enzyme activity [DiPaulo, 1965], and incorporation of labeled precursors [Freshney, Paul, and Kane, 1975] have all been used to measure response to potentially toxic stimuli. Although these are often interpreted as viability or survival assays, they are not and should be interpreted solely as metabolic responses, specific to the parameter measured. Application to cell survival is limited, but comparison of relative population responses is possible if culture is continued after removal of drug for two to three population doublings (see below).

Protocols for measuring precursor uptake and total protein or DNA are given in Chapter 18.

Long-Term Tests—Survival

While short-term tests are convenient and usually quick and easy to perform, they only reveal cells which are dead (i.e., permeable) at the time of the assay. Frequently, cells subjected to toxic influences, e.g., antineoplastic drugs, will only show an effect several hours or even days later. The nature of the tests required to measure viability in these cases is necessarily different since, by the time the measurement is made, the dead cells may have disappeared. Long-term tests are often used to demonstrate the metabolic or proliferative capacity of cells *after* rather than during exposure to a toxic influence. The objec-

tive is to measure survival rather than short-term toxicity, which may be reversible.

The ability of cells to survive a toxic insult has been the basis of most cytotoxicity assays. In this context survival implies the retention of regenerative capacity and is usually measured by plating efficiency as would be the case with bacteria or other microorganisms. Unfortunately, many animal cells have poor plating efficiencies, particularly normal cells, freshly isolated, so a number of alternatives have been devised for assaying cells at higher densities, e.g., in microtitration plates. None of these tests measures survival directly. Instead the net increase in cell number (growth curve), the increase in total protein or DNA, or the residual ability to synthesize protein or DNA is determined. "Survival" in these cases is defined as the retention of metabolic or proliferative ability by the cell population as a whole; such assays cannot discriminate between a reduction in metabolic or proliferative activity per cell and a reduced number of cells.

Plating efficiency, as described in Chapter 18, is the best measure of survival and proliferative capacity, provided that the cells plate with a high enough efficiency that the colonies can be considered representative. Though not ideal, anything over 10% is usually acceptable.

Since the colony number may fall at high toxic concentrations, it is usual to compensate by seeding more cells so that approximately the same number of colonies form at each concentration. This removes the risk of cell concentration influencing survival and improves statistical reliability. In addition cells should be plated on a preformed feeder layer the density of which ($5 \times 10^3/cm^2$) greatly exceeds that of the cloning cells, where the plating efficiencies of controls is <100%.

A typical survival curve is prepared as follows:

Outline

Treat cells with experimental agent at a range of concentrations for 24 h. Trypsinize, seed at low cell density, and incubate for 1–3 wk. Stain and count colonies.

Materials

    25 cm² flasks
    6- or 9-cm petri dishes
    PBSA
    0.25% trypsin
    growth medium
    hemocytometer or cell counter

agent to be tested at 5 × maximum concentration to be used, dissolved in growth medium—check pH and osmolality, adjust if necessary

HBSS

methanol

crystal violet, 1% (BDH)

Protocol

1.

Prepare a series of cultures in 25-cm^2 flasks, three for each of six agent concentrations, and three controls. Seed cells at 5×10^4/ml in 4-ml growth medium and incubate for 48 hr, by which time the cultures will have progressed into log phase (see above).

2.

Prepare 50 ml of agent to be tested in regular growth medium. Check pH and osmolality and adjust if necessary. Dilate agent 1:5 by adding 1 ml to first flasks, mix, transfer 1 ml to second flasks, mix, and so on, completing serial dilution. Remove 1 ml from lowest concentration and discard.

3.

Return to incubator.

4.

If the agent is slow-acting or partially reversible repeat steps 2 and 3 twice; i.e., expose cultures to the agent for 3 d, replacing the agent daily by changing the medium. With fast-acting agents, one hour's exposure will be sufficient.

5.

Remove medium from each group of three flasks in turn, trypsinize cells, and seed at required density for clonal growth (see Chapter 11), diluting all cultures by the same amount as the control. If toxicity is expected, increase the seeding density at higher agent concentrations to keep the number of colonies forming in the same range. In addition, plate cells on to a feeder layer (see Chapter 11) of the same cells treated with mitomycin C, if plating efficiency of controls is substantially less than 100%.

6.

Incubate until colonies form: fix, in absolute menthanol or 1% glutaraldehyde, stain for 10 min in 1% crystal violet (stain may be reused).

7.

Wash in tap water, drain, and dry inverted.

8.

Count colonies >32 cells (five generations).

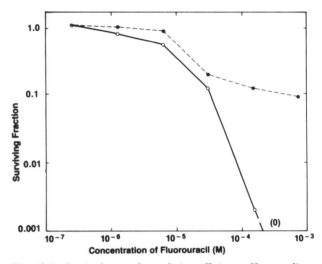

Fig. 19.1. *Survival curve from plating efficiency. Human glioma cells were plated out in the presence (dotted line) and absence (solid line) of a feeder layer after treatment with various concentrations of 5-fluorouracil. A 10% resistant fraction is apparent at 10^{-4} M drug only in the presence of a feeder layer. In the absence of the feeder layer, the small number of colonies constituting the resistant fraction was unable to survive alone.*

Analysis. 1. Plot relative plating efficiency (plating efficiency as a percentage of control) against drug concentration (Fig. 19.1) (*survival curve*).

2. Determine ID_{50} or ID_{90}: the concentration promoting 50% or 90% inhibition of colony formation.

3. Complex survival curves may be compared by calculating the area under the curve.

Variations

Concentration of agent. A wide concentration range, in log increments, e.g., 10^{-6} M, 10^{-5} M, 10^{-4} M, 10^{-3} M, 0, should be used for the first attempt and a narrower range (log or linear), based on the indications of the first, for subsequent attempts.

Invariate agent concentrations. Some conditions tested cannot easily be varied, e.g., testing the quality of medium, water, or an insoluble plastic. In these cases, the serum concentrations should be varied, as serum may have a masking effect on minor toxic effects.

Duration of exposure to agent. Some agents act rapidly; others, more slowly. Exposure to ionizing radiation, for example, need only be a matter of minutes, sufficient to achieve the required dose, while testing some antimetabolic drugs may take several days for a measurable effect.

Duration of exposure (T) and drug concentration (C) are related although C × T is not always a constant. Prolonging exposure can increase sensitivity beyond that predicted by C × T due to cell cycle effects and cumulative damage.

Time of exposure to agent. Where the agent is soluble and expected to be toxic, the above procedure should be followed; but where the quality of the agent is unknown, stimulation is expected, or only a minor effect is expected (e.g., 20% inhibition rather than 100-fold or more), the agent may be incorporated during clonal growth rather than at preincubation.

Cell density. The density of the cells during exposure to an agent can alter its response, e.g., HeLa cells are less sensitive to the alkylating agent mustine at high cell densities [Freshney et al., 1975]. In this kind of experiment, the cell density should be varied in the preincubation phase, during exposure to drug.

Colony size. Some agents are cytostatic but not cytotoxic, and during continuous exposure, may reduce colony size without reducing colony number. In this case the size of the colonies should be determined by densitometry [McKeehan et al., 1977], automatic colony counting or image analysis, or visually counting the number of cells per colony (see also Chapter 18).

For colony counting, the threshold number of cells per colony (e.g., 32 as above) is purely arbitrary, and assumes that most of the colonies are greatly in excess of this.

For colony sizing, harvest earlier before growth rate in larger colonies has slowed down, and score all colonies.

Solvents. Some agents to be tested above have low solubilities in aqueous media, and it may be necessary to use an organic solvent. Ethanol, propylene glycol, and dimethyl sulphoxide have been used for this purpose but may themselves be toxic. Use the minimum concentration of solvent to obtain solution. The agent may be made up at a high concentration in, for example, 100% ethanol, then diluted gradually with BSS, and finally diluted into medium. The final concentration of solvent should be < 0.5%, and a solvent control must be included.

Take care when using organic solvents with plastics or rubber. It is better to use glass with undiluted solvents and only to use plastic where the solvent concentration is < 10%.

While plating efficiency is one of the best methods for testing survival, it should only be used where the cloning efficiency is high enough for colonies that form to be representative of the whole cell population. Ideally this means that controls should plate at 100% efficiency. In practice this is seldom possible and control plating efficiencies of 20% or less are often accepted.

Plating efficiency tests are also time consuming to set up and analyze, particularly where a large number of samples is involved, and the duration of each experiment may be anything from 2 to 4 wk.

Microtitration. The introduction of multiwell plates revolutionized the approach to replicate sampling in tissue culture. They are economical to use, lend themselves to automated handling, and can be of good optical quality. The most popular is the 96-well microtitration plate, each well having 32 mm^2 growth area and capacity for 0.1 or 0.2 ml medium and up to 10^5 cells.

They may be used for cloning, for antibody, virus, and drug titration, for cytotoxicity assays of potential toxins, and for numerous other applications. The following example illustrates the use of microtitration plates in the assay of anticancer drugs but would be applicable with minor modification to any cytotoxicity assay.

Outline

Microtitration plate cultures are exposed to a range of drug concentrations during the log phase of growth and viability determined, several days after drug removal, by measuring incorporation of [^{35}S]-methionine (Fig. 19.2).

Materials

    culture of cells
    PBSA
    0.25% trypsin
    growth medium
    hemocytometer or cell counter
    96-well microtitration plates
    test solution
    Mylar film (Flow Labs)
    medium containing 0.1 μCi/ml [^{35}S]-
      methionine
    HBSS
    methanol
    10% trichloroacetic acid
    scintillation fluid
    x-ray film (Kodak Royal)
    dark box
    silica gel
    black light-tight bag
    −70°C freezer
    photographic developer and fixer

Protocol

1.

Trypsinize cells (see Chapter 10). Seed microtitration plates at 10^3 cells/well, 0.1 ml/well ($\sim$3,000

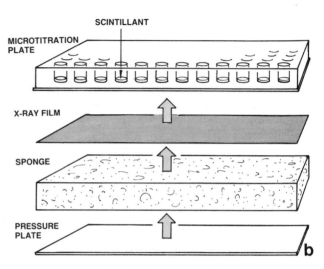

Fig. 19.2. *(Continued)*

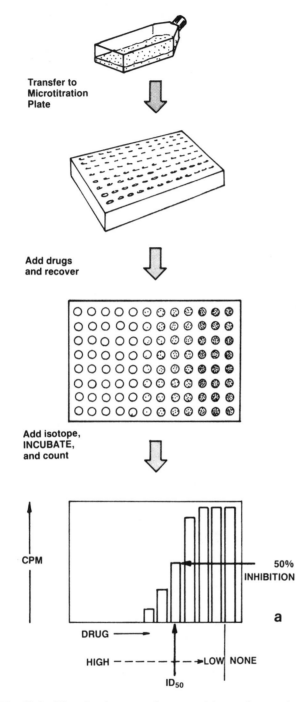

Fig. 19.2. *Microtitration assay for cytotoxicity. a. Stages of assay. b. Measurement of incorporated isotope by autofluorography.*

cells/cm^2). Set up one duplicate plate for cell counting.

2.
Place plate in CO_2 incubator with loose-fitting lid for 30 min to equilibrate with CO_2 (see Chapter 7).

3.
Prepare drug dilutions in 1.0-ml aliquots. Dilute in complete medium and prepare similar aliquots of medium with no drug. The number of dilution steps and the volume required in each aliquot will depend on the way the plate is subdivided (Fig. 19.3). One milliliter is sufficient for duplicates, changed daily for 3 d, 0.1 ml per well.

4.
Remove from incubator. Return to aseptic area, and quickly seal plate with self-adhesive Mylar film. Return plates to incubator for 48–72 hr.

5.
Remove cell count plate from incubator and swab with 70% alcohol.

6.
Cut Mylar film round eight wells and peel off film. Remove medium completely from these wells and add 0.1 ml of trypsin. Incubate for 15 min at 36.5°C and then disperse the cells in the trypsin. Pool the cell suspensions from four wells, dilute to 20 ml, and count on cell counter.

If a cell counter is not available, do not trypsinize; rinse the cells in BSS and fix in methanol. At the end of the experiment the cells may be stained and counted by eye on a microscope, or an estimate of cell number made by densitometry (see above).

Note. The cells must remain in exponential growth throughout (see Chapter 10). If the cell growth curve shows that the cells are moving into stationary phase, proceed directly to 16.

7.

Change medium in remaining wells and return plate to incubator, equilibrate with CO_2 for 30 min, and reseal.

8.

Remove drug plate from incubator, swab with 70% alcohol, and gently peel off Mylar film.

9.

Remove medium, two rows at a time, and add drug dilutions, the second row duplicating the first. There are a number of different ways of dividing up the plate to give different numbers of dilutions and replicates (see Fig. 19.3).

10.

Equilibrate with CO_2 in incubator for 30 min as in steps 2 and 3 above, reseal with Mylar film, and incubate for 24 hr.

11.

Repeat steps 8, 9, and 10 twice more to give three complete days exposure to the drug. Do cell counts and change the medium on accompanying plate each time the drugs are renewed.

12.

After drug exposure, remove drug medium and wash wells by gently adding and removing 0.1 ml medium three times ("dumping" the medium by inverting the plate can detach cells and increases the chance of contamination) and finally leave 0.1 ml of fresh medium in each well.

13.

Count samples from accompanying plate and feed remaining wells.

14.

Incubate for a further 5 d, changing the medium on the second or third d.

15.

Change medium and count samples on cell count plate at the time the medium is changed on the drug plate and at the end.

16.

Remove medium (drug plate only) and add 0.1 ml medium containing 0.1 μCi/ml [^{35}S]-methionine. (The specific activity is unimportant as this is controlled by the methionine in the culture medium.)

17.

Incubate for 3 hr.

18.

Remove isotope and wash plate by submerging it in BSS, rubbing the wells with a gloved finger or comb to promote entry of BSS into wells. Rapid removal of medium by pouring off or flicking the plate will dislodge cells.

19.

Repeat BSS wash twice; do not pour off previous wash from wells.

20.

Immerse plate in 100% methanol and rub as in step 18.

21.

Repeat twice in fresh methanol and leave for 10 min in final bath of methanol.

22.

Pour off methanol and dry plate at fan.

23.

Add ice-cold 10% trichloroacetic acid (TCA) to the plate from a wash bottle. Fill the wells and stand on ice for 5 min. Remove TCA and repeat twice more with fresh TCA.

24.

Wash in methanol and dry.

25.

Add 50 μl scintillation fluid (e.g., Ecoscint) and dry down in a flat film onto the cells by centrifuging the plate for 1 hr at 20°C.

26.

Bind dry plate with x-ray film (see Fig. 19.2b) (under dark-red safelight) and seal in dark box with desiccant such as silica gel.

27.

After 2–14 d, open and remove film, under safelight conditions, develop for 10 min in D19, wash in tap water, fix in photographic fixer for 5 min, wash, and dry.

Analysis. If the titration point is obvious, the plate may be read by eye. If not, scan plate on a densitometer, and determine ID_{50} (Fig. 19.4).

Variations

Duration. As for plating efficiency (see above), some agents may act more quickly, and the exposure period and recovery may be shortened.

Sampling. When trying the assay at first, it may be desirable to sample ([^{35}S]-methionine labeling and autofluorograph) on each day of drug exposure and recovery (Fig. 19.5). If a stable ID_{50} is reached earlier then the assay may be shortened.

End point. [^{35}S]-methionine labeling and autofluorography were chosen for speed and ease of analysis, and because active protein synthesis implies that the cells are still alive. Other alternatives are possible,

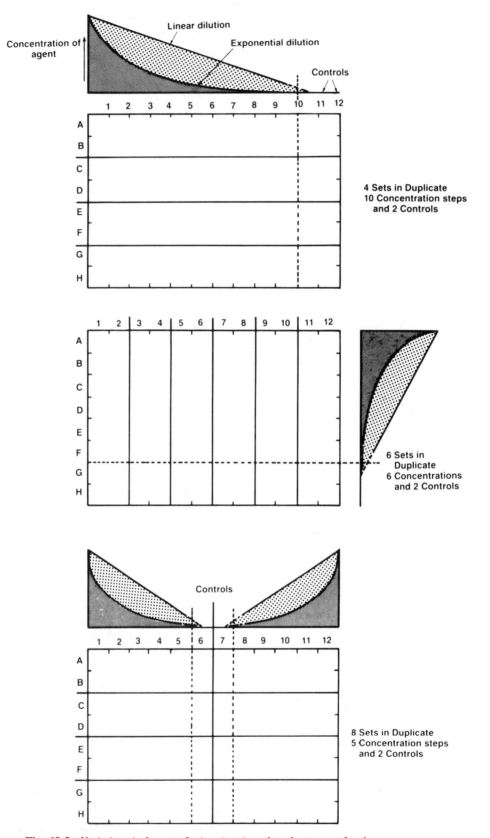

Fig. 19.3. *Variations in layout of microtitration plate for assay of a dose-response curve. The graphs represent agent concentrations in linear (stippled) or exponential (shaded) dilutions.*

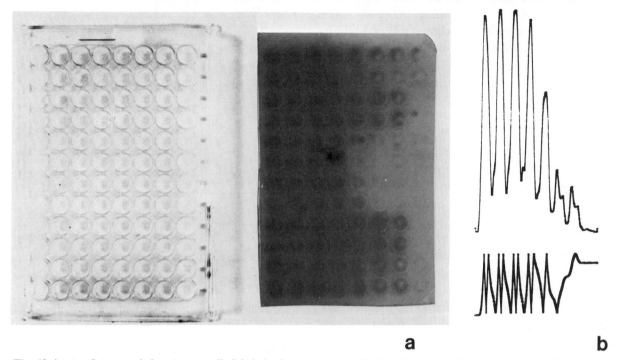

a   **b**

Fig. 19.4. *Autofluorograph from isotopically labeled cultures in a microtitration plate (a) and densitometer scan (b) from one row.*

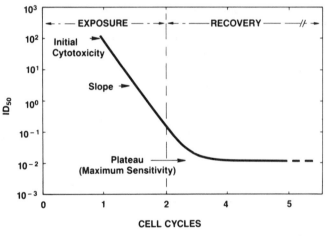

Fig. 19.5. *Time course of the fall in ID_{50}. Idealized curve for an agent with a progressive increase in cytotoxicity with time but eventually reaching a maximum effect after three cell cycles. Not all cytotoxic drugs will conform to this pattern.*

however, including direct staining and cell counting *in situ* or by densitometry, fluorimetric DNA assay [Kissane and Robbins, 1958; Brunk et al, 1979], measurement of dehydrogenase activity [DiPaulo, 1965], or labeling with [³H]-thymidine (DNA synthesis), [³H]-uridine (RNA synthesis), or other isotopes and analysis by scintillation counting or autofluorography. One per-

cent sodium salicylate may be substituted for conventional scintillant.

A recent innovation in microtitration assay has been the use of reduction of tetrazolium to a coloured formazen as an indicator of viability [Mossman, 1983]. This has been applied to the assay or anticancer drugs by Cole [1986] and Carmichael et al. [1987a], who have also applied the technique to the measurement of radiosensitivity [Carmichael et al., 1987b]. Plates are set up and treated as described above with a range of drugs and/or concentrations for the appropriate time. At the time of assessment of viability at the end of the assay, when, for example, a radioisotopically-labeled precursor would be added in the previous protocol, 3-[4,5-dimethylthiazol-2-yl]-2,5-diphenyl-tetrazolium bromide (MTT) is added directly to the medium in the wells and incubated for 2–4 hr. An insoluble formazan is formed in the cells in proportion to the amount of dehydrogenase activity. As this appears to be fairly constant for each cell type, this gives an approximation of the number of cells left in the well.

After removal of the medium, the formazan can be solubilized in DMSO and the plate read on an automatic plate reader (Biorad, Dynatek, Flow, Biotech) at 570 nm (see Chapter 4). As these plate readers can be linked directly to a microcomputer, titration curves, ID_{50}s, etc., can be generated automatically.

Around 0.5 mg/ml (final concentration) of MTT is required but it is preferable to determine the optimal concentration for each cell type (Plumb, personal communication). Likewise, the time of incubation in MTT may also be varied depending on cell number and enzyme activity per cell.

Microtitration offers a method whereby large numbers of samples may be handled simultaneously but with relatively few cells per sample. The whole population is exposed to the agent and viability determined metabolically, by staining or by counting cells from a few hours to several days later.

In practice it may not matter which criterion is used for viability or survival at the end of the assay; it is the design of the assay (drug exposure, recovery, cell density, growth rate, etc.) that is most important.

A variety of automated handling techniques are available (autodispensers, diluters, cell harvesters, and densitometers (see Fig. 6.2), reducing the time required per sample. The volume of medium required per sample is less than one-tenth of that required for cloning, though the cell number is approximately the same.

Microtitration, however, is unable to distinguish between differential responses between cells within a population and the degree of response in each cell—e.g., a 50% inhibition could mean that 50% of the cells respond or that each cell is inhibited by 50%.

A comparison of ID_{50}s derived by microtitration and plating efficiency assays showed a strong correlation between the two methods (Fig. 19.6) for the assay of antineoplastic drugs.

Metabolic Tests

The distinction between a metabolic test and a survival test is that a survival test examines the number of cells or amount of cellular activity remaining after prolonged exposure to an agent or short exposure and prolonged recovery in the absence of the agent. What is measured is the ability of the cell to survive, and in most cases, to continue to proliferate. In a metabolic test, on the other hand, the direct effect of the agent on one or more metabolic pathways is being measured. If, in the above protocol for microtitration assay, the labeling with ^{35}S-methionine is carried out during the first 24 hr of drug exposure, this becomes a metabolic test. There are many such tests, ranging from simple observation of the inhibition of pH depression to more specific tests depending on precursor incorporation or measurement of enzyme activity. They may be short-term (30 min or so) or long-term (several days).

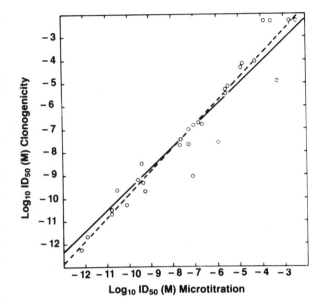

Fig. 19.6. *Correlations between microtitration and cloning in the measurement of the ID_{50} (solid line) of a group of five cell lines from human glioma and six drugs (vincristine, bleomycin, VM-26 epidophyllotoxin, 5-fluorouracil, methyl CCNU, mithramycin). Most of the outlying points were derived from one cell line which later proved to be a mixture of cell types. The dotted lines is the regression with the data points from the heterogeneous cell line omitted [from Freshney et al., 1982].*

In the context of cytotoxicity and viability testing, it should be kept in mind that effects on metabolism measured in the presence of an agent must be interpreted only as such. To establish an irreversible effect on cell survival, culture must be continued in the absence of the agent.

Drug Interaction

The investigation of cytotoxicity often involves the study of the interaction of different drugs, readily investigated by microtitration systems where several different ratios of interacting drugs can be examined simultaneously. Analysis of drug interaction can be performed using an *isobologram* to interpret the data [Szumiel and Nias, 1980]. A rectilinear plot implies an additive response while a curvilinear plot implies synergy if the curve dips below the predicted line and antagonism if above.

ANTICANCER DRUG SCREENING

The measurement of viability by clonal growth analysis and microtitration has been described in detail above and in Chapter 18. One of the major applications of such tests is in the development of new anticancer agents where comparison of survival curves in clonogenicity assays with L1210, P388, Hep-2, and more recently, with cell lines derived from human bronchial carcinoma and other human tumors, can give comparative figures for relative cytotoxicity. Drug testing *in vitro* does not allow for the modification of drugs by liver metabolism en route to the target tissue, so some workers have included liver microsomal enzyme preparations in the culture medium to activate drugs such as cyclophosphamide [Sladek, 1973].

Predictive Testing

The possibility has often been considered that measurement of the chemosensitivity of cells derived from a patient's tumor might be used in designing a chemotherapeutic regime for the patient. This has never been exhaustively tested, although small-scale trials have been encouraging [Limburg and Heckman, 1968; Hamburger and Salmon, 1977; Berry et al., 1975; Kauffmann et al., 1980; Thomas et al., 1985; Dendy, 1976; Bateman et al., 1979]. What is required now is the development of reliable and reproducible culture techniques for the common tumors such as breast, lung, and colon, such that cultures of pure tumor cells capable of cell proliferation over several cell cycles may be prepared routinely. Assays might then be performed in a high proportion of cases, within two weeks of receipt of the biopsy. So far this test has not been possible, but recent developments with new defined media (see Chapter 7) may have brought this closer [Carney et al., 1981].

The major problem is, however, logistic. The number of patients whose tumors will grow *in vitro* sufficiently to be tested, who can be expected to respond, and whose response can be followed up is extremely small. Hence it has proved difficult to use any *in vitro* test as a predictor of response or even to prove its reliability. Correlation of insensitivity *in vitro* with nonresponders is high, but few clinicians would withhold chemotherapy because of an *in vitro* test, particularly when the agent in question would probably not be used alone.

Culture Systems

Most of the procedures described above depend on the exposure of cells in exponential growth in conventional monolayer culture. There are, however, a number of variations that can be made to the culture system depending on whether maximum sensitivity, chronic exposure, or *in vivo* simulation is required. These types of culture systems must be considered along with the duration of exposure (see above) and stability of the agent, as this will influence sensitivty and *in vivo* simulation (Table 19.1). At one extreme, cloned cultures exposed continuously to a stable toxin will demonstrate higher sensitivity than a high-density culture pulsed with a brief exposure to an unstable toxin. The first may be appropriate, for example, for monitoring trace toxins in tap water where the keynote is chronic exposure and the effects minor, while the second may approximate more closely to intravenous bolus injection of a pharmaceutical, such as an anticancer drug.

The types of culture systems that have been used for cytotoxicity assay are almost as numerous as the workers in the field, and as a result a major problem has

TABLE 19.1. Types of Culture Systems for Cytotoxicity Assay

System	Exposure	Recovery period	End point	Use
Monolayer clones	Continuous	None	Colony no. and size	Chronic exposure, low grade toxin, e.g., tap water impurities or pharmaceuticals presumed to be nontoxic
Exponential monolayer, e.g., in microtitration plates or flasks	1–24[h] pulse, repeated if necessary	Approx. 3 cell cycles (cells must remain in exponential growth in controls)	^{3}H-leucine incorporation; ^{3}H-thymidine incorporation (but not with base analogues)	Large-scale screening of potential toxins, anticancer drugs, drug combinations, different doses
Spheroids	Pulsed or continuous	None	Spheroid volume (growth delay); trypsinize and clone	Drug penetration, solid tumor modeling, homologous cell interaction
Filter wells or mixed capillary bed perfusion	Pulsed or continuous (perfused)	None	Metabolic (e.g., isotopic precursor incorporation; clonogenicity)	Solid tissue simulation, complex tissue responses, heterologous cell interactions

emerged in comparing data between laborororatories. A number of suggestions are made in Table 19.1 based on practical examples.

Most tests to date have relied on cell survival, cell growth, or reductions in DNA, RNA, or protein sythesis as indicators of toxicity. However, there are many other ways that toxicity may be expressed *in vivo* and not recognised *in vitro*. One of the major *in vivo* indicators is inflammation; as yet there is no satisfactory *in vitro* model for the inflammatory response. However, as more is learned about primary cellular products which may elicit inflammatory or other more general responses, it may become possible to assay such products directly (e.g., by immunoassay) or indirectly by an appropriate bioassay with, for example, bone marrow stem cells. Until such times as this situation can be achieved, a substantial amount of toxicity testing will continue *in vivo*.

MUTAGENICITY

The determination of mutagenicity is a task more amenable to culture as it represents a cellular response. Metabolism by enzymes of the liver and the gastrointestinal tract may still be required, however ("S9 mix") [Venitt et al., 1984].

Assaying for mutagenicity can be preformed by a variety of standard genetic techniques [Venitt and Parry, 1984], e.g., an increase in the frequency of occurrence of such characterized mutants as the absence of thymidine kinase (TK⁻), which makes cells resistant to bromodeoxyuridine, or hypoxanthine guanine phosphoribosyl transferase deficiency (HGPRT⁻), making the cell resistant to 8-azaguanine [Littlefield, 1964].

As TK⁻ or HGPRT⁻ cells will not clone in HAT medium (hypoxanthine, aminopterin, and thymidine) but revertant mutants will, the reversion rate of TK or HGPRT deficiencies assayed by cloning in HAT medium can also be used to assay for frame-shift mutations.

Double strand breaks in DNA can lead to exchange of chromatid fragments between homologous chromatids. These exchanges, sister chromatid exchanges or SCEs can be detected by incorporating BudR into the chromatids and subsequently staining with Giemsa and Hoechst 33258 [Dean and Danford, 1984].

CARCINOGENICITY

The potential for *in vitro* testing for carcinogenesis is considerable for this is one area where *in vivo* testing is far from adequate. The models are poor and the tests often take weeks or even months to perform. However, the development of a satisfactory *in vivo* test is hampered (1) by the lack of a universally acceptable criterion for malignant transformation *in vitro*, and (2) by the inherent stability of human cells used as targets.

The most generally accepted tests so far assume that carcinogenesis, in most cases, is related to mutagenesis. This is the basis of the Ames tests (Ames, 1980), where bacteria are used as targets and activation can be carried out using liver microsomal enzyme preparations. This test has a high predictive value but nevertheless dissimilarities in uptake, susceptibility, and the type of cellular response have led to the introduction of alternative tests using mammalian and human cells as targets.

Some of these tests are mutagenesis assays also, using suspensions of L5178Y lymphoma cells as targets and the induction of mutations or reversion, or cytological evidence of sister chromatid exchange, as evidence of mutagenesis. Others [Styles, 1977] have used transformation as an end point, assaying clonogenicity in suspension (anchorage independence; see Chapter 15) as a criterion for transformation. Critics of these systems say that both use cells which are already transformed as targets; even the BHK21-C13 cell used by some workers is a continuous cell line and may not be regarded as completely normal. Furthermore, the bulk of the common cancers arise in epithelial tissues, which have so far been difficult to grow routinely, and not in connective tissue cells.

It would appear that at present the Ames test may be an appropriate first line test for screening potential carcinogens. The next step to improve on this is not obvious and may need the demonstration of a reliable and reproducible culture system with epithelial cells before carcinogenicity screening can be transferred routinely to mammalian cells.

Guillouzo et al. have proposed [1981] that such a system should incorporate normal hepatocytes in co-culture for their metabolic activity.

The final hurdle is the general acceptance of an appropriate criterion for malignant transformation (see Chapter 15). No one criterion may be sufficient and two or three may be required. Demonstration of increased oncogene expression, amplification, or the presence of increased or altered oncogene products may provide reliable criteria, in some cases functionally related to the carcinogen. However, as the proposals gain in complexity the resistance to relinquishing the Ames test increases and *in vitro* carcinogenesis assays remain the experimental tool of those studying the mechanism of carcinogenesis.

Chapter 20

It will be apparent from the discussion in previous chapters that the expression of specialized functions in culture is controlled by the nutritional constitution of the medium, the presence of hormones and other inducer or repressor substances, and the interaction of the cells with the substrate and other cells. Reviews of specialized culture techniques for specific cell types can be found Barnes et al. [1984a–d], in Jakoby and Pastan [1979], Kruse and Patterson [1973], Willmer [1965], Defendi [1964], Ursprung [1968], and Sato [1978, 1981] (see also Table 2.2). My main purpose here is to present an outline of some of the techniques that are available for culturing different tissue or cell types and to exemplify the diversity of cell types that can be cultured. It is useful to classify these anatomically, although there is considerable overlap in the techniques used.

A number of specialized procedures have now been devised and some representative examples have been contributed by experts in each area. It is assumed in the following protocols that the basic prerequisites of the cell biology laboratory, as specified in Chapter 4, will be available. Consequently items such as inverted microscopes, bench centrifuges, and waterbaths will not be reiterated in each Materials section.

EPITHELIAL CELLS

Epithelial cells are often responsible for the recognized functions of an organ, e.g., controlled absorption in kidney and gut, secretion in liver and pancreas, and gas exchange in lung. They are also of interest as models of differentiation and stem cell kinetics (e.g., epidermal keratinocytes) and are amongst the principal tissues where the common cancers arise. Consequently, culture of various epithelial cells has been a focus of attention for many years. The major problem in the culture of pure epithelium has been the overgrowth of the culture by stromal cells such as connective tissue fibroblasts and vascular endothelium. Most of the variations in technique are aimed at preventing this, by nutritional manipulation of the medium or alterations in the culture substrate (Table 20.1).

Factors contained in serum, many of them derived from platelets, have a strong mitogenic effect on fibroblasts, and tend to inhibit epithelial proliferation by inducing terminal differentiation. Consequently, one of the most significant developments in the isolation and propagation of specialized cell cultures has been the development of selective, serum-free media, supplemented with specific growth factors as appropriate.

Isolation of epithelial cells from donor tissue is best performed with collagenase (see Chapter 9) as this disperses the stroma but leaves the epithelial cells in small clusters, which favors their subsequent survival.

Epidermis

Rheinwald and Green [1975] showed that murine and human epidermal keratinocytes could be cultured selectively on feeder layers of irradiated 3T3 cells and could mature to form differentiated squames [Green, 1977]. Basal cell carcinoma can also be cultured by this method [Rheinwald and Beckett, 1981]. Alteration of the constituents of the medium enabled Peehl and Ham [1980] and Tsao et al. [1982] to culture keratinocytes from human foreskin selectively without feeder layers or serum and others have shown that reductions in pH [Eisinger et al., 1979], Ca^{2+} [Peehl and Ham, 1980], and temperature [Miller et al., 1980] may all contribute to improved selective growth of epidermal keratinocytes. Addition of hydrocortisone (10 μg/ml) [Peehl and Ham, 1980], 10^{-10} M cholera toxin, or 10^{-6} M isoprenaline (isoproterenol) and epidermal growth factor (10 ng/ml) [Rheinwald and Green, 1977] to the medium has made continued serial subculture possible over many cell generations [Green et al., 1979]. When mouse sublingual epidermal cultures were grown on collagen rafts (see also under "liver") at the gas-liquid interface, complete histological maturation was possible [Lillie et al., 1980].

The following protocol for the culture of epidermal cells has been contributed by Norbert E. Fusenig,

TABLE 20.1. Inhibition of Fibroblastic Overgrowth

Method	Agent	Tissue	Reference
Selective detachment	Trypsin	Fetal intestine, cardiac muscle, epidermis	Owens et al. [1974], Polinger [1970], Milo et al. [1980]
	Collagenase	Breast carcinoma	Lasfargues [1973]
Selective attachment and substrate modification	Polyacrylamide	Various tumors	Jones and Haskill [1973, 1976]
	Teflon	Transformed cells	Parenjpe et al. [1975]
	Collagen (pig skin)	Epidermis	Freeman et al. [1976]
Confluent feeder layers	Mouse 3T3 cells	Epidermis	Rheinwald and Green [1975]
	Fetal human intestine	Breast epithelium, normal and malignant	Stampfer et al. [1980] Freshney et al. [1982]
		Colon carcinoma	Freshney et al. [1982]
Selective media	D-valine	Kidney	Gilbert and Migeon [1975, 1977]
	Cis-OH-Proline	Cell lines	Whei-Yang Kao [1977]
	Ethylmercurithio-salicylate	Neonatal pancreas	Braaten et al. [1979]
	Phenobarbitone	Liver	Fry and Bridges [1979]
	MCDB 153	Epidermis	Boyce and Ham [1983]
	Antimesodermal antibody	Various carcinomas	Edwards et al. [1980]
	MCDB 170	Breast	Hammond et al. [1984]

Institute of Biochemistry, German Cancer Research Center, Im Neuenheimer Feld 280, 6900 Heidelberg, F.R.G.

Principle

Protease digestion of skin samples leads to a separation between epidermis and dermis with the split level being partly beneath and partly above the basal cell layer depending on tissue sample and incubation temperature. Single cells are obtained by mechanical dispersion.

Keratinocytes of animal and human skin have been grown in primary culture and for limited numbers of subcultures using a variety of substrata, culture media, and additives including feeder layers [for review see Karasek, 1983; Holbrook and Hennings, 1983; Fusenig, 1986]. Cultures tend to stratify forming differentiating multilayers when grown at normal (1.4 mM) Ca^{++} concentrations, but stay essentially as "undifferentiated" monolayers in media with low (below 0.1 mM) Ca^{++} concentrations. Maintenance at a high cell density or at low density in the presence of confluent feeder cells (e.g., 3T3) (see Chapter 11) helps to reduce fibroblast contamination.

Outline

Separate dermis from epidermis by cold trypsin digestion, collect keratinocytes by pipetting epidermal and scraping dermal layers, and propagate in low Ca^{2+} medium.

Materials

Tissue culture media:
(1) Eagle's MEM with a 4 × concentration of vitamins and amino acids (essential and nonessential) (4 × MEM) and 15% heat-inactivated fetal calf serum (FCS) [Fusenig and Worst, 1975]
(2) the same 4 × MEM medium with Ca^{++}-free HBSS, Chelex treated FCS [Hennings et al., 1980] and the Ca^{++} concentration adjusted by added CaCl$_2$ or normal FCS to a final concentration of 0.08 and 0.22 mM for mouse and human cells, respectively
(3) FAD-medium [Wu et al., 1982] consisting of a mixture of 1 part medium F-12 and 3 parts of DMEM, enriched with adenine (1.8 × 10^{-4} M), cholera toxin (10^{-10} M), EGF (10 ng/ml), hydrocortisone (0.4 μg/ml) and 5% FCS. All media contain antibiotics (penicillin 100 U/ml and streptomycin 50 μg/ml)
trypsin (1:250)
HBSS
DBSS (HBSS + antibiotics; see Reagent Appendix)
PBSA

100 mm bacteriological grade petri dishes

scalpels, curved forceps

Protocol

1.

Wash 5–10 times in DBSS (see Reagent Appendix), changing the instruments at least once during rinsing.

2.

To provide better and consistently good access of the enzyme to the epithelial-mesenchymal border zone, the skin specimens have to be dissected into pieces of equal size (approximately 1×2 cm) with scalpels. These specimens are placed epidermis-side-down into dry bacterial plastic petri dishes and irrigated with a few drops of PBS. The subcutaneous and lower dermal tissue is cut off as much as possible with curved scissors.

3.

The tissue samples are rinsed again (five to ten times) in sterile Ca^{++}- and Mg^{++}-free PBS and floated on ice-cold 0.2% crude trypsin (1:250) in Ca^{++}- and Mg^{++}-free PBS (pH 7.2). Usually five to eight pieces are floated on 10–15 ml trypsin in 100-mm (diameter) plastic petri dishes which are placed in a 4°C refrigerator under sterile conditions for 15 to 48 hr. The pH of the trypsin has to be controlled by added indicator dye (e.g., phenol red) to prevent pH shift leading to altered enzyme activity and loss in cell viability.

Alternatively, tissue strips are floated on 0.2% trypsin for 1 to 3 hr at 37°C in the incubator. At this temperature incubation time has to be carefully controlled to prevent cell damage by the protease (see Chapter 9).

4.

When the first detachment of epidermis is visible at the cut edges of skin samples, the pieces are placed (dermis-side down) in 100-mm plastic petri dishes irrigated with 5 ml complete culture medium including serum. With two fine curved forceps the epidermis is gently peeled off and pooled in a 50-ml centrifuge tube containing 20 ml complete culture medium. Viable keratinocytes are detached from the epidermal parts by vigorous pipetting and sieving through nylon gauze (100-μm mesh).

5.

The remaining dermal part is gently scraped with curved forceps on its epidermal (upper surface) to remove loosely attached basal cells. The isolated cells from dermal and epidermal parts are combined, washed twice in culture medium by centrifugation at 100 g for 10 min, and counted for total and viable (trypan blue excluding) cells (see Chapter 19).

6.

Cells are plated at 37°C in medium (1) or (3) at $1–5 \times 10^5$ cells/cm^2 and left undisturbed for 1 to 3 days, depending on species and isolation procedure, for attachment and spreading.

7.

When cells have attached, cultures are extensively rinsed with medium to eliminate nonattached dead and differentiated cells and cultivation is either continued in medium (1) or (3) for long-term growth of stratified cultures or changed to medium (2) and continued at low Ca^{++} concentration.

8.

The shift to low Ca^{++} culture medium (2) is done by rinsing 1–3-day-old primary cultures twice with low Ca^{++} medium and further incubating cultures in this medium. Remaining cell aggregates or differentiated cells sticking to the monolayer detach in this medium within a few days so that subsequent cultures are essentially monolayer.

9.

For subcultivation, cultures in medium (1) and (3) are first incubated in 0.05–0.1% EDTA to initiate cell detachment for 10–30 min until cells start to round up, visible by the enlargement of intercellular spaces. Final detachment is achieved by incubation in 0.1% trypsin and 0.05% EDTA and pipetting; the latter procedure alone is sufficient for cultures in medium (2).

10.

Human and mouse cells can be grown in all three media for several months. While mouse cells can only be subcultured once or twice human cells can be passaged for four and seven times in media (2) and (3), respectively [see Boukamp et al., 1986].

11.

Cultured cells have to be characterized for their epidermal (epithelial) nature to exclude contamination by mesenchymal cells. This is best achieved using cytokeratin-specific antibodies for epithelium and antivimentin for mesenchyme-derived cells. Further identification criteria for other epidermis-derived cells in keratinocyte cultures are available by histochemical and immunocytochemical methods [for review see Fusenig, 1986].

Variations

Keratinocytes can also be grown at clonal density cocultured with x-irradiated 3T3 cells [Rheinwald and Green, 1975], at reduced Ca^{++} concentrations with fibroblast conditioned medium [Yupsa et al, 1981] and in defined serum-free medium [Tsao et al., 1982].

In order to provide more *in vivo*-like growth conditions, "organotypic" culture systems have been developed by seeding the cells on collagen gels or pieces of dermis and lifting these supports to the air-medium interface [for review see Prunieras et al., 1983; Fusenig, 1986]. Under these improved growth conditions keratinocytes express many aspects of growth and differentiation of the epidermis *in vivo* which are less absent or pronounced in submerged cultures on plastic.

Growth under mesenchymal influence can be provided by transplantation of cell suspensions or intact cultures *in vivo* or by recombining cultures with mesenchyme *in vitro* [for review see Fusenig, 1983]. This leads to an almost complete expression of growth and differentiation characteristics of normal epidermis [Bohnert et al, 1986].

Alternative methods have been described for cell isolation in large quantities and further purification with Ficoll density gradients [Fusenig and Worst, 1975] or with Percoll gradients [Brysk et al., 1981], methods which are particularly useful for newborn rodent epidermis. For descriptions of other procedures see Yuspa et al. [1980] and Fusenig [1986].

Breast

Milk [Buehring, 1972; Ceriani et al., 1979] and reduction mammoplasty are suitable sources of normal ductal epithelium from breast, the first giving purer cultures of epithelial cells. Growth on confluent feeder layers of fetal human intestine [Stampfer et al., 1980; Freshney et al., 1982] repress stromal contamination with both normal and malignant tissue (see Fig. 11.8), and optimization of the medium [Stampfer et al., 1980; Smith et al, 1981, Hammond et al., 1984] enables serial passage and cloning of the epithelial cells. Cultivation in collagen gel allows three-dimensional structures to form; these correlate well with the histology of the original donor tissue [Yang et al., 1979, 1980, 1981].

As with epidermis, cholera toxin [Taylor-Papadimitriou et al., 1980] and EGF [Osborne et al., 1980] stimulate the growth of epithelioid cells from breast *in vitro*.

The hormonal picture is more complex. Many epithelial cells survive better with insulin added to the culture (1–10 IU/ml) in addition to hydrocortisone ($\sim 10^{-8}$ M). The differentiation of acinar breast epithelium in organ culture requires hydrocortisone, insulin, and prolactin [Stockdale and Topper, 1966], and requirements for estrogen, progesterone, and growth hormone have been demonstrated in cell culture [Klevjer-Anderson et al., 1980].

The following protocol for the culture of cells from human milk was contributed by Joyce Taylor-Papadimitriou, Imperial Cancer Research Fund, Lincoln's Inn Fields, London, UK.

Principle

Early lactation and postweaning milk, which give the highest cell yield, contain clumps of epithelium which can proliferate in culture [Buehring, 1972; Taylor-Papadimitriou et al., 1980]. Primary cultures, grown in hormone supplemented human serum containing medium, give cell lines of limited lifespan but clonogenic [Stoker et al., 1982]. These are eventually overtaken by nonepithelial "late milk" cells [McKay and Taylor-Papadimitriou, 1981].

Outline

Cells centrifuged from early lactation milk are grown in the presence of endogenous macrophages in an enriched medium. These may be subcultured with a mixed protease chelating solution.

Materials

Nunc plastic dishes

sterile universals or 20–50-ml centrifuge tubes

medium RPMI 1640 (Gibco)

fetal calf serum (FCS, Flow Labs)

human serum (HuS outdate pooled serum from blood blanks; Australian antigen negative)

stock solutions:

insulin (Sigma), 1 mg/ml in 6 mM HCl

hydrocortisone, 0.5 mg/ml in physiological saline

cholera toxin (Schwartz-Mann), 50 µg/ml in physiological saline

serum and stock solutions of insulin, hydrocortisone, cholera toxin, pancreatin, and trypsin should be kept at $-20°C$

trypsinization solution (TEGPED):

10 ml 0.5% EGTA (Sigma, ethylene glycol-bis-(β-aminoethylether) N'N' tetraacetic acid) in PBSA

4 ml 0.02% EDTA (Sigma, diaminoethane tetraacetic acid) in PBSA

4 ml 0.2% trypsin (Difco) in Hanks' balanced salt solution (HBSS)

2 ml 1.0% pancreatin (Difco) in HBSS

growth medium:

RPMI 1640 containing 15% FCS, 10% HuS, cholera toxin 50 ng/ml, hydrocortisone 0.5 μg/ml, insulin 1 μg/ml

Protocol

Milk collection: Milk (2–7 days postpartum) can best be collected on hospital wards. The breast is swabbed with sterile H_2O and the milk manually expressed into a sterile container. Five to 20 ml are usually obtained per patient. The milks are pooled and diluted 1:1 with RPMI 1640 medium to facilitate centrifugation.

Primary cultures:

1.

Spin diluted milk at 600–1000 g for 20 min. Carefully remove supernatant leaving some liquid so as not to disturb the pellet.

2.

Wash the pelleted cells two to four times with RPMI containing 5% FCS until supernatant is not turbid.

3.

Resuspend the packed cell volume in growth medium and plate 50 μl packed cells in 5-cm dishes (Nunc), in 6 ml growth medium. Incubate at 37°C in 5% CO_2.

4.

Change medium after 3–5 days and thereafter twice weekly. Colonies appear around 6–8 days and expand to push off the milk macrophages, which initially act as feeders.

Subculture of milk cells: Incubate in TEGPED (1.5 ml per 55-cm plate) at 37°C for 5–15 min, depending on the age of the culture, to produce a single-cell suspension which can be diluted one third in fresh medium for replating (after washing free of enzyme mixture).

Modifications of the technique: It is convenient to use the macrophages which are already present in the milk as feeders, and these are gradually lost as the epithelial colonies expand. However, macrophages can be removed by absorption to glass and in that case other feeders must be added. Irradiated or mitomycin-treated 3T6 cells (see Chapter 11) show the best growth-promoting activity [Taylor-Papadimitriou et al., 1977]. Analogues of cyclic AMP can be used to replace the cholera toxin [Taylor-Papadimitriou et al., 1980], although this is not possible with macrophage feeders, which are killed by the analogues.

Uses and application of milk epithelial cell cultures: Milk cultures provide cells from the fully functioning gland and allow definition of phenotypes by immunological markers [Chang and Taylor-Papadimitriou, 1983]. They have been successfully transformed by SV40 virus [Chang et al., 1982] and provide an important source of normal cells for comparison with breast cancer cell lines and for transfection with oncogenes.

Cervix

A modification of the epidermal culture technique can be used for the propagation of cervical epithelium [Stanley and Parkinson, 1979].

The following protocol for the culture of epithelial cells from cervical biopsy samples has been contributed by Mary Freshney, The Beatson Institute for Cancer Research, Garscube Estate, Switchback Road, Bearsden, Glasgow G61 1BD, Scotland. It is based on a protocol devised by Stanley and Parkinson [1979] and personal communications from Dr. Margaret Stanley.

Principle

Epithelial cell cultures may be established from biopsies of the cervix [Stanley and Parkinson, 1979] by culturing cells on a feeder layer of irradiated 3T3 cells in the presence of growth factors [Rheinwald and Green 1975, 1977].

Outline

Punch biopsy samples of tissue are diced and after initial growth are seeded onto irradiated 3T3 cells at first passage.

Materials

Dissection BSS—see Appendix

transport medium—Leibovitz L15 with

10% FCS: 0.5 μg/ml hydrocortisone,
50 U/ml penicillin,
50 μg/ml streptomycin

trypsin solution for disaggregation of epithelial cells: 0.1% trypsin + 0.01% EDTA in PBSA

epidermal growth factor (EGF from BRL) ($\times$ 200 stock): 100 μg dissolved in 5 ml sterile water; dilute to 1 μg/ml

cholera enterotoxin (Schwartz-Mann) 856011 from Becton Dickinson: mol wt 60,000; make 1,000 $\times$ stock solution of 10^{-7} M

hydrocortisone: Sigma H4001 ($\times$ 1000 stock): weigh 25 mg, dissolve in 3 ml absolute alcohol, add 47 ml water; filter sterilize, aliquot, and keep dark at 4°C

growth medium of primary explants: Leibovitz L15 with 10% FCS, 50 U/ml pencillin, 50 μg/ml streptomycin, 0.05 g/ml hydrocortisone, 10^{-10} M cholera toxin (EGF 5 ng/ml added after explants attach; EGF inhibits plating)

tissue culture flasks 25 cm^2 and 75 cm^2 growth area

9-cm petri dishes

sterile forceps sterile scalpels with Beaver blades number 65-mini

sterile pipettes

universal containers or conical centrifuge tubes

hemocytometer

irradiation source, e.g., ^{60}Co

growth medium for 3T3 cells and epithelial cells on irradiated
3T3 cells—MEM + 10% FCS
PE (see Appendix)

3T3 Swiss fibroblasts

These cells should be maintained at low density. After trypsinzation an F75 should be reseeded at 5 $\times$ 10^4–10^5 cells/flask and a 175^2-cm flask seeded at 2 $\times$ 10^5 cells/flask. Feed after 3 d and then subculture 3–4 d later when an F75 will contain 3–5 $\times$ 10^6 cells and a 175^2-cm flask will contain 15–20 $\times$ 10^6 cells.

To irradiate 3T3 cells

Trypsinize flasks of cells, resuspend in medium—MEM + 10% FCS—and count cells. Irra- diate cells in a universal container giving 60 Gy (6,000 rads) from a cobalt source. Inoculate F75 flasks with 1.5 $\times$ 10^6 irradiated cells/flask and incubate until the cells are attached (from 4 hr to overnight). Replace medium with MEM + 10% FCS + H.C. + CET (don't add EGF until epithelial cells have attached).

Protocol

1.
Collect biopsy samples in transport medium and store 1–4 days at 4°C until convenient to set up cultures.

2.
Take five 9-cm petri dishes and put 15 ml DBSS into three, 5 ml into one, and leave one empty.

3.
Transfer sample with transport medium into empty petri dish.

4.
Transfer sample to first dish of BSS and rinse.

5.
Transfer sample to second dish. Using two scalpels with Beaver blades number 65-mini, pull away any connective tissue that may be attached to the sample. This is obvious as it is thin and stringy.

6.
Transfer sample to third dish of BSS and rinse.

7.
Transfer sample to dish with 5 ml DBSS and start chopping. This is easier if done on a dry part of the dish. Cut through the tissue as cleanly as possible without pulling it apart until the pieces are about 2 mm across. Let the BSS wet the pieces.

8.
Using a pipette transfer the pieces to a universal container or centrifuge tube. (Wet the inside of the pipette first with DBSS to prevent the pieces from sticking to it.) Allow the pieces to settle.

9.
Withdraw the supernatant and discard; add 10—15 ml fresh DBSS to the container and allow pieces to settle.

10.
Repeat once using growth medium.

11.
Discard medium and add about 1 ml of fresh medium.

12.
Label and date two to four F25 flasks (depending on the number of explants you have) and add 2.5 ml growth medium to each.

13.

Add 0.25 ml of medium containing approximately ten pieces to each flask. (Remember to wet the inside of the pipette.)

14.

Incubate at 37°C until some of the explants attach. This may take up to 10 days.

15.

Add EGF to 5 ng/ml final concentration. If many cells are growing out from the explants, replace medium with medium containing EGF.

16.

Continue to feed the cultures every 3–4 d until they become one-half to two-thirds confluent, when they will be ready to be subcultured onto irradiated fibroblasts.

17.

To trypsinize primary cultures, wash with PE (see Appendix). Add 2.5 ml 0.1% trypsin 0.01 EDTA in PBS. Leave for 1 min and remove. Incubate flask at 37°C until cells loosen.

18.

Resuspend cells in medium and count using a hemocytometer.

19.

Inoculate flasks containing irradiated 3T3 cells with 10^4–10^5 cervical cells/flask.

20.

Incubate and feed every 3–4 d until epithelial cells have grown.

21.

Subculture epithelial cells onto fresh irradiated 3T3 cells before they become confluent.

Variations

The epithelial cell layer may be separated from the feeder layer by treatment with Dispase II (Protease, Boehringer Mannheim 165 859). Stock solution: 5 g dissolved in 20 ml PBSA (gives a cloudy solution, but sediment will settle out or can be filtered off). Sterilize by filtration, aliquot, and store at −20°C.

When cultures are confluent add Dispase II to 1.2 U/ml (in medium), incubate at 37°C for 1 hr, when the epithelial cells should have rolled up and come off. Remove medium and cells and spin at 100 g for 10 min. Discard medium and wash cells with PE, spin, and discard PE.

Gastrointestinal Tract

Culture of normal gut lining epithelium has not been extensively reported although there are numerous re-

ports in the literature of continuous lines from human colon carcinoma [Tom et al., 1976; Van der Bosch et al., 1981; Kim et al., 1979; Noguchi et al, 1979; Bergerat et al., 1979]. Colorectal carcinoma cells plated on confluent feeder layers of FHI (see below and Chapter 11) form colonies which apparently disappear but reappear 8–10 wk later as nodules in the monolayer (Fig. 20.1). These nodules will increase and can be subcultured with or without feeder layers and, in several cases, have given rise to continuous cell lines, some of which produce carcinoembryonic antigen (CEA) and sialomucin, markers of neoplasia in human colon [Freshney et al., 1982].

Owens et al. [1974] were able to culture cells from fetal human intestine as a finite cell line (FHS 74 Int). Similar results have been obtained in the author's laboratory, where more vigrorous growth was observed in the epithelial cell component of these cultures and fibroblastic cells were eventually diluted out, giving rise to a finite cell line, FHI.

Although no protocol is available for the propagation of normal cell lines from the gastrointestinal tract, at the time of writing it seems probable that the promising approach adopted by Ham and colleagues (see Chapter 7) will lead to a selective medium for gut epithelium in the near future. Such media as MCDB 153 or LHC 9 (see Tables 7.6, 7.7) would seem to be an appropriate starting point.

Liver

One of the major objectives of the 1960s was the culture of cell lines of normal functional liver parenchyma. This has not yet been fully realized, but there

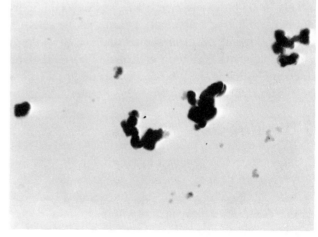

Fig. 20.1. *Colonies of colonic epithelium shed into medium after several weeks of cultivating colon carcinoma cells on a confluent feeder layer (FHI). (Courtesy of Dr. J.M. Russell.)*

are now many examples of epithelial cells cultured from rat liver. Although these cultures do not express all the properties of liver parenchyma, there is little doubt that the correct lineage of cells may be cultured even if the correct environment cannot be re-created for full functional expression [Guillouzo et al., 1981].

The development of the correct conditions for perfusing liver *in situ* with collagenase and hyaluronidase [Berry and Friend, 1969] (see Chapter 9) provided a technique whereby a good viable suspension of liver parenchymal cells could be plated out with high purity and form a viable monolayer, and reports such as those of Malan-Shibley and Iype [1981] suggest that some epithelial cultures may be propagated.

Some of the most useful continuous liver cell lines were derived from Reuber H35 [Pitot et al., 1964] and Morris [Granner et al., 1968] minimal deviation hepatomas of the rat. Induction of tyrosine aminotransferase in these cell lines with dexamethasone proved to be a valuable model for studying the regulation of enzyme adaptation in mammalian cells [Reel and Kenney, 1968; Granner et al., 1968].

Pitot and others have demonstrated that, as with epidermis, greater functional expression can be obtained by culturing liver parenchymal cells on free-floating collagen sheets [Michalopoulos and Pitot, 1975; Sirica et al., 1980]. The cells are seeded in medium onto a preformed collagen gel on the base of a petri dish. After the cells have attached, the gel is released from the base of the dish with a bent Pasteur pipette or spatula, allowing it to float freely in the medium. This permits access of nutrients to the cell from above and below. It is possible that diffusion of nutrients and metabolites via the collagen layer is analogous to the situation *in vivo* where epithelial cells usually lie on basement membrane, and may be important in establishing a necessary polarity in the cells.

The following protocol for the culture of isolated adult hepatocytes has been contributed by Christiane Guguen-Guillouzo, Hôpital de Pontchaillou, INSERM U49, Rue Henri le Guilloux, 35033 Rennes, France.

Principle

Low cytotoxic proteolytic enzymes such as collagenase, when perfused into the liver through the vessels and capillaries at an adequate flow rate, will disrupt intercellular junctions, and will digest the connective framework within 15 min if the liver is previously cleared of blood and depleted of Ca^{++} by washing with calcium-free buffer.

Hepatocytes are selected from the cell suspension by two or three differential centrifugations.

Outline

Introduce a cannula in the portal vein or a portal branch, wash the liver with a calcium-free buffer (15 min), perfuse with the enzymatic solution (15 min), collect and wash the cells, and count the viable hepatocytes.

Materials

Peristaltic pump (10 to 200 rpm.)
waterbath
sterile Tygon tube (ID 3.0 mm; OD 5.0 mm), disposable scalp vein infusion needles, 20 gauge (Dubernard Hospital Labratory, Bordeaux, France), and sewing thread for cannulation
chronometer and sterile graduated bottles and petri dishes
sterile surgical instruments (sharp, straight, and curved scissors and clips)
2 × 1-ml disposable syringes
Heparin (Roche)
Nembutal (Abbot, 5%)
calcium-free Hepes buffer pH 7.65: 160.8 mM NaCl; 3.15 mM KCl; 0.7 mM Na$_2$HPO$_4$12H$_2$O, 33 mM Hepes—sterlization by filtration on 0.22-μm Millipore filters and storage at 4°C (2 months)
collagenase (Sigma grade I; Worthington CLS; Boehringer 103578)
collagenase solution: 0.025% collagenase; 0.075% CaCl$_2$2H$_2$O in calcium-free Hepes buffer pH 7.65—preparation and sterilization by filtration just before use.
L$_{15}$ Leibovitz medium

Protocol for isolation of rat hepatocytes (Guguen-Guillouzo and Guillouzo, 1986)

1.
Warm the washing Hepes buffer and collagenase solution in a waterbath (usually approximately 38–39°C to achieve 37°C in the liver). Oxygenation is not necessary.
2.
Set the pump flow rate at 30 ml/min.
3.
Anesthetize the rat (180–200 g) by intraperitoneal

injection of nembutal (100 μl/100 g) and inject heparin into the femoral vein (1,000 IU).

4.

Open the abdomen, place a loosely tied ligature around the portal vein approximately 5 mm from the liver, insert the cannula up to the liver, and ligate.

5.

Rapidly incise the subhepatic vessels to avoid excess pressure and start the perfusion with 500 ml calcium-free Hepes buffer, at a flow rate of 30 ml/min; verify that the liver whitens within a few sec.

6.

Perfuse 300 ml of the collagenase solution at a flow rate of 15 ml/min for 20 min. The liver becomes swollen.

7.

Remove the liver and wash it with Hepes buffer; after disrupting the Glisson capsule, disperse the cells in 100 ml L_{15} Leibovitz medium.

8.

Filter through two-layer gauze or 60–80-μm nylon mesh, allow the viable cells to sediment for 20 min (usually at room temperature), and discard supernatant (60 ml) containing debris and dead cells.

9.

Wash three times by slow centrifugations (50 g for 40 sec) to remove collagenase, damaged cells, and non-parenchymal cells.

10.

Collect the hepatocytes in Ham's F12 or Williams' E medium enriched with 0.2% bovine albumin (grade V, Sigma) and 10 μg/ml bovine insulin (80–100 ml).

Analysis

Determine cell yield and viability by the well preserved-refringent shape or the trypan blue exclusion test (0.2% w/v) (usually 400 to 600 $\times$ 10^6 viable cells with a viability of more than 95%).

Isolation of hepatocytes from other species

The basic two-step perfusion procedure [Seglen, 1975] can be used for obtaining hepatocytes from various rodents, including mouse, rabbit, guinea pig, or woodchuck by adapting the volume and the flow rate of the perfused solutions to the size of the liver. The technique has been adapted for the human liver [Guguen-Guillouzo et al., 1982] by perfusing a portion of the whole liver (usually at

1.5l Hepes buffer and 1l collagenase solution 70 and 30 ml/min respectively) or biopsies (15 to 30 ml/min depending on the size). A complete isolation into a single cell suspension can be obtained by an additional collagenase incubation at 37°C under gentle stirring for 10 to 20 min (especially for human liver) [Guguen-Guillouzo and Guillouzo, 1986]. Fish hepatocytes can be obtained by cannulating the intestinal vein and incising the heart to avoid excess pressure. Perfusion is performed at room temperature at a flow rate of 12 ml/min.

Applications

Isolated parenchymal cells can be maintained in suspension for 4–6 h and used for short-term experiments. They survive for a few days when seeded in the nutrient medium supplemented with 10^{-6} M dexamethasone on plastic culture dishes (7 $\times$ 10^5 viable cells/ml). A survival of several weeks is obtained by seeding the cells onto a biomatrix [Rojkind et al., 1980]. However, they rapidly lose their specific functions. A high stability (2 months) can be obtained by coculturing hepatocytes with rat liver epithelial cells presumed to derive from primitive biliary cells [Guguen-Guillouzo et al., 1983].

Pancreas

There has not been the same effort expended on culture of pancreas as in liver. Pahlman et al. [1979] and Lieber et al. [1975] described neoplastic cell lines from exocrine pancreas; and Wallace and Hegre [1979] produced epithelial monolayers from fetal rat pancreas by a primary explant technique. These cultures remained free of fibroblasts for several days, and contained many endocrine cells (see also under Endocrine Cells, this chapter).

The following protocol for the culture of pancreatic acinar cells has been contributed by Robert J. Hay, American Type Culture Collection, 12301 Parklawn Drive, Rockville, MD.

Principle

Fractionated populations of pancreative acinar epithelia are caused to aggregate by gyration, the three-dimensional complexes are inoculated onto collagen-coated culture dishes, and two-dimensional aggregated colonies are allowed to develop for subsequent study [Ruoff and Hay, 1979; Hay, 1979].

Outline

Dissociate guinea pig pancreatic tissue, filter the mixed cell suspension through nylon sieves, and layer the suspension over a BSA solution. After three sequential centrifugation/fractionation steps, collect the cell pellet and aggregate by gyration in a waterbath (2–24 hr). Inoculate onto collagen-coated culture vessels, incubate, and observe.

Materials

> Shaker-incubator, Model G-76 (New Brunswick)
>
> conical flasks, 25-ml (Bellco), siliconized (see Chapter 9)
>
> nylon mesh, 40 and 20-μm pore size (Tetco)
>
> dialysis tubing
>
> HBSS
>
> HBSS without Ca^{++} and Mg^{++} (HBSS-DVC)
>
> F-12K tissue-culture medium with 20% bovine calf serum (F12K-CS20)
>
> collagenase type 1 (Sigma): dissolve collagenase in HBSS-DVC (pH 7.2–7.4) to give 1,800 U/ml and dialyze overnight at 4°C against 20 × its volume with HBSS-DVC and a 12,000 MW exclusion membrane—filter, sterilize, and store in 10–20-ml aliquots at −70°C or below
>
> trypsin, 1:250 (ICN), 0.25% in citrate buffer, 3 g/l trisodium citrate, 6 g/l NaCl, 5 g/l glucose, adjusted to pH 7.6 with NaOH
>
> bovine serum albumin, fraction V, (Sigma)
>
> pipettes: 5- or 10-ml, wide-bore (Bellco)
>
> collagen-coated culture dishes (see Chapter 7)

Protocol

1.

Make up the dissociation fluid proper just before use by mixing 1 part of trypsin solution with 1 part of collagenase.

2.

Aseptically remove the entire pancreas (0.4–1.0 gm) and place in HBSS, trim away mesenteric membranes and other extraneous matter, and mince into 1–3 mm-^{3} fragments.

3.

Transfer to a siliconized, 25-ml conical flask in 5 ml of prewarmed, dissociation fluid and agitate at about 120 rpm for 15 min at 37°C in a shaker bath. Allow larger fragments to settle and decant the supernate over a sterile (autoclaved) nylon filter (40-μm pore size) fitted to a Buchner funnel. Repeat these dissociation steps two or three times with fresh fluid until most of the tissue has been dispersed.

4.

Pool the cell suspensions and pass through a second filter/funnel combination using 20-μm nylon mesh in this case. It may be necessary to apply light suction by inserting the funnel into a vacuum filter flask during each filtration step.

5.

Dispense the resulting fluid to 40-ml centrifuge tubes, take an aliquot for counting, and sediment the cells by centrifugation at 250 g for 5 min. Generally, 5–10 × 10^7 cells are obtained at this stage and 30 to 40% can be identified as acinar cells on the basis of zymogen droplet inclusions. These are readily visible by light and phase microscopy.

6.

Resuspend the cell pellet in 10 ml HBSS-DVC and carefully layer 5 ml over each of two columns consisting of 35 ml of 4% BSA in HBSS-DVC in centrifuge tubes. Centrifuge at 100 × g for 5 min, discard the supernate, and repeat this fractionation step twice more, pooling and resuspending the pellets after each step.

7.

Collect the cell pellets in 5 of 10 ml F-12K CS20. Yields of 0.9–3 × 10^7 viable cells (90%) are obtained, with 80 to 96% of the total being acinar.

8.

Transfer the suspension to 25-ml siliconized conical flasks, 5 ml per flask; equilibrate with 5% CO_2 in air and place in a gyration-type water bath (shaker-incubator) at 37°C at 80 rpm for 2–24 hr.

9.

After this period, allow the aggregates to settle for 3 to 5 min and discard the supernate. Resuspend the aggregates in F-12K-CS20 using a wide-bore pipette and take an aliquot for counting. Inoculation densities can be varied from 10^2–10^5 aggregates per cm^2 depending upon experimental design. Aggregates adhere and form colonies within 24–48 hr.

Variations

The method has been applied for studies with guinea pig and human (transplant donor) tissues. Addition of human lung irradiated feeder-fibroblasts produces a marked stimulation (up to 500%) in ^{3}H-thymidine incorporation and prolongs survival at least by a factor of two.

Kidney

The separation of tubular epithelium from stroma has been one of the simpler problems in the field of epithelial cell culture. As these cells have D-amino acid oxidase, they will grow in D-valine while the stromal cells do not [Gilbert and Migeon, 1975]. Kidney tubular epithelium has also proved amenable to separation from stroma by curtain electrophoresis and velocity sedimentation [Kreisberg et al., 1977]. One of the simpler approaches is to treat the finely chopped kidney with collagenase (see Chapter 9) overnight and then, following dispersal of the tissue by gentle pipetting, to allow the undisaggregated fragments of tubule and glomeruli to sediment through the more finely dispersed stroma. If the tubules are washed two or three times by repeating this differential sedimentation, a culture highly enriched for tubular epithelium and glomeruli may be obtained.

A continuous line of dog kidney epithelium has been described [e.g., MDCK, Taub and Saier, 1979; Rindler et al., 1979], and numerous monkey kidney primaries and cell lines are in regular use in vaccine laboratories. BSC-1 is an apparently normal epithelial line from African green monkey kidney [Hopps et al., 1963], but human lines are rare. The success in growing kidney epithelium from many species (see Trade Index, American Type Culture Collection catalogue) may derive from the fact that they are of mesodermal origin; this may confer a different growth factor requirement from endodermally or ectodermally derived epithelium, one more likely to be found in serum.

Bronchial and Tracheal Epithelium

There are a number of reports of primary culture of alveolar, bronchial, and tracheal epithelium [e.g., Steele et al., 1978; Fraser and Venter, 1980; Stoner et al., 1980] including the use of floating collagen [Geppert et al., 1980] and pigskin [Yoshida et al., 1980], and Steele et al. [1978] were able to produce nontumorigenic continuous cell lines by treating tracheal epithelium with a phorbol ester. More recently Lechner and LaVeck [1985] have developed a low serum medium for clonal growth of normal lung (see Chapter

20) and Carney et al. [1981] have developed a serum-free medium supplemented with hydrocortisone, insulin, transferrin, estrogen, and selenium (HITES medium) for small cell carcinoma of lung and with modifications for large cell and adenocarcinoma (see Chapters 7 and 21). These media are selective and do not support stromal cells.

The following protocol for the isolation and culture of normal human bronchial epithelial cells from autopsy tissue has been contributed by Moira A. LaVeck and John F. Lechner, National Institutes of Health, Bethesda, MD.

Principle

In the presence of serum, normal human bronchial epithelial (NHBE) cells cease to divide and, furthermore, terminally differentiate. The serum-free medium which has been optimized for NHBE cell growth does not support lung fibroblast cell replication, thus permitting the establishment of pure NHBE cell cultures [Lechner and LaVeck, 1985].

Outline

This procedure involves first explanting fragments of large airway tissue in a serum-free medium (LHC-9) for initiating and subsequently propagating fibroblast-free outgrowths of NHBE cells; four subculturings and 30 population doublings is routine.

Materials

Culture medium (sterile, filtered):
L-15 (Gibco)
LHC-9 [Lechner and LaVeck, 1985] (see Table 7.6)
HB [Lechner and LaVeck, 1985] (see Reagent Appendix)
bronchial tissue from autopsy of noncancerous donors
plastic tissue culture dishes (60 and 100 mm, sterile)
mixture of human fibronectin (Collaborative Research)/collagen (Vitrogen 100, Collagen Corp.)/crystallized bovine serum albumin (Miles Biochemical) in LHC basal medium (Biofluids Inc.) (FN/V.BSA) [Lechner and LaVeck, 1985].
gloves (human tissue can be contaminated with biologically hazardous agents)

humidified CO_2 incubator at 36.5°C

scalpel no. 1621 Becton Dickinson (sterile)

surgical scissors (sterile)

half-curved microdissecting forceps (sterile)

high O_2 as mixture (50% O_2, 45% N_2, 5% CO_2)

controlled atmosphere chamber (Bellco no. 7741)

rocker platform (Bellco no. 7740)

0.02% Trypsin (Cooper Biomedical)/ 0.02% EGTA (Sigma)/1% polyvinyl-pyrrolidine (U.S. Biochemical Corp.) solution (sterile, filtration) [Lechner and LaVeck, 1985]

pipettes (10 and 25 ml, sterile)

phase contrast inverted microscope

Protocol

1.

Coat culture dish with 1 ml of FN/V/BSA mixture per 60-mm dish and incubate in a humidified CO_2 incubator at 36.5°C for at least 2 hr (not to exceed 48 hr). Vacuum aspirate the mixture and fill the dish with 5 ml culture medium.

2.

Aseptically dissected lung tissue from noncancerous donors autopsied within the previous 12 hr is placed into ice-cold L-15 medium for transport to the laboratory where the bronchus is further dissected from the peripheral lung tissues.

3.

Before culturing, scratch a square centimeter area at one edge of the surface of the culture dishes using a scalpel blade.

4.

Open the airways (submerged in L-15 medium) with surgical scissors and cut (slice, not saw) with a scalpel into two pieces, 20 × 30 mm.

5.

Using a scooping motion to prevent damage to the epithelium, pick up the moist fragments and place epithelium-side-up onto the scratched area of the 60-mm dish. Remove medium and incubate at room temperature for 3 to 5 min to allow time for fragments to adhere to the scratched areas of the dishes.

6.

Add 3 ml of HB medium to each dish and place in

a controlled atmosphere chamber. Flush chamber with a high O_2 gas mixture and place on a rocker platform. Rock chamber at 10 cycles per minute, causing the medium to intermittently flow over the epithelial surface. Incubate rocking tissue fragments at 36.5°C, changing the medium and atmosphere after 1 d and at 2-d intervals for 6–8 d. This step improves subsequent explant cultures by reversing ischemic damage to the epithelium that occurred from time of death of the donor until the tissue was placed in ice-cold L-15 medium.

7.

Before explanting, scratch seven areas of the surface of each 100-mm culture dish with a scalpel. Coat the culture dish surfaces with the FN/V/BSA mixture and aspirate as before.

8.

Cut the moist ischemia-reversed fragments into 7 × 7-mm pieces and explant epithelium-side-up on the scratched areas. Incubate at room temperature without medium for 3 to 5 min as before.

9.

Add 10 ml of LHC-9 medium to each dish and incubate explants at 36.5°C in a humidified 5% air/CO_2 incubator. Replace spent medium with fresh every 3 to 4 d.

10.

After 8 to 11 d of incubation, when epithelial cell outgrowths radiate from the tissue explants more than 0.5 cm, transfer the explants to new scratched and FN/V/BSA coated culture dishes to produce new outgrowths of epithelial cells. This step can be repeated up to seven times with high yields of NHBE cells.

11.

Incubate the postexplant outgrowth cultures in LHC-9 medium for an additional 2 to 4 d before trypsinizing (with trypsin/EGTA/PVP solution) for subculture or for experimental use.

Prostate

Propagation of prostatic cell lines was not reported before the introduction of serum-free selective media (see below). There have been reports of primary culture systems [Webber et al., 1974; Franks, 1980], but these were not successfully subcultured.

The following protocol for the primary culture of rat prostate epithelial cells has been contributed by W.L. McKeehan, W. Alton Jones Cell Science Center, Lake Placid, New York.

Principle

Isolated normal rat prostate epithelial cells are a valuable model to study the cell biology of maintenance, growth, and function of normal prostate under investigator-controlled and defined conditions [McKeehan et al., 1984; Chaproniere and McKeehan, 1986]. Normal cells serve as the control cell type out of which prostate adenocarcinoma cells arise. Conditions similar to those described here have also been useful for primary culture of epithelial cells for transplantable rat tumors. Key features of this method are specific support of epithelial cells by improved nutrient medium and hormone-like growth factors, while concurrently inhibiting fibroblast outgrowth due to deficient or inhibitory properties of the medium.

Outline

Remove and prepare prostates for cell culture (30 min). Incubate with collagenase (1 hr) and collect single cells and small aggregates of cells (1 hr). Inoculate and culture cells to monolayer (7 d).

Materials

10–12-week-old male rats
shaking waterbath at 37°C
centrifuge
hemocytometer or Coulter counter
Sterile
60-mm petri dishes (glass or plastic)
type I collagenase (Sigma)
MSS (media salt solution) [McKeehan et al., 1984]
penicillin
kanamycin
scissors
syringe and 14-g cannula
25-ml Erlenmeyer flasks
1-mm wire mesh screen made to fit a 50-ml conical centrifuge tube
fetal calf or horse serum
50-ml plastic conical centrifuge tubes
nylon screen filters of 253, 150, 100, and 41 µm to fit 50-ml conical tubes
nutrient medium WAJC 404 [McKeehan et al., 1984; Chaproniere and McKeehan, 1986]
25-well plastic tissue culture dishes
cholera toxin
dexamethasone
epidermal growth factor (EGF)
ovine or rat prolactin

insulin
partially purified [McKeehan et al., 1984; Chaproniere and McKeehan, 1986] or purified [Crabb et al., 1986] prostatropin (prostate epithelial cell growth factor)

Protocol

1.
Aseptically remove desired lobes of the prostate and place in a sterile 60-mm petri dish.
2.
Trim fat from lobes and weigh.
3.
Add 2 ml collagenase at 675 U/ml in MSS containing 100 U/ml penicillin and 100 µg/ml kanamycin.
4.
Mince with scissors to approximately 1-mm pieces, small enough to fit through a 14-g cannula.
5.
Using a syringe and 14-g cannula, transfer minced tissue fragments to a sterile 25-ml Erlenmeyer flask. Add collagenase to 1 ml per 0.1 g original wet tissue weight.
6.
Incubate for 1 hr at 37°C on a shaking water bath.
7.
Aspirate the digested suspension three times through a 14-g cannula and then pass the suspension through a coarse (1-mm) wire mesh screen fitted to 50-ml plastic conical tubes to remove debris and undigested material. Rinse through with an equal volume of MSS containing 5% whole fetal calf serum or horse serum.
8.
Collect cells by centrifugation at 100 g for 5 min at 4°C.
9.
Resuspend pellet in 5 ml of MSS plus 5% serum and pass the suspension successively through nylon screen filters of mesh sizes 253, 150, 100, and 41 µm. Wash each screen with 5 ml of MSS plus 5% serum.
10.
Collect cells by centrifugation and resuspend in 5 ml of nutrient medium WAJC 404.
11.
Count cells and adjust concentration to 4×10^6 cells/ml.
12.
Inoculate 50 µl containing 2×10^5 cells into each well of a 24-well plate (area = 2 cm²) containing

1 ml medium WAJC 404 and 10 ng/ml cholera toxin, 1 μM dexamethasone, 10 ng/ml EGF, 1 μg/ml prolactin, 5 μg/ml insulin, 10 ng/ml prostatropin, 100 U/ml penicillin, and 100 μg/ml streptomycin. Partially purified sources of prostatropin can be substituted as described in McKeehan et al. [1984] and Chaproniere and McKeehan [1986].

13.

Incubate in a humidified atmosphere of 95% air and 5% CO_2 at 37°C. Change medium at days 3 and 5. Cells should be near confluent by day 7.

Variations

This procedure can be applied to culture of human prostate epithelial cells with modifications described in Chaproniere and McKeehan [1986].

MESENCHYMAL CELLS

I will include here those cells which are derived from the embryonic mesoderm, but exclude the hemopoietic system, which will be discussed below. This group includes the structural and vascular cells.

Connective Tissue

These cells are generally regarded as the weeds of the tissue culturist's garden. They survive most mechanical and enzymatic explantation techniques and may be cultured in many of the simplest media, such as Eagle's basal medium.

Although cells loosely called fibroblasts have been isolated from many different tissues and assumed to be connective tissue cells, the precise identity of cells in this class remains somewhat obscure. Fibroblast lines, e.g., 3T3 from mouse, produce types I and III collagen and release it into the medium [Goldberg, 1977]. While collagen production is not restricted to fibroblasts, synthesis of type I in relatively large amounts is characteristic of connective tissue. However, 3T3 cells can also be induced to differentiate into adipose cells [Kuriharcuch and Green, 1978]. It is possible that cells may transfer from one lineage to another under certain conditions, but such transdifferentiation has rarely been confirmed. It is more likely that mouse embryo fibroblastic cell lines are primitive mesodermal cells [Franks, personal communication] which may be induced to differentiate in more than one direction.

Human, hamster, and chick fibroblasts are morphologically distinct from mouse fibroblasts as they assume a spindle-shaped morphology at confluence, producing characteristic parallel assays of cells distinct from the pavement-like appearance of mouse fibroblasts. The spindle-shaped cell may represent a more highly committed precursor and may be more correctly termed a fibroblast. NIH3T3 cells may become spindle shaped if allowed to remain at high cell density.

It has also been suggested that fibroblastic cell lines may be derived from vascular pericytes, connective tissue-like cells in the blood vessels, but in the absence of the appropriate markers, this is difficult to confirm.

It is clearly possible to cultivate cell lines, loosely termed fibroblastic, from embryonic and adult tissues, but these should not be regarded as identical or classed as fibroblasts without confirmation with the appropriate markers. Collagen, type I, is one such marker. Thy I antigen has also been used [Raff et al., 1979], although this may also appear on some hemopoietic cells.

Cultures of fetal or adult fibroblasts can be prepared by any of the techniques described in Chapter 9.

Adipose Tissue

Although it may be difficult to prepare cultures from mature fat cells, differentiation may be induced in cultures of mesenchymal cells (mouse 3T3) by maintenance of the cells at a high density for several days [Kuriharcuch and Green, 1978]. An adipogenic factor in serum appears to be responsible for the induction.

Muscle

Myoblasts from the three main categories of muscle may be grown in culture. Skeletal and cardiac myoblasts may be prepared from chick embryo as described in Chapter 9 (see also Konigsberg, 1979). Yaffe [1968] and others have described the stages of differentiation in these cultures.

Cardiac myoblast cells will also progress through differentiation *in vitro* and can be seen to contract rhythmically a few days after explantation from the embryo, although they tend to lose this capacity with continued subculture. Polinger [1970] used the differential rate of attachment to reduce fibroblastic contamination of primary cultures.

Smooth muscle cells may be cultured from blood vessels following disaggregation in trypsin or collagenase [Ross, 1971; Burke and Ross, 1977]. Gospodarowicz et al. [1976] described cloned cell lines from bovine aorta derived from scraped or collagenase-treated tissue. Yasin et al [1981] obtained cell lines from adult skeletal muscle and showed that they retained specialized markers.

Muscle cells may be identified by a number of antigenic markers including myosin and tropomyosin. Actin is not a good marker as it can be found in most cells. Creatine phosphokinase activity increases as muscle cells differentiate [Richler and Yaffe, 1970; Yaffe, 1971]. The most obvious property of all is spontaneous contraction, which is observed in both skeletal and cardiac muscle.

In common with other cells with excitable membranes (and some hemopoietic cells), muscle cells may be stained selectively with the fluorescent dye merocyanine 540 [Easton et al., 1978].

The following protocol for the culture of skeletal muscle has been contributed by Jane Plumb, Dept. of Medical Oncology, University of Glasgow, 1 Horselethill Road, Glasgow G12 9LX, Scotland.

Principle

Skeletal muscle cultures can be established from either embryonic or newborn animal tissues. The muscle is dissected free of skin and bone and disaggregated by enzymatic digestion. Myoblasts are plated out at a high density to minimize fibroblast growth and fusion of the myotubes occurs within 5 to 6 days.

Outline

Remove muscle tissue from animals, mince, and incubate with collagenase for 24 hr. Disperse, remove collagenase, add medium, and seed cultures.

Materials

scissors
forceps
scalpels
universal containers or centrifuge tubes
petri dishes
25-cm^2 culture flask
Primaria petri dishes, 50 mm (Falcon)
plastic box
centrifuge
PBSA + antibiotics (see Chapter 9)
serum-free medium (Ham's F10: DMEM, 50:50)
culture medium: as above plus 10% fetal bovine serum
collagenase (2,000 U/ml) (Worthington, CLS, Sigma, 1A)
3–4-day-old rats
pipettes

Protocol

1.
Kill rats by cervical dislocation and wash thoroughly with 70% alcohol. Transfer to a sterile paper towel.
2.
Use sterile instruments to remove skin. Start with a transverse cut around the body in the abdominal region. Then peel skin back over the fore and hind limbs.
3.
Cut off fore and hind feet and remove fore and hind limbs at shoulder and hip joints respectively. Place limbs in PBSA + antibiotics in a sterile container and keep on ice. Repeat procedure for up to 4 rats at a time.
4.
Transfer container to a laminar flow hood. Remove PBSA and place limbs in a fresh sterile container. Wash four times with 20 ml PBSA and antibiotics.
5.
Place the limbs in a sterile petri dish and add 5 ml of serum-free medium. Place 5 ml of serum-free medium in a second petri dish.
6.
Remove fat from limbs and dissect muscle tissue away from the bone. Transfer tissue to the second petri dish.
7.
Mince muscle tissue with crossed scalpels into pieces about 1 mm^3 and transfer medium and tissue to a sterile container.
8.
Allow tissue to settle for about 5 min and then remove medium. Add 10 ml of serum-free medium and pipette medium plus tissue into a 25-cm^2 culture flask. Add collagenase (2,000 U) and incubate for 24 hr at 37°C.
9.
Pipette vigorously to disperse tissue and transfer suspension to a sterile universal container or centrifuge tube. Centrifuge for 5 min at 500 g.
10.
Remove supernatant and resuspend pellet in culture medium containing fetal calf serum (10%) and antibiotics (not streptomycin since this may inhibit myoblast fusion) [Moss et al., 1984]. Seed cultures at a high density in Primaria petri dishes. Tissue from four rats is sufficient for 30 × 50-mm dishes.

11.

Put dishes in a plastic box and place in a CO_2 incubator at 36.5°C. Feed daily.

Characterization of cultures

Fusion of the myoblasts occurs after 4 to 5 d and can be observed with the aid of a phase-contrast microscope. Provided that the cells are plated at a high density, fibroblast contamination is less than 20% of the total cell population. Fusion can be quantified by monitoring the production of the myosin heavy chain [Moss and Strohman, 1976].

Variations

Cultures of skeletal muscle from chick embryos can be established in a similar manner [Konigsberg, 1979; Walker et al., 1979]. Fibroblast contamination can be reduced further by preplating the culture for 1 hr in a standard tissue culture flask prior to culture in Primaria petri dishes. This procedure reduces the overall yield of muscle cells. Human muscle cultures can be derived from primary explant cultures prepared from muscle biopsies of both normal and diseased tissues [Wilkouski et al., 1976].

Cartilage

Coon and Cahn [1966] described a technique (see also Chapter 11) for the cultivation of cartilage-synthesizing cells from chick embryo somites. Cahn and Lasher [1967] later used this system for analysis of the involvement of DNA synthesis as a prerequisite for cartilage differentiation. Chondrocytes respond to stimulation of growth by both EFG and FGF [Gospodarowicz and Mescher, 1977] but ultimately lose their differentiated function [Benya et al., 1978].

The following protocol for the culture of human chondrocyte cultures has been contributed by Edith Schwartz, Departments of Orthopedic Surgery and Physiology, Tufts University School of Medicine, 136 Harrison Avenue, MA 02111.

Principle

Chondrocytes are embedded in a dense matrix of proteoglycans which must be digested by sequential enzymatic treatment to release the relatively small cellular compartment of the tissue. The cells are then cultured in slightly alkaline Ham's medium with serum and increased Mg^{2+}.

Outline

Finely chopped fragments of cartilage are treated twice with hyaluronidase, trypsin, and collagenase to remove the matrix and make them available to more prolonged collagenase digestion. The cells which are then released are propagated by conventional monolayer techniques in slightly alkaline Ham's F12 medium with an elevated Mg^{2+} concentration.

Materials

Gey's balanced salt solution (GBSS) pH 7.0; if contamination becomes a problem, add antibiotics (100 U/ml penicillin, 100 μg/m streptomycin or 50 μg/ml gentamycin, with or without 100 μg/ml mycostatin) to digestion mixture

Ham's F12 with 12% fetal bovine serum, 2.3 mM Mg^{2+}, 100 U/ml penicillin, 100 μg/ml streptomycin SO_4, pH 7.6

serum-free medium: as above, without serum

0.5% testicular hyaluronidase in GBSS with 100 U/ml penicillin and 100 μg/ml streptomycin

0.2% collagenase in GBSS

0.2% trypsin (Sigma type IX) (for tissue digestion) in GBSS

0.25% trypsin (Sigma type XI) in Ca^{2+}- and Mg^{2+}-free Tyrode's (to be used for cell passage only)

Protocol

1.

Transfer cartilage to a 50-cm sterile glass petri dish.

2.

Cover cartilage with GBSS.

3.

With the use of two scalpels (one no. 20 blade and one no. 11 blade), cut cartilage into 1–2-mm^3 segments.

4.

Transfer cut segments to a second glass petri dish. Cover with GBSS.

5.

When all segments have been cut and combined, remove GBSS and cover tissue with 4 ml of hyaluronidase for 5 min at room temperature.

6.

Remove the hyaluronidase solution and replace with 8 ml fresh hyaluronidase solution for an additional 10 min at room temperature.

7.

Remove the hyaluronidase solution at the end of this time and wash the tissue fragments two times with 5 ml of GBSS.

8.

At the end of this procedure, use GBSS to transfer the cartilage pieces to a 25-ml Corex screw-top tube.

9.

Remove the GBSS and wash the tissue with 2 ml of trypsin solution. Discard wash.

10.

Wash the tissue again with an additional 2 ml of trypsin solution and discard wash.

11.

Incubate the tissue with 4 ml of trypsin at 37°C for 30 min with stirring.

12.

Remove and discard the trypsin solution after 30 min and wash the tissue fragments twice with 5 ml GBSS at room temperature.

13.

Wash the tissue fragments for 5 min with 2 ml of 0.2% collagenase dissolved in GBSS. Discard collagenase wash.

14.

Incubate the tissue with 4 ml of collagenase for 30 min at 37°C.

15.

Remove and discard the supernatant.

16.

Add an additional 4 ml of collagenase to the tissue sample and incubate at 37°C for 90 min.

17.

Remove the supernatant and centrifuge at 600 g for 8 min to pellet the cells.

18.

Resuspend the cells in 8 ml Ham's F12 medium with 20% FCS and centrifuge at 180 g for 1 min to sediment undigested matrix particles.

19.

Remove the supernatant to another tube and centrifuge it at 600 g for 10 min to sediment the cells.

20.

Resuspend the cells in 8 ml of F12 medium with 12% FBS and inoculate a 75-cm^2 flask.

21.

Add an additional 4 ml of F12 medium with 12% FBS to bring the final volume to 12 ml.

22.

Repeat steps 16–21 two or three more times as warranted by the amount of tissue in the sample.

Bone

Although bone is mechanically difficult to handle, thin slices treated with EDTA and digested in collagenase [Bard et al., 1972] give rise to cultures of osteoblasts which have some functional characteristics of the tissue. Antiserum against collagen has been used to prevent fibroblastic overgrowth without inhibiting the osteoblasts [Duksin et al., 1975]. Propagated lines have been obtained from osteosarcoma [Smith et al., 1976; Weichselbaum et al., 1976] but not from normal osteoblasts.

The following protocol for the culture of bone cells has also been contributed by Edith Schwartz.

Principle

Bone culture suffers from the inherent problem that the hard nature of the tissue makes manipulation difficult. However, conventional primary explant culture or digestion in collagenase and trypsin releases cells which may be passaged in the usual way.

Materials

Ham's F12 medium with 12% fetal bovine serum, 2.3 mM Mg^{2+}, 100 U/ml penicillin, and 100 μg/ml streptomycin SO$_4$

trypsin solution to be used for cell passage: dissolve 125 mg trypsin (Sigma type XI) in 50 ml of Ca^{2+}- and Mg^{2+}-free Tyrode's solution. Adjust to pH 7 and filter sterilize. Place aliquots of 5 and 10 ml in Pyrex tubes and store at −20°C.

digestion solution for the isolation of osteoblasts:

to prepare solution A, dissolve 8.0 g NaCl, 0.2 g KCl, 0.05 g NaH$_2$PO$_4$ H$_2$O in 100 ml distilled water

to prepare (collagenase-trypsin solution) solution B, dissolve 137 mg Collagenase (type I, Worthington Biochemicals), and 50 mg trypsin (Sigma, type

III) in 10 ml solution A. Adjust to pH 7.2 and then bring to 100 ml with distilled water. Filter sterilize and distribute into 10-ml aliquots which are stored at $-20°C$

scalpels

forceps

9-cm petri dishes

25- or 75-cm^2 flasks (Corning, Falcon, Nunc)

Explant cultures

Outline

As described in Chapter 9, small fragments of tissue are allowed to adhere to the culture flask by incubation in a minimal amount of medium. The adherent explants are then flooded and the outgrowth monitored.

Protocol

1.

Obtain bone specimens from the operating room.

2.

Rinse the tissue several times at room temperature with sterile saline.

3.

If the bone cannot be used immediately, cover the bone specimen with sterile Ham's F12 medium containing 12% fetal calf serum and penicillin and streptomycin sulfate. The bone may be stored overnight at 4°C.

4.

The next morning, rinse the tissue with Tyrode's solution containing penicillin (100 U/ml) and streptomycin sulfate (100 μg/ml).

5.

Place the bone in a petri dish and with the use of scalpel and forceps, remove trabeculae and place in a second petri dish.

6.

Add 10 ml Tyrode's solution over the excised trabeculae. Rinse the trabeculae several times with Tyrode's solution until blood and fat cells are removed.

7.

To initiate explant cultures, prepare a 25-cm^2 flask by preincubating with 2 ml of complete medium for 20 min to equilibrate the medium with the gas phase. Adjust the pH as necessary with CO_2 or 4.5% $NaHCO_3$, or HCl.

8.

Cut the trabeculae into fragments of 1–3 mm^3.

9.

Remove the preincubation medium from the flask and add 2.5 ml of fresh Ham's F12 medium to the flask. Transfer between 25 and 40 fragments of trabeculae to the flask.

10.

With the flask in an upright position, slide the explant pieces with the aid of an inoculating loop along the base of the flask and distribute the explant pieces evenly.

11.

Permit the flask to remain upright for 15 min at 37°C.

12.

Slowly restore the flask into a normal horizontal position. The explant pieces will stick to the bottom of the flask.

13.

Leave the flask in the horizontal position at 37°C for 5 to 7 d. After this period, check for outgrowth and replace the medium of the flask with fresh medium.

To avoid detachment of the explant pieces, lift the flask slowly into the vertical position before carrying from the incubator to the hood or to the microscope.

14.

To maintain cultures, change medium two times per wk.

15.

When confluency is reached, the explant pieces are removed, the cell layer is trypsinized, and the cells are isolated by centrifugation and seeded into flasks or wells.

Monolayer cultures from disaggregated cells

Outline

Trabecular bone is dissected down to 2–5 mm^3 and digested in collagenase and trypsin. Suspended cells are seeded into flasks in F12 medium.

Protocol

1.

Wash trabecular bone specimens repeatedly with Tyrode's solution to remove the fat and blood cells. The trabeculae are excised with scalpel and forceps under sterile conditions.

2.

After collecting as much bone as possible, wash the remaining blood and fat cells away by rinsing three times with Tyrode's solution

3.

Wash the cut trabeculae with F12 medium containing fetal calf serum.

4.

Place the bone pieces in a small sterile bottle with a magnetic stirrer and add 4 ml of digestion solution (this should cover the bone specimens).

5.

Stir the solution containing bone fragments at room temperature for 45 min.

6.

Remove the suspension of released cells and discard since these cells are most likely to contain fibroblasts.

7.

Add a second aliquot of 4 ml of digestion solution to the bone fragments and stir the mixture at room temperature for 30 min.

8.

Collect the digestion solution from bone fragments and centrifuge for 2 min at 580 g at room temperature.

9.

After removing the supernatant, suspend the cells in 4 ml of Ham's F12 medium with 20% fetal calf serum and count the cells.

10.

Centrifuge at 580 g for 10 min and resuspend the cells in 4 ml of complete medium. This will become the inoculum.

11.

Preincubate 75-cm^2 flasks for 20 min with 8 ml of complete F12 medium to equilibrate with the gas phase.

12.

Remove the preincubation medium and add 2 ml of complete F12 medium.

13.

Add 4 ml of medium containing the cell suspension. The inoculum should contain 6–10,000 cells per cm^2 of surface area.

14.

Finally, add an additional 6 ml of Ham's F12 medium to give a total volume of 12 ml.

15.

In the interim, add an additional 4 ml of digestion solution to the remaining bone pieces and repeat the digestion for 30 min. The released cells are harvested and, if necessary, the digestion step is repeated several more times. With large amounts of bone, the digestion period can be increased to 1–3 hr. Cell counts are performed after each digestion period and the released cells are used to inoculate a different flask.

Passage of cells in culture

1.

Remove the explant pieces.

2.

Remove the medium and rinse the cell layer with PBS, 0.2 ml per cm^2.

3.

Add trypsin to the flask, 0.1 ml/cm^2, and incubate at 37°C until the cells have detached and separated from one another. This is monitored under the microscope. In general, a 10-min incubation is sufficient.

4.

Transfer the released cells to a centrifuge tube with an equal volume of Ham's F12 medium with 20% fetal calf serum (FCS).

5.

Centrifuge at 600 g for 5 min.

6.

Discard the supernatant and resuspend the cells in complete medium by gentle repeated pipetting.

7.

Set one aliquot aside for the determination of cell concentration and another for DNA determination (see Chapter 18).

8.

Inoculate the remaining cells into culture flasks or wells which have previously been equilibrated with medium. The cells should reattach within 24 hr.

Endothelium

Endothelium has been successfully cultured by collagenase perfusion of bovine aorta (Fig. 13.1e) [Gospodarowicz et al., 1976, 1977, 1978; Schwartz, 1978] and human umbilical vein [Gimbrone et al., 1974], trypsinization of white matter from rat cerebral cortex [Phillips et al., 1979], and microdissection of adrenal cortex [Folkman et al., 1979]. In the author's laboratory, an endothelial cell line has been developed from a human anaplastic astrocytoma by collagenase digestion. The astrocytoma cells were overgrown during serial passage.

Endothelium can be characterized by the presence of factor VIII antigen [Booyse et al., 1975], type IV collagen [Howard et al., 1976], Weibel-Palade bodies [Weibel and Palade, 1961] and, sometimes, the formation of tight junctions, although the last feature is not always demonstrated readily in culture.

Endothelial cultures are good models for contact inhibition and density limitation of growth as cell proliferation is strongly inhibited after confluence is reached [Haudenschild et al., 1976].

Much interest has been generated in endothelial cell culture because of the potential involvement of endothelial cells in vascular disease, blood vessel repairs, and angiogenesis in cancer. Folkman and Haudenschild [1980] described the development of three-dimensional structures resembling capillary blood vessels derived from pure endothelial lines *in vitro*. Growth factors, including angiogenesis factor derived from Walker 256 cells *in vitro*, play an important part in maintaining proliferation and survival, so that secondary structures can be formed.

The following protocol for the culture of large vessel endothelial cells has been contributed by Bruce Zetter, Department of Surgery, Harvard Medical School, Boston, MA.

Principle

Endothelial cells comprise a single cell layer at the inner surface of all blood vessels. The vessels most commonly used to obtain cultured endothelial cells are the bovine aorta [Booyse et al., 1975], bovine adrenal capillaries [Folkman et al., 1979], rat brain capillaries [Bowman et al., 1981], human umbilical veins [Jaffe et al., 1973], and human dermal [Davison et al., 1983] and adipose [Kern et al., 1983] capillaries. Although all endothelia share some properties, significant differences exist between the endothelial cells of large and small blood vessels [Zetter, 1981].

Outline

Endothelial cells released by collagenase incubated within the blood vessel are cultured on a gelatin-coated substrate in medium supplemented with mitogens (human and bovine capillary) or without mitogens (bovine large vessel).

Materials

aseptically isolated blood vessels, pref-

erably in 10-cm sections, approximately 5-mm diameter. If asepsis cannot be guaranteed, clamp both ends and dip in 70% alcohol for 30 sec

collagenase: 0.25% crude collagenase, approximately 200 U/mg, in PBSA containing 0.5% bovine serum albumin

Dulbecco's modified Eagle's medium (see Chapter 7) supplemented with 10% calf serum and antibiotics (see Chapter 9)

10-cm tissue culture grade petri dishes or 75-cm² flasks coated with 1.5% gelatin in PBSA: incubate overnight in gelatin solution, remove gelatin, add medium with serum, and incubate until cells are ready

two clamps or hemostats, 25 mm

sharp scissors, 50 mm

Protocol: isolation of endothelial cells

1.
Ligature one end of a 10-cm section of blood vessel 2–10-mm diameter to a 5-ml plastic syringe.
2.
Wash vessel with 20 ml PBSA to remove blood.
3.
Run in collagenase solution until it appears at bottom end, clamp lower end with hemostat, and incubate at room temperature for 10 min.
4.
Cut vessel above clamp with sharp scissors and collect the collagenase in a 10-cm Petri dish.
5.
Rinse lumen of vessel with 10 ml PBSA and add to collagenase.
6.
Centrifuge pooled digest and wash at 100 *g* for 5 min.
7.
Wash twice by resuspending in medium and centrifuging.
8.
Resuspend final pellet in growth medium and seed into gelatin coated dishes or flasks, approximately one 10-cm section of blood vessel, 5-mm diameter, per 75-cm flask or 10-cm diameter dish.
9.
Subculture by conventional trypsinization (see Chapter 10).

Identification

This is proved by production of factor VIII [Hoyer et al., 1973], angiogensin converting enzyme [Del Vecchio and Smith, 1981), uptake of acetylated low-density lipoprotein [Voyta et al., 1984], presence of Weibel-Palade bodies [Weibel and Palade, 1964] and expression of endothelial specific cell surface antigens [Parks et al., 1985].

Variations

Human endothelial cells are cultured in medium 199 with 10–20% human serum, 25 μg/ml hypothalamus derived endothelial mitogen (BTI, Stoughton, MA) and 90 μg/ml heparin [Thornton et al., 1983]. Capillary endothelial cells also require direct addition of an endothelial mitogen or conditioned medium from a tumor culture [Folkman et al., 1979].

NEURECTODERMAL CELLS

Neurones

Nerve cells appear more fastidious in their choice of substrate than most other cells [Nelson and Lieberman, 1981]. They will not survive well on untreated glass or plastic but will demonstrate neurite outgrowth in collagen [Ebendal and Jacobson, 1977; Ebendal, 1979] and poly-D-lysine [Yavin and Yavin, 1980]. Neurite outgrowth is encouraged by a polypeptide nerve growth factor (NGF) [Levi-Montalcini, 1964, 1979] and a factor secreted by glial cells [Barde et al., 1978; Lindsay, 1979] immunologically distinct from NGF.

Cell proliferation has not been found in cultures of neurons even with cells from embryonic stages where mitosis was apparent *in vivo*. Much of the work on nerve cell differentiation has, therefore, been performed on neuroblastoma cell lines [Augusti-Tocco and Sab, 1969; Lieberman and Sachs, 1978; Littauer et al., 1979] or on glial-neuronal hybrids [Minna et al., 1972; Minna and Gilman, 1973]. This remains an intriguing area with many unsolved problems.

The following protocol for the monolayer culture of cerebellar neurons has been contributed by Bernt Engelsen and Rolf Bjerkvig, The Gade Institute, Department of Pathology, University of Bergen, N-5016 Haukeland Hospital, Norway.

Principle

Cerebellar granule cells in culture provide a well-characterized neuronal cell population suited for morphological and biochemical studies [Messer, 1977; Kingsbury et al., 1985; Drejer et al., 1983]. The cells are obtained from the cerebella of 7- or 8-day-old rats, and initial growth inhibition of nonneuronal cells is obtained by a short addition of cytosine arabinoside to the cultures.

Outline

The cerebella from four to eight neonates are cut into small cubes and trypsinized for 15 min at 37°C in Hanks' balanced salt solution. The cell suspension is seeded in poly-L-lysine coated culture wells/flasks.

Materials

Dulbecco's modification of Eagle's medium containing:
10% heat inactivated fetal calf serum
30 mM glucose
L-glutamine 293.2 mg/l
24.5 mM KCl
100 mU/l insulin (Sigma I-1882)
7 μM p-aminobenzoic acid (Sigma A-3659)
100 μg/ml gentamycin
Hanks' balanced salt solution with 3 g/l BSA (HBSS)
poly-L-lysine, mol wt > 300,000 (Sigma P-1524)
cytosine arabinoside (Cytostar, Upjohn; powder)
trypsin (type II) (0.025% in HBSS)
silicone (Aquasil, Pierce 42799)
35-mm tissue culture petri dishes
scalpels, scissors, and surgical tweezers (sterile)
waterbath
Pasteur pipettes and 10-ml pipettes (sterile)
12- and 50-ml sterile test tubes

Protocol

1.
Siliconization of Pasteur pipettes: dilute the Aquasil solution in distilled deionized water to a 0.1–1% concentration. Dip the pipettes into the solution or flush on the inside. Air dry for 24 hr, or for several min at 100°C, and sterilize.

2.
Poly-L-lysine treatment of culture dishes: dissolve the poly-L-lysine in distilled water (10 mg/l) and sterilize by filtration. Add 1 ml of poly-L-lysine

solution to each of the 35-mm petri dishes. Remove the poly-L-lysine solution after 10–15 min and add 1–15 ml of culture medium. Place the culture dishes in the incubator (minimum 2 hr) until seeding the cells.

3.

Preparation of cells:

   a.

   Dissect out the cerebella aseptically and place them in HBSS. Mince the tissue with scalpels into small cubes approximately 0.5 mm^3.

   b.

   Transfer to test tubes (12 ml) and wash three times in HBSS. Allow the tissue to settle to the bottom of the tubes between each washing.

   c.

   Add 10 ml of 0.025% trypsin (in HBSS) to the tissue and incubate in a waterbath for 15 min at 37°C.

   d.

   Transfer the trypsinized tissue to a 50-ml test tube and add 20 ml of growth medium to stop trypsin action.

   e.

   Shear tissue by trituration through a siliconized Pasteur pipette, until a single cell suspension is obtained.

   f.

   Let the cell suspension stay in the test tube for 3–5 min allowing small clumps of tissue to settle to the bottom of the tube. Remove these clumps with a Pasteur pipette.

   g.

   Centrifuge the single cell suspension at 200 *g* for 5 min and aspirate off the supernatant.

   h.

   Resuspend the pellet in growth medium and seed the cells at a concentration of 2.5–3.0 × 10^6 cells/dish.

   i.

   After 2–4 d (best results usually after 2 d), incubate the cultures with 5–10 μM cytosine arabinoside for 24 hr. Then change to ordinary culture medium.

Analysis

Neurons can be identified by immunological characterization using neuron specific enolase antibodies or by using tetanus toxin as a neuronal marker. Astrocyte contamination can be quantified by using glial fibrillary acidic protein as a marker.

Variations

A single cell suspension can be obtained by mechanical sieving through nylon meshes of decreasing diameter, or by sequential trypsinization (i.e., 3 × 5-min trypsin treatment). Instead of HBSS, Puck's solution, Krebs, or other buffers with glucolse can be used.

The following protocol for aggregating cultures of brain cells has been contributed by Rolf Bjerkvig, The Gade Institute, Department of Pathology, University of Bergen, N-5016 Heukeland Hospital, Norway.

Principle

Aggregating cultures of fetal brain cells have been extensively used to study neural cell differentiation [Seeds, 1971; Trapp et al., 1981; Bjerkvig et al., 1986a]. The aggregating cells follow the same developmental sequence as observed *in vivo*, leading to an organoid structure consisting of mature neurons, astyrocytes, and oligodendrocytes. A prominent neuropil is also formed. In tumor biology, the aggregates can be used to study brain tumor cell invasion *in vitro* [Bjerkvig et al., 1986b].

Outline

Brains from fetal rats at day 17–18 of gestation are removed and prepared as a single cell suspension. Brain aggregates are formed by overlay cultures in agar coated multiwells. The cells in the aggregates will form a mature organoid brain structure during a 20-day culture period.

Materials

   Dulbecco's modification of Eagle's medium containing:

   10% heat inactivated newborn calf serum

   four times the prescribed concentration of non-essential amino acids

   L-glutamine 293.2 mg/l

   penicillin 100 U/ml

   streptomycin 100 μg/ml

   phosphate buffered saline (PBS) with Ca^{2+} and Mg^{2+}

   trypsin type II (0.025% in PBSA)

   waterbath

   agar (Difco)

   multiwell tissue culture dishes (24 wells; Nunc)

   10-cm sterile petri dishes

12-ml sterile test tubes

Pasteur pipettes and 5-ml pipettes (autoclaved)

scalpels, scissors, and surgical tweezers (sterile)

2 sterile 100-ml Erlenmeyer flasks

Protocol

1.

Medium-agar coating of microwells: prepare a 3% stock solution (3 g agar in 100 ml PBSA) in an Erlenmeyer flask. Heat the flask in boiling water until the agar is dissolved. Place an empty Erlenmeyer flask in boiling water and add 10 ml of hot agar solution. Then slowly add complete growth medium to the flask until a medium-agar concentration of 0.75% is reached. Add 0.5 ml of hot medium-agar solution to each well in the multiwell dish. Wait until the agar has cooled. The multiwell dishes can be stored in a refrigerator for 1 wk.

2.

Preparation of cells:

a. Dissect out aseptically the whole brains from a litter of fetal rats at day 17–18 gestation. Place the tissue in a 10-cm petri dish containing PBSA. Mince the tissue with scalpels into small cubes—0.5 cm^3.

b. Transfer the tissue to a test tube and wash three times in PBSA. Allow the tissue to settle to the bottom of the tube between each washing.

c. Add 5 ml of trypsin solution and incubate in a waterbath for 5 min at 37°C.

d. Shear tissue by trituration through a Pasteur pipette approximately 20 times. Allow to settle for 3 min and transfer the clump-free milky cell suspension to a test tube containing 5 ml of growth medium. Five ml of new trypsin is added to the undissociated tissue and the trypsinzation and dissociation procedure is repeated two more times.

e. Spin the cell suspensions at 200 *g* for 5 min. Aspirate supernatant, resuspend, and pool cells in 10 ml of growth medium. Count the cells.

f. Add 3×10^6 cells to each agar coated well. The volume of the overlay suspension should be 1 ml. Place the multidish in a CO_2 incubator for 48 hr.

g. Remove the aggregates to a sterile 10-cm petri dish. Add 10 ml growth medium. Trans-

fer larger aggregates individually to new agar coated multiwells by using a Pasteur pipette.

h. Change medium every third day by carefully removing and adding new overlay medium. During 20 days in culture, the aggregates will become spherical and develop into an organoid structure.

Analysis

The next step is fixation and embedding in paraffin or epon for histological or electron microscopic evaluation. Oligodendrocytes, astrocytes, and neurons are identified by transmission electron microscopy or by immunohistochemical localization of myelin basic protein, glial fibrillary acidic protein, and neuron specific enolase, respectively.

Variations

A single cell suspension can be obtained by mechanical sieving through steel or nylon meshes [Trapp et al., 1981]. Reaggregation cultures can also be obtained using a gyratory shaker. A speed (about 70 rpm) is selected such that the cells are brought into vortex, thereby greatly increasing the number of collisions between cells. This movement also prevents cell attachment to the culture flasks.

Glia

Greater success has been obtained in culturing glial cells from avian, rodent, and human brain. Embryonic and adult brain give cultures by trypsinization [Pontén and McIntyre, 1968], collagenase digestion (see Chapter 9), and primary explant [Bornstein and Murray, 1958] which closely resemble glia. Astrocytic markers can be demonstrated for several subcultures, although there is only one report that cell lines from human adult normal brain lines express the most specific marker, glial fibrillary acidic protein (GFAP) [Gilden et al., 1976]. It is our experience that while some glial properties remain (high-affinity γ-aminobutyric acid and glutamate uptake, glutamine synthetase activity), GFAP is lost [Frame et al, 1980]. Oligodendrocytes do not readily survive subculture, but Schwann cells from optic nerve have been subcultured using cholera toxin as a mitogen [Brockes et al., 1979; Raff et al., 1978].

Cultures of human glioma can also be prepared by mechanical disaggregation, trypsinization, or collagenase digestion [Pontén, 1975; Freshney, 1980] (see Fig.

13.1c). The right temporal lobe from human males appears to be marginally better than other regions of the brain [Westermark et al., 1973], but most give a good chance of success. The glia/glioma system provides a good model for comparing normal and neoplastic cells under the same conditions.

There is good evidence that the cell lineage is the same, particularly between embryonic normal cells and tumor cells, but the position of cells within the lineage, as with many cultured cells, is still debatable.

The primary explant technique and collagenase digestion (see Chapter 9) have both been found suitable for preparing glial cultures from normal fetal and adult brain. Satisfactory cultures may also be obtained from fetal and newborn rodent brain by the cold trypsinization techniques (see Chapter 9).

A number of gliomas have been cultured from rodents among which the C_6 deserves special mention [Benda et al., 1968]. This cell line expresses the astrocytic marker, glial fibrillary acidic protein, in up to 98% of cells [Freshney et al., 1980a] but still carries the enzymes glycerol phosphate dehydrogenase and 2'3' cyclic nucleotide phosphorylase [Breen and de Vellis, 1974], both of which are oligodendrocytic markers. This appears to be an interesting example of a precursor cell tumor which can mature along two distinct phenotype routes simultaneously.

Linser and Moscona [1980] separated the Müller cells of the neural retina from pigmented retina and neurons and demonstrated that full functional development could not be achieved unless the Müller cells (astroglia) were recombined with neurons from the retina. Neurons from other regions of the brain were ineffective.

Endocrine Cells

The problems of culturing endocrine cells [O'Hare et al., 1978] are similar to the culture of any other specialized cell but accentuated because the relative number of secretory cells may be quite small. Sato and colleagues [Sato and Yasamura, 1966; Buonassisi et al., 1962] cultured functional adrenal and pituitary cells from rat tumors by mechanical disaggregation of the tumor [Zaroff et al., 1961] and regular monolayer culture. The functional integrity of the cells was retained by intermittent passage of the cells as tumors in rats [Buonassisi et al., 1962; Tashjian et al., 1968]. These lines are now fully adapted to culture and can be maintained without animal passage [Tashjian, 1979], and in some cases in fully defined media [Hayashi and Sato, 1976].

Fibroblasts have been reduced in cultures of pancreatic islet cells by treatment with ethylmercurithiosalicylate [Braaten et al., 1974] and have also been purified by density gradient centrifugation [Prince et al., 1978]. These cells apparently produce insulin but not as propagated cell lines.

Pituitary cells, which continue to produce pituitary hormones for several subcultures, have been isolated from the mouse [DeVitry et al., 1974], but in our experience, normal human pituitary cells do not survive well and even pituitary adenoma cells gradually lose the capacity for hormone synthesis.

Melanocytes

Pigment cells were cultured successfully by Coon [Coon and Cahn, 1966] from chick pigmented retina and propagated over many generations. As with the chick embryo cartilage cells, a fraction derived from embryo extract was required for the function differentiation of these cells.

Until recently, other normal pigment cells have proven difficult to culture although cultures were obtained from human uveal melanocytes [Meyskens et al., 1980]. Pigment cells from skin do not survive readily without the appropriate growth factors (see below) although cultures can be obtained from melanomas with a reasonable degree of success [Creasey et al, 1979; Mather and Sato, 1979a,b]. Primary melanomas are often contaminated with fibroblasts, but since they can be cloned on confluent feeder layers of normal cells (see Chapter 11; Fig. 15.2) [Freshney et al., 1982; Creasey et al., 1979] purification may be possible. In general, however, greater success is obtained with secondary growth from lymph nodes, or from distant metastatic recurrences. Sato [1979] has described conditions for serum-free culture of cell lines from human and murine melanoma.

The following protocol for the culture of human melanocytes has been contributed by Barbara A. Gilchrest, US Department of Agriculture Human Nutrition Research Center on Aging at Tufts University, 711 Washington Street, Boston, MA 02111.

Principle

The greater substrate dependency of cultured keratinocytes has been utilized to obtain preferential melanocyte attachment and growth in a hormone-supplemented medium containing a potent, previously undescribed melanocyte growth factor (MGF) extracted from bovine hypothalamus [Wilkins et al., 1985]. The system has been modified to obviate the problem of keratinocyte contamination while supporting good melanocyte proliferation and pigment production in the absence of tumor promoters or chemotherapeutic agents with minimal serum supplementation [Gilchrest et al., 1984; Wilkins et al., 1985]. With conventional serum supplementation (5–20%), melanocyte growth is far better than reported in other systems.

Outline

Epidermis, stripped from small fragments of skin following cold trypsinization, is dissociated in EDTA and cultured in serum-free medium supplemented with MGF.

Materials

humidified incubator (37°C, 8% CO_2)
waterbath
Coulter counter or hemocytometer
tissue culture dishes (Falcon, Becton Dickinson), 100-, 60-, and 35-mm diameter
sterile pipettes
15- and 50-ml centrifuge tubes
PBSA
fetal bovine serum
hormone-supplemented medium:
   medium 199 (Gibco item 400-1200), 93.1 ml
   EGF (Collaborative Research), 0.1 ml, 10 μg/ml stock concentration, 10 ng/ml final concentration
   transferrin (Sigma), 0.1 ml, 10 mg/ml stock, 10 μg/ml final
   insulin (Sigma), 0.1 ml, 10 mg/ml stock, 10 μg/ml final
   triiodothyronine (Collaborative Research), 0.1 ml, 10^{-6} M stock, 10^{-9} M final
   hydrocortisone (Calbiochem), 0.5 ml, 200 × stock, 1.4×10^{-6} M final
   Cholera toxin (Calbiochem), 1.0 ml, 10^{-8} M stock, 10^{-9} M final
   MFG (see below), 5.0 ml, 2 mg/ml stock, 100 μg/ml final

Preparation of melanocyte growth factor (MGF)

1.
Wash bovine hypothalami (Pel Freeze) in dH_2O (2 l/kg tissue) until grossly free of blood.

2.
Homogenize in Waring blender for 3 min in 0.15 M Cl (1 l/kg tissue).

3.
Extract homogenate for 2 hr at 4°C in beaker with magnetic stirrer bar at moderate speed.

4.
Centrifuge at 10,000 g for 30 min at 4°C in 500-ml aliquots.

5.
Decant supernatant (containing MGF) and delipidate with streptomycin sulphate. Prepare streptomycin sulphate 7.5 g/kg tissue in 50 ml chilled dH_2O, pH chilled 8.0, and add to extract while stirring over a 5-min period. Maintain extraction preparation at pH 6.5–7.5 at all times.

6.
Centrifuge at 20,000 g for 30 min at 4°C in 50-ml tubes. Decant supernatant (containing MGF).

7.
Calculate protein concentration (A_{280}) and store as a lyophilized powder at −20°C in 100-mg aliquots in plastic scintillation vials.

8.
Reconstitute in 10 ml medium 199 for dialysis. Wash dialysis tubing (MW cut-off 6–8,000) extensively with ddH_2O. Transfer resuspended extract into dialysis tubing with Pasteur pipette and seal tubing. Dialyse overnight in 500 ml 200 mM Tris-HCl buffer, pH 7.2, at 4°C. Replace buffer next morning with 500 ml fresh Tris-HCl and continue to complete 24-hr dialysis.

9.
Filter sterilize with a 0.22-μm filter.

10.
Using Tris-HCl as a blank, obtain A_{280} for a 1:10 dilution of the dialyzed extract. Bring concentration to 2 mg per ml with medium 199. Store at 4°C.

Protocol: establishing the primary culture
Day 1.
Rinse skin specimen in 70% ethanol, then twice in PBSA. Transfer tissue to a sterile 100-ml dish, epidermal-side-down. With dissecting scissors, excise subcutaneous fat and deep dermis. Cut remaining tissue into 5 × 5-mm pieces with a scalpel, rolling blade over tissue (do not use a sawing action). Transfer tissue fragments to cold 0.25% trypsin in a 15-ml centrifuge tube. Incubate tube at 4°C for 18–24 hr.

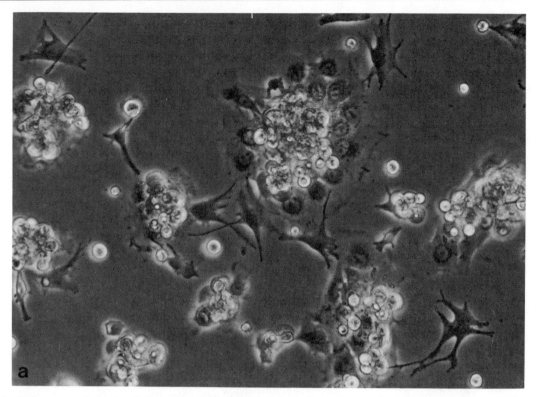

Fig. 20.2. *Melanocyte cultures. a. Culture of newborn foreskin-derived melanocytes 48 hr after inoculation. Note multiple keratinocyte colonies with central stratification and tightly apposed epithelial cells at the periphery. Melanocytes are relatively small dark dendritic cells, most of them in contact with the keratinocyte colonies via dendritic projections (phase contrast, × 640). b.* *Secondary cultures of newborn and adult, c, epidermal melanocytes. Newborn cells tend to be polygonal under these culture conditions, while adult cells are more dendritic. Note larger cell size compared to melanocytes immediately after establishment of the culture (a). (phase contrast, × 640).*

Day 2.

Gently tap centrifuge tube to dislodge fragments settled in the bottom, then rapidly pour tube contents into a 60-mm dish. Using forceps, transfer tissue fragments individually to a dry 100-mm dish, epidermal-side-down. Gently roll each tissue fragment against the dish. The epidermal sheet should adhere and allow clean separation of the dermis with forceps. Discard dermal fragments. Transfer all epidermal sheets to a sterile 15-ml centrifuge tube containing 5 ml 0.02% EDTA, taking care to place each epidermal sheet in the EDTA solution, not on the plastic wall of the centrifuge tube. Gently vortex tube to disintegrate epidermal sheets into a single cell suspension. Centrifuge cells 5 min at 350 g, then aspirate supernatant. Resuspend pellet in serum-free melanocyte medium. Determine cell count using a hemocytometer and inoculate 10^6 cells (approximately 2–4 × 10^4 melanocytes) per 35-mm dish in 2 ml of hormone supplemented medium containing 2% serum to facilitate attachment. Refeed fresh serum-free medium twice weekly.

Days 3–30.

At 24 hr, cultures will contain primary keratinocytes with scattered melanocytes (Fig. 20.2a). Keratinocyte proliferation should cease within several days and colonies should begin to detach during the second week. By the end of the third week, only melanocytes should remain (Fig. 20.2b,c). In most cases, cultures attain near confluence and are ready to passage within one month.

Subcultivation

1.

Gently rinse culture dish twice with 0.02% EDTA. Add 1 ml 0.25% trypsin/0.1% EDTA (Gibco item 610-5050) and incubate at 37°C. Examine dish under phase microscopy every 5 min to detect cell detachment.

2.

When most cells have detached, inactivate trypsin with soybean trypsin inhibitor or 1 ml medium containing 10% serum. (Melanocytes maintained under serum-free conditions greatly benefit from a "serum kick" at the time of cultivation.)

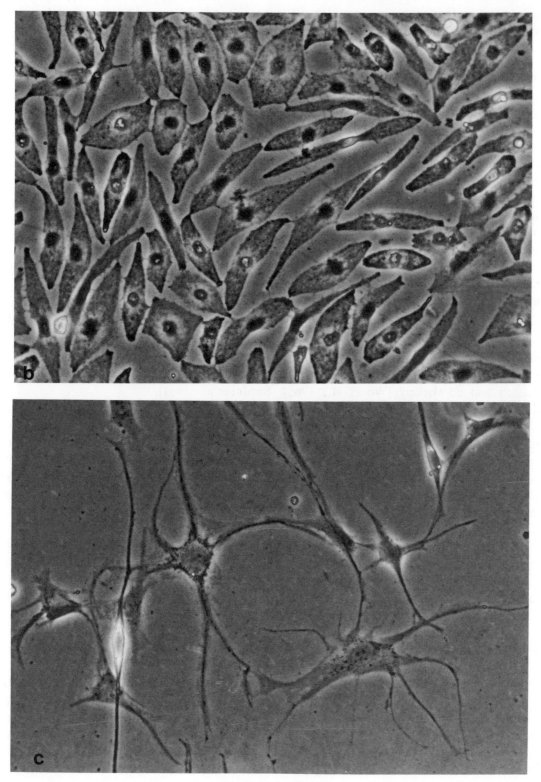

Fig. 20.2. *(continued)*

3.

Pipette dish contents to ensure complete melano-cyte detachment. Aspirate and centrifuge for 5 min at 350 *g*. Aspirate supernatant, resuspend cells in melanocyte medium containing 2% serum, and replate at 2–4 × 10⁴ cells per 35-mm dish. Serum is required for good melanocyte attachment (> 75%) to plastic dishes, but can be omitted if dishes are coated with fibronectin or type I/III collagen [Gilchrest et al., 1985]. Refeed twice weekly with serum-free or serum-containing melanocyte medium, dependent on experimental protocol.

Confirmation of melanocytic identity

Melanocyte cultures may be contaminated initially with keratinocytes and at any time by dermal fibroblasts. Both forms of contamination are rare in cultures established and maintained by an experienced technician/investigator, but are common problems for the novice. Melanocytic identity of the cultured cells can be confirmed with moderate certainty by frequent examination of cultures under phase microscopy, presuming familiarity with the respective cell morphologies (Fig. 20.2B,C). More definitive identification is provided by electron microscopic examination, dopa staining, or immunofluorescent staining with S 100 antibody.

HEMOPOIETIC CELLS

There have been three major milestones in this area. Bradley and Metcalf [1966], Pluznik and Sachs [1965], and McCulloch and co-workers [Wu et al., 1968] developed techniques for cloning normal hemopoietic precursor cells in agar or Methocel (see also Chapter 11), in the presence of colony stimulating factor(s) [Burgess and Metcalf, 1980]. The colonies matured during growth and could not be subloned, implying that the colony-forming unit (CFU) was a precursor cell which was not regenerated in culture. This cell, the CFU-C ("colony-forming unit-culture") is distinct from the CFU-S (spleen colony-forming unit), which is a pluripotent stem cell present in colonies forming in the spleens of sublethally irradiated mice after bone marrow reconstitution [Wu et al., 1968].

Hence, suspension colonies, which contain cells of only one lineage, survive only as primary cultures which lose repopulation efficiency and cannot be subcultured. Granulocytic colonies are the most common; but under the appropriate conditions (see below), lymphoid [Choi and Bloom, 1970] and erythroid [Stephenson and Axelrad, 1971] colonies can be produced.

Golde and Cline [1973] obtained survival in a liquid culture system of normal and neoplastic leukocytes at high cell densities but with abundant medium by placing the cells in a small diffusion chamber immersed in medium. Dexter has also demonstrated, in a liquid culture system, that lymphoid, granulocytic, and erythroid stem cells could be propagated from bone marrow if a bone marrow culture was first prepared and allowed to form a monolayer and to act as a feeder layer for a later, second bone marrow primary culture [Dexter et al., 1977, 1979].

The third major development which occurred over several years between the earlier suspension cloning and Dexter's liquid culture system was the development of a number of functional cell lines from hemopoietic cells. Human lymphoblastic cell lines in both B and T cell lineage were developed by Moore et al. [1967] and subsequently Epstein-Barr virus has been found to be implicated in the ability of these cell lines to become permanent. A number of myeloid cell lines have also been developed from murine leukemias [Horibata and Harris, 1970] and, like some of the human lymphoblastoid lines [Collins et al., 1977], have been shown to make globulin chains, and in some cases, complete α- and γ-globulins (see Chapter 23, Production of Monoclonal Antibodies). Some of these lines can be grown in serum-free medium [Iscove and Melchers, 1978]. T-cell lines require T-cell growth factors (e.g., Interleukin IL-2) [Burgess and Metcalf, 1980; Gillis and Watson, 1981] and B-cell growth factors have also been described [Howard et al., 1981; Snedni et al., 1981].

Originally human lymphoblastoid cell lines were derived by culturing peripheral lymphocytes from blood at very high cell densities ($\sim 10^6$/ml), usually in deep culture (> 10 mm) [Moore et al., 1967]. A monolayer culture appeared in the cell pellet at the bottom of the culture tube and eventually cells were shed into suspension and started to proliferate. This could be detected by the pH drop and the cells were then subcultured. The cell concentration was kept high initially, but eventually these cells adapted to regular culture conditions and could be passaged at 10^5 cells/ml or less. More recently the development of cell lines has become easier by the use of irradiated spleen cells, antigenic stimulation (for T-cell lines), and T- and B-cell growth factors [Paul et al., 1981; Schnook et al., 1981].

Erythroid cell lines have also been cultured from the mouse. Rossi and Friend [1967] demonstrated that a mouse RNA virus (the "Friend virus") could cause splenomegaly and erythroblastosis in infected mice. Cell cultures taken from minced spleens of these animals could, in some cases, give rise to continous cell lines of erythroleukemia cells. All of these cell lines are transformed by what is now recognized as a complex of defective and helper virus derived from Moloney sarcoma virus [Ostertag and Pragnell, 1978, 1981] Some cell lines can produce virus which is infective *in vivo* but not *in vitro*, and the cells can also be passaged as solid tumors or ascites tumors in DBA2 or BALB-C mice.

Treatment of cultures of Friend cells with a number of agents, including DMSO, sodium butyrate, isobutyric acid, and hexamethyl-bis-acetamide, promotes erythroid differentiation [Friend et al., 1971; Leder and Leder, 1975]. Untreated cells resemble undifferentiated proerythroblasts while treated cells show nuclear condensation, reduction in cell size, and an accumulation of hemoglobin to the extent that centrifuged cell pellets are red in color. Evidence for differentiation can also be demonstrated by staining for hemoglobin with benzidine, isolating globin-specific messenger RNA, and fluorescent antibody detection of spectrin, a specific cell surface constituent of erythrocytes, on the surface of stimulated cells [Eisen et al., 1977].

Anderson et al. [1979] have shown that the human leukemic cell line K562 can also be induced to differentiate with sodium butyrate and hemin, though not with DMSO.

Macrophages may be isolated from many tissues by collecting the cells that attach *during* enzymatic disaggregation. The yield is rather low, however, and a number of techniques have been developed to obtain larger numbers of macrophages. Mineral oil or thioglycollate broth [Adams, 1979] may be injected into the peritoneum of a mouse, and 3 days later the peritoneal washings contain a high proportion of macrophages.

If necessary, macrophages may be purified by their ability to attach to the culture substrate in the presence of proteases, as above. They can only be subcultured with difficulty because of their insensitivity to trypsin. Methods have been developed using hydrophobic plastics, e.g., Petriperm dishes (Heraeus).

There are some reports of propagated lines of macrophages mostly from murine neoplasia [Defindi, 1976]. Normal mature macrophages do not proliferate although it may be possible to culture replicating precursor cells by the method of Dexter (see below).

The following protocol for the long-term culture of bone marrow has been contributed by T.M. Dexter, Paterson Laboratories, Christie Hospital, Manchester, England.

Principle

By culturing whole bone marrow the relationship between the stroma and stem cells is main-tained and, in the presence of the appropriate hemopoietic cell and stromal cell interactions, proliferation of stem cells and specific progenitor cells can be maintained over several weeks [Dexter et al., 1984]. Progenitor cells from fresh marrow or long-term cultures may be assayed by clonogenic growth in soft agar [Metcalf, 1985a], or in mice [Till and McCulloch, 1961].

Outline

Marrow is aspirated into growth medium and maintained as an adherent cell multilayer for at least 12 wk and up to 30 wk. Stem cells, maturing, and mature cells are released from the adherent layer into the growth medium. Granulocyte/macrophage progenitor cells can be assayed in soft gels.

Materials

(All reagents must be pretested to check their ability to support the growth of the cultures)

Fischer's medium (Gibco) supplemented with 50 U/ml penicillin and 50 μg/ml streptomycin and containing 16 mM (1.32 g/l) NaHCO$_3$

growth medium: as above, 100-ml aliquots supplemented with 10^{-6} M hydrocortisone sodium succinate and 20% horse serum (Flow Labs) (hydrocortisone sodium succinate made up as 10^{-3} M stock in Fischer's medium and stored at $-20°$C)

five mice—strain differences exist [Greenberger, 1980]

1-ml sterile syringes with 21-gauge needles

gauze, swabs, scissors, forceps

25-cm^2 tissue culture flasks (Nunc, Costar, Corning, Sterilin, but not Falcon)

Protocol: long-term bone marrow cultures

1.

Kill donor mice by cervical dislocation. Wet the fur with 70% alcohol, remove both femurs. Collect ten femurs in a petri dish on ice containing Fischer's medium. One femur contains 1.5–2.0 $\times$ 10^7 nucleated cells.

2.

In a laminar flow cabinet, clean off any remaining muscle tissue using gauze swabs. Hold the femur with forceps and cut off the *knee* end. The 21-

gauge needle should fit snugly into the bone cavity. Cut off the other end of the femur as close to the end as possible. Insert the tip of the bone into a 100-ml bottle of growth medium and aspirate/depress the syringe plunger several times until all the bone marrow is flushed out of the femur. Repeat with the other nine bones.

3.

Disperse to a fine suspension by pipetting the large marrow cores through a 10-ml pipette. There is no need to disaggregate small cell clumps. Dispense 10-ml aliquots of the cell suspension into 25-cm^2 tissue culture flasks, swirling the suspension often to ensure even distribution of the cells in the ten cultures. Gas the flasks with 5% CO_2 in air and tighten the caps. Incubate the cultures horizontally at 33°C.

4.

Feed the cultures weekly. Agitate the flasks *gently* to suspend the loosely adherent cells. Remove 5 ml of growth medium including the suspension cells, taking care not to touch the layer of adherent cells with the pipette. Add 5 ml of fresh growth medium to each flask; to avoid damage do not dispense the medium directly onto the adherent layer. Gas the cultures and replace in the incubator.

Analysis

Cells harvested during feeding can be investigated by a range of methods including morphology, CFC-assays (see below), and the *in vivo* CFU-S assay for stem cells [Till and McCulloch, 1961].

Variations

Human long-term cultures have been grown [Gartner and Kaplan, 1980].

The following protocol for GM-CFC Assay has also been contributed by T.M. Dexter.

Principle

Hemopoietic progenitor cells may be cloned in suspension in semisolid media, in the presence of the appropriate growth factor(s) (Table 20.2) [Metcalf, 1985a]. Pure or mixed colonies will be obtained depending on the potency of the stem cells isolated.

Outline

Fresh bone marrow or supernatant cells from long-term cultures are diluted in a growth factor enriched plating medium, mixed with melted agar,

and plated out. Colonies form in suspension in the gelled agar.

Materials

Fischer's medium, horse serum, mice, syringes, 21-gauge needles, scissors, swabs, and forceps

granulocyte/macrophage colony stimulating factor (GM-CSF) or interleukin 3 (IL-3). GM-CSF is prepared by conditioning medium for 2 d with mouse lung tissue and IL-3 by conditioning medium with WEHI-3B cells for 3–6 d. Both these growth factors have been molecularly cloned and recombinant material will soon be commercially available [Metcalf, 1985b]

3.3% Noble agar (Difco) in double distilled water sterilized by boiling

white cell diluting fluid (WCFD): 3% glacial acetic acid (nonsterile) colored with gentian violet

sterile bottles in which to mix plating medium: capacity > 6 ml

35-mm tissue culture grade plastic petri dishes

Protocol: GM-CFC Assay

1.

Flush the marrow cells into Fischer's medium as described in steps 1 and 2 above. Count the nucleated cells in a hemocytometer using WCFD and adjust to the appropriate concentration. Fresh bone marrow cells should be plated at about 5×10^4/ml to produce 50–150 colonies/plate.

2.

The plating mixture is made as follows: cells ($\times$ 10 final desired concentration), 0.5 ml; horse serum (final concentration 20% v/v), 1.0 ml; WEHI-CM 10% v/v or lung-CM (final concentration 10% v/v), 0.5 ml; Fischer's medium to make final volume to 5.0 ml, 2.5 ml. Warm the plating mixture to above 20°C.

3.

Melt the agar in a boiling waterbath and add 0.5 ml per plating mixture.

4.

Rapidly mix the plating mixture thoroughly by pipetting to ensure the agar is evenly distributed. Dispense 1-ml aliquots into triplicate 35-mm petri dishes. Swirl the plates gently so the agar mixture covers the base of the plate.

TABLE 20.2. Growth Factors in Myelopoiesis*

Growth[†] factor	Synonyms	Species	Target cells								
			CFU-S	CFC-MIX	GM-CFC	BFU-E	EOS-CFC	Meg-CFC	Mast	CFU-E	Mature myeloid
IL-3	Multi-CSF HCGF BPA	Mouse Human Gibbon	+ (insufficient data)	+	+	+	+	+	+	±	+
GM-CSF	MGI-GM Plutipoietin α	Mouse Human	−	+	+	+	+	+	−	−	+
G-CSF	MGI-G Pluripoietin	Mouse Human	− NA	− (+)	+(G) +(G)	− (+)	− (+)	− (+)	− −	− −	+ +
M-CSF	MGI-M CSF-1	Mouse Human	−	−	+(M)	−	−	−	−	−	+
EOS-DF	IL-4	Mouse	−	−	−	−	+	−	−	⊤	+
Erythro-poietin		Human Monkey Mouse	NA	−	−	−	−	−	−	+	−

*These growth factors have all been molecularly cloned, from at least one species. Other "factors" have been described which have not yet been cloned; −, no reported effect; +, direct stimulation; (+), indirect stimulation; G, granulocyte; M, Macrophage; NA, not applicable.

[†] General references: Burgess and Nicola, 1983, Metcalf, 1985, 1986, Whetton and Dexter 1986; IL-3, Hapel et al, 1985, Kindler et al, 1986, Lord et al, 1986, Metcalf et al, 1986, Yang et al, 1986; GM-CSF, Gabrilove et al, 1986, Gough et al, 1984, Nicola et al, 1983, Sachs, 1982; G-CSF, Nicola et al, 1985, Sachs, 1982, Souza et al 1986. M-CSF, Das and Stanley, 1982, Kawasaki et al, 1985, Sachs, 1982, Stanley and Guilbert 1981; EOS-DF, O'Garra et al, 1986; Erythropoietin, Jacobs et al, 1985.

5.

Place the agar cultures in the refrigerator for about 5 min to set the agar.

6.

Place the cultures at 37°C in a humidified atmosphere of 5% CO_2 in air.

7.

After 7 d growth the colonies should be ready for scoring. Colonies can be scored using a stereomicroscope (magnification about 25). A colony is classed as a group of more than 50 cells. It may have a very compact form, a diffuse form, or it may have a compact center with a diffuse halo. These colony types are likely to be granulocytic, monocytic, and mixed granulocyte/monocyte respectively. However, in order to classify colony types correctly, the colonies must be picked out, disaggregated, cytocentrifuged onto slides, and stained.

Variations

CFC assays exist for mouse multipotent-CFC, erythroid-CFC, megakaryocyte-CFC, B-cell-CFC, eosinophil-CFC, and "fibroblast"-CFC (a component of the hemopoietic environment [Golde, 1984; Dexter et al., 1984; Metcalf, 1985a]. A similar range of human CFC can also be grown [Testa, 1985].

GONADS

Culture of germ cells has on the whole been disappointing. Ovarian granulosa cells can be maintained and are apparently functional in primary culture [Orly et al., 1980], but specific functions are lost on subculture. A cell line started from Chinese hamster ovary (CHO-K1) [Kao and Puck, 1968] has been in culture for many years but its identity is still not confirmed. Although epithelioid at some stages of growth, it undergoes a fibroblasticlike modification when cultured in dibutyryl cyclic AMP [Ilsie and Puck, 1971].

Cellular fractions from testis have been separated by velocity sedimentation at unit gravity, but prolonged culture of these has not been reported. The TM4 is an epithelial line from mouse testis although its differentiated features have not been reported, and Sertoli cells have also been cultured from testis [Mather, 1979].

MINIMAL DEVIATION TUMORS

Several cell lines have been derived from the Reuber and Morris hepatomas of the rat [Pitot et al., 1964; Granner et al., 1968] (see above) adrenal cortex and pituitary [Sato and Yasamura, 1966] (as described above) and provide a valuable, if rare, source of continuous cell lines with differentiated properties.

TERATOMAS

When cells from an embryo are implanted into the adult, e.g., under the kidney capsule, these can give rise to tumors known as teratomas. Teratomas also arise spontaneously when groups of embryonic cells or single cells are carried over into the adult, often at an inappropriate site.

Artificially derived teratomas have been used extensively to study differentiation [Martin, 1975, 1978; Martin and Evans, 1974], as they may develop into a variety of different cell types (muscle, bone, nerve, etc.). Growth of teratoma cells on feeder layers of, for example, SCI mouse fibroblasts, will proliferate but not differentiate, whereas when grown on gelatin without feeder layer, or in nonadherent plastic dishes, nodules form which eventually differentiate.

Culture of cells from tumors, particularly spontaneous human tumors, presents similar problems to the culture of specialized cells from normal tissues. The tumor cells must be separated from normal connective tissue cells, preferably by provision of a selective medium which will sustain tumor cell growth but not that of normal cells. While the development of such media for normal cells has advanced considerably (see Chapters 7 and 20), progress in tumor culture has been limited by variation both among and within samples of tumor tissue, even from the same tumor type. It is often surprising to find that tumors which grow *in vivo*, largely as the result of their apparent autonomy from normal regulatory controls, fail to grow *in vitro*.

There are many possible reasons for the failure of some tumor cultures to survive. The nutritional balance may be wrong, or attempts to remove stroma may actually deprive the tumor cells of a matrix, or of nutritional or informational stimulus necessary for survival. Alternatively, dilution of tumor cells to provide sufficient nutrient per cell may dilute out autocrine growth factors produced by the cells. Strictly speaking, truly autocrine factors should be independent of dilution if they are secreted onto the surface of the cell and are active on the same cell, but it is possible that some so-called autocrine growth factors are in fact often paracrine, i.e., they act on adjacent cells, not only on the cell releasing them. Hence a closely interacting population is required. Interaction with certain types of stromal cell may substitute if the stroma are able to make the requisite growth factors either spontaneously or in response to the tumor cells.

Finally, it is probable that most cells in a tumor have a limited life-span due to genetic aberration, terminal differentiation, or natural senescence. Dilution into culture may reduce the numbers, and their interaction with other cells, such that survival is impossible. Cells from multicellular animals, unlike prokaryotes, do not survive readily in isolation. Even a tumor is still a multicellular organ and may require continuing cell interaction for survival. The lethality of the tumor to the host lies in its uncontrolled infiltration and uncontrolled colonial growth, but the origin of the bulk of the cell population may reside in a relatively small population of transformed stem cells, so small that its dilution subsequent to explantation deprives it of some prerequisites for survival.

In sum, the problem is either to create the correct, defined nutritional and hormonal environment, or, failing that, to provide a sustaining environment as yet undefined but nevertheless able to permit the survival of the appropriate or representative population.

SAMPLING

Isolation of cell cultures from normal tissue presents a sampling problem, more in evidence in subsequent culture, in producing cultures representative of a particular cell type. Tumor cell culture in addition to selecting the appropriate cell type, be it gastric epithelium or neuroblast, must also separate this cell type from its normal equivalent and prevent the overgrowth of connective tissue or vascular cells, which tend to predominate in conventional culture systems. Furthermore, while any section of gastric epithelium may be regarded as representative of that particular zone of the gastric mucosa, tumor tissue, dependent as it is on genetic variation and natural selection for its development, is usually heterogeneous and composed of a series of often diverse subclones displaying considerable phenotypic diversity. Ensuring representativity in cultures derived from this heterogeneous population is difficult and can never be guaranteed unless sampling is totally representative and survival is 100%. Since this is practically impossible the average tumor culture is a compromise. Assuming representative subpopulations have been retained, and able to interact, the corporate identity may be similar to the original tumor.

The problem of selectivity is accentuated when sampling is carried out from secondary metastases, which often grow better, but may not be typical either of the primary or all other secondaries.

In view of these almost overwhelming problems facing tumor culture, it is almost surprising that the field has produced any valid data whatsoever. In fact, it has and this may result from: (1) the aforementioned autonomy of tumor populations, which may have allowed their proliferation under conditions where normal cells would not multiply; (2) the increased size of the proliferative pool in tumors, which is larger than that of most normal tissues; and (3) the propensity of malignantly transformed cells to give rise to continuous immortalized cell lines more frequently than normal cells. This last feature, more than any other, has allowed extensive research to be carried out on tumor cell populations, even on supposedly "normal" differentiation processes, in spite of the uncertainty of their relationship to the tumor from which they were derived.

The uncertainty of the status of continuous cell lines remains, but nevertheless they have provided a valuable source of human cell lines for molecular and virological research. The question of whether they represent advanced stages of progression of the tumor whose development has been accelerated in culture, or a cryptic stem cell population, or a purely *in vitro* artefact is still to be resolved. They are certainly distinct from most early passage tumor cultures and should be regarded with caution, though certainly not disregarded, until their true status is fully recognised.

DISAGGREGATION

Some tumors such as human ovarian carcinoma, some gliomas, and many transplantable rodent tumors, are readily disaggregated by purely mechanical means such as pipetting or sieving (see Chapter 9), which may also help to minimize stromal contamination. With many of the common human carcinomas, however, the tumor incorporates large amounts of fibrous stroma, making mechanical disaggregation difficult. In these cases, enzymatic digestion has proved preferable. Trypsin has often been used although its effectiveness against fibrous connective tissue is limited and its toxicity to some epithelial tumor cells may be high.

In the author's laboratory, crude collagenase has been used successfully with several different types of tumor although disaggregation also releases many stromal cells, requiring selective culture techniques for their elimination. Collagenase exposure may be car-

ried out over several hours or even days in complete growth medium.

Necrosis is also a problem of tumors not usually encountered with normal tissue. Usually attachment of viable cells allows necrotic material to be removed on subsequent feeding but if the amount is large and not easily removed at dissection, it may be advisable to use a Ficoll/Hypaque separation (see Chapter 23) to remove necrotic cells.

PRIMARY CULTURE

Removal of protease allows viable cells to attach although some cells (particularly macrophages) can attach in collagenase, and at this stage it is necessary to apply selective culture conditions to promote tumor cell survival and block stromal growth (see below).

Physical separation techniques have also been used [e.g., Green et al., 1980; Sykes et al., 1970](see also Chapter 12), but in general these are only suitable if the cells are to be used immediately as stromal overgrowth usually follows in the absence of selective conditions.

Cloning is a method which suggests itself but there are several limitations. Tumor cells in primary culture often have poor plating efficiencies. Furthermore, because of the heterogeneity of tumor cell populations several clones must be isolated to be at all representative. By the time a clone has grown to sufficient numbers to be of potential analytical value it may have changed considerably and even become heterogeneous itself due to genetic instability. Cloned isolates from a tumor should be studied collectively and even in co-culture for a meaningful interpretation.

There has also been some difficulty in propagating cell lines from primary clones, particularly by the suspension method. It may be that although they are "clonogenic," few of them are really stem cells, or, if they are they mature spontaneously due to the suspension mode of growth and lose their regenerative capacity. Nevertheless cloned lines from tumors would be valuable material for studying tumor clonal diversity and interaction, and they represent a key area of study for future investigation.

CHARACTERIZATION

Isolation of cells from tumors may give rise to several different types of cell line. Besides the neoplastic cells, connective tissue fibroblasts, vascular endothelium and smooth muscle cells, infiltrating lymphocytes, granulocytes and macrophages, and elements of the normal tissue in which the neoplasia arose, can all survive explantation. The hemopoietic components seldom form cell lines although this has been demonstrated with small cell carcinoma of the lung, where the confusion is serious since this carcinoma also tends to produce suspension cultures. Macrophages and granulocytes are so strongly adherent that they generally remain in the primary culture vessel at subculture. Smooth muscle does not propagate readily without the appropriate conditions, so the major potential contaminants of tumor cultures are fibroblasts, endothelium, and normal equivalents to the neoplastic cells.

Of these contaminants the major problem lies with the fibroblasts, which grow readily in culture in any case, but may also respond to tumor-derived factors and grow even faster. Similarly, endothelial cells, particularly in the absence of fibroblasts, may respond to tumor-derived factors and proliferate readily. The role of normal equivalent cells is harder to define as their similarity to the neoplastic cells has made the appropriate experiments difficult to analyze.

Characterization must therefore define lineage. Endothelium is factor VIII positive, contact inhibited, and sensitive to density limitation of growth. Fibroblasts have a characteristic spindle-shaped morphology, are density limited for growth (though less so than endothelium), have a finite lifespan of 50 generations or so, make type II collagen, and are rigidly diploid.

The normal tissue, equivalent to the tumor cells, is harder to identify and eliminate. In general it will be diploid, though some normal epithelia become aneuploid with time in culture (in excess of 1 year). It is usually anchorage dependent and will have a finite life-span (although again there are cases of normal epithelial cell lines becoming continuous) [Christensen et al., 1984]. Normal cells will have the same lineage markers and general morphological characteristics, so behavioral and some biochemical parameters will have to be employed to distinguish them and separate them from their neoplastic derivatives.

From a behavioral aspect, the ability of neoplastic cells to grow on a preformed monolayer of the same cell type is both a good criterion for cell recognition and a potential model of separation. Glioma, for example, will grow readily (better than on plastic in some cases) on a preformed monolayer of normal glial cells [Macdonald et al., 1985] (see Fig. 11.9), and the

same may be true for hepatoma cells and skin carcinoma.

Normal cells tend to have a low growth fraction (labeling index with ^3H thymidine exposure; see Chapters 15 and 18) at saturation density, while neoplastic cells continue to grow faster postconfluence. Maintenance of cultures at high density can sometimes provide conditions for overgrowth of the neoplastic cells.

Many of these properties have been reviewed in Chapter 15; the major objective here is to provide criteria which might favour selection as well as recognition. One of the more useful among these is the capacity for growth on confluent homologous or even heterologous feeder layers in media supplemented by a lower serum concentration. One of the most successful methods of culturing carcinoma cells has been to seed these on to a preformed monolayer of normal cells, usually mouse embryo fibroblasts (See Chapter 11), although fetal human intestinal epithelium has been very successful for breast epithelium [Lan, et al., 1981; Freshney et al., 1982] Subsequent identification of cell tissues derived from the colonies that form is probably best achieved by chromosome analysis, measuring their angiogenic capacity, urokinase-like plasminogenic activator activity, and transforming ability for normal rodent [Todaro and DeLarco, 1978] (see Chapter 15).

A major problem in characterization of many tumor lines is that the criteria employed are often negative, e.g., lack of differentiation markers; hence the appearance of reliable malignancy markers becomes doubly important.

DEVELOPMENT OF CELL LINES

Propagation of primary cultures into cell lines is often difficult. Primary cultures of carcinoma cells do not always take readily to trypsin passage and many of the cells in the primary culture may not be capable of propagation due to genetic or phenotypic aberrations, terminal differentiation, or nutritional insufficiency. Nevertheless some tumor cultures can be subcultured and this often opens up major possibilities. It implies that the neoplastic cell has not been overgrown, probably has a faster growth rate than contaminating normal cells, and, if necessary, is available for cloning or other selective culture methods (see Chapters 11 and 12).

One of the major advantages of subculture is amplification. Replicate cultures can be prepared for characterization and assay of specific parameters such as chemosensitivity or invasiveness. Disadvantages include evolution away from the phenotype of the tumor due to inherent genetic instability and selective adaptation to the culture environment.

Continuous Cell Lines

One major criterion for the neoplastic origin of a culture is its capacity to form a continuous cell line. The constituent cells of this line are normally aneuploid, heteroploid, insensitive to density limitation of growth, less anchorage dependent, and often tumorigenic. The relationship to the primary culture and the parent tumor is still difficult to assess as such cells are not always typical of the tumor proliferation (and need not even be tumorigenic). The two conflicting views that they are (1) either a peculiar and specific adaptation to culture, made possible by the genotypic characteristics of tumor cells, or (2) a specific subset or stem cell population, are still not readily resolved. Certainly, the capacity to form continuous cell lines is a useful criterion of a malignant origin and some authors maintain that their characteristics (tumorigenicity, type of tumor, chemosensitivity, etc.) remain constant [Minna et al., 1983; Tveit and Pihl, 1981]. In any event they provide useful experimental material although the time required for their evolution makes immediate clinical application difficult.

GENERAL METHOD

There are too many methods for the culture of tumors for them to be reported here in detail, so one frequently successful method will be presented with reference to other specific techniques.

Principle

Since trypsin is often toxic and yet many tumors require prolonged enzymatic digestion, collagenase is employed in completed growth medium for prolonged periods. The disaggregation is gentle, nonmechanical, and can be carried out in complete growth medium. Prolonged incubation facilitates disaggregation of the tumor, but may restrict survival of normal anchorage dependent cells. Collagenase is active in the presence of serum which may be a necessary additive to the culture medium, but is less effective in disaggregating epithelium than stroma and hence allows for potential separation by sedimentation of partially disaggregated epithelial cell clusters.

Outline

Tumor tissue is chopped finely with crossed scalpels and incubated in collagenase in complete growth medium for 1–5 d; collagenase is removed by centrifugation and fractions cultured after separation by sedimentation at unit gravity.

This technique has been described in detail in Chapter 9 under "Collagenase Alone."

SELECTIVE CULTURE

Four main approaches have been adopted to select tumor cells in primary culture: selective media (Table 7.7), selective substrates, confluent feeder layers, (see also Chapter 11), and suspension cloning (see Chapters 11 and 15).

Selective Media

There are only a few media which have been developed as selective agents for tumor cells, due to the problems of variability and heterogeneity described above. HITES medium [Carney et al., 1981; Table 7.6] is one exception and may owe its success to the production of peptide growth factors by small cell lung cancer, for which the medium was developed. A proportion, but not all, small cell lung cancer biopsies will grow in pure HITES; others will survive with a low serum supplement (2.5%). HITES medium is modified RPMI 1640 with hydrocortisone, insulin, transferrin, estradiol, and selenium. Of these selenium, insulin, and transferrin are probably the most important.

The same group [Brower et al., 1986] have also produced a selective medium for adenocarcinoma, reputedly suitable for lung, colon, and potentially many other adenocarcinoma. It is also based on RPMI 1640 supplemented with selenium, insulin, and transferrin but also incorporates hydrocortisone EGF, triiodotyrosine, BSA, and sodium pyruvate (Table 21.1).

A simplified version of this medium, RPMI 1640 plus selenium, insulin, and transferrin supplemented with 2.5% fetal bovine serum, may be suitable for a number of tumors with minimal stromal overgrowth. A selective, serum-free medium has also been reported for urinary bladder epithelium [Messing et al., 1982].

Other types of selective media depend on metabolic inhibition of fibroblastic growth and are not specifically optimized for any particular tumor type (see Chapter 11). They have not been found to be generally effective, with the exception of the use of a monoclonal antibody against fibroblasts by Edwards et al.

TABLE 21.1. Serum-Free Medium for Non-Small Cell Lung Cancer Cells

Basal Medium*	RPMI 1640
Supplements	
Insulin	20 μg/ml
Transferrin	10 μg/ml
Hydrocortisone	5.0 E-8
Sodium selenite	2.5 E-8
EGF	10 ng/ml
BSA (optional)	5.0 mg/ml
Na pyruvate (in addition to basal medium)	5.0 E-4
Glutamine	2.0 E-3
Triiodothyronine	1.0 E-10

*Precoating of the substrate with collagen and fibronectin (see Chapter 7) is recommended with this medium

[1980]. This has proved useful in establishing cultures from laryngeal cancer.

Selective Substrates

As transformed cells are less anchorage dependent for growth (see Chapter 15), substrates with reduced adhesive properties have been employed for the selective culture of tumor cells (see also Chapter 11). Jones and Haskill [1973, 1976] found repression of fibroblastic overgrowth on polyacrylamide as did Parenjpe et al. [1975] using Teflon.

Culture in HITES medium often gives rise to aggregate cells growing in suspension which, in itself, helps to separate tumor from stroma. Dr. Morag McCallum of the Victoria Hospital, Glasgow, has observed a similar situation with cultures of breast carcinoma, separated from the biopsy by vigorous manual shaking. These tend to grow in suspension as aggregates when cultured in supplemented MCDB 170 (see Table 7.6). The proportion of stromal cells released is less than by enzymatic digestion and those that are released tend to adhere to the culture vessel, allowing subsequent separation from the floating tumor aggregates. The only disadvantage of this technique is that it may select for more anaplastic tumors, and against very fibrous tumors.

Confluent Feeder Layers

This technique (see Figs. 11.3, 11.8), perhaps more than any other, has been applied successfully to many types of tumor. Smith and others [Lan et al., 1981] used confluent feeder layers of fetal human intestine, FHS-74Int, to grow epithelial cells from mammary carcinoma, using media conditioned by other cell lines although later reports from this and Ham's laboratory suggest that selective culture in MCDB 170 is a more

suitable approach [Hammond et al., 1984]. In the author's laboratory this technique has been used successfully with breast, colon, and lung carcinoma, employing human fetal intestinal epithelium or mouse 3T3 or STO embryonic fibroblasts. Basal cell carcinoma has also been grown on a similar feeder layer system as for epidermis. [Rheinwald and Beckett, 1981].

This technique relies on the prevention of fibroblastic overgrowth by a preformed monolayer of other contact inhibited cells. It is not selective against normal epithelium as normal epidermis and normal breast epithelium both form colonies on confluent feeder layers. Results from glioma, [Macdonald et al., 1985], however, suggest that selection against equivalent normal cells may require a homologous feeder layer. Glioma grown on normal glial feeder layers should lose any normal glial contaminants. By the same argument, breast carcinoma seeded on confluent cultures of normal breast epithelium, e.g., milk cells (see Chapter 20), could become free of any contaminating normal epithelium.

Confluent feeder layers can be highly selective and may even remove some elements of the neoplastic population. Colon carcinoma seeded on normal human fetal intestinal epithelium feeder layers virtually disappeared during early primary culture, but if maintained for 2–3 months gave rise to colonies of apparently transformed cells (CEA producing, aneuploid, and heteroploid). Hence only a very small proportion of the initial inoculum gave rise to the eventual culture. Although this raises questions of representativity, it is interesting to speculate that these may be neoplastic stem cells.

Outline (Fig. 21.1)

Feeder cells are treated in mid-exponential phase with mitomycin C and reseeded to give a confluent monolayer. Tumor cells, dissociated from the biopsy by collagenase digestion, or from a primary culture with trypsin, are seeded onto the confluent layer. Colonies may form in 3 weeks to 3 months from epithelial tumors. Fibrosarcoma and gliomas do not always form colonies but may infiltrate the feeder layer.

Materials (All sterile)

Feeder cells (e.g., 3T3, STO, 10T.1/2 or FHS-74Int)
mitomycin C (Sigma) 10 μg/ml
growth medium

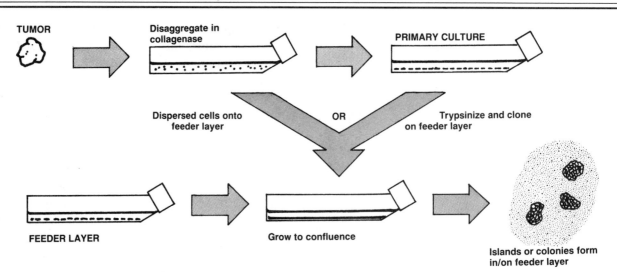

Fig. 21.1. *Selective growth on confluent feeder layers. Colonies of epithelial cells will form on confluent fetal human intestinal epithelium or on confluent normal human glia.*

collagenase, 2,000 U/ml, CLS grade (Worthington) or equivalent

trypsin, 0.25% in PBSA

tumor biopsy or primary culture

forceps, scalpels, petri dishes for dissection as for primary culture

Protocol

1.

Grow up feeder cells to 50% confluence in six 75-cm^2 flasks.

2.

Add mitomycin C to give 5 μg/10^6 cells).

It is advisable to do a dose response curve with mitomycin C when using feeder cells for the first time to confirm that this dose allows the feeder layer to survive for 2–3 wk but does not permit further replication in the feeder layer after about two doublings at most. Treat the cells in 25-cm^2 flasks as above, and trypsinize and reseed entire contents into 75-cm^2 in 20 ml fresh medium, grow for 3 wk, feeding two times per wk. Stain and check for surviving colonies.

3.

Incubate overnight (18 hr) in mitomycin C, remove, wash monolayer, and grow for a further 24–48 hr.

4.

Trypsinize and reseed in 25-cm^2 flasks at 5 × 10^5 cells/ml (10^5/cm^2) and leave 24 hr.

5.

If using biopsy material, biopsy should be dis-

sected and placed in collagenase at the same time as step 2.

6.

Seed cell suspension from biopsy, approximately 20–100 mg/flask, into two of the 25-cm^2 flasks in 6 ml total. Remove 1 ml and add to 4 ml medium in second pair of flasks. The third pair of flasks is kept as a control to guard against feeder cells surviving the mitomycin C treatment.

If using a primary culture, trypsinize or dissociate in collagenase, 200 U/ml final, and seed onto feeder layer at 10^5/ml in two flasks and 10^4/ml in two flasks.

If glioma or fibrosarcoma, colonies may not appear and surviving tumor will only be confirmed by subculturing without feeder layer (by which time contaminating normal cells should have been eliminated).

It is essential to confirm the species of origin of any cell line derived by this method to guard against accidental contamination from resistant cells in the feeder layer. This can be done by chromosomal analysis and lactate dehydrogenase isoenzyme electrophoresis (see Chapter 13).

Suspension Cloning

Transformation of cells *in vitro* leads to an increase in their clonogenicity in agar [Macpherson and Montagnier, 1964]; tumorigenicity has also been shown to correlate with cloning in methocel [Freedman and Shin, 1974]. Since cells may be cloned in suspension directly from disaggregated tumors [Hamburger and Salmon, 1977], or at least colonies may grow (they may not be clones) in preference to normal stromal cells, this would seem to be a potentially selective technique. However, the colony forming efficiency is very low (often < 0.1%) and it is not easy to propagate colonies isolated from agar. On the whole this has not been a successful method for generating cell lines and the method has had greater exploitation in the assay of primary cultures from tumor biopsies (see Chapter 19).

Suspension cloning is described in detail in Chapter 11 and the Hamburger and Salmon modification in Chapter 19. These methods are also reviewed in Dendy and Hill [1983]. However, as this has not been a major method used in establishing cell lines it will not be discussed further here.

Histotypic Culture

Organ culture and histotypic culture in general will be discussed in Chapter 22. Apart for organ culture itself, which is not fundamentally different for tumor tissue, other methods with particular application to tumor culture are spheroid culture and filter with culture.

Spheroids

The technique for generating spheroids will be described in Chapter 23 and their use in drug assay in Chapter 19. Their particular relevance to this chapter stems from the inability of normal stroma to form spheroids or even to become incorporated in tumor-derived spheroids. Hence cultures from tumors allowed to form spheroids on nonadhesive substrates like agarose will tend to overgrow their stromal component.

Some cultures from breast and small cell carcinoma of lung often generate spheroids or other nonspheroidal cellular aggregates which float off into suspension (see above). These may be collected from supernatant medium, leaving the stroma behind. These spheroids or organoids do not always appear soon after culture and can sometimes take weeks or even months to form, suggesting a derivation from a minority cell population in the tumor.

Spheroid generation does not arise in all tumor cultures, but has also been described in neuroblastoma, melanoma, and glioma. Its potential has not been fully explored, however, as the bulk of attention has been given to either forming attached monolayers or suspension colonies in agar for assay purposes. The selectivity described above for PTFE and polyacrylamide might be extended to simpler substrates such as nontissue culture grade polystyrene to favor the selection of anchorage independent cells.

Xenografts

When cultures are derived from human tumors, scarcity of material and the infrequency of rebiopsy means that it is difficult to make several attempts with the same tumor, using different selective techniques. Some tumors can be made to grow in immune-deprived animals, which makes much greater amounts of tumor available. It has sometimes been found that cultures can be initiated more easily from xenografts than from the parent biopsy but whether this is due to the availability of more tissue, progression of the tumor, or modification by the heterologous host (e.g., by c-type virus) is not clear.

Two main types of host are in current use, the genetically athymic nude mouse [Giovanella et al., 1974] or neonatally thymectomized animals subsequently irradiated and treated with cytosine arabinoside [Selby et al., 1980; Fergusson et al., 1986]. The first are expensive to buy and difficult to rear but maintain the tumor for longer. Thymectomized animals are more trouble to prepare but cheaper and easier to provide in large numbers. They will, however, regain immune competence and ultimately reject the tumor after a few months.

If access to a nude mouse colony is available, or facilities exist for neonatal thymectomy and irradiation, this is a step to be considered. Although only a small proportion of tumor may "take," the resulting tumor will probably be easier to culture and repeated attempts may be made.

However, particular care must be taken, as with isolation from mouse feeder layers, to ensure that the cell line ultimately surviving is human and not mouse by proper characterization with isoenzyme and chromosome analysis.

Freezing

It is often difficult to take advantage of a large biopsy and utilize all the valuable material that it provides. It is possible in these cases to preserve the tissue by freezing.

Outline
Chop tumor, expose to DMSO, and freeze.

Materials
Biopsy
instruments (scalpels, forceps, dishes, etc. as for primary culture)

dissection BSS
growth medium with antibiotics (see above)
dimethyl sulphoxide.

Protocol

1.
Chop tumor (after removing necrotic, fatty, and fibrous tissue) into about 3–4-mm pieces and wash in DBSS as for primary culture.

2.
Place four or five pieces in each ampule.

3.
Add 1 ml growth medium + 10% DMSO and leave for 30 min at room temperature.

4.
Freeze at 1°C/min (see Chapter 17) and transfer to liquid N_2.

5.
To thaw, place in 37°C water (with appropriate precautions; see Chapter 17).

6.
Swab thoroughly in alcohol, open ampule and remove half of medium.

7.
Replace medium slowly with fresh medium without DMSO.

8.
Gradually replace all of medium with DMSO free medium, transfer to petri dish and proceed as for regular primary culture, but allowing twice as much material per flask.

When tissue culture was first developed it was based on the explantation of whole fragments of tissue with a view to studying them as tissues in isolation. However, it was observed that cells often grew out from these tissues in a sheet to form a monolayer on the supporting glass substrate. This divergence in growth properties–growth as a migratory and potentially proliferative monolayer versus residence within the original explant—set the pattern for future divergence in approach to the culture of animal cells. One school of thought believed in the retention of histological structure and the possibility of organotypic function while the other looked towards the biology of individual cells. The former required retention of histotypic structure, cell interaction, and histological characteristics and enabled the study of developmental problems such as embryonic induction, *in vitro* modeling of malignant invasion, and hormonal control of morphogenesis and differentiation. The latter gave rise to propagated cell lines which laid down the basis for most of modern cellular and molecular biology and its insight into the regulation of gene expression.

Now, although the potential for propagated cell lines is far from exhausted, many people are reverting to the notion that nutritional completeness and hormonal supplementation are inadequate in themselves to recreate full structural and functional competence in a given cell population. The vital missing factor is cell interaction and the signaling capacity that it entails. A hormone may activate a specific pathway in cell A, and it or a different hormone activate a different commitment in cell B. Both, alone, may lead to individual modifications in phenotypic expression, but if allowed to interact, a cascade of interactions may occur dependent on the cells being in association. Alveolar cells of the lung will only synthesize and release surfactant in response to hormonal stimulation of adjacent fibroblasts [Post et al., 1984]; prostate epithelium response to stromal signals is in turn activated by androgen binding to the stroma [Cunha, 1984]. Epithelium differentiates in response to matrix constituents often determined jointly by the epithelium on one side and connective tissue stroma on the other, as may be the case with the interaction between epidermis and dermis *in vitro* [Bohnert et al., 1986]. Hence the whole integrated tissue may easily, and understandably, respond differently to simple ubiquitous signals, not because of the specificity of the signal or the receptor capacity but because of the response encoded in the juxtaposition of one cell type with a specific correspondent. As in human society, the response of one individual to an exogenous stimulus will be dictated as much by the spatial and temporal relationship with other individuals as by the endogenous make-up. Likewise a primitive neural crest cell may become a neurone, an endocrine cell, or a teratoma, dependent on its ultimate location, its interaction with adjacent cells and its response, mediated by neighboring cells, to hormonal stimuli.

In essence this preamble establishes that while some cell functions, such as cell proliferation, glycolysis, respiration, and gene transcription, proceed in isolation, their regulation, as related to a functioning multicellular organism, must depend ultimately on the interaction between cells of the appropriate lineage, the appropriate stage in that lineage, and in the appropriate intracellular regulatory environment. This suggests that if you want to study cell biology cell lines are fine, but if you want to learn something of the integrated function, or dysfunction, of whole organisms a histotypic or organotypic model will be required.

There are two major ways to approach this goal. One is to accept the cellular distribution of the tissue, explant it, and maintain it as an organ culture. The second is to purify and propagate individual cell lineages, recombine them, and study their interactions. This has given rise to three major types of technique:

Organ culture, where whole organs, or representative parts, are maintained as small fragments in culture and retain their intrinsic distribution, numerical and spatial, of participating cells; *histotypic culture*, where propagated cell lines are grown to high density in a 3D matrix alone; or *organotypic culture*, in which cells of different lineages are recombined in experimentally determined ratios and spatial relationships to recreate a component of the organ under study.

Organ culture seeks to retain the original structural relationship of different and similar cells and hence their interactive function, in order to study the effect of exogenous stimuli on further development [Lasnitzky, 1986]. This may even be achieved by separating the constituents and recombining them as in the now classical experiments of Grobstein and Auerbach and others in organogenesis [Grobstein, 1953; Auerbach and Grobstein, 1958; Cooper, 1965]. Organotypic culture represents the synthetic approach whereby a three-dimensional, high-density culture is regenerated from isolated (and preferably purified and characterized) lineages of cells that are then recombined, their interaction studied, and, in particular, their response to exogenous stimuli characterized.

"Exogenous stimuli" may be regulatory hormones, nutritional conditions, or environmental toxins. In each case the response, and the justification of this approach, will be different from the responses of a pure cell type in isolation.

Several types of systems have been described to study isolated, whole, undisaggregated tissue, or recombinations of tissues, or purified cell lineages. As these may provide models for quite distinct types of investigation each will be described separately.

ORGAN CULTURE

Gas and Nutrient Exchange

When cells are cultured as a solid mass of tissue, gaseous diffusion and the exchange of nutrients and metabolites becomes limiting. The dimensions of individual cells cultured in suspension or as a monolayer are such that diffusion is rapid but aggregates of cells beyond about 250 μm (5,000 cells) start to become limited by diffusion and at or above 1.0-mm diameter ($\sim 2.5 \times 10^5$ cells) central necrosis is often apparent. To alleviate this problem organ cultures are usually placed at the interface between the liquid and gaseous phases to facilitate gas exchange while retaining access to nutrients. Most systems achieve this by positioning the explant on a raft or gel exposed to the air but explants anchored to a solid substrate can also be aerated by rocking the culture, exposing it alternately to liquid medium or gas phase [Nicosia et al., 1983; see also LaVeck and Lechner and protocol Chapter 20], or by using a roller bottle or tube (see Chapter 23) to the same end.

Anchorage to a solid substrate can lead to the development of an outgrowth from the explant and resultant alterations in geometry although this can be minimized using a nonwettable surface. One of the advantages of culture at the gas-liquid interface is that the explant retains a spherical geometry if the liquid is maintained at the correct level: too deep and gas exchange is impaired; too shallow and surface tension will tend to flatten the explant and promote outgrowth.

Increased permeation of oxygen can also be achieved by using increased O_2 concentrations up to pure oxygen or by using hyperbaric oxygen. Certain tissues, e.g., thyroid [de Ridder and Marcel, 1978] and prostate trachea and skin [Lasnitzky, 1986] particularly from newborn or adult may benefit from elevated O_2 tension but often this is at the risk of O_2-induced toxicity. This may have to be determined for each tissue type under study.

Increasing the O_2 tension will not facilitate CO_2 release or nutrient metabolite exchange, so the benefits of increased oxygen may be overidden by other limiting factors.

Structural Integrity

Structural integrity, above other considerations, was and is the main reason for adopting organ culture as an *in vitro* technique in preference to cell culture. While cell culture utilizes cells dissociated by mechanical, enzymic techniques or spontaneous migration, organ culture deliberately maintains the cellular associations found in the tissue. Initially this was selected to facilitate histological characterization but ultimately it was discovered that certain elements of phenotypic expression were only found if cells were maintained in close association.

It is now recognized that associated cells do exchange signals via junctional communications ("gap" junctions) and via paracrine hormones and surface information exchange. This is most striking during organogenesis, but is probably also required for the maintenance of fully mature tissues.

Hence maintenance of the structural integrity of the original tissue may preserve the correct homologous and heterologous cellular interactions present in the original tissue, and maintain the correct chemical configuration of the extracellular matrix.

A major deficiency in tissue architecture in organ culture is the absence of a vascular system, limiting the size (by diffusion) and potentially also the polarity of the cells within the organ culture.

Growth and Differentiation

It has been stated previously (Chapter 14) that there appears to be a relationship between growth and differentiation such that differentiated cells no longer proliferate. It is also possible that cessation of growth may in itself contribute to the induction of differentiation if only by providing a permissive phenotypic state receptive to exogenous inducers of differentiation.

Because of the rules of density limitation of growth, and the physical restrictions imposed by their geometry, most organ cultures do not grow, or if they do, proliferation is limited to the outer cell layers. Hence the status of the culture is permissive to differentiation and given the appropriate cellular interactions and soluble inducers (see Chapter 14) should provide an ideal environment for differentiation to occur.

Limitations of Organ Culture

In view of the advantages described above it is perhaps surprising that organ culture is not more popular. The reasons relate largely to the development of biochemical and molecular criteria for *in vitro* behavior, particularly the monitoring of differentiation. These criteria have been adopted because they are more readily quantified and generally more objective than histological criteria although they lack the resolution of histological techniques where local response in a minority of cells can be detected.

Biochemical monitoring requires reproducibility between samples which is less easily achieved in organ culture than in propagated cell lines. This is due to the sampling variation in preparing an organ culture, to minor differences in handling and geometry, and to variations in cell type heterogeneity between cultures.

Organ cultures are also more difficult to prepare than replicate cultures from a passaged cell line and do not have the advantage of a characterized reference stock to which they may be related. Preparation is labor intensive and as a result the yield of usable tissue is often too low to be of value in biochemical or molecular assays. Furthermore, as the reacting cell population may be a minor component it is difficult to analyze the biochemical nature of the response and attribute it to the correct cell type other than by autoradiographic, histochemical, or immunocytochemical techniques that tend to be more qualitative than quantitative.

Organ cultures cannot be propagated and hence each experiment requires recourse to the original donor tissue.

Organ culture is essentially a technique for studying the behavior of integrated tissues rather than isolated cells. It is precisely in this area that the future understanding of the control of gene expression (and ultimately of cell behavior) in multicellular organisms may lie, but the limitations imposed by the organ culture system are such that recombinant systems between purified cell types may contribute more at this particular stage. However, there is no doubt that organ culture has contributed a great deal to our understanding of developmental biology and tissue interactions and that it will continue to do so in the absence of adequate synthetic systems.

Types of Organ Culture

As the technique has been dictated largely by the requirement to place the tissue at a location allowing optimal gas and nutrient exchange, most techniques place the tissue at the gas-liquid interface. This has been achieved by placing fragments of tissue on semisolid gel substrates of agar [Wolff and Haffen, 1952] or clotted plasma [Fell and Robison, 1929] or on a raft of microporous filter, lens paper, or rayon supported on a stainless steel grid or adherent to a strip of Perspex or Plexiglas [see Fig. 23.1 and Lasnitzky, 1986].

Two techniques have successfully departed from this method, the use of stirred cultures of small tissue fragments [Mareel et al., 1979; Bjerkvig et al., 1986a,b] (see Chapter 21), and the use of rocking or rolled cultures where the tissue is anchored to a substrate and subjected alternately to liquid culture medium and the gas phase [see Lechner, Chapter 20; Nicosia et al., 1983]. Stirred suspensions of tissue fragments tend to be restricted to embryonic or newborn tissue.

Several techniques have been described for gas-liquid interface culture, but one which has become popular because of the provision of suitable disposable plasticware is the grid technique derived from that of Trowell [1954, 1959].

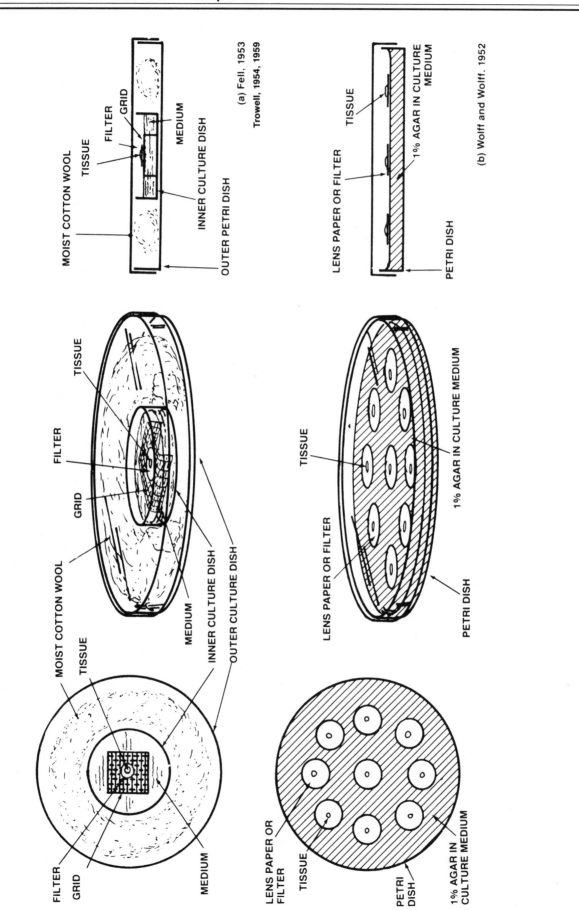

MOIST COTTON WOOL
TISSUE
FILTER
GRID
MEDIUM
INNER CULTURE DISH
OUTER PETRI DISH

(a) Fell, 1953
Trowell, 1954, 1959

TISSUE
LENS PAPER OR FILTER
1% AGAR IN CULTURE MEDIUM
PETRI DISH

(b) Wolff and Wolff, 1952

TISSUE
FILTER
GRID
MOIST COTTON WOOL
MEDIUM
INNER CULTURE DISH
OUTER CULTURE DISH

TISSUE
LENS PAPER OR FILTER
1% AGAR IN CULTURE MEDIUM
PETRI DISH

FILTER
GRID
MOIST COTTON WOOL
TISSUE
MEDIUM
INNER CULTURE DISH
OUTER CULTURE DISH

LENS PAPER OR FILTER
TISSUE
PETRI DISH
1% AGAR IN CULTURE MEDIUM

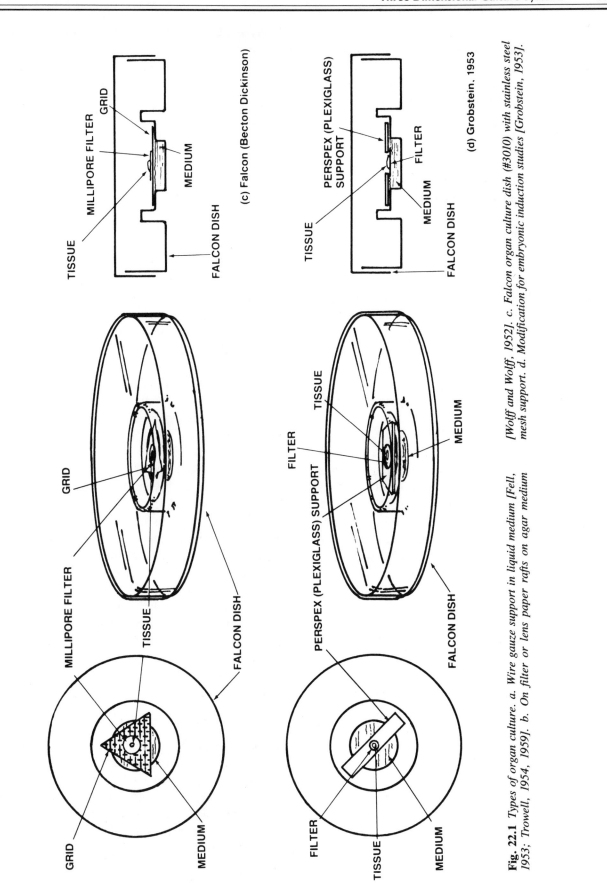

Fig. 22.1 *Types of organ culture. a. Wire gauze support in liquid medium [Fell, 1953; Trowell, 1954, 1959]. b. On filter or lens paper rafts on agar medium [Wolff and Wolff, 1952]. c. Falcon organ culture dish (#3010) with stainless steel mesh support. d. Modification for embryonic induction studies [Grobstein, 1953].*

(c) Falcon (Becton Dickinson)

(d) Grobstein. 1953

Outline

Dissect out organ or tissue, reduce to 1 mm^3, or to thin membrane or rod, place on support at gas (air)/ medium interface, and incubate in humid CO_2 incubator, changing the medium as required.

Materials

Instruments
stainless steel grids
sterile filters, 0.5 μm Nuclepore (sterilize by autoclaving)
medium
organ culture dishes (Falcon #3010)

Protocol

1.
Prepare grids (Falcon) with sterile filters (Nuclepore) in position (see Fig. 22.1) in dishes (Falcon).
2.
Add enough medium to wet the filter but not so much that it floats (~ 1.1 ml).
3.
Place dishes in humid CO_2 incubator to equilibrate at 36.5°C.
4.
Prepare tissue or dissect out whole embryonic organs, e.g., 8-day femur or tibiotarsus of chick embryo (see Chapter 9 for dissection). Tissue must not be more than 1 mm thick, preferably less, in one dimension, e.g., 8-day embryonic tibiotarsus is perhaps 5 mm long but only 0.5–0.8 mm in diameter. A fragment of skin might be 10 mm^2 but only 200 μm thick. Tissue like liver or kidney which must be chopped down to size should be no more than 1 mm^3.
For short dissections, HBSS is sufficient, but for longer dissections, use 50% serum in HBSS buffered with HEPES to pH 7.4.
5.
Take dishes from incubator and transfer tissue carefully to filters. A pipette is usually best and can be used to aspirate any surplus fluid transferred with the explant.
6.
Check level of medium, making sure tissue is wetted, and return dishes to incubator.
7.
Incubate for 1–3 wk, changing medium every 2 or 3 days.

Analysis. Usually by histology, autoradiography, or immunocytochemistry, but assay of total amounts of cellular constituents or enzyme activity is possible, although variation between replicates will be high.

Variations. Most variations are in:

(1) Type of medium; 199 or CMRL 1066 may be used with or without serum, and BJG [Biggers et al., 1961] for cartilage or bone.

(2) Type of support (see Fig. 22.1). The Grobstein technique has a number of advantages. Different types of tissue may be combined on the opposite sides of the filter to study their interaction (see above). Furthermore, the well formed on the top side of the filter assembly generates a meniscus of medium with a large surface area available for gas exchange. It is also possible to alter the configuration of the tissue by raising or lowering the level of medium in the dish, and thereby in the well; deeper medium gives a spherical explant and shallower medium flattens the explant. This system does have the disadvantage that the filter assemblies are not commercially available.

(3) O_2 tension: embryonic cultures are usually best kept in air, but late-stage embryos, newborn, and adult tissue are better kept in 95% O_2 [Trowell, 1959; de Ridder and Mareel, 1978].

Organ cultures are useful in the demonstration of processes such as embryonic induction [e.g., Grobstein, 1953; Cooper, 1965], where the maintenance of the integrity of whole tissue is important. However, they are slow to prepare and present problems of reproducibility between samples. Growth is limited by diffusion (although growth is perhaps not necessary and may even be undesirable) and mitosis is nonrandomly distributed throughout the explant. Mitosis occurs round the periphery only, while the centers of explants frequently become necrotic (Fig. 22.2). It has been argued that this type of geometry makes organ cultures good models of tumor growth, where peripheral cell division is often accompanied by central necrosis.

HISTOTYPIC CULTURE

Various attempts have been made to regenerate tissuelike architecture from dispersed monolayer cultures. Kruse and Miedema [1965] demonstrated that perfused monolayers could grow to more than ten cells deep and organoid structures can develop in multilayered cultures if kept supplied with medium [Schneider et al., 1963; Bell et al., 1979]. Green [1978] has shown that human epidermal keratinocytes will form dematoglyphs (friction ridges) if kept for several weeks without transfer, and Folkman and Haudenschild [1980] were able to demonstrate formation of capillary tubules in cultures of vascular endothelial cells cultured in the presence of endothelial growth factor and medium conditioned by tumor cells.

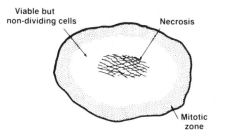

Fig. 22.2 *Diagrammatic representation of the expected distribution of mitoses (stippled area) and necrosis (shaded central area) in an organ culture explant.*

Sponge Techniques

Leighton first demonstrated that cells would penetrate cellulose sponge [Leighton et al., 1968]. Both normal and malignant cells can do this and it does not seem to reflect malignant behaviour. Collagen coating of the sponge may facilitate occupation and Gelfoam (a gelatin sponge matrix used in reconstructive surgery) may be used in place of cellulose [Sorour et al., 1975]. These systems require histological analysis and are limited in dimensions, like organ cultures, by gaseous and nutrient diffusion.

Collagen gel (native collagen as distinct from denatured collagen coating) provides a matrix for the morphogenesis of primitive epithelial structures. Yang et al. [1979, 1980, 1981] have shown that breast epithelium forms rudimentary tubular and glandular structures grown in collagen. Analysis is again by histology.

Capillary Bed Perfusion

Since medium supply and gas exchange become limiting at high cell densities, Knazek et al. [1972; Gullino and Knazek, 1979] developed a perfusion chamber from a bed of plastic capillary fibers. This is now available commercially from Amicon. The fibers are gas- and nutrient-permeable and support cell growth on their outer surfaces. Medium, saturated with 5% CO_2 in air, is pumped through the centers of the capillaries, and cells are added to the outer chamber surrounding the bundle (Fig. 22.3; see also Fig. 7.3). The cells attach and grow on the outside of the capillary fibers fed by diffusion from the perfusate and can reach tissue-like cell densities. There is an option between two types of plastic and different ultrafiltration properties giving molecular weight cut-off points at 10,000, 50,000, or 100,000 daltons, regulating the diffusion of macromolecules from the medium to the cells.

It is claimed that cells in this type of high-density culture behave as they would *in vivo*. Choriocarci-

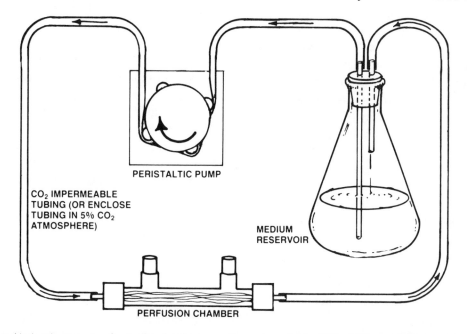

Fig. 22.3. *Vitafiber (Amicon) apparatus for perfused culture on capillary bundles of permeable plastic. Medium is circulated from a reservoir to the culture chamber by a peristaltic pump. As silicone tubing is gas permeable, the apparatus should be enclosed in an atmosphere of 5% of CO_2. Alternatively, the length of tubing between the pump and the culture chamber may be enclosed in a polyethylene bag purged with CO_2.*

noma cells release more human chorionic gonadotrophin [Knazek et al., 1974] and colonic carcinoma cells produce elevated levels of CEA [Rutzky et al., 1979; Quarles et al., 1980]. There are considerable technical difficulties in setting up the chambers, however, and they are costly.

Sampling cells from these chambers and determination of the cell concentration are also difficult. However, they appear to present an ideal system for studying the synthesis and release of biologically generated compounds and are now being exploited on a semi-industrial scale (Endotronics).

Reaggregation and Spheroids

When dissociated cells are cultured in a gyratory shaker, they may reassociate into clusters. Dispersed cells from embryonic tissues will sort during reaggregation in a highly specific fashion [Linser and Moscona, 1980], e.g., Müller cells of the chick embryo retina reaggregated with neuronal cells from the retina were inducible for glutamine synthetase; but those reaggregated with neurons from other parts of the brain were not. Cells in these heterotypic aggregates appear to be capable of sorting themselves into groups and forming tissue-like structures. This property is less easily demonstrated in adult cells, although some results suggest that it may be possible for adult cells to form organoid structures [Douglas et al., 1976, 1980; Bell et al., 1979].

Homotypic reaggregation also occurs fairly readily, and spheroids generated in gyratory shakers or by growth on agar have been used as models for chemotherapy *in vitro* [Twentyman, 1980] and for the characterization of malignant invasion [Mareel et al., 1980]. As with organ cultures, growth is limited by diffusion and a steady state may be reached where cell proliferation in the outer layers is balanced by central necrosis.

The following protocol for generating spheroid cultures has been contributed by Tom Wheldon, Institute of Radiotherapeutics and Oncology, Belvidere Hospital, Glasgow, Scotland.

Principle

Multicellular tumor spheroids provide a proliferating model for avascular micrometastases. The three-dimensional structure of spheroids allows the experimental study of aspects of drug penetration and resistance to radiation or chemotherapy dependent on intercellular contact. Human tumor spheroids are more easily developed from established cell lines or from xenografts than from primary tumours [Sutherland et al., 1981].

Outline

From single cell suspension (trypsinized monolayer or disaggregated tumor) cells are inoculated into flasks, previously base-coated with agar, and incubated to allow formation of small aggregates over 3–5 days. Aggregates may be transferred to fresh flasks, to magnetic stirrer vessels or to multiwell plates where continued growth over 2–4 weeks yields spheroids of maximum size, about 1,000 μm [Yuhas et al., 1977].

Materials

> Noble agar (Difco)
> growth medium
> distilled water (sterile)
> trypsin 0.25% in PBS
> 25-cm^2 flasks or 24-well plates
> 9-cm petri dishes
> Note: where agar coating is used flasks, plates, and dishes should be sterile but not necessarily tissue culture grade.

Protocol

1.

Agar coating. In 25-cm^2 flasks: add 1 g Noble agar (Difco) to 20 ml distilled water in a 150-ml Erlenmeyer flask and boil gently until the agar has completely dissolved (about 10 min). Add contents immediately to 60 ml of growth medium, previously heated to 37°C, and put 5-ml aliquots into each flask. Ensure that the agar is free from bubbles. It will set (at room temperature) in about 5 min giving a 1.25% agar coated flask. In multiwell plates: add 0.5 g agar to 10 ml distilled water, heat as above and then add 40 ml of distilled water. Place 0.5 ml of the resultant solution in each well of a 24-well plate (Corning 25820) to give a base coat of 1% agar. Accuracy and careful placement is important to ensure easy well-to-well focus of the microscope in subsequent viewing of spheroids.

2.

Spheroid initiation. Trypsinize confluent monolayer (established lines) (see Chapter 10) or disaggregate (solid tumors) (see Chapter 9) to give single cell suspension. Neutralize trypsin with medium containing serum (if necessary). Count cells (Coulter counter or hemocytometer). Place 5 × 10^5 cells in each agar-coated 25-cm^2 flask in 5 ml of growth medium and incubate. If the cells are capable of spheroid formation, small aggregated clumps (about 100–300 μM diameter) will form spontaneously in 3–5 days.

3.

For subsequent growth, spheroids should be transferred to new vessels. If growth is to be continued in 25-cm^2 flasks, the contents of the original flasks should be transferred to conical centrifuge tubes or Universal containers, the speroids allowed to settle, and single cells removed with the supernatant. Spheroids may then be resuspended in fresh medium and transferred to new agar-coated flasks, where growth will proceed by division of cells in the outer layer.

4.

For growth in wells, decant the contents of a 25-cm^2 flask into a 9-cm petri dish and, under laminar flow conditions, select individual spheroids of chosen dimensions under low power magnification ($\times$ 40). Using a Pasteur pipette and a Pi-Pump (Shuco International) transfer selected spheroids of similar diameter individually to agar-coated wells or a 24-well plate, each containing 0.5 ml medium. Place plates in CO_2 incubator. Replace medium once or twice weekly (exchanging 0.5 ml each time) or add 0.5 ml medium (without removal) once or twice weekly giving 2–4 wk for a 2-ml well).

5.

Spheroid growth in wells or flasks may be quantified by regular measurement (e.g., two to three times weekly) of diameter using a microscope graticule or, better, by measurement of cross-sectional area using an image analysis scanner. The most accurate growth curves are obtained when spheroids are grown in wells and individually monitored.

Variations

1. Some spheroids can be grown only when subjected to continuous agitation. Transfer single cells in suspension, or preformed aggregates, to siliconized glass culture vessels and agitate on a magnetic stirrer (Techne).

2. As normal stromal cells are usually excluded from spheroids, spheroids formed from a disaggregated tumor cell suspension will usually contain only tumor cells. Monolayer cultures of tumor cells may then be established by placing spheroids

in uncoated tissue culture petri dishes, to which they will attach. Monolayers will grow out from each attached spheroid and may be harvested by trypsinization [Bruland et al., 1985].

Applications

Spheroids have wide applications in assessment of cytotoxic treatment. End-points include treatment-induced growth delay, proportion of spheroids sterilized ("cured") by treatment, and colony formation in monolayer following disaggregation of treated spheroids. Some (not all) spheroid types may be "stripped" by exposure to proteolytic enzymes; exposure for varying periods followed by sequential removal of cells liberated allows assessment of cell survival or drug penetration in progressively deeper layers of the spheroid [Freyer and Sutherland, 1980].

FILTER WELLS

Several attempts have been made to generate dense populations of cells supported on a filter, either perfused from below by aerated medium [Dickson and Suzangar, 1976] or located at the gas-liquid interface or near to it. Mauchamp demonstrated the development of polarity and functional integrity in thyroid epithelium explanted on a collagen coated filter in a specially constructed mount [Chambard et al., 1983] (see Chapter 14). Others have used the system to study invasion by granulocytes or malignant cells [Elvin et al., 1985; McCall et al., 1981].

One of the major advantages of the system is that it allows the recombination of cells at very high tissue-like densities, with ready access to medium and gas exchange, but in a multireplicate form. McCall et al., 1981] constructed chambers from disposable 2-ml syringes which utilized multiwell plates for support, and Millipore has now produced the Millicell-HA chambers in two sizes, 12-mm and 30-mm diameter (Fig. 22.4a), suitable for 24-well plates, 6-well plates, or larger dishes. These filters are available in HA grade Millipore filters of different porosities but as yet are not transparent. The author has used transparent polycarbonate filters (Nuclepore) with defined and regular pores (1 μm or 8 μm are the most useful) in polypropylene mounts (Hendley Engineering) (Fig. 22.4b). These are useful for invasion studies (Fig. 22.4c) and cell interacting studies but require further treatment to promote cell adhesion. Similar filters are produced by Costar, already mounted in polystyrene (Transwells).

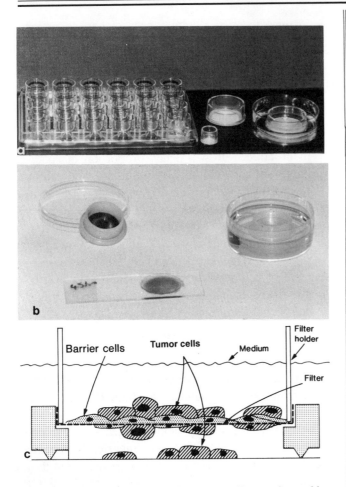

Fig. 22.4. *Filter well units for culture. a. Millipore, disposable. b. Hendley Engineering, Ltd., reusable polypropylene with Nuclepore polycarbonate filter. c. Transverse section of Hendley filter holder used as a model for tumor invasion. Medium level shown at its deepest, but can be used with level at or near the level of the filter. (Reproduced from Freshney, 1985, by permission of the publisher.)*

Outline

Cells are seeded into filter chambers and cultured in excess medium in deep petri dishes or sample pots (Sterilin, Macom).

Materials

Approximately 2×10^6 cells per filter
growth medium, 20 ml per filter
filter holders (Hendley Engineering)
filters (polcarbonate 1-μm or 8-μm porosity, Nuclepore treated by Flow Laboratories—filter wells are assembled and sterilized by autoclaving, ethylene oxide, or γ-irradiation; autoclaving tends to distort the filter; must

be left 2–3 wk after ethylene oxide treatment before use; irradiation preferable (5×10^4 Gy, 5 Mrads)
containers for filter assembly—Corning 9-cm petri dishes or deep sample pots (Sterilin or Macom)

Protocol

1.
Place filter wells in dish or container.
2.
Add medium, tilting dish to allow medium to occupy space below filter and displace air with minmum entrapment. Add medium until level with filter (15 ml in 9-cm petri dish).
3.
Level dish and add 2×10^6 cells, 10^6/ml in 2 ml medium to top of filter, taking care not to perforate filter.
4.
Place in humid CO_2 incubator in protective box (see Chapter 11). It is critical to avoid shaking the box, and the cultures should not be moved in the incubator to avoid spillage and resultant contamination.
5.
Monolayer (or multilayer) should become established in about 5 d. It forms before then but complete integrity (formation of matrix, cell contacts, polarity, etc.) takes several days.
6.
Culture may be maintained indefinitely, replacing medium every 3–5 d or transferring to fresh dish. A second cell type may be added on top if it is desired to study interactions, or to the other side of the filter by inverting the filter holder.

Analysis

1. Penetration of cells through filter. Trypsinize and count each side of filter in turn (trypsinized cells will not pass through even an 8-μm filter as their spherical diameter in suspension exceeds this), or fix, embed, and section by electron microscope or conventional histology. Visualization is possible in whole mounts by mounting the fixed, stained (Giemsa) filter on a slide in DePeX under a coverslip under pressure to flatten it. Differential counting can then be performed by focusing on each plane alternately.

2. Detachment of cells from filter to bottom of dish. Count by trypsinization or scanning.

3. Partition above and below filter (chemotaxis or invasion) either by counting as in step 1 above or by prelabeling the cells with rhodamine or fluorescein isothiocyanate (5 μg/ml for 30 mins in trypsinized suspension) and measuring the fluorescence of solubilized cells (0.1% SDS in 0.3N NaOH for 30 min) trypsinized from either side of the filter.

4. Invasion. Precoat filter with cell layer (normal fibroblasts, MDCK, etc.), ensure confluence is achieved by microscopic examination , and then seed EDTA dissociated test cells on top of preformed layer (10^5–10^6 cells per filter). If test cells are RITC-or FITC-labeled, fluorescent measurements will reveal appearance of cells below filter.

Variations

1. Type of filter. Millipore filters (preformed chambers) may be used. These are presterilized but opaque and of the mesh type filter matrix. Cells can be grown directly on the filter or after coating with gelatin or collagen.

2. Filter porosity. One-μm filters allow cell interaction and contact without transit of filter. Eight-μm filters allow live cells to cross. Filters of 0.2 μm probably do not allow cell contact. Low-porosity filters may be used to study cell interaction without intermingling.

3. Transfilter combinations. Invert filter and load underside first with 0.5 ml, 2×10^6/ml of cell suspension. After 2 d invert filter and load well as above.

Chapter 23

It is intended in this chapter to provide detailed information on some specialized techniques referred to in the text but so far not described. In some cases, these are not tissue culture techniques *per se*, but rather techniques associated with tissue culture and which might be used in a number of different tissue culture-based experiments, e.g., autoradiography. Other techniques involving tissue culture directly, but of a very specialized nature, e.g., monoclonal antibody production, are also described.

MASS CULTURE TECHNIQUES

Mass culture might be defined as encompassing from 10^9 cells to semi-industrial pilot plant 10^{11} or 10^{12} cells (1–1000 l). The method employed depends on whether the cells proliferate in suspension or require to be anchored to the substrate.

Suspension Culture

Increasing the bulk or "scale-up" of suspension cultures is relatively simple since only the volume need be increased. Above about 5-mm-deep agitation of the medium is necessary, and above 10 cm sparging with CO_2 and air is required to maintain adequate gas exchange (Fig. 23.1). Stirring of such cultures is best done slowly with a large surface area paddle or large diameter magnetic stirrer bar. The stirring speed should be between 30 and 100 rpm, sufficient to prevent cell sedimentation but not so fast as to grind or shear the cells. If a bar is used, it must be kept off the base of the culture vessel with a collar or be suspended from above. Antifoam (Dow Chemical Co.) must be included where the serum concentration is above 2% particularly if the medium is sparged. In the absence of serum, it may be necessary to increase the viscosity of the medium with (1–2%) carboxymethyl cellulose (molecular weight $\sim 10^5$).

The procedure for setting up a 4-1 culture of suspended cells is as follows:

Outline

Grow pilot culture of cells and add to prewarmed, pregassed aspirator of medium. Stir slowly with sparging until required cell concentration is reached and harvest.

Materials

Medium with antifoam
pilot culture
prepared aspirator (Fig. 23.1)
magnetic stirrer
supply of 5% CO_2
bunsen
gas lighter
counting fluid
cell counter or hemocytometer

Protocol

1.
Prepare a standard 5- or 10-1 aspirator as in Figure 23.1, with a one-holed silicon rubber stopper at the top and a two-holed stopper at the bottom. The top stopper carries a glass tube with a cotton plug and the bottom stopper has (a) a glass tube and silicone rubber connection to an inlet port closed with a silicone membrane closure ("skirted cap") and (b) a screw cap fixed to a tube leading from the stopper with a screw cap vial inserted in the cap. The hole for the bottom stopper should be about 15 mm above the base of the aspirator bottle and the stopper should fill it completely and leave no crevices for cells to lodge.

A Teflon-coated bar magnet is placed in the aspirator. It should be as large as possible, while still able to turn freely in the bottom of the aspirator ($\sim$ 9–12 cm) and have a central collar to raise it up from the base of the aspirator to avoid grinding cells below the bar.

Sterilize by autoclaving 100 kPa (15 lb/in²) for 20 min.

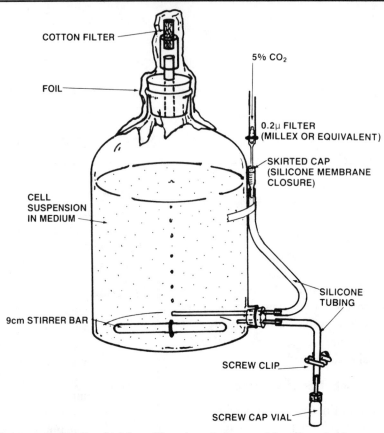

COTTON FILTER

5% CO₂

FOIL

0.2μ FILTER
(MILLEX OR EQUIVALENT)

SKIRTED CAP
(SILICONE MEMBRANE
CLOSURE)

CELL
SUSPENSION
IN MEDIUM

SILICONE
TUBING

9cm STIRRER BAR

SCREW CLIP

SCREW CAP VIAL

Fig. 23.1. *Bulk culture of cells in suspension. Standard 5- or 10-l aspirators may be modified as illustrated. Optimum mixing is achieved by using a large stirrer bar, with a central collar, and a slow stirring speed (~ 60 rpm). (Apparatus developed by the staff of the Beatson Institute for Cancer Research, Glasgow, Scotland.)*

2.

Set up "starter culture," using a standard screw-capped 1-l reagent bottle, or equivalent, with a Teflon-coated bar magnet in the bottom (see Fig. 10.2). Add 400 ml medium and seed with cells at 5×10^4–10^5/ml. Place on magnetic stirrer (see Chapter 4) rotating at 60 rpm and incubate until 5×10^5–10^6 cells/ml is reached.

3.

Add 4 l medium and 0.4 ml antifoam to sterile aspirator. The antifoam should be added directly to the aspirator using a disposable pipette or syringe. Place on magnetic stirrer in 36.5°C room or incubator. Connect 5% CO_2 air line via fresh, sterile, 25-mm, 0.2-μm Millex filter to sterile disposable hypodermic needle and insert needle into skirted cap. Turn on gas at a flow rate of approximately 10–15 ml/min and stir at 60 rpm. Incubate for about 2 hr to allow temperature and CO_2 tension to equilibrate.

4.

Bring aspirator and starter culture back to laminar flow hood, remove top stopper, keeping aluminum foil in place, and starter bottle cap, taking care to keep stopper sterile. Flame neck of aspirator and starter culture bottle and pour starter culture into aspirator.

5.

Replace stopper in aspirator and return to incubator or hot room. Reconnect 5% CO_2 line and restart stirrer at 60 rpm. Adjust gas flow to 10–15 ml/min.

6.

Incubate for 4–7 d, sampling (see below) every day to check cell growth. When cell concentration reaches desired level, disconnect aspirator, run off cells into centrifuge bottles, and centrifuge at 100 g for 10 min.

Sampling. Open screw clip and run 5–10 ml into vial to remove cells and medium which have been stagnant in the delivery line. Discard vial and contents and replace with fresh vial. Collect second 5–10 ml, perform a cell count, and check viability by dye exclusion.

Analysis. For best results, cells should not show a

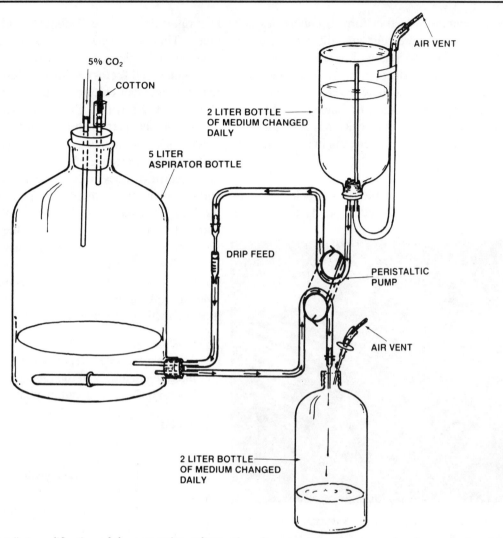

5% CO_2

COTTON

2 LITER BOTTLE
OF MEDIUM CHANGED
DAILY

AIR VENT

5 LITER
ASPIRATOR BOTTLE

DRIP FEED

PERISTALTIC
PUMP

AIR VENT

2 LITER BOTTLE
OF MEDIUM CHANGED
DAILY

Fig. 23.2. *"Biostat." A modification of the suspension culture vessel of Figure 23.1, with continuous matched input of fresh medium and output of cell suspension. (Based on a system developed by John Paul and George Lanyon.) The objective is to keep the culture conditions constant rather than to produce large numbers of cells. Bulk culture, per se, is best performed in batches in the apparatus in Figure 23.1.*

lag period of more than 24 hr and should still be in exponential growth when harvested. Plot cell counts daily and harvest at approximately 10^6 cells/ml.

Variations

Continuous culture—"Biostat." If it is required that the cells be maintained at a set concentration, e.g., at mid-log phase, cells may be removed and medium added daily or cells may be run off and medium added continuously using the skirted cap entry port to add medium and the screw cap outlet to collect into a larger reservoir (Fig. 23.2). The flow rate of medium may be calculated from the growth rate of the culture [Griffiths, 1986] but is better determined experimentally by serial cell counting at different flow rates of medium.

Production of cells in bulk, in the 1–20 l range, is best done by the "batch" method outlined first. The "steady-state" method is required for monitoring metabolic changes related to cell density but is more expensive in medium and is more likely to lead to contamination. However, if the operation is in the 50–1,000-l range, more investment and time is spent in generating the culture, and the batch method becomes more costly in time, materials and "down time."

Suspension cultures can also be grown in bottles rotating on a special rack as for monolayer cultures (see below).

Monolayer cells. Anchorage-dependent cells cannot be grown in liquid suspension except on microcarriers (see below), but transformed cells, e.g., virally transformed or spontaneously transformed continuous cell lines, can. Because these cells are still capable of attachment, the culture vessels will require to be coated with a water-repellent silicone (e.g., Repelcote) and the calcium concentration may need to be reduced. MEMS medium is a variation of Eagle's MEM with no calcium in the formulation, which has been used for the culture of HeLa-S_3 and other cells in suspension.

Monolayer Culture

For anchorage-dependent monolayer cultures, it is necessary to increase the surface area of the substrate

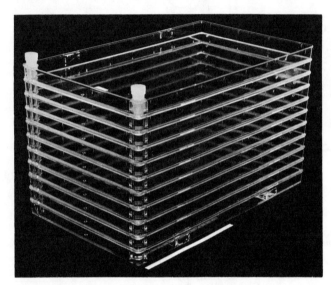

Fig. 23.3. *Nunclon "Cell Factory."*

in proportion to the cell number and volume of medium. This requirement has prompted a variety of different strategies, some simple, others complex.

Nunclon cell factory. The simplest system for scaling up monolayer cultures is the Nunclon Cell Factory (Figs. 23.3, 23.4) (see Table 7.1). This is made up of ten rectangular petri dish-like units, total surface area 6,500 cm^2, interconnected at two adjacent corners by vertical tubes. Because of the positions of the apertures in the vertical tubes, medium can only flow between compartments when the unit is placed on end. When the unit is rotated and laid flat, the liquid in each compartment is isolated, although the apertures in the interconnecting tubes still allow connection of the gas phase. The cell factory has the advantage that it is not different in the geometry or the nature of its substrate from a conventional flask or petri dish. The recommended method of use is as follows:

Outline

Prepare a cell suspension in medium and run into the chambers of the unit. Lay the unit flat and gas with CO_2. Seal and incubate.

Materials

> Monolayer cells
> medium
> 0.25% crude trypsin
> PBS
> hemocytometer or cell counter and
>   counting fluid

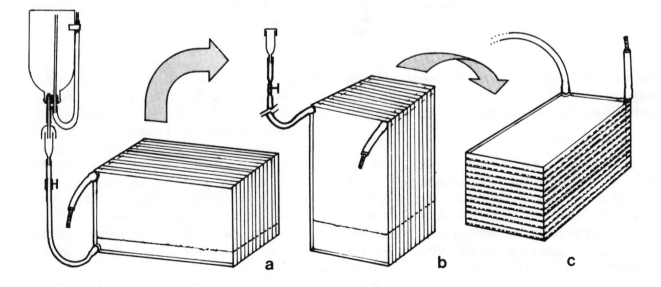

Fig. 23.4. *Filling Nunclon Cell Factory. a. Run medium in. b. Rotate onto short side away from inlet. c. Lay down flat, seal inlet, or connect to 5% CO₂ line.*

culture chamber
silicone tubing and connectors

Protocol

1.

Trypsinize cells (Chapter 10), resuspend, and dilute to 2×10^4 cells/ml in 1,500 ml medium.

2.

Place chamber on long edge with supply tube to the bottom (see Fig. 23.4) and run cells and medium in through supply tube. Medium in all chambers will reach the same level.

3.

Clamp off supply tube and disconnect from medium reservoir.

4.

Rotate unit through 90° in the plane of the monolayer so that it lies on a short edge with the supply tube at the top.

5.

Rotate unit through 90° perpendicular to the plane of the monolayer so that it now lies flat on its base with the culture surfaces horizontal. To transport to incubator, tip medium away from supply port.

6.

If it is necessary to gas the culture, loosen clamp on supply line and purge unit with 5% CO_2 in air for 5 min, then clamp off both supply and outlet. The unit may be gassed continuously if desired.

7.

To change medium (or collect medium), reverse step 5 and then 4, flame clamped line, open clamp, and drain off medium.

8.

Replace medium as in steps 2–6.

9.

To collect cells, remove medium as in step 7, add 500 ml PBS, and remove. Add 500 ml trypsin at 4°C and remove after 30 s. Incubate, add medium after 15 min, and shake to resuspend cells. Run off cells as in step 7.

10.

The residue may be used to seed the next culture, although this does make it difficult to control the seeding density. It is better to discard the chamber and start fresh.

Analysis. Following growth in these chambers is difficult, so a single tray or chamber is supplied to act as a pilot culture. It is assumed the single tray will behave as the multichamber unit.

The supernatant medium can be collected repeatedly for virus or cell product purification. Collection of cells for analysis depends on the efficiency of trypsinization.

This technique has the advantage of simplicity but can be expensive if the unit is discarded each time cells are collected. It was designed primarily for harvesting supernatant medium but is a good method for produc-

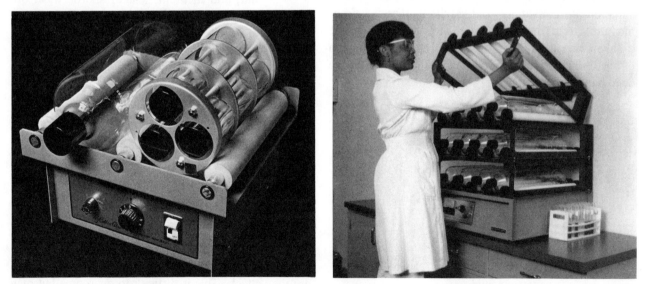

Fig. 23.5A   Fig. 23.5B

Fig. 23.5. *Roller culture bottles on racks. A. Small, bench top rack (Bellco). B. Large benchtop or free standing extendable rack (photograph courtesy of New Brunswick Scientific).*

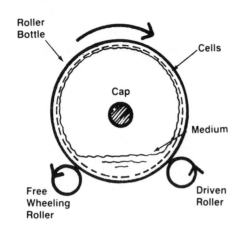

Fig. 23.6. *Roller bottle culture. Cell monolayer (dotted line) is constantly bathed in liquid but only submerged for about one-fourth of the cycle, enabling free gas exchange.*

ing large numbers of cells (3×10^8–3×10^9) for a pilot run, or on an intermittent basis.

Roller culture. If cells are seeded into a round bottle or tube which is then rolled around its long axis, the medium carrying the cells runs around the inside of the bottle (Figs. 23.5, 23.6). If the cells are nonadhesive, they will be agitated by the rolling action but remain in the medium. If the cells are adhesive they will gradually attach to the inner surface of the bottle and grow to form a monolayer. This system has three major advantages over static monolayer culture: (1) the increase in surface area; (2) the constant, but gentle, agitation of the medium; and (3) the increased ratio of medium surface area to volume, allowing gas exchange to take place at an increased rate through the thin film of medium over cells not actually submerged in the deep part of the medium.

Outline

Seed cell suspension in medium into round bottle and rotate slowly on a roller rack.

Materials

Medium and medium dispenser
PBSA
0.25% crude trypsin
monolayer culture
hemocytometer or cell counter and
    counting fluid
roller bottles
supply of 5% CO_2
roller apparatus

Protocol

1.
Trypsinize cells and seed at usual density.

Note. The gas phase is large in a roller bottle so it may be necessary to blow a little 5% CO_2 into the bottle (e.g., 2s at 10 1/min). If medium is CO_2/HCO_3^- buffered, then the gas phase should be purged with 5% CO_2 (30 s to 1 min at 20 1/min depending on the size of the bottle; see Chapter 10).

2.
Rotate bottle slowly around its axis at 20 rev/hr until cells attach (24–48 hr).

3.
Increase rotational speed to 60–80 rev/hr as cell density increases.

4.
To feed or harvest medium, take bottles to sterile work area and draw off medium as usual and replace with fresh medium. A transfusion device (see Fig. 4.13) is useful for adding fresh medium, provided that the volume is not critical. If the volume of medium is critical, it may be dispensed by pipette or metered by a peristaltic pump (Camlab, Jencons) (see Chapter 4).

5.
To harvest cells, remove medium, rinse with 50–100 ml PBSA and discard PBSA, add 50–100 ml trypsin at 4°C, roll for 15 s, by hand or on rack at 20 rpm. Draw off trypsin, incubate the bottle for 5–15 min, add medium, shake, and/or wash off cells by pipetting.

Analysis. Monitoring cells in roller bottles can be difficult, but it is usually possible to see cells on an inverted microscope. With some microscopes, the condensor needs to be removed, and with others, the bottle may not fit on the stage. Choose a microscope with sufficient stage accommodation.

For repeated harvesting of large numbers of cells or for repeated collection of supernatant medium, the roller bottle system is probably most economical, although it is labor intensive and requires investment in a bottle rolling unit ("roller rack"; Fig. 23.5).

Variations

Aggregation. Some cells may tend to aggregate before they attach. This is difficult to overcome but may be improved by reducing the initial rotational speed to 5 or even 2 rev/hr or trying a different type or batch of serum.

Size. A range of bottles, both disposable and reusable, are available (see Table 7.1, Figs. 23.5, 23.7).

Volume. Medium volume may be varied. A low volume will give better gas exchange and may be better for untransformed cells. Transformed cells, which are more anaerobic and produce more lactic acid, may be better in a larger volume. The volumes given in Table 7.1 are mean values and may be halved or doubled as appropriate.

Mechanics. The system where bottles are supported on rollers (see Fig. 23.5) is now the most popular because it is most economical in space. Roller drums (see Fig. 23.8), once popular, demand more space. They are still used for smaller bottles or tubes.

Spiral propagator (Sterilin). [House et al., 1972; see also House, 1973] (Fig. 23.9). The Sterilin propagator comprises a spiral coil of tissue culture-treated polystyrene sheet inside a plastic container. Cells adhere to and grow on both surfaces of the sheet giving a total area of 8,500 cm^2. Mixing of the medium and adequate gas exchange is achieved by sparging with 5% CO_2 in air.

Outline

Cells are suspended in medium, and run into the chamber, which is first rolled around its long axis to allow the cells to adhere and then placed upright and sparged for the remainder of the culture period.

Materials

    Medium
    dispenser
    PBSA
    0.25% crude trypsin
    monolayer culture
    hemocytometer or cell counter and
        counting fluid
    Sterilin spiral culture vessel and
        connectors
    roller rack capable of 4 rev/hr (it may
        be necessary to make rings to go round
        the culture vessel if the distance
        between the rollers is too great)

Protocol

1.
Trypsinize cells and dilute to required concentration in 1,500 ml CO_2-buffered medium with antifoam (e.g., 2×10^4/ml, 3×10^7 cells total for 3T3).

2.
Run medium plus cells into chamber with bottle and dispenser device (see Fig. 4.13).

3.
Place screw caps in place and tighten.

4.
Rotate slowly (4 rev/hr) around long axis for 24 hr.

5.
Stand flask on end and connect center inlet to 5% CO_2 line and cotton-plugged outlet to outer narrow opening. Bubble gas through at 20–50 ml/min (~2 bubbles per s).

6.
Replacement of medium is not recommended; but if necessary, pour off from outer narrow outlet and refill as in step 2 above.

7.
To harvest cells, pour off medium, add 500 ml trypsin and run around inside of coils, checking by eye that all surface is covered.

8.
Incubate on roller for 10–15 min.

9.
Pour off cells and centrifuge to remove trypsin.

10.
To reseed, rinse out with 1 l of BSS, pour off, and repeat from step 1. Reseeding may increase the risk of contamination.

The Sterilin spiral can give a higher yield of cells than the Nunclon Cell Factory because of its greater surface area, but it is limited in the types of cells which will grow in it satisfactorily. Human diploid fibroblasts, for example, tend to form aggregates before they attach and grow slowly from the aggregates.

Microcarriers. Monolayer cells may be grown on plastic microbeads of approximately 100 μm diameter, made of polystyrene (Biosilon, Nunc), Sephadex (Supabeads, Flow and Cytodex, Pharmacia), polyacrylamide (Biorad), collagen (Pharmacia), gelatin (KC Biologicals), or polyacrylamide (Biorad) [Griffiths, 1986]. Culturing monolayer cells on microbeads gives maximum ratio of surface area to medium volume and has the additional advantage that the cells may be treated as a suspension. While the Nunclon Cell Factory gives an increase in scale with conventional geometry, microcarriers require a significant departure from usual substrate design. This has relatively little effect at the microscopic level as the cells are still

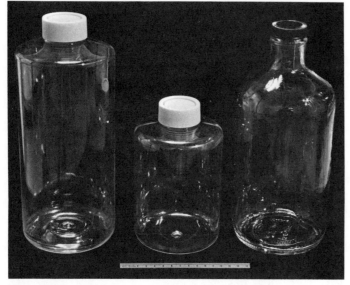

Fig. 23.7. *Examples of roller culture bottles. Center and left, disposable plastic (Falcon, Corning); right, glass.*

Fig. 23.8. *Roller drum apparatus for roller culture of large numbers of small bottles or tubes (New Brunswick Scientific).*

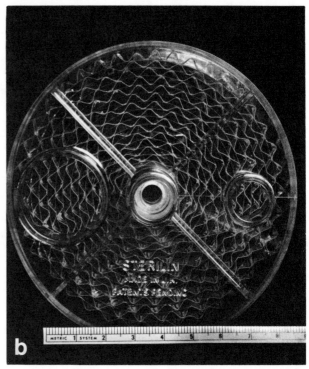

Fig. 23.9. *Sterilin spiral culture vessel, developed in collaboration with the Imperial Cancer Research Fund Laboratories, Lincoln's Inn Fields, London. The vessel is filled with cell suspension in medium and rolled slowly (2–5 rev/hr) overnight to allow cells to attach to both sides of spiral film. It is then incubated vertically and 5% CO_2 pumped through slowly to mix and gas the medium. a. Side view with gassing connections. b. Top view showing spiral.*

growing on a smooth surface at the solid liquid inter-face. The major difference created by microcarrier systems is in the mechanics of handling [Thilly and Levine, 1979; Clark et al., 1979; Griffiths, 1986]. Efficient stirring without grinding the beads is essential and a paddle system (Fig. 23.10) rotating at 30 rpm is preferable to a plain cylindrical magnet. Techne has produced a rotating pendulum system that is becoming increasingly popular. Technical literature is available from microcarrier suppliers to assist in setting up sat-isfactory cultures.

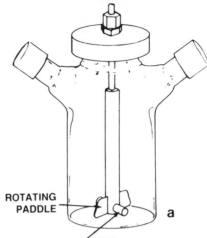

MICROCARRIER CULTURE FLASK

ROTATING PADDLE

MAGNETIC FOLLOWER

a

Outline

Cells are seeded at a high cell and bead concen-tration, diluted, stirred, and sampled as required.

Materials

Microcarriers (Gelatin, KC Biologicals)
growth medium
donor culture
culture vessel (Techne, Bellco)
magnetic stirrer (Techne, BTL, Bellco)

Protocol

1.
Suspend beads in one-third of the final volume of medium required.
2.
Trypsinize cells, count, and seed at three to five times the normal seeding concentration into the bead suspension.
3.
Stir at 10–25 rpm for 8 hr.
4.
Increase stirring speed to approximately 60 rpm.
5.
If pH falls, feed by switching off stirrer for 5 min, allowing beads to settle, and replacing one-half to two-thirds of medium.
6.
To harvest cells, remove medium and wash by settling, digest beads with trypsin EDTA, spin down, and wash cells.

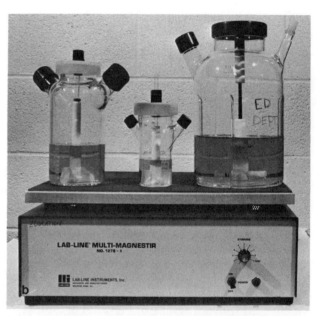

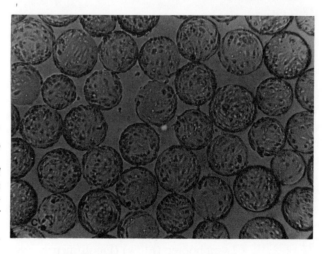

Fig. 23.10. *Microcarrier culture. a. The apparatus is similar to that used for suspension culture as the principle is the same. Modifications include a larger surface area for stirring, provided by a paddle, to enable slower stirring speeds to be used, and a suspended stirrer bar. Both modifications are designed to mini-mize shearing and grinding of the cells. b. A range of microcarrier culture vessels (Bellco) on a multiplace stirrer rack (Lab-Line). c. Vero cells growing on microcarriers (courtesy of Flow Laborato-ries, Irvine, Scotland).*

Analysis. Cell counting on beads can be difficult, so growth rate should be checked by determination of DNA (see Chapter 18), or protein if nonproteinaceous beads are used, or dehydrogenase activity using the MTT assay (see Chapter 19).

Variation. Most variations on the method arise from the choice of bead or design of the culture vessel and stirrer [Griffiths, 1986].

Many other mass culture techniques exist [McLimans, 1979] but they are of such specialized application that they will not be described in detail here. Linbro produced a multiplate system, similar to the Sterilin Chamber, but with plates at right angles to the long axis of the chamber. This resembled the multiplate system of Schleicher [1973] but was smaller. Amicon and Endotronics supply larger perfusion chambers in a similar style to the Vitafiber system (see above). The potential of these systems for large-scale high-density culture has yet to be explored, but they may be valuable in recreating high tissue-like cell densities both for production of natural substances and for synthesizing large numbers of cells in a tissue-like matrix.

Millipore has introduced a large-scale culture system (MCCS) for adherent cells based on a filter membrane as a support, which also allows for perfusion of the culture. As long as current restrictions on the use of transformed cells in biotechnology exist, there will be a need for mass culture systems for anchorage-dependent cells.

LYMPHOCYTE PREPARATION

There is a variety of methods for the preparation of lymphocytes, but flotation on a combination of Ficoll and sodium metrizoate (e.g., Hypaque) is still most widely used [Boyum, 1968a,b; Perper et al., 1968].

Outline

Whole citrated blood or plasma depleted in red cells by dextran accelerated sedimentation is layered on top of a dense layer of Ficoll and sodium metrizoate. After centrifugation most of the lymphocytes are found at the interface between the Ficoll/metrizoate and the plasma.

Materials

Blood sample
clear centrifuge tubes or universal
    containers
Dextraven 110 (Fisons)
PBSA
Lymphoprep (Flow) (Ficoll/metrizoate,

adjusted to 1.077 g/cc (Pharmacia,
    Nygaard))
centrifuge
syringe or Pasteur pipette
serum-free medium
hemocytometer or cell counter

Protocol

1.
Add Dextraven 110 to blood sample to final concentration of 10% and incubate at 36.5°C for 30 min to allow most of the erythrocytes to sediment.
2.
Collect supernatant plasma, dilute 1:1 with PBSA and layer 9 ml onto 6 ml Lymphoprep or other Ficoll/sodium metrizoate mixture. This should be done in a wide transparent centrifuge tube with a cap such as the 25-ml Sterilin or Nunclon Universal Container, or the clear plastic Corning 50-ml tube, using double the above volumes.
3.
Centrifuge for 15 min at 400 g (measured at center of interface).
4.
Carefully remove plasma/PBSA without disturbing the interface.
5.
Collect the interface with a syringe or Pasteur pipette and dilute to 20 ml in serum-free medium (e.g., RPMI 1640 [Moore et al., 1967]).
6.
Centrifuge at 70 g for 10 min.
7.
Discard supernatant fluid and resuspend pellet in 2 ml serum-free medium. If several washes are required, e.g., to remove serum factors, resuspend cells in 20-ml serum-free medium, and centrifuge two or three times more, and finally resuspend pellet in 2 ml.
8.
Count cells on hemocytometer (count only nucleated cells) or on electronic counter.

Lymphocytes will be concentrated in the interface, along with some platelets and monocytes. Granulocytes will be found mostly in the Ficoll/metrizoate and in the 4 hr pellet, and erythrocytes will pellet at the bottom of the tube. Removal of monocytes and residual granulocytes can be achieved by their adherence to glass (beads or flask surface) or to nylon mesh. If purer preparations are required, fractionation on density gradients of metrizamide (Nygaard) or Percoll (Pharmacia) or by centrifugal elutriation (see Chapter 12) may be attempted. Alternatively, specific subpopulations of lymphocytes may be purified on antibody or lectin-bound affinity columns (Pharmacia).

Blast Transformation [Hume and Weidemann, 1980]

Lymphocytes in purified preparations, or in whole blood, may be stimulated with mitogens such as phytohemagglutinin (PHA), pokeweed mitogen (PWM), or antigen [Berger, 1979]. The resultant response may be used to quantify the immunocompetence of the cells. PHA stimulation is also used to produce mitosis for chromosomal analysis of peripheral blood [Kinlough and Robson, 1961; Rothvells and Siminovitch, 1958].

Materials

Medium + 10% FBS or autologous serum
phytohemaglutinin (PHA), 50 μg/ml
test tubes or universal containers
microscope slides
Colcemid, 0.01 μg/ml in BSS
0.075 M KCl

Protocol

1.
Using the washed interface fraction from step 7 above, incubate 2×10^6 cells/ml in medium, 1.5–2.0 cm deep, in HEPES or CO_2-buffered DMEM, CMRL 1066, or RPMI 1640 supplemented with 10% autologous serum or fetal bovine serum.
2.
Add PHA, 5 μg/ml (Final), to stimulate mitosis from 24 to 72 hr later.
3.
Collect samples at 24, 36, 48, 60, and 72 hr and prepare smears or cytocentrifuge slides to determine optimum incubation time (peak mitotic index).
4.
Add 0.001 μg/ml (final concentration) Colcemid for 2 hr when peak of mitosis is anticipated [Berger, 1979].
5.
Centrifuge cells after Colcemid treatment, resuspend in 0.075 M KCl for hypotonic swelling, and proceed as for chromosome preparation in Chapter 13.

AUTORADIOGRAPHY

The following description is intended to cover autoradiography of any small molecular precursor into a cold acid-insoluble macromolecule such as DNA, RNA, or protein. Other variations may be derived from this or found in the literature [Rogers, 1979; Stein and Yanishevsky, 1979].

Isotopes suitable for autoradiography are listed in Table 23.1. A low energy emitter, e.g., ^{3}H or ^{55}Fe, in combination with a thin emulsion, gives high intracellular resolution. Slightly higher energy emitters, e.g., ^{14}C and ^{35}S, give localization at the cellular level. Still higher energy isotopes, e.g., ^{131}I, ^{59}Fe, and ^{32}P, give poor resolution at the microscopic level but are used for autoradiographs of chromatograms and electropherograms where absorption of low energy emitters limits detection. Low concentrations of higher energy isotopes (^{14}C and above) used in conjunction with thick nuclear emulsions produce tracks useful in locating a few highly labeled particles, e.g., virus particles infecting a cell.

Tritium is used most frequently for autoradiography at the cellular level because the β-particles released have a mean range of about 1 μm, giving very good resolution. Tritium-labeled compounds are usually less expensive than the ^{14}C- or ^{35}S-labeled equivalents and have a long half-life. Because of the low energy of emission, however, it is important that the radiosensitive emulsion is positioned in close proximity to the specimen, with nothing between the cell and the emulsion. Even in this situation only the top 1 μm of the specimen will irradiate the emulsion.

β-particles entering the emulsion produce a latent image in the silver halide crystal lattice within the emulsion at the point where they stop and release their energy. The image may be visualized as metallic silver grains by treatment with an alkaline reducing agent (developer) with subsequent removal of the remaining unexposed silver halide by an acid fixer.

TABLE 23.1 Isotopes Suitable for Autoradiography

Isotope	Emission	Energy (mV) (mean)	T½
^{3}H	β^-	0.018	12·3 yr
^{55}Fe	X-rays	0.0065	2·6 yr
^{125}I	X-rays	0.035	60d
		0.033	
^{14}C	β^-	0.155	5568
^{35}S	β^-	0.167	87d
^{45}Ca	β^-	0.254	164d

The latent image is more stable at low temperature and in anhydrous conditions, so sensitivity may be improved by exposing in a refrigerator or freezer over desiccant. This will reduce background grain formation by thermal activity.

Outline

Cultured cells are incubated with the appropriate isotopically labeled precursor (e.g., [³H]thymidine to label DNA), washed, fixed, and dried (Fig. 23.11). Any extractions necessary, e.g., to remove

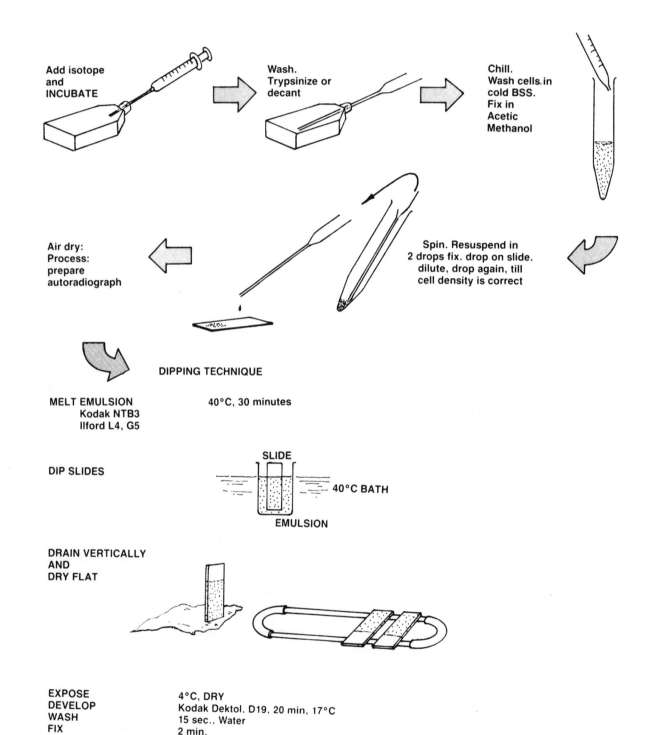

Fig. 23.11. *Steps in preparing an autoradiograph from a cell culture.*

unincorporated precursors, are performed, and the specimen is coated with emulsion in the dark and left to expose. When subsequently developed in photographic developer, silver grains can be seen overlying areas where radioisotope was incorporated (Fig. 23.12).

Materials

(Sterile where indicated, otherwise nonsterile)

Setting up culture:
cells (sterile)
PBSA (sterile)
trypsin (sterile)
medium (sterile)
hemocytometer or cell counter
counting fluid if using automatic cell counter
coverslips or slides and petri dishes (may be nontissue culture grade if cover slips or slides are used) or plastic bottles (sterile)

Labeling with isotope and setting up autoradiographs:
isotope (sterile)
HBSS (sterile)
protective gloves
containers for disposal of radioactive pipettes
container for radioactive liquid waste
acetic methanol (1:3, ice-cold, freshly prepared)
DPX 10% TCA
emulsion (Kodak NTB2, Ilford G5, diluted 1 + 2 in distilled or deionized water
light-tight microscope slide boxes (Clay Adams, Raven)
silica gel
dark vinyl tape
black paper or polyethylene

Processing:
D19 developer (Kodak)
photographic fixer (Kodak, Ilford)
hypoclearing agent (Kodak)
coverslips (00)
Giemsa stain
0.01 M phosphate buffer (pH 6.5)

Protocol

1.

Prepare culture. Monolayer cells may be grown on coverslips (Lux Thermanox or Polystyrene),

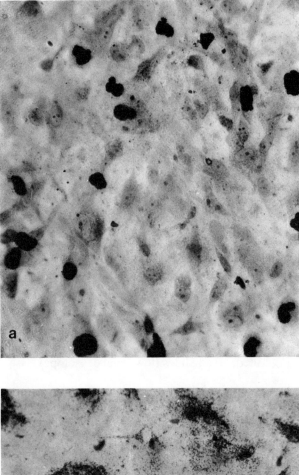
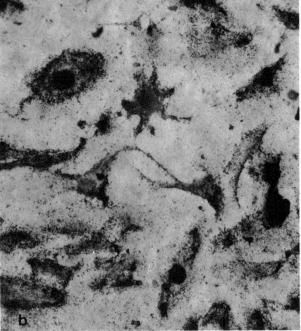

Fig. 23.12. *Autoradiograph. This is an example of [³H]thymidine incorporation into a cell monolayer. Normal glial cells were incubated with 0.1 μCi/ml (200 Ci/mMol) [³H]thymidine for 24 hr, washed, and processed as in text. a. Typical densely-labeled nuclei, suitable for determining labeling index (see Chapter 18). b. Similar culture infected with mycoplasma; cytoplasm is now labeled also.*

slides (Lab-Tek, Bellco Nuclon), or in conventional plastic bottles or petri dishes.

2.

Add isotope (usually in the range 0.1–10 μCi/ml, 100 Ci/mmol) for 0.5–48 hr as appropriate.

◊ Follow local rules for handling radioisotopes. Since these vary, no special recommendation will be made here.

3.

Remove medium containing isotope, wash cells carefully in BSS, discarding medium and washes ◊Radioactive! ^{3}H-nucleosides are highly toxic due to their ultimate localization in DNA (see Chapter 6).

All glassware must be carefully washed and free of isotopic contamination. Plastic coverslips should be used in preference to glass to minimize radioactive background. Be particularly careful with spillages; mop up right away. Wear gloves and change regularly, e.g., when you move from incubation (high level of isotope) to handling washed, fixed slides (low level of isotope).

4.

Fix cells in ice-cold acetic methanol for 10 min. Coverslips should be mounted on a slide with DPX or Permount, cells uppermost. Cell suspensions may be centrifuged onto a slide (Cytospin or Cytobuckets) or drop preparations made (see Chapter 13). Prepare several extra control slides for use later to determine correct duration of exposure. All preparations will be referred to as "slides" from now on.

5.

Extract acid-soluble precursors (when labeling DNA, RNA, or protein) with ice-cold 10% TCA (3 × 10 min), and perform any other control extractions, e.g., with lipid solvent or enzymatic digestion.

6.

Wash slide in distilled water and methanol and dry.

7.

Take to darkroom [see also Kopriwa, 1963] and under dark red safelight, melt emulsion, in water bath at 40°C and dilute 50:50 with deionized distilled water. It is convenient to place aliquots of the diluted emulsion in containers suitable for the number of slides to be handled at one time. If sealed in a dark box, these may be stored at 4°C until required.

8.

Still under safelight, dip slides in emulsion, making sure that the cells are completely immersed, withdraw, blot the end of the slide, drain vertically for 5 sec, and allow to dry flat.

9.

When dry ($\sim$ 30 min), transfer to light-tight microscope slide boxes (Clay Adams, Raven Scientific) with a desiccant, such as silica gel, and seal with dark vinyl tape (e.g., electrical insulation tape).

10.

Wrap in black paper and place in refrigerator. Make sure that this refrigerator is not used for storage of isotopes.

11.

Leave at 4°C for 24 hr to 2 wk. The time required will depend on the activity of the specimen and can be determined by processing one of the extra slides at intervals.

12.

To develop, return to darkroom (dark red safelight), unseal box, and allow slides to come to atmospheric temperature and humidity ($\sim$ 2 min).

13.

Place slides in developer (e.g., Kodak D19) for 10 min with gentle intermittent agitation.

14.

Wash briefly in distilled deionized water.

15.

Transfer to photographic fixer for 3–5 min.

16.

Rinse in deionized water and place in hypoclearing agent (Kodak) for 1 min.

17.

Wash in deionized water, five changes over 5 min.

18.

Dry slides and examine on microscope. Phase contrast may be used by mounting a thin glass (00) coverslip in water. Remove coverslip when finished before water dries out or it will stick to the emulsion.

19.

If staining is desired, immerse dry slide in neat Giemsa stain for 1 min then dilute 1:10 in 0.01 M, pH 6.5, phosphate buffer for 10 min. Rinse thoroughly under running tap water until color is removed from emulsion but not from cells [see also Thurston and Joftes, 1963]. Staining solution should be removed by upward displacement with water. The slides should not be withdrawn or the stain poured off or the scum which forms on top of the stain will adhere to the specimen.

Analysis

Qualitative. Determine specific localization of grains, e.g., over nuclei only, or over one cell type rather than another.

Quantitative. (1) Grain counting. Count number of grains per cell, per nucleus, etc. This requires a low grain density, about five to 20 grains per nucleus, ten to 50 grains per cell, no overlapping grains, and a low uniform background.

(2) Labeling index. Count number of labeled cells as a proportion of the total. Grain density should be higher than in (1) to ease the recognition of labeled cells. If the grain density is high (e.g., ~ 100 grains per nucleus), set the lower threshold at, say, ten grains per nucleus or per cell; but remember that low levels of labeling, significantly over background, may yet contain useful information.

Autoradiography is a useful tool for determining the distribution of isotope incorporation within a population, but it is less suited to total quantitation of isotope uptake or incorporation, when scintillation counting is preferable.

Variation. Autoradiographic localization of water-soluble precursors is possible with rapidly frozen specimens which have been freeze-dried or freeze-substituted to remove the water. These may be mounted dry, clamped to the radiosensitive film, or with a minimal amount of moisture (obtained by brief condensation on a cold slide) to promote adhesion of the emulsion [Novak, 1962; Hassbroek et al., 1962]. Both processes require Kodak AR-10 stripping film (see below).

Isotopes of two different energies, e.g., ^{3}H and ^{14}C, may be localized in one preparation by coating the slide first with a thin layer of emulsion, coating that with gelatin alone, and finally coating the gelatin with a second layer of emulsion [Baserga, 1962; Kempner and Miller, 1962; Rogers, 1979]. The weaker β-emission from ^{3}H is stopped by the first emulsion and the gelatin overlay, while the higher-energy β-emission from ^{14}C, having a longer mean path length of around 20 μm, will penetrate the upper emulsion.

Soft β-emitters may also be detected in electron microscope preparations using very thin films of emulsion or silver halide sublimed directly on to the section [Salpeter, 1974; Rogers, 1979].

Adams [1980] described a method for autoradiographic preparations from petri dishes or flasks where

liquid emulsion is poured directly onto fixed preparations without the necessity for trypsinization.

Radiographic emulsion may be applied to slides in the form of a film stripped from a glass plate (Kodak, AR-10)[Rogers, 1979]. The film is made up of a 5-μm thick layer of sensitive emulsion on a 10-μm gelatin backing. It is applied to the slide by inverting the film, peeled off the plate in 30 mm × 40 mm-rectangles, after drying the prescored plate in a desiccator, on to the surface of warm water, and bringing the slide up under the film and allowing the film to drape over and around the slide. After drying, it is treated in the same way as dipping emulsion for exposure and development.

AR-10 gives a very reproducible emulsion thickness and is good for high resolution work and quantitation. As the emulsion tends to detach during processing, coat the slide before adding cells with 0.5% gelatin, 0.05% chrome alum, and avoid prolonged washing in low ionic strength.

CULTURE OF CELLS FROM POIKILOTHERMS

The approach to the culture of cells from cold-blooded animals (poikilotherms) has been similar to that employed for warm-blooded animals largely because the bulk of present-day experience has been derived from birds and mammals. Thus, the dissociation techniques for primary culture employ proteolytic enzymes such as trypsin and EDTA as a chelating agent. Fetal bovine serum appears to substitute well for homologous serum or hemolymph (and is more readily available), but modified media formulations may improve growth. A number of these media are available through commercial suppliers and the procedure is much the same as for mammalian cells—try those media and sera which are currently available, assessing for growth, plating efficiency, and specialized functions (see Chapter 7). Since the development of media for many invertebrate cell lines is in its infancy it may prove necessary to develop new formulations if an untried class of invertebrates is examined. Most of the accumulated experience so far relates to insects and molluscs.

Two reviews cover some aspects of the field [Maramorosch, 1976; Vago, 1971, 1972], but since this is a rapidly expanding area it is to be hoped that a fundamental methodological review text will be forthcoming in the near future. Culture of vertebrate cells other than birds and mammals has also followed procedures for warm-blooded vertebrates, and so far there has not been a major divergence in technique.

Since this is a developmental area, certain basic parameters will still need to be considered to render culture conditions optimal, and if a new species is being investigated, optimal conditions for growth may need to be established, e.g., pH, osmolality (which will vary from species to species), nutrients, and mineral concentration. Temperature may be less vital but it should be fixed within the appropriate environmental range at $\pm 0.5°C$.

CELL SYNCHRONY

The percentage-labeled mitosis method for determining the duration of the stages of the cell cycle has been described in Chapter 18. In order to follow the progression of cells through the cell cycle, a number of techniques have been developed whereby a cell population may be fractionated or blocked metabolically so that on return to regular culture they will all be at the same phase.

Cell Separation

Techniques for this have been described in Chapter 12. Sedimentation at unit gravity (Figs. 12.2, 12.3) is the simplest [Shall and McLelland, 1971; Shall, 1973], but centrifugal elutriation is preferable if a large number of cells ($> 5 \times 10^7$) is required [Meistrich et al., 1977a,b] (see Figs. 12.5, 12.6). Fluorescence-activated cell sorting (see Figs. 12.13, 12.14, 12.15) can also be used in conjunction with a nontoxic, reversible DNA stain such as Hoechst 33342. The yield is lower than unit gravity sedimentation ($\sim 10^7$ cells or less) but the purity of the fractions is higher.

One of the simplest techniques for separating synchronized cells is mitotic "shake off." Mitotic cells tend to round up and detach when the flask is shaken. This works well with CHO cells [Tobey et al., 1967; Petersen et al., 1968] and some sublines of HeLa-S$_3$. Placing the cells at 4°C for 30 min to 1 hr a few hours previously enhances the yield at shake off [Newton and Widly, 1959; Sinclair and Morton, 1963; Lesser and Brent, 1970; Miller et al., 1972].

Blockade

Two types of blocking have been used:

DNA synthesis inhibition (S-phase). Thymidine, hydroxyurea, cytosine arabinoside, aminopterin, etc. [Stubblefield, 1968]: the effects of these agents are variable because many are toxic as blocking cells in cycle at phases other than G, tends to lead to deterioration. Hence, the culture will contain nonviable cells,

cells blocked in S but viable, and cells which have escaped the block.

Nutritional deprivation (G$_1$ phase). In these cases serum [Chang and Baserga, 1977] or isoleucine [Ley and Tobey, 1970] is removed from the medium for 24 hr and then restored, whereupon transit through cycle is resumed in synchrony.

A high degree of synchrony (e.g., $> 80\%$) is only achieved in the first cycle; by the second cycle it may be $< 60\%$ and by the third cycle, close to random. Chemical blockade is often toxic to the cells and nutritional deprivation does not work well in many transformed cells. Physical fractionation techniques are probably most effective and do less harm to the cells.

TIME-LAPSE CINEMICROGRAPHY

This is a technique whereby living cultures may be filmed and their behavior (e.g., cell membrane ruffling, mitosis, migration) accelerated for viewing [Riddle, 1979]. A typical apparatus is depicted in Figure 23.13. It consists of: (1) an inverted microscope with phase-contrast, interference-contrast or surface-interference-contrast optics (e.g., Nikon or Reichert Biovert); (2) a perfusion slide (Fig. 23.14) to present the cells in optical clarity but still in optimal physiological conditions; (3) an incubator chamber or warm air curtain (Sage) to keep the specimen at 36.5°C; (4) a time-lapse control unit (a) to determine the frequency of exposure (i.e., the lapsed time between exposures), (b) to switch on the light before each exposure and switch off the light when it is completed, or to open and close a magnetic shutter in the light path if the light is left on continuously, and (c) to activate the camera shutter and advance film; (5) an exposure meter linked to the time-lapse control unit and camera to set the duration of each exposure; and (6) a still camera, preferably Polaroid, to take record shots at intervals without interrupting filming as this would upset the time base of the film.

Regular film may be replaced with a television camera and video recorder. This reduces the need for exposure control and gives an instant result, but poorer resolution and limited acceleration of movement.

The following protocol for time-lapse recording has been contributed by John Lackie, Department of Cell Biology, University of Glasgow, Scotland.

Principle

By recording images frame by frame with a variable lapse interval and replaying at normal speed, movements of a few microns per hour can be observed and analyzed. Film or videorecording

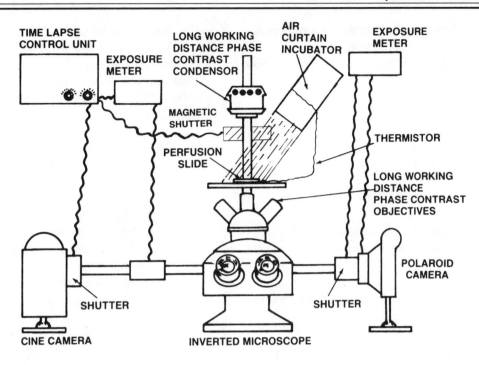

Fig. 23.13. *Suggested layout for time-lapse cinemicrography. Microscope should be mechanically isolated from cameras, control units, and air curtain incubator.*

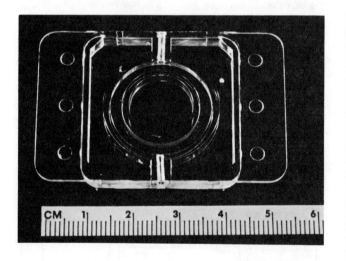

Fig. 23.14. *Perfusion slide for use in cytological observation such as time-lapse cinemicrography. Medium can be perfused in and out via the holes at top and bottom (Sterilin).*

can be used; both involve considerable capital investment. Any image can be recorded, although phase-contrast or differential interference contrast microscopy are generally used. Techniques vary depending upon the specimen and the equipment available.

There are two reasons for "time-lapsing": 1) to observe behavior over a long period and 2) to quantify parameters of movement. Time-lapse videotaping will serve for both, but for high resolution optical images, 16-mm film is preferable. The ultimate resolution comes from Allen Video Enhanced Contrast (AVEC) systems (US patent 4412246) [Allen and Allen, 1983], but these are very expensive ($30,000 for the Hamamatsu AVEC system).

Outline

Cells are seeded in sealed flasks or chambers and viewed microscopically with appropriate optics on a warm-stage. The image is recorded on film or tape and replayed at a faster rate.

Videotaping: example, division of BHK cells

Equipment

Inverted phase-contrast microscope with camera tube.

Low-light level monochrome camera (e.g., Newvicon, Chalnicon; cheap Vidicon tubes are unsuitable) fitted to microscope without eyepiece or intermediate lens, linked to National Panasonic time-lapse videorecorder (time-date generator is an essential "extra"), and connected to monochrome monitor.

Air curtain incubator (fan heater with feedback thermistor on stage: local workshop or Sage Instruments). Temperature controlled room preferable if room temperature fluctuates overnight.

Cost (excluding microscope) around $3,500 minimum.

Protocol

1.
Put sealed tissue culture flask on prewarmed microscope stage and select field.
2.
Fasten flask in place (plasticine).
3.
Adjust light level to get adequate contrast (on monitor) with minimum illumination (to prevent local heating).
4.
Start recording—the longer the proposed observation time, the slower the speed.
5.
Check focus (on the monitor) until the system settles to thermal stability and often thereafter.
6.
Note the starting position on the videocassette, details of optics, etc.

Hints. Overnight stability of focus is essential. Use a slow recording rate for trial runs. Use a heat filter between lamp and stage; a green filter sometimes improves contrast. A short recording of a stage scale gives calibration.

Advantages. Immediate replay once recording is complete, reusable cassettes, capital and running cost lower than for film.

Disadvantages. Loss of resolution, difficult to analyse step by step, difficult to show elsewhere (incompatible with domestic videorecorders). Photographs from monitor unsatisfactory.

Time-lapse filming: example, analysis of fibroblast movement

Equipment (Fig. 23.13)
(1) Microscope with phase-contrast optics and long-working distance condenser. (Microscope sets resolution limits.)

16-mm motor-driven cine camera (e.g., Bolex H16) linked to the microscope through a focusing beam splitter with exposure meter and compensating eyepiece. Camera control through pulse generator (i.e., variable timer to initiate exposure of a single frame).

Temperature regulation (as for videotaping).

(2) Stop-action projector (e.g., L & W 224A Mark IV) and digitizing tablet and microcomputer for numerical analysis.

Cost (excluding microscope): (1) $5,000, (2) $3,500 + $2,500. Total: $11,000 (very approximately).

Materials

Film: choice is between black and white negative (Kodak Plus-X or faster, grainier Tri-X) or color (processing of reversal film impossible in U.K.); best resolution is obtained from black and white negative; may lose resolution on prints

Stainless steel slides 75 × 40 mm: 0.3–0.7-mm-thick with 15-mm-diameter central hole (sterile) (Fig. 23.15)

32 × 32-mm coverslips (sterile)

32 × 32-mm coverslips with cells

sterile medium (easier if not bicarbonate buffered)

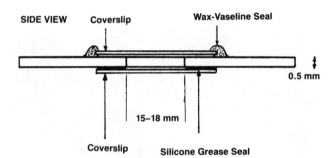

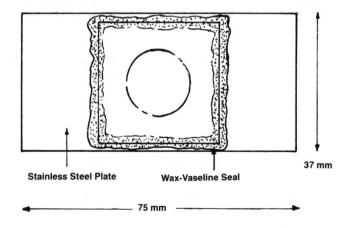

Fig. 23.15. *Filming chamber.*

silicone grease (e.g., vacuum grease) in
   syringe (autoclaved)
sterile Pasteur pipettes
watchmaker's forceps (sterile)
paraffin wax/vaseline (petroleum jelly)
   mix (3:2) (molten: 60°C) and cheap
   artist's paintbrush
tissues
distilled water
70% alcohol

Protocol

1.
Load film, switch on stage heater and wax melter.
2.
Prepare filming chamber aseptically (see Fig.
23.15):
   a. Affix bottom coverslip (dry) over hole in steel
      slide with silicone grease (avoid smearing
      "optical" surfaces) to form chamber.
   b. Slightly overfill chamber with medium (about
      300 μl).
   c. Invert coverslip with cells over chamber
      (avoiding air bubbles).
   d. Draw off excess medium so coverslip rests
      on steel slide.
   e. Seal with hot wax/vaseline applied with
      paintbrush.
   f. Gently clean upper surface with water and
      alcohol.
3.
Put chamber on stage with cells on chamber
ceiling.
4.
Set up microscope, adjust light levels to give correct exposure. Focus through beam splitter.
5.
Choose lapse interval—depends on time-resolution
required: for fibroblasts, about a 23-s interval.
6.
Check light switched through to camera and start
filming.
7.
Take a few frames of a stage scale.

Hints. One hundred ft (35 m) of film = 4,000
frames. Minimum useful sequence is probably 2–
300 frames. Leave a few blank frames between
sequences. If still pictures are needed, use separate
35-mm camera, preferably connected through separate prism housing so still shots can be taken
without interrupting filming, to preserve time base.
Printing from 16-mm film is possible. Comments

about videorecording all apply. Chamber is designed to minimize the specimen thickness; by
filming inverted very high numerical aperture objectives can be used and the light path is very
clean. Much variation is possible.

Advantages. Maximum resolution (without
computer enhanced optics), no magnification
anomalies at edges. Can be projected onto paper,
digitizing tablet, or screen anywhere.

Disadvantages. Delay between completion and
viewing. Large capital outlay. Image cannot be
digitized directly.

Variations. Any microscope system can be
used. Lapse intervals, exposure times, and film
speeds can be varied to get contrast (or color).

Analysis of movement. A complex problem:
see Wilkinson et al. [1982] and Noble and Levine
[1986]. It is essential to have a time base—either
from known frame interval or from time-date
generator.

AMNIOCENTESIS

Inborn metabolic abnormalities may be identified by
culturing cells collected from the amnion during early
pregnancy [Valenti, 1973] (Fig. 23.16). The amniotic
fluid is centrifuged, the supernatant removed (it may
be used for biochemical analysis of proteins and enzyme activity), and the cells cultured as a monolayer.
When the monolayer starts to proliferate, it is used for
chromosome analysis (see also Chapter 13).

The following culture method for amniotic fluid
cells has been contributed by Marie Ferguson-Smith,
Department of Medical Genetics, Yorkhill Hospital,
Glasgow, Scotland.

Principle

Method of culturing amniotic fluid cells from
fluid samples obtained by amniocentesis for chromosome analysis using the *in situ* method in an
open system (petri dishes).

Outline

Cell suspension is set up in 5% CO_2, 95%
humidity incubator on coverslips in petri dishes.
Sufficient colonies are established in 10–14 d for
chromosome analysis.

Materials

Equipment required:
   5% CO_2 incubator
   inverted microscope
   centrifuge

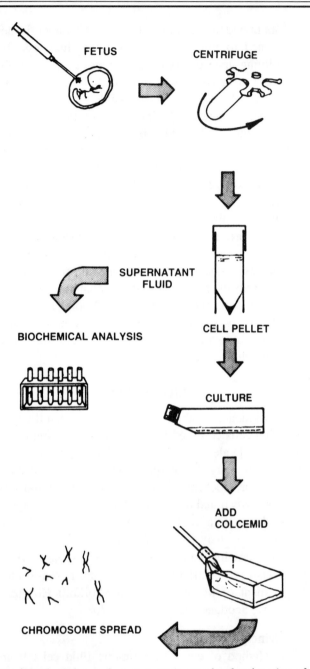

Fig. 23.16. *Culture of amniocentesis samples for detection of chromosomal abnormalities.*

rubber teats
syringes and needles
measuring cylinders
Sterile materials:
  complete medium Ham's F10 (Gibco)
  glutamine, working concentration 4
    μM/ml
  penicillin/streptomycin (10,000 U/ml
    Gibco), working conc. 100 U/ml

fetal bovine serum (Gibco), working
  conc. 5%
ultra ser G serum substitute (LKB),
  working conc. 1%
amniotic fluid samples, 15–20 ml obtained by amniocentesis at 16–18 wk
  gestation, collected in sterile universal
  (cone shaped base) containers or
  centrifuge tubes
35-mm plastic petri dish (Nunc or
  Falcon)
22-mm glass coverslip (dry-heat
  sterilized)
25-cm² plastic flask (best three-
  position screw cap)
Pasteur or 2-ml pipettes or syringes
  with needles
colcemid 10 mg/ml (Gibco) (final concentration for harvesting 0.05
  mg/ml)
Non-sterile materials:
  hypotonic solution 0.05 M KCL (for
    harvesting)
  fixative: 3 parts methanol:1 part
    glacial acetic acid
  forceps

Protocol: setting up and cell culture
1.
Collect amniotic fluid in sterile, cone shaped Universal containers or centrifuge tubes.
2.
Spin at 120 *g* for 10 min.
3.
Remove supernatant leaving 1 ml above cell pellet and mix to disperse pellet.
4.
Add 10–15 ml of complete medium, resuspend cells.
5.
Label four petri dishes, place one coverslip in each with sterile forceps.
6.
Add 2 ml of cell suspension on top of the coverslip in each petri dish. Incubate in 5% CO_2 humid incubator.
7.
Put rest of cell suspension into 25-cm² plastic flask. Adjust cap of flask to allow gaseous perfusion, incubate in 5% CO_2 humid incubator.

8.

After 5–7 d culture should be inspected. Some cells should have settled. Change media (about 2 ml).

9.

Cultures are then inspected and fed two times a wk. They are ready for harvest when actively growing colonies are observed.

10.

Twenty-four hr prior to harvesting remove coverslip, place into fresh petri dish, and feed with fresh culture media. Petri dish should have between 1.8 to 2.0 ml of fluid.

Harvesting

1.

Add colcemid: 0.05 mg/ml final concentration to petri dish for 2–3 hr. (Colcemid can be diluted in culture media and this can be added to petri dish making total fluid volume to about 2.5 ml.)

2.

Taking care not to upset rounded mitotic cells, remove media gently.

3.

Flood coverslip gently with 2 ml of 0.05 M KCl heated to 37°C. Place culture into incubator for 8–10 min (time depends on size and activity of colonies).

4.

Fixation: 6 drops of cold fixative (3 methanol:1 acetic acid) are dropped gently into petri dish with KCl. Allow fixative to disperse.

5.

Remove solution from petri dish, flood coverslip with fresh fixative, leave for 10 min.

6.

Remove coverslip from petri dish and air dry. Preparation is ready for staining (see Chromosome Preparation, Chapter 13).

Variations

Remaining cells in original petri dish and flask can be subcultured either onto a new coverslip or made into drop preparations (see Chapter 13). These cells can also be trypsinized or scraped into a pellet for biochemical analysis or DNA extraction.

SOMATIC CELL FUSION

For many years mammalian, and particularly human, genetic analysis was hampered by the limitations of the duration of the breeding cycle and difficulties in performing breeding experiments. The discovery by Barski et al. [1960] and Sorieul and Ephrussi [1961] that somatic cells would fuse in the presence of Sendai virus led to a burst of activity that has developed into the field of somatic cell genetics.

Briefly, somatic cells fuse if cultured with inactivated Sendai virus, or with polyethylene glycol (PEG) [Pontecorvo, 1975; Milstein, 1979]. A proportion of the cells that fuse will progress to nuclear fusion, and a proportion of these will progress through mitosis such that both sets of chromosomes replicate together and a hybrid is formed. In some interspecific hybrids, e.g., human-mouse, one set of chromosomes (the human) is gradually lost [Weiss and Green, 1967]. Thus, genetic recombination is possible *in vitro* and, in some cases, segregation as well.

Since the proportion of viable hybrids is low, selective media are required to favor the survival of the hybrids at the expense of the parental cells. TK^- and $HGPRT^-$ mutants (see below) of the two parental cell types are used, and the selection is carried out in HAT medium (Fig. 23.17) [Littlefield, 1964]. Only cells formed by the fusion of two different parental cells (heterokaryons) survive, as the parental cells, and fusion products of the same parental cell type (homokaryons) are deficient in either thymidine kinase or hypoxanthine guanine phosphoribosyl transferase. They cannot, therefore, utilize thymidine or hypoxanthine from the medium, and since aminopterin blocks endogenous synthesis of purines and pyrimidines, they are unable to synthesize DNA.

The following protocol for somatic cell fusion has been contributed by Ivor Hickey, Department of Genetics, Queen's University, Belfast, Northern Ireland.

Principle

Although many cell lines will undergo spontaneous fusion, the frequency of such events is very low. In order to produce hybrids in significant numbers cells are treated with either inactivated Sendai virus [Harris and Watkins, 1965] or, more commonly, the chemical fusogen polyethylene glycol (PEG) [Pontecorvo, 1975]. Selection systems which kill parental cells but not hybrids are then used to isolate clones of hybrid cells.

Outline

Cells to be fused are brought into close contact either in suspension or in monolayers before treatment with PEG, which is brief to reduce cell killing. Usually cells are given a 24-hr period to recover before selection for hybrids is exercised.

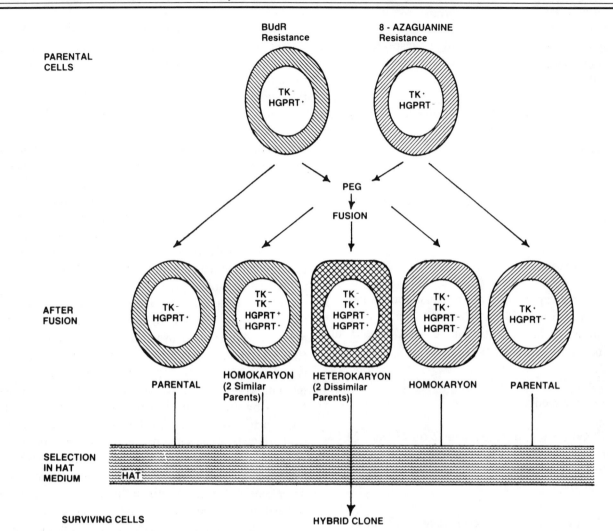

Fig. 23.17. *Somatic cell hybridization. Selection of hybrid cells after fusion (see text).*

Materials

  PEG 1,000 (BDH)
  complete growth medium
  serum-free growth medium
  1.0 M NaOH
  50-mm tissue culture dishes
  plastic universal containers (Sterilin)

Protocol

Monolayer fusion:

1.
Autoclave PEG 1,000. This both liquifies and sterilizes the PEG. Allow it to cool to 37°C and mix with an equal volume of serum-free medium prewarmed to the same temperature. Adjust the pH to approximately 7.6–7.9 with 1.0 M NaOH. This solution can be stored at 4°C for up to 2 wk.

2.
Inoculate equal numbers of the cells to be fused into 50-mm tissue culture dishes. Between 2.5×10^5 and 10^6 of each parental cell line per dish is usually sufficient. The mixed culture is incubated overnight.

3.
Warm the PEG solution to 37°C. It may be necessary at this point to readjust the pH with NaOH.

4.
Remove the medium thoroughly from the cultures and wash once with serum-free medium. Add 3.0 ml of the PEG solution and spread over the monolayer of cells.

5.
Remove the PEG after exactly 1.0 min, and rinse the monolayer three times with 10.0 ml volumes

of serum-free medium before returning to complete medium.

6.

Culture the cells overnight before adding selection medium.

Suspension fusion:

1.

Prepare PEG as above.

2.

Centrifuge a mixture of 4×10^6 cells of each of the two parental cell lines at 150 g for 5 min at room temperature. Centrifugation and subsequent fusion are conveniently carried out in 30-ml plastic universal containers or centrifuge tubes.

3.

Resuspend the pellet in 15.0 ml of serum-free medium and repeat the centrifugation.

4.

Aspirate off all of the medium and resuspend the cells in 1.0 ml PEG solution by gentle pipetting. After 1.0 min dilute with 9.0 ml of serum-free medium and transfer half of the suspension to each of two universal containers or centrifuge tubes containing a further 15.0 ml of serum-free medium. Centrifuge as before, remove the supernatant, and resuspend in complete medium.

5.

After overnight incubation clone the cells in selection medium.

Variations

A large number of variations of the PEG fusion technique have been reported. While the procedures described here work well with a range of mouse, hamster, and human cells in interspecific and intraspecific fusions, they are unlikely to be optimal for all cell lines. Inclusion of 10% DMSO in the PEG solution has the advantage of reducing its viscosity, and has been reported to improve fusion [Norwood et al., 1976]. The molecular weight of the PEG used need not be 1,000. Preparations with molecular weights from 400 to 6,000 have been successfully used to produce hybrids. Although now largely superseded by PEG as a fusogen, for reasons of convenience, Sendai virus fusion remains a reliable method. If a source of virus is available the method of Harris and Watkins [1965] can be used.

Selection of hybrid clones

The method of selection used in any particular instance depends on the species of origin of the two parental cell lines, their growth properties, and whether selectable genetic markers are present in either or both cell lines. Hybrids are most frequently selected using the HAT system: 10^{-4} M hypoxanthine, 6×10^{-7} M aminopterin and 1.6×10^{-6} M thymidine [Littlefield, 1964]. This can be used to isolate hybrids made between pairs of mutant cell lines deficient in the enzymes thymidine kinase (TK$^-$) and hypoxanthine guanosine phosphoribosyl transferase (HGPRT$^-$), respectively. TK$^-$ cells are selected by exposure to BUdR and HGPRT$^-$ by exposure to thioguanine by the procedures described by Biedler in Chapter 11. Where only one parent carries such a mutation HAT, selection can still be applied if the other cell line does not grow, or grows poorly in culture, e.g., lymphocytes or senescing primary cultures.

Differential sensitivity to the cardiac glycoside ouabain is an important factor in the selection of hybrids between rodent cells and cells from a number of other species including human. Rodent cells are resistant to concentrations of this antimetabolite up to 2.0 mM, while human cells are killed at 10^{-5} M ouabain. The hybrids are much more resistant to ouabain than the human parental cells. If a rodent cell line which is HGPRT deficient is fused to unmarked human cells, then the hybrids can be selected in medium containing HAT and low concentrations of ouabain.

Although many other selection systems have been reported, only complementation of auxotrophy [Kao et al., 1969] has been widely used.

It must be stressed that whatever method is used to isolate clones of putative hybrid cells, confirmation of the hybrid nature of the cells must be obtained. This is usually done using cytogenetic or biochemical techniques. In certain cases comparison of numbers of hybrids with frequencies of revertants may be the only way of doing this.

GENE TRANSFER

In addition to fusion of whole cells, fusion of isolated nuclei, individual chromosomes, and even purified genes or gene fragments with whole cells or enucleated cytoplasts is now possible (Fig. 23.18) [Shows and Sakaguchi, 1980]. Enucleation is performed by centrifuging cytochalasin-B-treated cells such that the nuclei detach from an anchored monolayer and pellet at the bottom of the tube. This gives cytoplasmic residues without nuclei (cytoplasts) and nuclei with only some residual plasma membrane sur-

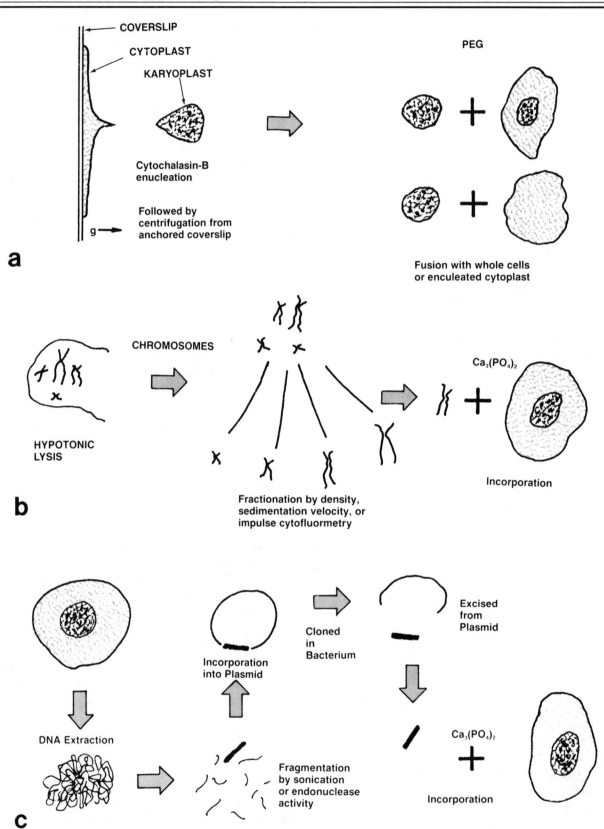

Fig. 23.18. *Gene transfer techniques. a. Whole nuclei extruded by treatment with cytochalasin B hybridized with whole cells or enucleated cytoplast. b. Chromosomes isolated from cells in mi-totic arrest and fractionated by density gradient centrifugation or flow cytophotometry added to whole cells. c. Isolated DNA fragments, amplified by gene cloning techniques, added to whole cells.*

rounding them (karyoplasts). Incubation of karyoplasts with cytoplasts, or whole cells, in the presence of polyethylene glycol results in fusion.

Chromosomes may be isolated from metaphase cells by hypotonic lysis and incubation of these with whole cells after coprecipitation with calcium phosphate results in their incorporation into the nucleus. The chromosomes may be fractionated by density centrifugation or flow cytophotometry and individual chromosome pairs inserted into recipient cells.

At a still higher level of resolution, DNA extracted from one cell can be incorporated into another by a technique similar to that used for whole chromosomes. The DNA may also be cut by restriction endonuclease, the fragments cloned in plasmids, and released by further nuclease treatment. The purified genes or fragments so produced can then be incorporated into recipient cells and their effect on gene expression determined (see Fig. 15.1b).

The following protocol for gene transfer into mammalian cells by the calcium phosphate technique has been contributed by Demetrios A. Spandidos, The Beatson Institute for Cancer Research, Garscube Estate, Switchback Road, Bearsden, Glasgow, Scotland.

Principle

The calcium phosphate technique for introducing genes into mammalian cells was first described by Graham and Van der Eb [1973] and is the most widely used method in current practice.

Exogenous DNA is mixed with calcium chloride and is then added to a solution containing phosphate ions. A calcium phosphate-DNA coprecipitate is formed which is taken up by mammalian cells in culture resulting in expression of the exogenous gene. This method can be used to introduce any DNA into mammalian cells for transient expression assays or long-term transformation.

Outline

Transfect cells with the appropriate DNA carrying a selectable marker, i.e., the aminoglycoside phosphotransferase (aph) gene, and apply selection to eliminate cells which have not taken up and expressed the exogenous gene.

Materials

> Geneticin (Gibco)
> methocel MC4000 CP (Fluka)
> plasmids carrying the aph gene [see Spandidos and Wilkie, 1984a, for a variety of aph recombinant plasmids]
> carrier salmon sperm DNA (Sigma)
> growth medium: Ham's SF12, supplemented with Eagle's MEM amino acids (Flow Laboratories SF12), with 15% fetal bovine serum (Sterile Systems Inc.)

Liquid medium selection procedure

1.

2 × HEPES-buffered saline (2 × HBS):
> 1.63 g NaCl (Analar, BDH)
> 1.19 g HEPES (Sigma)
> 0.023 g Na_2HPO_4-$2H_2O$ (Analar, BDH)
> distilled water to 100 ml

Adjust the pH to 7.1 with 0.5 M NaOH. Filter sterilize the solution and store it at 4°C. The final composition of 2 × HBS is 0.28 M NaCl, 1.5 mM Na_2HPO_4, 50 mM HEPES, pH 7.1.

2.5 M $CaCl_2$: dissolve 10.8 g of $CaCl_2.6H_2O$ in 20 ml (final volume) of distilled water. Filter sterilize the solution and store it at 4°C.

Tris-EDTA buffer (TEB): 0.1 mM EDTA, 1.0 mM Tris-HCl, pH 8.0. Mix 50 μl of 0.2 M EDTA, pH 8.0 and 100 μl of 1.0 M Tris-HCl, pH 8.0, with distilled water to 100 ml final volume. Filter-sterilize the solution and store it at 4°C.

Methocel selection procedure

Methocel medium:

1.

Mix 3 g of methocel with 200 ml of distilled water and autoclave. The methocel dissolves to yield a clear solution which can be stored at 4°C for at least 6 months (see Reagent Appendix).

2.

Just before use, warm the medium to 37°C and add:
> 22.0 ml 10 × Ham's SF12 medium
> 4.0 ml 50 × MEM essential amino acids
> 4.0 ml 0.1 M sodium pyruvate
> 2.5 ml 0.2 M glutamine
> 5.0 ml 7.5% sodium bicarbonate

The stock solutions can be obtained from Flow Laboratories Inc.

3.

Next add 100 ml of serum and the appropriate concentration of the drugs to be used for selection. Thus, for methocel medium containing geneticin (200 μg/ml), add 3.4 ml of 20 mg geneticin/ml water (stock solution, filter sterilized).

The final concentration of methocel and serum are 0.9% and 30%, respectively. Note that the composition of the methocel medium in step 2 depends upon the cell line being used while the nature of the components in step 3 depends upon the selection marker being used.

Protocol

Liquid medium selection procedure for attached cells:

1.

Harvest exponentially growing cells (i.e., mouse LATK$^-$ cells) by trypsinization (see Chapter 10).

2.

Replate the cells at a density of 5×10^5 per flask (25-cm^2 growth area) containing 5 ml of SF12 medium containing 15% fetal bovine serum. Incubate at 37°C for 24 hr.

3.

Into a plastic bijoux vial place 0.5 ml of donor DNA plus carrier DNA at a concentration of 80 μg/ml in TEB. Add 0.4 ml TEB and 0.1 ml of 2.5 M CaCl$_2$. Mix.

4.

Add this DNA solution slowly (about 30 s) with continuous mixing to 1.0 ml of 2 × HBS already in a second bijoux vial. Mix immediately by vortexing and leave the solution at room temperature for 30 min. The DNA concentration at this stage is 20 μg/ml.

5.

After the incubation, a fine precipitate will have formed. Add 0.5 ml of this DNA-calcium phosphate suspension to each flask containing cells in 5 ml of growth medium.

6.

Incubate the flasks at 37°C for 24 hr to allow absorption of the DNA-calcium phosphate coprecipitate by the cells.

7.

Preselection expression stage. Replace the medium with fresh, prewarmed medium and incubate at 37°C for a further 24 hr to allow expression of the transferred gene(s) to occur.

8.

Selection stage. Replace the medium with an appropriate selection medium, in this case SF12 medium containing 15% serum and 200 μg/ml geneticin. (Cultured cell lines differ in their sensitivity to geneticin and the most suitable concentration of geneticin to use must be empirically determined.)

9.

Renew the selection medium every 2–3 d for up to 2–3 wk when colonies are routinely counted.

Analysis

To pick individual colonies, remove the growth medium and then cut off the top of the flask using a heated scalpel. Place a stainless steel cloning ring (diameter 0.4–0.8 cm) over each colony and detach the cells by incubation for 2–3 min at room temperature with 0.1 ml of 0.25% trypsin. After the cells have detached, use a sterile Pasteur pipette to transfer them to flasks (25-cm^2 growth area) containing fresh prewarmed growth medium. If a permanent record is desired, the cells in some of the flasks can be fixed in ice-cold methanol for 15 min, then stained with Giemsa (see Chapter 13).

Methocel selection procedure for anchorage-independent cells

1.

Start transformation as described in the protocol for liquid medium selection procedure, steps 1–8. After allowing for preselection expression (step 8), trypsinize and then count the cells.

2.

Mix 0.2 ml of cells with 20 ml of methocel medium in a plastic universal container and plate on bacteriological plates (9-cm diameter). Up to 2×10^6 cells per plate can be plated, the choice of plating density depending on the transformation frequency expected.

3.

Incubate the plates at 37°C for 7–10 d depending on the doubling time of the recipient cell line.

Analysis

Count the colonies using an inverted microscope. If required, pick individual colonies using a Pasteur pipette and grow these in a suitable growth medium (5 ml per flask of 25-cm^2 growth area or 6-cm diameter dish).

PRODUCTION OF MONOCLONAL ANTIBODIES

One of the most exciting developments of somatic cell fusion arises from the demonstration that sensitized plasma cells from the spleen of an immunized mouse can be fused to continuous lines of mouse myeloma cells. Some of the resultant fusion products are capable of synthesizing immunoglobulins; if

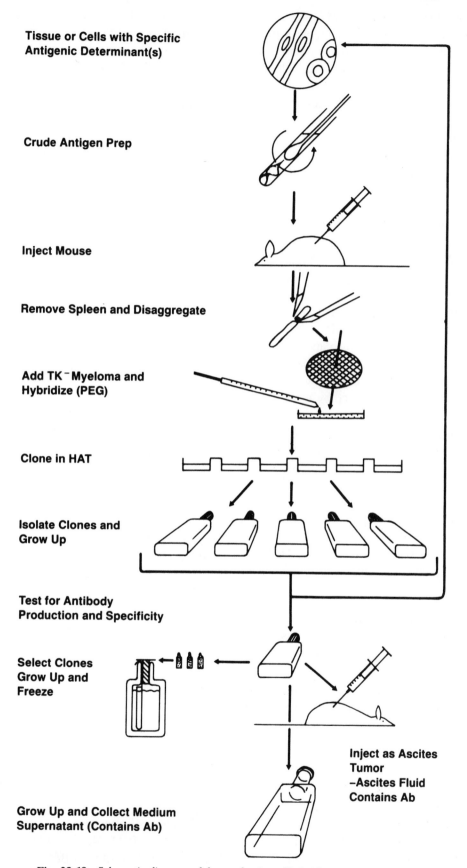

Tissue or Cells with Specific Antigenic Determinant(s)

Crude Antigen Prep

Inject Mouse

Remove Spleen and Disaggregate

Add TK⁻ Myeloma and Hybridize (PEG)

Clone in HAT

Isolate Clones and Grow Up

Test for Antibody Production and Specificity

Select Clones Grow Up and Freeze

Inject as Ascites Tumor —Ascites Fluid Contains Ab

Grow Up and Collect Medium Supernatant (Contains Ab)

Fig. 23.19. *Schematic diagram of the production of hybridoma clones capable of secreting monoclonal antibodies. (Drawing by David Tallach.)*

cloned, each clone produces a single specific monoclonal antibody [Milstein et al., 1979]. The steps in the technique are outlined in Figure 23.19. A mouse is first immunized with crude antigen, and later the spleen is removed. It is then minced and placed in culture with TK^- mouse myeloma cells in the presence of polyethylene glycol. The cells fuse, and the fusion products are cloned in HAT medium. The parental cells or homokaryons will not grow in HAT medium because the myeloma is TK^- and the spleen cells are unable to grow *in vitro*. The clones are then grown and their specificity tested by immunoassay of their supernatant medium with appropriate target cells or antigen, anchored to microtitration plates. Where a specific antigen-antibody complex forms, it can be detected autoradiographically with ^{131}I-labeled or enzyme conjugated antimouse immunoglobulin.

The required clones are recloned, retested, and then frozen. For subsequent antibody production, the cells may be propagated in suspension culture and the supernatant medium used, or they may be passaged in mice as an ascites tumor. Ascitic fluid gives very high yields of antibody although contaminated by globulins from the host animal.

The potential of this technique is enormous for the study of the immune system, production of cell type specific markers, performance of structural analysis of proteins, in the production of vaccines, and in purification of proteins by immunoaffinity chromatography.

The following protocol for the preparation of hybridomas has been contributed by Chris Morris, European Collection of Animal Cell Cultures, PHLS, CAMR, Porton Down, Salisbury, England.

Principle

The production of heterokaryons is a result of fusion between myeloma cells and immunocompetent spleen cells. The resulting hybrids are selected for in HAT medium and subsequently screened for the production of a specific (monoclonal) antibody. Stable secreting hybrids are established by recloning the antibody producing cells.

Outline

Immunize rats or mice with the desired antigen from whole or partially purified cells. It may be necessary to conjugate the antigen to increase antigenicity, e.g., with hemocyanin. Spleen cells are fused with myelomas using polyethylene glycol (PEG) and stable hybrids are selected by diluting the cells into 24- or 96-well plates in medium containing HAT medium.

Culture supernatants are assayed for specific antibody activity between 7–15 d after fusion. Cells from positive wells are cloned by dilution in 96-well trays and rescreened prior to expansion. The positive cells should be cloned twice more; frozen stock should be made from amplified cultures at each cloning and stored in liquid nitrogen.

Materials

Fluorescence microscope
24- and 96-well tissue culture trays
screened NBCS and FCS
DMEM (4.5 g/l glucose) or RPMI 1640
0.34 M sucrose
15-ml and 50-ml sterile conical tubes (Falcon, Corning)
sterile coverslips (20 × 20 mm)
mountant suitable for uv illumination
hemocytometer (improved Neubauer)
trypan blue solution (see Chapter 19)
myelomas, e.g., NS/1, Sp2/0-Ag14 (mouse); y/o (rat) (European Culture Collection or ATCC)
antibody screening reagents, e.g., ELISA, RIA, immunofluorescence
Pristane (2,6,10,14-tetramethyl pentadecane) (Sigma)
mice (balb/c) or rats (Lou)

HT stock (100 x)

Hypoxanthine (6-hydroxypurin, Sigma no. 9377) 136.1 mg (10^{-2} M) and thymidine (Sigma no. T5018) 37.8 mg (1.6×10^{-3} M) dissolved in 100 ml d H_2O. Heat to 50°C until dissolved. Filter at 0.2 μm and aliquot. Store at -20°C. Keep working stocks at 4°C.

A stock (100 x)

Aminopterin (4-aminofolic acid; 4-aminopteroyl-glutamic acid, Sigma no. 1784) 1.76 mg (4×10^{-5} M is added to 90 ml d H_2O. Add 0.5 ml 1N NaOH. When dissolved add 0.5 ml 1N HCl. Make up to 100 ml, 0.2 μm filter and aliquot. Store at -20°C. Protect working stock (4°C) from light.

Myeloma growth medium

Dulbecco's MEM (4.5 g/l glucose) or RPMI 1640 with 10% serum (FBS, or a mixture of FBS + NBCS, e.g., 5 + 5) with 10 ml of HT per l, 10 ml of 200 mM L-glutamine per l. Prescreen sera by cloning the myelomas in 96-well trays, e.g., at 3,1, and 0.3 cells/well concentrations, and selecting the one giving the highest number of wells with growing cells. All serum used should be *held* at 56°C for 40 min prior to use.

Hybridoma growth medium

As for myeloma medium, with an additional 5% serum, e.g., 10% FBS + 5% NBCS. Aminopterin is added as required.

PEG solution (50% v/v)

Two molecular weights have been used successfully by the author, i.e., 1,500 and 4,000. It may be necessary to try several brands for the best results, e.g., BDH, Merck, Roth.

Melt 2 × 10 g by autoclaving, and cool to 50°C. With a glass pipette heated in a flame, measure accurately the PEG volume of one (this PEG is then discarded). Add an equivalent volume of DMEM (no serum), heated to 50°C, to the other container of PEG and mix rapidly. If any of the PEG solidifies, discard, and start again with fresh PEG. The solution is kept at 4°C and should go alkaline (purple color) after 2–3 d. Use only alkaline solutions.

Protocol

A fusion should not be attempted *until* a reliable screening method for specific antibodies has been devised. The essentials of a successful screening method are speed, ease, and reliability, e.g., ELISA, radioimmunoassay.

1.

Culture myelomas in HT medium and maintain between 2–6 × 10^5 cells/ml for 7–10 d prior to a fusion. The cells should be from a reliable source (known to give *stable* fusion products) and must be screened for mycoplasma contamination.

2.

One to 3 d prior to the fusion, aseptically prepare peritoneal macrophages from 4–8-week-old normal healthy mice by injecting 5 ml of ice-cold 0.34 M sucrose. Count the cells and seed at 10^5/well for 24-well trays (1 ml) and 2 × 10^4/well for 96-well trays (0.2 ml) in HAT medium.

3.

After the final antigen injection, screen, using your chosen system, the immunized animals' blood for specific antibody production (tail or eye bleeding).

Successfully immunized animals are sacrificed by neck dislocation or CO_2 asphyxiation. Aseptically remove the spleen. Transfer to a laminar flow cabinet and place in a petri dish with 5 ml of growth medium (HT). Tease spleen apart with surgical needes and mix with a pipette. Transfer the dispersed cells in medium to a 15-ml tube and allow 1–2 min for large debris to settle. Transfer supernatant to another tube and count cells. Centrifuge (80 g for 5 min).

4.

Aliquot enough myeloma cells into 50-ml tubes to give a final ratio of one myeloma cell for every two to three spleen cells, e.g., 5 × 10^7 myelomas to 10^8 spleen cells. Centrifuge (80 g for 5 min), resuspend both myeloma and spleen cell pellets in 20–30 ml of serum-free medium in one 50-ml tube. Centrifuge (80 g for 5 min). Pour off supernatant and drain pellet by inverting the tube onto sterile tissue. The cells must be fully drained to avoid diluting the PEG.

5.

Hold the tube in water at 37°C and add *dropwise* 1 ml of PEG solution onto the wall of the tube just above the pellet, over a 5–10 s period. Gently vibrate the base of tube with finger or thumb while adding PEG, but do not break up the pellet. Hold the tube for a further 50 s at 37°C and mix the contents gently every 10 s with a pipette tip.

Remove the tube from 37°C and immediately start to add dropwise, with gentle swirling, 5 ml of serum-free medium at the rate of 1 ml per min. A further 5 ml is added at 2 ml per min. Stand the tube for 2–3 min (capped). Centrifuge gently (60–70 g for 5 min).

Decant the medium. Resuspend the cells by gently adding HAT medium to give a final myeloma concentration of 1.5 × 10^6/ml for 24-well or 5 × 10^5/ml for 96-well plates. Add 0.1 ml per well to macrophage plates. Transfer to 7–8% CO_2 chambers at 37°C.

7.

Start to examine wells for colonies after 3–4 d. Maintain expanding cultures in HAT for at least 10–14 d and medium change (50%) growing wells at least once prior to screening. Screen wells when more than 50% of the base is covered, expand positives to at least 3 × 10^6 cells, then clone in 96-well trays (HAT medium). Rescreen them using HT medium for subsequent steps, expand, and reclone. Repeat once more to establish stable hybrids. Expand selected positive hybrids at each cloning step and freeze cells at $5-6 \times 10^6$ per ampule. Store in liquid nitrogen.

8.

Ascites production: hybridomas grown in peritoneal cavity will grow to very high densities, thus producing high concentrations of antibody (up to 100 times higher than in cell culture). To facilitate *in vivo* growth the mouse is given an intraperitoneal injection of 0.5 ml Pristane, 7–10 d before injecting hybrids (minimum 3×10^6/mouse). Tumors usually take 7–14 d to grow. Remove ascites fluid from swollen mice with a syringe and centrifuge. If taken aseptically, the cells can be cultured *in vitro* or, alternatively, reinjected into a new mouse. If second or third drainings of ascites are taken, the fluid usually becomes contaminated with blood. It is considered humane at this point to kill the mouse.

9.

Cryopreservation (see Chapter 17).

IN SITU MOLECULAR HYBRIDIZATION

Like transformation and cloning, hybridization is a word with many meanings. In this context it implies the pairing or matching of two complementary molecules of nucleic acid, single-strand DNA with DNA, or with RNA, the complementarity induced by the base sequences in the nucleic acid polymer. In hybridization *in situ* a purified tritiated DNA molecule is used to probe for complementary RNA sequences in fixed cells, and binding is revealed by autoradiography [Wilkes et al., 1978; Conkie et al., 1974; Harrison et al., 1974]. The ^{3}H-DNA is prepared from a specific purified messenger RNA (mRNA), by incubating the mRNA with ^{3}H-nucleotides and the enzyme reverse transcriptase (Fig. 23.20). This enzyme can be extracted from avian myeloblastosis virus, and catalyzes the synthesis of DNA from nucleotides using RNA as a template. Hence the DNA that is synthesized (cDNA) is a complementary copy of the mRNA and has a high affinity for similar mRNA in the cell.

The isolation of purified proteins by monoclonal antibodies may be extended to the isolation of polysomes carrying nascent protein chains from the cytoplasm of a cell, using a specific antibody to the protein being synthesized. On dissociation, the isolated poly-

somes give ribosomes, nascent polypeptide chains, and the mRNA for that protein. This mRNA can then be used to make a cDNA probe as above. Hence it may be possible to make a wide range of cDNA probes for the characterization of specific cell types, or their stage in differentiation.

The following protocol for *in situ* hybridization has been contributed by David Conkie, The Beatson Institute for Cancer Research, Garscube Estate, Switchback Road, Bearsden, Glasgow, Scotland.

Principle

Molecular hybridization *in situ* is the annealing of nucleic acid probes, single stranded and highly radioactive, to complementary regions of nucleic acid fixed in conventional cytological preparations. Specifically, the technique described permits localization of RNA transcripts within individual cells and is readily accomplished without the necessity for high specific activity probes or signal enhancement.

Outline

Apply the radiolabeled probe directly on to fixed cells at 43°C for 18 hr. Wash in 2 × SSC at 55°C and then at 18°C. Detect molecular hybrids by autoradiography; for cytological discrimination, stain the target tissue.

Materials

Water bath at 43°C and then at 55°C

glass slide carriers with lid

methanol

10% trichloroacetic acid

70% ethanol

gelatin chrome alum (gelatin 5 g, chrome alum 0.5 g, distilled water 1 l).

gelatin-coated glass slides (immerse slides in chromic acid, wash thoroughly in water, dip into gelatin chrome alum freshly prepared, and allow to drain and dry dust-free; store at 4°C)

16-mm diameter glass coverslips (immerse in chromic acid, then wash thoroughly in water; store in 70% ethanol)

2 × SSC (2 X standard saline citrate; sodium chloride 17.5 g, sodium citrate 8.8 g, distilled water 1 l, adjust to pH 7.0)

3 × SSC (sodium chloride 26.25 g, so-

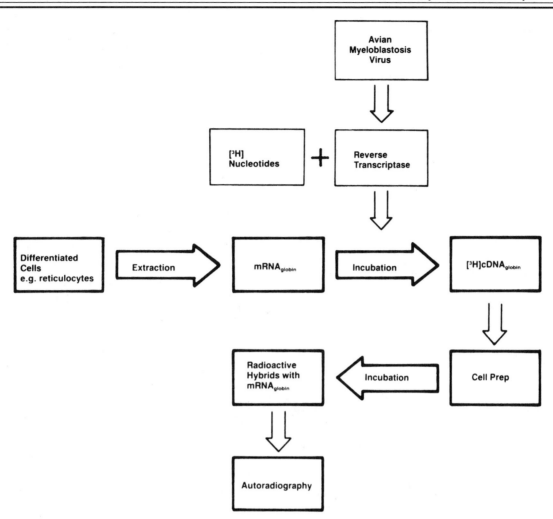

Fig. 23.20. In situ *hybridization of radioactive complementary DNA, synthesized on a messenger RNA template by the action of reverse transcriptase, with intracellular messenger RNA.*

dium citrate 13.2 g, distilled water 1 1, adjust to pH 7.0

60–80° petroleum ether

rubber solution (Cow Proofings Ltd.; reduce viscosity by adding 1/10th volume of petroeum ether)

hybridization buffer (3 × SSC containing 40% formamide (Fluka) adjusted to pH 6.5)

hybridization probe: radioactive nucleic acid probes may be prepared by reverse transcriptase or nick translation in the presence of ^{35}S- or tritium-labeled precursors; incorporation of a single ^{35}S- nucleotide precursor results in a specific activity of up to 10^9 dpm/μg using reverse transcriptase or 3×10^8 dpm/μg by nick translation (kits containing essential reagents and

detailed protocol for reverse transcription and nick translation are available from Amersham or NEN Division of DuPont); dissolve the labeled nucleic acid probe in the hybridization buffer at about 5×10^7 dpm/ ml; if double stranded, denature at 70°C for 10 min immediately before use

autoradiography materials as in Autoradiography section above

DPX mountant

Protocol

1.

Centrifuge suspension cultures onto gelatin-coated glass slides using a cytocentrifuge 10^6 cells/cm^2. Alternatively, prepare thin cryostat sections, tissue imprints, or monolayer cultures grown on glass or

Thermanox (Flow Laboratories).

2.

Fix the cytological preparations in methanol for 5 min followed by three immersions in 10% trichloracetic acid at 4°C.

3.

Dehydrate in 70% ethanol and air dry. At this stage coverslip cultures should be fixed to a glass slide with cells uppermost, using a drop of DPX. Fixed preparations may be stored at 4°C.

4.

Warm the fixed cells to 43°C by placing the slide on a metal tray floating in a water bath at 43°C.

5.

Place 5 μl of the hybridization probe directly onto the fixed cells.

6.

Cover with a 16-mm diameter coverslip.

7.

Apply rubber solution to seal the edges of the coverslip.

8.

Incubate the slides for 18 hr on the flat metal tray at 43°C floating in a water bath with lid. The high humidity prevents dehydration of the hybridization buffer if the seal is incomplete.

9.

After incubation cool the slides on ice and then peel the rubber seal from the coverslip using fine forceps.

10.

Dislodge the coverslip either under the surface of 2 × SSC or by using a jet of 2 × SSC.

11.

Immerse the slides in a glass slide carrier containing 2 × SSC.

12.

Seal the lid with tape and partially immerse the slide carrier in a 55°C water bath for 30 min.

13.

Wash the slides four times in 2 × SSC at 18°C.

14.

Dehydrate in 70% ethanol and air dry.

15.

Detect molecular hybrids by autoradiography using stripping film or liquid emulsion as in the Autoradiography section in this Chapter.

16.

Stain the cells through the autoradiographic emulsion (see above).

Analysis

Microscopy reveals the localization of silver grains over specific cells which contain the target RNA transcripts.

Variations

Alternative nonradioactive techniques for detecting molecular hybrids in cytological preparations have been devised. These methods provide a faster detection time with enhanced resolution and signal-to-noise ratio but may be more difficult than conventional autoradiographic detection methods to use successfully.

Uridine triphosphates containing a biotin molecule attached through "linker" allylamine groups can be incorporated into DNA probes by nick translation [Langer et al., 1981]. These probes have been used in the *in situ* hybridization technique by Langer-Safer et al. [1982] and are detected by either: (1) immunological methods using antibiotin antibodies and a second antibody tagged with a fluorescent, enzymetic, or electron-dense reagent or (2) affinity labeling with avidin conjugated to rhodamine or biotinylated derivatives of peroxidase or alkaline phosphatase.

Biotinylated dUTP for nick translations, avidin, and biotinylated polyalkaline phosphatase are available from Bethesda Research Laboratories.

VIRUS PREPARATION AND ASSAY

Many of the mass propagation methods described above were developed to produce large quantities of virus for analytical and preparative purposes. The current status of our understanding of viruses and their use in the manufacture of vaccines owes much to the development of tissue culture [Habel and Salzman, 1969; Kuchler, 1974,1977; Petricciani, 1979]. The corollary is also true that the development of large-scale culture techniques and serum-free media was prompted in part by the requirements of virology.

Viral assays are of two main types: (1) cytopathic and (2) transforming. Cytopathic viruses may be assayed by their antimetabolic effects in microtitration plates or by the formation of characteristic plaques in monolayers of the appropriate host cell. A viral suspension is serially diluted and added to monolayer culture plates. The number of plaques forming at the limiting dilution is taken as equivalent to the number of infectious particles in the supernatant medium, allowing the concentration of virus in the initial sample to be calculated. Characterization of the virus may be performed with specific antisera, measuring inhibition of the cytopathic effect of the virus, or by radioimmunoassay.

Transforming viruses may be assayed by the selective growth of transformed clones in suspension [Macpherson and Montagnier, 1964], or by looking for transformation foci in monolayer cultures [Temin and Rubin, 1985] (see Chapter 15).

The following protocol for virus preparation and assay has been contributed by Joan C.M. Macnab, Medical Research Council Institute of Virology, University of Glasgow, Scotland.

Principle

Viruses require a susceptible host cell system to replicate. Virus particles adsorb to cultured cells at physiological temperatures. The virus penetrates, uncoats, replicates its nucleic acid in the nucleus or cytoplasm, and after transcription and translation assembles progeny virions which egress either by cell lysis or budding. The number of progeny virus particles released from one cell is referred to as the burst size. Virus is titrated by adsorbing serial tenfold dilutions to cells and counting the resultant plaques after appropriate incubation at the permissive temperature.

Outline

Herpes simplex virus (HSV) will be used as the example.

Permissive cell cultures are infected with an aliquot of virus (1 hr), incubated at 37°C for 2 d and the resultant virus harvested by (1) sonicating and extracting the cells and (2) pelleting virus from the supernatant. Virus is titrated by adsorbing dilutions of virus to cells (1 hr) and counting resultant plaques (37°C for 2 d).

Materials

> Suitable gassed incubator or warm room roller culture facilities if available
> culture vessels, containing required cells just subconfluent, e.g., BHK C13 cells
> 80-oz (2-l) autoclaved round glass bottles or plastic (Falcon) bottles suitable for roller culture or Nuncon cell factory or large plastic flasks (Nunclon, Falcon)
> 50-mm plastic culture dishes (Nunclon, Falcon)
> medium with 5% serum for culturing

> virus
> PBSA
> 1-, 5-, and 10-ml sterile glass pipettes
> Eppendorf pipettes
> autoclaved Eppendorf tips
> bath sonicator, e.g., Cole Palmer
> autoclaved bijoux bottles for dilutions
> black cup vial or ampules
> ice
> storage space at −70°C or −186°C
> stock virus of *known* titer—refer to virus assay (see below) if titer not known
> sterile scraper suitable for removing virus infected cells
> optional—human serum (pooled human serum with HSV antibodies) instead of calf serum to prevent virus spread and secondary plaques

Protocol

Virus preparation
1.
Remove medium from cells.
2.
Infect cells with 0.001 plaque forming units per ml (pfu/ml) virus in enough medium (without serum) to just cover cell monolayer.
3.
Incubate 37°C 1–1½ hr with gassing if necessary.
4.
Add medium with 5% serum—17ml for large culture flask and 20–30 ml for roller bottle.
5.
Incubate 37°C for 2 d.
6.
Check for virus cytopathic effect (cpe). Cells should be rounded, giant, syncytial, and coming off substrate into medium. Shake or scrape remaining cells into medium.
7.
Reserve supernatant fluid, pellet cells by low speed centrifugation (50–100 g, 5–10 min).
Two virus stocks [(1), supernatant virus and (2), cell associated virus] are now prepared as below.
1.
Remove all but 20–30 ml supernatant to sterilized centrifuge bottles. Pellet virus at 12,000 g for 2–3 hr. Resuspend virus pellet in 10 ml supernatant fluid, sonicate for 10 min on ice in 60-s bursts,

aliquot to ampules (0.5 ml each), and freeze at −70°C. Call this preparation (1) supernatant

2.
Add 10 ml supernatant fluid to cells pelleted at low speed.

3.
Sonicate 10 min on ice in 60-s bursts until cells are obviously disrupted.

4.
Pellet cell debris (100 g, 10 min).

5.   —
Collect supernatant and reserve as cell associated virus extraction (2a).

6.
Aliquot and freeze.

7.
Re-extract cell pellet as in steps 2–6. Call this cell associated virus extraction (2b).

Virus assay
Titrate virus preparations separately (the titers will be different).

1.
To sterile bijoux add 0.9 ml PBSA + 5% calf serum (to improve viability of virus).

2.
Add 0.1 ml and, using fresh Eppendorf tip or pipette for each dilution, dilute virus in serial ten-fold steps.

3.
Remove medium from 50-mm dishes. (Each preparation is usually titerd in duplicate).

4.
Carefully add 0.1 ml virus dilution at rim of plate and gently run this over all the surface of the dish. (Do not pipette roughly onto cell sheet or cells will detach).

5.
Incubate at 37°C for 1–1½ hr in gassed incubator.

6.
Overlay with 4.0 ml medium with 5% calf serum or 5% human serum (if desired). Human serum contains HSV antibodies and prevents secondary plaques.

7.
Incubate at 37°C in gassed incubator 2 d.

8.
Fix monolayer after careful removal of medium.

9.
Stain with Giemsa.

10.
Count number of plaques (these usually appear as holes in the cell monolayer) with characteristic virus cpe, i.e., cell rounding, syncytia, or giant cells.

Analysis
When countable plaques are visualized, then note dilution. An average of eight plaques at 10^{-7} dilution represents the titer of 0.1 ml virus stock. Stock titer is therefore 8×10^8 pfu/ml (assuming that one plaque arises from one infectious particle).

Variations
If poor titers of virus are obtained this may be due to poor yield of infectious particles per cell, i.e., burst size. Defective noninfectious virus may be present in high amounts. This may be calculated by determining the total number of virus particles by the electron microscope (EM) and calculating the number of those which are infectious particles. This gives the particle:infectivity ratio and should preferably not exceed 50:1. High ratios of particles (seen in EM) to infectious particles frequently interfere with replication of infectious virions. Some viruses release few particles into the supernatant, e.g., HSV-2 and most virus is cell associated.

Virus stock should be checked for contamination with bacteria, fungi, and mycoplasma by plating on suitable medium or infecting cells.

Most viruses are titrated by using an overlay of 0.6% purified agarose, Noble (Difco) agar or methyl cellulose to prevent spread of virus to secondary plaques. However, HSV is very sensitive to impurities in agarose and this step is generally omitted unless plaque purified stocks are being made.

Herpes simplex virus replicates in a wide variety of human and animal cell types, wheres other viruses, e.g., human cytomegalovirus (CMV) replicates only in human fibroblast cells [Oram et al., 1982]. Some Bunyaviridae replicate in amphibian cell lines [Watret et al., 1985]. The tumor virus SV40 grows in monkey (BSC-1) cells whereas polyoma virus grows in mouse embryo cells. The best culture system for a specific virus may be found by consulting reference textbooks such as *Comprehensive Virology* [Fraenkel-Conrat and Wagner, 1979] or the *Molecular Biology of the*

Tumour Viruses [Weiss et al., 1982, 1985].

Note. Some important viruses cannot, as yet, be cultured in tissue culture systems *in vitro*, e.g., papilloma virus and the AIDS virus (HTLVIII).

IN CONCLUSION

Tissue culture is not limited to mammalian and avian systems, though much of modern technology and our current understanding of cellular and molecular biology have derived from these cultures. Using appropriate culture conditions, it is possible to culture cells from cold-blooded vertebrates such as reptiles, amphibia, and fish and many cell lines are now available (commercial catalogues, e.g., Flow, Gibco, and American Type Culture Collection; see Trade Index and list of Cell Banks, Repositories, and Indices).

To cover these and many other fascinating aspects of this field would take many volumes and defeat the objective of this book. It has been more my intention to provide sufficient information to set up a laboratory and prepare the necessary materials with which to perform basic tissue culture and to develop some of the more important techniques required for the characterization and understanding of your cell lines. This book may not be sufficient on its own, but with help and advice from colleagues and other laboratories, it may make your introduction to tissue culture easier and more profitable than it otherwise might have been.

Reagent Appendix

Acetic/Methanol
1 part glacial acetic
3 parts methanol
Make up fresh each time used and keep at 4°C.

Agar 2.5%
2.5 g agar
100 ml water
Boil to dissolve.
Sterilize by autoclaving.
Store at room temperature.

Amino Acids—Essential
See Eagle's MEM: "Amino Acids," Chapter 7.
    (Available as 50 × concentrate in 0.1N HCL from commercial
    suppliers such as Flow Laboratories and GIBCO)
Sterilize by filtration.
Store at 4°C in the dark.

Amino Acids—Nonessential

Ingredient	g/liter in a 100 × solution
L-alanine	0.89
L-asparagine H_2O	1.50
L-aspartic acid	1.33
Glycine	0.75
L-glutamic acid	1.47
L-proline	1.15
L-serine	1.05
Water	1000 ml

Sterilize by Filtration.
Store at 4°C.
Use 1:100.

Antibiotics
See penicillin, streptomycin, kanamycin, gentamycin, mycostatin.

Antifoam (RD emulsion 9964.40)
Aliquot and autoclave to sterilize.
Store at room temperature.
Dilute 0.1 ml/l liter, i.e., 1:10,000.

Bactopeptone, 5%
5 g Difco bactopeptone dissolved in 100 ml Hanks' BSS
Stir to dissolve.
Dispense and autoclave.
Store at room temperature.
Dilute 1/10 for use.

Balanced Salt Solutions(BSS)
See Table 7.4.
Dissolve each constituent separately, adding $CaCl_2$ last, and
make up to 1 liter. Adjust to pH 6.5.
Hanks' BSS. Without phenol red: as regular recipe but omit
    phenol red.
Sterilize by autoclaving; mark liquid level before autoclaving.
    Store at room temperature. Make up to mark with sterile
    deionized distilled water before use, if evaporation has
    occurred.
Dissection BSS
Hanks' BSS without bicarbonate—previously sterilized by auto-
    claving
250 U/ml penicillin
250 μg/ml streptomycin
100 μg/ml kanamycin or 50 μg/ml gentamycin
2.5 μg/ml Amphotericin B
(All preparations sterile, see below)
Store at -20°C.

Broths
See manufacturers' instructions (Difco, microbiological Associ
-ates, Gibco) for preparation.
Sterilize by autoclaving.

Carboxymethylcellulose (CMC)
4 g CMC
90 ml Hanks' BSS
Weigh out 4 g CMC and place in beaker.
Add 90 ml Hanks' BSS, and bring to boil to "wet" the CMC.
Allow to stand overnight at 4°C to clear.
Make volume to 100 ml with Hanks' BSS.
Sterilize by autoclaving. The CMC will solidify again during
    autoclaving but will dissolve at 4°C.
For use, e.g., to increase viscosity of medium in suspension
    cultures, use 3 ml per 100 ml growth medium.

Chick Embryo Extract [Paul, 1975]
1. Remove embryos from eggs as described in Chapter 9 and
    place in 9-cm petri dishes.
2. Take out eyes using two pairs of sterile forceps.
3. Transfer embryos to flat or round bottomed containers—two
    embryos to each container.
4. Add an equal volume of Hank' BSS to each.
5. Using a sterile glass rod which has been previously heated and
    flattened at one end, mash the embryos in the BSS until they
    have broken up.
6. Stand for 30 min at room temperature.
7. Centrifuge for 15 min at 2,000 *g*.
8. Remove supernatant and after keeping a sample to check steril-
    ity (see Chapter 8), aliquot, and store at −20°C.
Extracts of chick and other tissues may also be prepared by ho-
    mogenization in a Potter homogenizer or Waring blender
    [Coon and Cahn, 1966].

1. Homogenize chopped embryos with an equal volume of Hanks' BSS.
2. Transfer homogenate to centrifuge tubes and spin at 1000 g for 10 min.
3. Transfer the supernatant to fresh tubes and centrifuge for a further 20 min at 10,000 g.
4. Check sample for sterility (see Chapter 8), aliquot remainder, and store at $-20°C$.

0.1 M Citric Acid/0.1% Crystal Violet

21.0 g citric acid
1.0 g crystal violet
Make up to 1,000 ml with deionized water.
Stir to dissolve.
To clarify, filter through Whatman No. 1 filter paper.

CMC
See "Carboxymethylcellulose"

CMF *See* "Calcium-Magnesium-Free Saline"

Colcemid, 100 × Concentrate

100 mg colcemid
100 ml Hanks' BSS
Stir to dissolve.
Sterilize by filtration.
Aliquot and store at $-20°C$.

Collagenase

2,000 U/ml in Hanks' BSS
100,000 units Worthington CLS grade collagenase, or equivalent (specific activity = 1500–2000 u/mg)
50 ml Hanks' BSS
To dissolve: stir at 36.5°C for 2 hr or at 4°C overnight.
Sterilize by filtration as for serum (see Chapter 8).
Divide into aliquots suitable for 1–2 wk use.
Store at $-20°$.

Collagenase-Trypsin-Chicken Serum (CTC) [Coon and Cahn, 1966]

	Volume	Final Concentration
Calcium-and magnesium-free saline [Moscona, 1952], sterile	85 ml	
Trypsin stock, 2.5%, sterile	4 ml	0.1%
Collagenase stock, 1%, sterile	10 ml	0.1%
Chick serum stock	1 ml	1.0%

Aliquot and store at $-20°C$.

Collection Medium (for tissue biopsies)

Growth medium + antibiotics	
Growth medium	500 ml
Penicillin	125,000 units
Streptomycin	125 mg
Kanamycin or	50 mg
Gentamycin	25 mg
Amphotericin	1.25 mg

Store up to 3 wk at 4°C or at $-20°C$ for longer periods.

Counting Fluid × 10 [Paul, 1975]

Sodium chloride	700g
Tri-sodium citrate ($2H_2O$)	170g

Distilled water up to 10 liters
Stir ingredients in water until dissolved.
Filter through Whatman Filter Paper No. 1, then dispense into Pyrex bottles.
Adjust pH to 7.2
Autoclave, and store at room temperature.
Dilute 1:10 to use, when osmolality should be 290 mOs mols/Kg.

Crystal Violet 0.1% in Water

Crystal violet	100mg
Water	100 ml

Available ready made from BDH

D19 (Kodak)
See manufacturers instructions.

Dexamethasone (Merck) 1 mg/ml (100 ×)

This comes already sterile in glass vials. To dissolve, add 5 ml water by syringe to vial, remove, and dilute to give a concentration of 1 mg/ml.
Aliquot and store at $-20°C$.
β-methasone (Glaxo) and methylprednisolone (Sigma) may be prepared in the same way.

Dissection BSS
See "Balanced Salt Solutions."

Ficoll, 20%

Sprinkle 20 g Ficoll (Pharmacia) on the surface of 80 ml water and leave overnight to settle and dissolve. Make up to 100 ml.
Sterilize by autoclaving.
Store at room temperature.

Fixative, Photographic (Kodak or Ilford)
See manufacturer's instructions.

Fixative for Tissue Culture
See "Acetic/Methanol", this Appendix.

Gentamycin (Flow, GIBCO, Schering)
Dilute to 50 μg/ml for use.

Gey's Balanced Salt Solution

NaCl	7.00 g/l
KCl	0.37 g/l
$CaCl_2$	0.17 g/l
$MgCl_2 \cdot 6H_2O$	0.21 g/l
$MgSO_4 \cdot 7H_2O$	0.07 g/l
$Na_2HPO_4 \cdot 12H_2O$	0.30 g/l
KH_2PO_4	0.03 g/l
$NaHCO_3$	2.27 g/l
Glucose	1.00 g/l
CO_2	5%

Giemsa in 0.01 M Phosphate Buffer pH 6.5

$NaH_2PO_4 \cdot 2H_2O$	0.01 M	1.38 g/liter
$Na_2HPO_4 \cdot 7H_2O$	0.01 M	2.68 g/liter

Combine to give pH 6.5.

Dilute prepared Giemsa concentrate (Gurr, BDH, Fisher) 1:10 in 100 ml buffer. Filter through Whatman No. 1 filter paper to clarify.

Use fresh each time (precipitates on storage).

Glucose, 20%

Glucose 20g

Dissolve in Hanks' BSS and make up to 100 ml. Sterilize by autoclaving.

Store at room temperature.

Glutamine 200 mM

| L-glutamine | 29.2g |
| Hanks' BSS | 1000 ml |

Dissolve glutamine in BSS and sterilize by filtration (see Chapter 8).

Aliquot and store at $-20°C$.

Glutathione

Make $100 \times$ stock, i.e., 0.10 M in BSS or PBSA and dilute to 1 mM for use.

Sterilize by filtration.

Aliquot and store at $-20°C$.

Growth Medium

That medium normally used to maintain the cells in exponential growth with the appropriate serum added if specified.

Ham's F12

See Chapter 7.

Hanks' BSS

See "Balanced Salt Solutions," Chapter 7.

HAT Medium

Drug	Concentration	Dissolve in	Molarity $100 \times$ final
Hypoxan-thine	136 mg/100 ml	0.05N HCl	1×10^{-2} M
Aminopterin	1.76 mg/100 ml	0.1 N NaOH	4×10^{-5} M
Thymidine	38.7 mg/100 ml	BSS	1.6×10^{-3} M

For use in the HAT selective medium mix equal volumes sterilize (filter) H, A., and T and add mixture to medium at 3%.

Store H and T at 4°C, A at $-20°C$.

HB Medium

Add the following to CMRL 1066 Medium:

Insulin	5 μg/ml
Hydrocortisone	0.36 μg/ml
β-retinyl acetate	0.1 μg/ml
Glutamine	1.17 mM
Penicillin	50 U/ml
Streptomycin	50 μg/ml
Gentamycin	50 μg/ml
Fungisone	1.0 μg/ml
Fetal bovine serum	1%

HBSS

See "Hanks' BSS," Chapter 7.

Hoechst 33258 [Chen, 1977]

(2-2(4-hydroxyphenol)-6-benzimidazolyl -6-(1-methyl-4-pierpa zyl)-benzimidalol-trihydrochloride

Make up 1 mg/ml stock in BSS without phenol red and store at $-20°C$.

For use dilute 1:20,000 (1.0 μl→20 ml) in BSS without phenol red at pH 7.0.

◇This substance may be carcinogenic. Handle with extreme care.

Kanamycin Sulphate "Kannasyn" (10 mg/ml)

4×1 g vials

Hanks' BSS 400 ml

Add 5 ml BSS from a 400-ml bottle of BSS to each vial.

Leave for a few minutes to dissolve.

Remove the BSS and kanamycin from the vials and add back to the BSS bottle.

Add another 5 ml BSS to each vial to rinse and return to BSS bottle. Mix well.

Dispense 20 ml into sterile containers and store at $-20°C$.

Test sterility: add 2 ml to 10 ml sterile medium, free of all other antibodies, and incubate at 36.5°C for 72 hr.

Use at 100 μg/ml

Lactalbumin Hydrolysate 5% (10×)

| Lactalbumin hydrolysate | 5g |
| Hanks' BSS | 100ml |

Heat to dissolve.

Sterilize by autoclaving.

Use at 0.5%.

McIlvaines Buffer pH 5.5

		To make 20 ml	100 ml
0.2 M Na$_2$HPO$_4$	28.4 g/l	11.37 ml	56.85 ml
0.1 M Citric Acid	21.0 g/l	8.63 ml	43.15 ml

Media

The constituent of some media in common use are listed in Chapter 7 with the recommended procedure for their preparation. For those not described see Morton [1970], manufacturers catalogues (Flow, GIBCO, Microbiological Associates, etc.), or the original literature references.

MEM

See "Eagle's MEM," Chapter 7.

2-Mercaptoethanol (M.W. 78) (Stock solution 5×10^{-3}M (100 ×))

3.9 mg in 10 ml HBSS

Sterilize by filtration in fume cupboard.

Store at $-20°C$ or make fresh each time.

Methocel

See "Methylcellulose."

Methylcellulose (1.6% in medium)

1. Add 8.0 gm methylcellulose (4,000 centipose) to a 500-ml medium storage bottle containing 250 ml distilled water at 80—100°C and a magnetic stirrer bar.
2. Shake by hand until the methylcellulose is wetted.
3. Place on a magnetic stirrer and stir until cooled to room temperature.
4. Remove to cold room and continue to stir overnight.

5. Autoclave the methocel, when it will form an opaque solid.
6. Allow to cool to room temperature and then cool to 4°C.
7. Add 250 ml of 2 × strength medium and stir overnight at 4°C.
8. Dispense into sterile 100-ml bottles and store at −20°C. Thawed bottles may be kept at 4°C for 1 month.

For use, dilute methylcellulose medium with appropriate volume of serum and add cell suspension in sufficient growth medium to give final concentration of 0.8% methylcellulose.

Mitomycin-C (Stock solution 10 μg/ml (50 ×))

2 mg vial
Measure 20 ml HBSS into a sterile container.
Remove 2 ml by syringe and add to vial of mitomycin.
Allow to dissolve, withdraw, and add back to container.
Store for 1 wk only at 4°C in the dark.
(Cover container with aluminum foil.)
For longer period store at −20°C.
Dilute to 2 μg/10^6 cells, 0.2 μg/ml, for use, but check for effective concentration when using each new cell line as a feeder layer

Mycostatin (Nystatin) (2 mg/ml, 100 ×)

Mycostatin   200 mg
Hanks' BSS   100 ml
Make up by same method as kanamycin. Final concentration 20 μg/ml.

Napthalene Black 1% in Hanks' BSS

Napthalene black   1 g
Hanks' BSS   100 ml
Dissolve as much as possible of the stain then filter through Whatman No. 1 filter paper.

Penicillin (e.g., Crystapen Benzylpenicillin (sodium)) 1,000,000 units per vial

Use 4 vials and 400 ml Hanks' BSS.
Make up as for kanamycin, stock concentration 10,000 U/ml.
Use at 50–100 U/ml.

Percoll (Pharmacia)

Made up and sterile and should be diluted with medium or HBSS until correct density is achieved.
Check osmolality and adjust to 290 mOsm/Kg by adding water or less medium.

Phosphate Buffered Saline PBS (Dulbecco 'A') (See Table 7.4, Chapter 7)

Oxoid tablets, Code BR 14a, 1 tablet per 100 ml distilled water
Dispense, then autoclave.
Store at room temperature, pH 7.3, osmolality 280 mOsm/Kg.
PBSB contains the calcium and magnesium and should be made up and sterilized separately. Mix with PBSA, if required, immediately before use.

Phosphate Buffered Saline/EDTA (PBS/EDTA) (1mM 0.372 g/ liter)

Make PBSA as above.
Add EDTA and stir.
Dispense, autoclave, and store at room temperature.

Phytohemagglutinin Stock (500 μg/ml, 100 ×. Lyophilized (Sigma))

Dissolve powder by adding HBSS by syringe to ampule.
Aliquot and store at −20°C.
Dilute 1:100 for use.

SF12

Eagle's MEM with 2x amino acid concentration and lacking thymidine

Sodium Citrate/Sodium Chloride (SSC)

Trisodium citrate (dihydrate)	0.03 M	8.82 g	0.09 M	26.46 g
Sodium chloride	0.3 M	17.53 g	0.9 M	52.60 g
Water				

SSC

See "Sodium Citrate/Sodium Chloride" above.

Streptomycin Sulphate (1 g per vial) (For method see Kanamycin)

Take 2 ml from a bottle containing 100 ml sterile Hanks' BSS and add to 1 g vial of streptomycin. When solution complete, return 2 ml to 98 ml Hanks'.
Dilute 1/200 for use.
Final concentration—50 μg/ml

Trypsin Diluent-Buffered

Sodium chloride	6.0g
Trisodium citrate	2.9g
Tricine (N-Tris (hydroxymethyl) methyl glycine)	1.79g
Phenol red	0.005g
Distilled water to	1,000 ml

Stir ingredients until dissolved, adjust pH to 7.8.
Filter through Whatman No. 1 filter paper.
Dispense and autoclave.
Osmolality—290 mOsm/Kg.

Trypsin Stock 2.5% in 0.85% (0.14 M) NaCl

Trypsin solutions can be bought commercially. Alternatively, to make up a 2.5% solution in 0.85% NaCl, stir for 1 hr at room temperature or 10 hr at 4°C. If trypsin does not dissolve completely, clarify by filtration through Whatman No. 1 filter paper.
Sterilize by filtration, aliquot and store at −20°C.
Note. Trypsin is available as crude (e.g., Difco 1:250) or purified (e.g., Sigma 3 × recrystalized) preparations. Crude preparations contain several other proteases which may be important in cell dissociation but may also be harmful to more sensitive cells. The usual practice is to use crude trypsin unless cell damage reduces viability or reduced growth is observed, when purified trypsin may be used. Pure trypsin has a higher specific activity and should therefore be used at a proportionally lower concentration, e.g., 0.05 or 0.05%.

Trypsin, Versene, Phosphate (TVP)

Trypsin (Difco 1:250)	25 mg (or 1 ml Flow or
Gibco 2.5%)	
Phosphate buffered saline	98 ml
Disodium EDTA (2H$_2$O)	37 mg
Chick serum (Flow)	1ml

Mix PBS and EDTA and autoclave, then add chick serum and trypsin.

If using powdered trypsin, sterilize by filtration before adding chick serum. Aliquot and store at $-20°$C.

Tryptose Phosphate Broth

10% in Hanks' BSS

Tryptose phosphate (Difco)	100 g
Hanks' BSS	1,000 ml

Stir until dissolved.

Aliquot and sterilize in the autoclave.

Store at room temperture.

Use 1:100 (final concentration 0.1%).

Tyrodes's Solution

NaCl	8.00g/l
KCl	0.20g/l
CaCl$_2$	0.20g/l
Mg$_2$ · 6 H$_2$O	0.10g/l
NaH$_2$ PO$_4$ · H$_2$O	0.05 g/l
glucose	1.00
Gluphase	Air

Viability Stain

See "Naphthalene Black."

Vitamins

See media recipes (Chapter 7).

Make up at 100 × concentrated.

Sterilize by filtration.

Store at $-20°$C in the dark.

Trade Index

SOURCES OF MATERIALS

Air Curtain Incubator	Sage Instruments	Bovine serum albumin	Sigma
Agar	GIBCO	BSS, Hanks', Earle's, etc.	Flow
	Difco		GIBCO
Amino acids	Sigma	Camera lucida	British-American Optical
Aminopterin	Sigma		E. Leitz (Instruments) Ltd.
Ampules—glass	Wheaton		Olympus
Ampules—plastic	Nunc, *see* GIBCO		Nikon
Ampule sealer	Kahlenberg-Globe		Carl Zeiss (Oberkochen) Ltd.
Anemometers	*See* Laminar flow cabinets	Carbosymethylcellulose	Fisher
Antibiotics	*See* individual antibiotics	(CMC)	BDH
Antibodies	Amersham	Cell counters	Coulter
	Boehringer		Grant Instruments
	DAKO		GIBCO
	Bethesda Research		Particle data
	Laboratories	Cellulose nitrate filters	Gelman
Antifoam RD emulsion	Dow-Corning		Millipore
9964.40	Hopkin & Williams		Sartorius GmbH
	Miles	Centrifugal elutriator	Beckman
Automatic glassware washing	John Burge (Equip.) Ltd.	Centrifuges	MSE
machines	(U.K.)		Fisher
	Camlab Ltd.		Beckman
	Vernitron (U.S.A.)		DAMON/IEC
Automatic pipettes	Becton Dickinson	Centrifuge—Cytospin	Shandon
	Camlab	Centrifuge tubes, tissue	Corning
	Gilson	culture	Falcon, *see* Becton Dickinson
	Boehringer	Channelyzer attachment	Coulter
	Flow	Chemicals	Aldrich
	Jencons		BDH
	Warner		Boehringer
Automatic pipette plugger	Bellco (A.R. Horwell, U.K.)		Calbiochem
	Volac, *see* Camlab		Fisher
Bactopeptone	Difco		Fluka
Bench centrifuges	MSE		Sigma
	Fisher	Chick plasma	GIBCO
	DAMON/IEC	Chicken serum	GIBCO
	Sorvall		Flow
	Beckman	Cine camera	Bolex
Biochemicals	Aldrich	Citifluor AFI	Citicorp
	U.S. Biochemical Co.	Cloning rings	Fisher
	Sigma	Closed-circuit TV	British-American Optical
	BDH		E. Leitz (Instruments) Ltd.
	Calbiochem		Olympus
	Fluka		Gallenkamp
	Boehringer		Nikon
Bottles	Corning		National (via local TV/video
	Bellco		agent or lab. supplier)
	Schott		Carl Zeiss (Oberkochen) Ltd.

CMC	*See* Carboxymethylcellulose Distillers Co.		Helena (M.I. Scientific, U.K.)
CO_2 automatic change-over unit for cylinders	Air Products L.I.P.		Gelman
CO_2 controllers	Laboratory Impex Ltd.	Density meter	Paar
	Gow Mac	DePeX, permount	*See* Stains
	Heinicke		Fishers
	Lab-Line	Detergents	Flow
	Forma		Decon
CO_2 incubators	Assab		Diversey
	Astell Hearson		Medical Pharmaceutical Dev. Ltd.
	Baird & Tatlock	Dexamethasone	Sigma
	Flow	Dextran	Calbiochem
	Napco		Sigma
	Gallenkamp	Dextraven 100	Fisons
	Heinicke	Diacetyl fluorescein	Fisher
	Heraeus (Tekmar Co. U.S.A.)		Sigma
	A.R. Horwell Ltd.	Dialysis tubing	laboratory suppliers
	Lab-Line (Burkard Scientific (Sales) Ltd., U.K.)	Disc filter assembly for sterilization	*See* Filters
	LEEC	Disinfectants (Chloros (1Cl))	laboratory suppliers
	New Brunswick Scientific	Dispase	Boehringer Mannheim
	Precision Scientific	DMSO	Merck
	Vindon	DNase	Sigma
Colcemid	Sigma	Ecoscint	National Diagnosis
	CIBA	EDTA	*See* Chemicals
Collagen (Vitrogen 100)	Collagen Corporation	Ehrlenmeyer flasks	*See* Glassware
	Flow	Electronic cell counter for cell sizing	Coulter
Collagenase	Worthington (Millipore)	Electronic thermometer	Comark
Collagenase	Sigma		Fisher
Colony counter	New Brunswick Scientific	Emulsion	*See* Photographic
	Dynatech (Artek)	Epidermal growth factor	Bethesda Research Laboratories
	Micromeasurements		Uniscience
Compupet	W.R. Warner		Collaborative Research
	General Diagnostics Ltd.	Equipment	Bellco
	Planer		Cole Parmer
Controlled rate coolers for liquid N_2 freezing	Union Carbide		Gallenkamp
	Cyro-Med.		New Brunswick
Coverslips, glass, plastic	Gallenkamp	Electrophoresis	BRL
	Fisher		Haake
	Lux		Camas
	L.I.P.		LKB
	M A Bioproducts		Pharmacia
Crystal violet (Gurr)	Searle Scientific Services		Shandon
	Fisher	Ethidium bromide	Sigma
Cytobuckets	DAMON/IEC	Fibronectin	Uniscience
Cytofluorograph	Ortho		Collaborative Research
D19	*See* Photographic	Fermentors	New Brunswick
DEAE dextran	Pharmacia		L.H. Fermentation
	Bio-Rad		Bethesda Research Laboratories
Deionizers	Fisons, *see* MSE	Ficoll	Pharmacia
	Elga	Filters	Millipore
	Corning		Sartorius
	Bellco		Gelman
Densitometers	Joyce-Loebl		Nuclepore, (Sterilin, U.K.)
	Beckman Instruments		Pall (Europe) Ltd.
	Gilford Instruments Ltd.		

Polaroid camera	Polaroid	Nu-Serum	Collaborative Research
Poly-D-lysine	Sigma	CLEX	Dextran Products Ltd.
Polystyrene flasks, petri dishes	*See* Tissue Culture Flasks, etc.	Sieves	Endecotts
		Silica gel	Fisher
Polyvinyl pyrrolidone, dextran	Calbiochem		BDH
	Sigma	Silicon tubing	ESCO
Pressure cooker, bench top autoclave	Astell Hearson	Silicone	Fisher
	Valley Forge	Slide containers for emulsion-Cyto-Mailer	Lab-Teck, *See* Fisher
	Napco		
Projection table	Bellco (A.R. Horwell Ltd., U.K.)	Slide Boxes, light light	Raven Scientific
			Clay Adams
Projectors	L.W. Photooptical	Stains	Searle Scientific Services
Pronase	Sigma		Fisher
Pumps, peristaltic	Pharmacia	Stainless steel mesh	Falcon, *See* Becton Dickinson
	Gilson		
	LKB	Streptomycin	Glaxo
	Millipore	Sterilizing and drying oven	Astell Hearson
	Watson-Marlowe		Baird & Tatlock
Pumps, vacuum	Millipore	Sterilizing ovens	Astell Hearson
	Gelman		Baird & Tatlock
Quinacrine dihydrochloride (Atebrin)	Searle Scientific Services		Camlab Ltd.
			Forma
Recording thermometer	Cambridge Instruments		Gallenkamp Ltd.
	Fisher		A.R. Horwell Ltd.
Refractometer	Townson & Mercer Ltd.		LEEC
	American Optical Co.		New Brunswick Scientific
Refrigerators	Fisher		Precision Scientific Co.
	Gallenkamp		Strands Scientific
	local laboratory suppliers		
Repelcote	Hopkin & Williams	Sterilizing tape (indicator)	Fisher
Reusable in line filter assembly	*See* Filters	Sterility indicators, Thermalog	Brownes Tubes
			Bennet & Co.
Reverse osmosis	Millipore		Propper
	Fisons	Stills	Fisons
	Elga		Jencons
Roller bottles (glass)	Bellco		Corning
Roller bottles (plastic)	*See* Tissue Culture Flasks, etc.	Storage tank for liquid N_2	BOC Cryo Products
			Minnesota Valley Engineering
Roller bottle rack	New Brunswick Scientific		Union Carbide Ltd.
	Bellco		
Rubber bulb with inlet and outlet valve	Jencons Scientific Ltd.	Supplements, e.g., tryptose (lactalbumin hydrolysate, bactopeptone)	Difco
Scalpels	*See* Instruments, dissecting		
Scintillation fluid	Packard	Syringes	local laboratory suppliers
	Econoscint, National Diagnostics	Temperature controllers	*See* Thermostats
		Temperature recorders	*See* Recording Thermometer
	New England Nuclear	Thermalog	Bennet (U.K.)
Scintillation vials, minivials	Packard	Thermostats, proportional controllers	Fisher
Semi-permeable nylon film	*See* Packaging		Jumo
Serum	Flow Laboratories		Napco
	Microbiological Associates	α-Thioglycerol	Sigma
	GIBCO	Thymidine	Sigma
	Seralab	Time-lapse cinemicrography	British-American Optical
	Northumbria Biological		E. Leitz Ltd.
Serum substitutes			Olympus
Jerxtend	NEN		Carl Zeiss Ltd.
Ventren	Ventrex		Bolex
Ultraser	LKB	Tissue culture flasks, etc.	Nunc (GIBCO)
Nutricyte	Brooks Labs		Bellco

	A.R. Horwell
	Corning
	Costar (Northumbria Biologicals, U.K.)
	Falcon (Becton Dickinson)
	Linbro Lux (Flow)
Tissue culture media	Flow
	GIBCO
	Microbiological Associates
Trypan blue	*See* Stains
Tryptophan	*See* Amino Acids
Tryptose phosphate broth	Difco
	Oxoid
Trypsin	Flow
	GIBCO
Trypsin inhibitor (soya bean)	Sigma

Tyrosine	*See* Amino Acids
Universal containers	Sterilin
	Nunc (GIBCO)
Washing machine	Vernitron/Betterbuilt (Burge, U.K.)
Vacuum pump	*See* Pumps
Vinblastin	Sigma
Vinyl tape	3M
Vitafiber	Amicon
Vitamins, solid	Sigma
Vitamins, solution	*See* Tissue Culture Media
Vortex mixer	Gallenkamp
	Fisher
	general laboratory suppliers
X-ray film	Eastman Kodak Company
	Fuji

ADDRESSES OF COMMERCIAL SUPPLIERS

United Kingdom and Europe

ADELPHI MANUFACTURING CO.
207 Duncan Terrace, London NI8
AGFA-GAVAERT
125 Whittinghame Drive, GI3 1NW
AMERSHAM INTERNATIONAL
White Lion Road, Aylesbury, Bucks HP20 2TP
AMICON LTD.
Amicon House, 2 Kingsway, Woking, Surrey, GU21 1UR
ANACHEM LTD.
20a North Street, Luton, Beds, Luton 37274
ASSAB LTD.
Bishops Walk Precinct, High Street, Tewkesbury, Glos., Tewkesbury (0684)
ASTELL HEARSON
172 Brownhill Road, Catford, London, SE6 2DL Unit 1, Brickfields, Huyton Industrial Estate, Huyton, Merseyside, L36 6HY
BDH CHEMICALS LTD.
Broom Rd., Poole Dorset, HB12 4NN
BAIRD & TATLOCK (LONDON) LTD.
P.O. Box 1, Romford, Essex, RM1 1HA
BECKMAN-RIIC LTD. ANALYTICAL INSTRUMENTS SALES AND SERVICE OPERATION
Cressex Industrial Estate, Turnpike Road, High Wycombe, Bucks
BECTON DICKINSON UK LTD.
Between Towns Rd., Cowley, Oxford OX4 3LY
BENNET & CO.
Unit 1A Taskers, Anna Valley, Andover SP11 7NF, Kent, England
BETHESDA RESEARCH LABORATORIES GmbH (BRL)
Offenbacher Strasse 113, D-6078 Neu-Isenburg 1, Federal Republic of Germany
BETHESDA RESEARCH LABORATORIES
P.O. Box 35, Paisley PA3 4EP

BIONIQUE (*See* NORTHUMBRIA BIOLOGICALS)
BIO-RAD LABORATORIES LTD.
27 Homesdale Road, Bromley, Kent
BIOTEC INSTRUMENTS, LTD.
Unit 1, Caxton Hill Extn. Rd., Caxton Hill, Hertford, SG13 7LS
BOC LTD., SPECIAL GASES
Deer Park Road, London, SW19 3UF
BOEHRINGER CORPORATION (LONDON) LTD., THE
Bell Lane, Lewes, East Sussex, BN7 1LG
BOEHRINGER MANNHEIM GmbH BIOCHEMICA
Postfach 310120, 6800 Mannheim 31, Federal Republic of Germany
BOOTS PHARMACEUTICALS
No. Higham, Nottingham
BOYDEN DATA PAPERS, LTD.
Parkhouse Street, Camberwell, London
BRITISH AMERICAN OPTICAL CO., LTD. (INSTRUMENT GROUP), REICHERT JUNG UK
820 Yeovil Road, Slough, Berks.
ALBERT BROWNE LTD.
Chancery Street, Leicester, LE1 5NA
BURGE (JOHN) (EQUIPMENT) LTD.
35 Furze Platt Road, Maidenhead, Berks.
CALBIOCHEM LTD.
79/81 South Street, Bishops, Stortford, Herts., CM23 3AL
CAMBRIDGE MEDICAL INSTRUMENTS LTD.
Rustat Road, Cambridge, CB1 3QH
CAMLAB LTD.
Nutfield Road, Cambridge CB4 1TH
CIBA
Horsham, Sussex
COMARK ELECTRONICS LTD.
Brookside Avenue, Rustington, West Sussex
CORNING LTD.
Stone, Staffordshire ST15 0BG

COSTAR, NORTHUMBRIA BIOLOGICALS LTD.
South Nelson Industrial Estates, Gramlington,
Northumberland, NE23 9HL
COULTER ELECTRONICS LTD.
Northwell Drive, Luton, Beds, LU3 3RH
COW PROOFINGS LTD.
Slough, Berkshire
CRYODIFFUSION
49 Ruede Verdun, F27690 Lery, France
CRYOSERVICE LTD.
Platts Common Industrial Estate, Hawshaw Lane, Hoyland,
Barnsley, Yorkshire
DAKO LTD.
22 The Arcade, The Octagon, High Wycombe, Bucks, HP11
2HT
DAMON/IEC (U.K.) LTD.
Unit 7, Lawrence Way, Brewers Hill Road, Dustable, Beds,
LU6 1BD
DECON LABORATORIES LTD.
Conway Street, Hove, E. Sussex BN3 3LY
DENLEY INSTRUMENTS LTD.
Daux Road, Billinghurst, Sussex, RH14 9SJ
DIFCO LABORATORIES
P.O. Box 14B, Central Avenue, East Molesey, KT8 0SE
DISTILLERS CO. (CARBON DIOXIDE) LTD., THE
Cedar House, 39 London Road, Reigate, RH2 9QE
DIVERSEY LTD.
Weston Favell Centre, Northhampton, NN3 4PD
DOW CHEMICAL CO.
Heathrow House, Bath Road, Hounslow, TW5 9QQY
DU PONT UK LTD.
2 New Road, Southampton, SO2 0AA
ELGA GROUP, THE
Lane End, High Wycombe, Bucks
ENVAIR LTD.
York Ave., Haslingden, Rossendale, Lancashire, BB4 4HX
ESCO (RUBBER) LTD.
43-45 Broad Street, Teddington, Middlesex, TW11 8QZ
EUROPEAN BUSINESS ASSOCIATES (eba)
69, rue de la Pétrusse, L-8084 Betrange, Luxembourg
EUROPEAN COLLECTION OF ANIMAL CELL CULTURES
PHLS CAMR, Porton Down, Salisbury SP4 0JG
EXPANDED METAL CO.
Hartlepool, England
FISONS SCIENTIFIC APPARATUS
Bishop Meadow Road, Loughborough, LE11 0RG
FLOW LABORATORIES LTD.
P.O. Box 17, Second Avenue Industrial Estate, Irvine,
Ayrshire, KA12 8NB, Scotland
FLOW LABORATORIES LTD.
Woodcock Hill, Harefield Rd., Rickmansworth,
Hertfordshire, WD3 1PQ
FLUKA A.G.
Buchs, Switzerland CH-9470
FLURECHEM LTD. (FLUKA agents)
Dinting Vale Trading Estate, Glassop, Derbyshire
GALLENKAMP (A.) & CO. LTD.
P.O. Box 290, Technico House, Christopher Street, London
EC2P 2ER

GENERAL DIAGNOSTICS, WILLIAM R. WARNER & CO.
LTD.
Chestnut Avenue, Eastliegh, SO5 3ZQ
GENZYME (IC CHEMIKALIEN GmbH)
Sohnckestrasse 17, 8000 Munchen 71, Federal Republic of
Germany
GIBCO EUROPE LTD.
P.O. Box 35, Trident House, Renfrew Road, Paisley, PA3
4EF
GILFORD INSTRUMENTS LTD.
46-48 Church Road, Teddington, Middlesex
GLAXO LABORATORIES LTD.
Greenford, Middlesex
GONOTEC
Gesellschaft für Mess- und Regeltechnik mbH, Ringstrasse
105, D-1000 Berlin 45, Federal Republic of Germany
GORDON-KEEBLE LTD.
Petersfield House, St. Peter's St., Duxford, Cambridge CB2
4RP
GRANT INSTRUMENTS (CAMBRIDGE) LTD.
Barrington, Cambridge CB2 5QZ
HEINICKE INSTRUMENTS
Postfach 1203, Friedrich-Ebert-Str. 10, D-8223 Trostberg/Alz,
Federal Republic of Germany
HEPAIRE MANUFACTURING LTD.
Station Road, Thatcham, Berks, RG12 4JE
W.C. HERAEUS GmbH, PEW
Postfach 1220, D-3360 Osterade am Harz, Federal Republic
of Germany
HOECHST (UK), LTD.
Hoechst House, Salisbury Road, Hounslow, Middlesex, TW4
6HJ
HOPKIN & WILLIAMS
P.O. Box 1, Romford, Essex, RM1 1HA
HORWELL (ARNOLD R.) LTD.
2 Grangeway, Kilburn High Road, London NW6 2BP
HOWORTH PARTICULATE DIVISION
Lorne Street, Farnworth, Bolton, BL4 7LZ
ILFORD LTD.
Ilford, London
JENCONS SCIENTIFIC LTD.
Mark Road, Hemel Hempstead, Herts.
JOBLING (GLASSWARE)
Stone, Staffs.
JUMO INSTRUMENT CO.
Hysol, Harlow, CM18 60Z
K C BIOLOGICALS EUROPE
11 Chemin de Ronde, 78110 Le Vesinet, France
KODAK LTD.
P.O. Box 66, Kodak House, Station Road, Hemel Hempstead,
Herts., HP1 1JU
KOR BIOCHEMCIALS
69, rue de la Pétrusse, L-8084 Betrange, CD Luxemburg
L.I.P. (EQUIPMENT & SERVICES) LTD.
111 Dockfield Road, Shipley, West Yorkshire, BD17 7AS
LAB-EQUIP
42 Babbacombe Road, Coventry, CV3 5PD
LABORATORY IMPEX LTD.
Lion Road, Twickenham, Middx.

L'AIRE LIQUIDE (U.K.) LTD.
   44 Hertford St., London W1Y 7TF
LAMB (RAYMOND A.)
   6 Sunbeam Road, London NW10 6JL
LEEC
   Private Road 7, Colwick, Nottingham, NG4 2AJ
LEITZ (E.) (INSTRUMENTS) LTD.
   Luton, Beds.
L H FERMENTATION
   Bells Hill, Stoke Poges, Slough, Bucks, SL2 4EJ
LKB INSTRUMENTS LTD.
   LKB House, 232 Addington Road, Selsdon, Croydon, CR2 8YD
LKB
   LKB-Produkter AB, Box 305, S-16126 Bromma, Sweden
MEDICAL PHARMACEUTICAL DEVELOPMENTS LTD.
   Ellen Street, Portslade-by-Sea, Sussex
MERCK
   Gottingen, Federal Republic of Germany
MICROFLOW LTD.
   Fleet Mill, Minley Road, Fleet, Hampshire, GU13 8RD
MICROMEASUREMENTS
   Shirehill Industrial Estate, Shirehill, Saffron, Walden, Essex
MILLIPORE (UK) LTD.
   Millipore House, 11–15 Peterborough Rd., Harrow, Middlesex, HA1 2BR
M.I. SCIENTIFIC
   Suite 7, Second Floor, Exchange Buildings, Quayside, Newcastle-Upon-Tyne
MSE SCIENTIFIC INSTRUMENTS
   Manor Royal, Crawley, West Sussex, Crawley
NEN (DUPONT LTD.)
   Wedgewood Way, Stevenage, Herts SG1 4QN
NEW BRUNSWICK SCIENTIFIC
   26-34 Emerald Street, London WC1N 3QA
NORTHUMBRIA BIOLOGICALS
   South Nelson Industrial Estate, Cranlington, Northumberland NE23 9HL
NUNC, (see GIBCO EUROPE LTD.)
NUNC
   P.O. Box 280, Kamstrup DK-4000, Roskilde, Denmark
NYEGAARD & CO. A/S
   Oslo, Norway
ORTHO DIAGNOSTICS (DIVISION OF ORTHO PHARMACEUTICAL LTD.)
   Saunderton, High Wycombe, Bucks.
OXOID LTD.
   Wade Road, Basingstoke, Hants, RG24 OPW
PACKARD INSTRUMENT LTD.
   13-17 Church Road, Caversham, Berks., RG4 7AA
PAESEL GmbH AND CO.
   Borsigallee 6, 6000 Frankfurt/Main 63, Federal Republic of Germany
PALL PROCESS FILTRATION LTD.
   Europa House, P.O. Box 62, Portsmouth, Hants., PO1 3PD
PHARMACIA (GREAT BRITAIN) LTD.
   Prince Regent Road, Hounslow, Middx., TW3 1NE
PLANER PRODUCTS LTD.
   Windmill Road, Sunbury on Thames, Middlesex, TW16 7HD

POLAROID (U.K.) LTD.
   Ashley Road, St. Albans, AL1 5 PR
POLYSCIENCES LTD.
   24 Low Farm Place, Moulton Park, Northampton, NN3 1HY
PORTLAND PLASTICS
   Portex Ltd., The Reachfields, Hythe, Kent
RÉACTIS IBF
   Société Chimique Pointet-Girard, 35 avenue Jean-Jaurès, 92390 Villeneuve-La-Garenne, France
REICHERT JUNG, U.K., LTD.
   820 Yeovil Road, Slough, Berks.
SARTORIUS GmbH
   P.O. Box 19, 3440 Göttingen, Federal Republic of Germany
SCHERING CHEMICALS LTD.
   Pharmaceutical Division, Burgess Hill, Sussex, RH15 9NE
SEARLE SCIENTIFIC SERVICES
   Coronation Road, Cressex, High Wycombe, Bucks.
SERA LAB
   Crawley Down, Sussex RH10 4LL
SIGMA CHEMICAL CO. LTD.
   Fancy Road, Poole, Dorset, BH17 7NH
SQUIBB AND SONS LTD.
   Regal House, Twickenham, Middlesex, TW1 3QT
STERILIN LTD.
   43-45 Broad Street, Teddington, TW11 8QZ
STERILIN LTD.
   Sterilin House, Clockhouse Lane, Feltham, Middlesex, TW14 8QS
TISSUE CULTURE SERVICES LTD.
   10 Henry Road, Slough, SL1 2QI
TOWNSON & MERCER LTD.
   101 Beddington Lane, Croydon, Surrey, CR9 4EG
UNION CARBIDE U.K. LTD., CRYOGENICS DIVISION
   Redworth Way, Aycliffe Industrial Estage, Aycliffe, Co., Durham
UNISCIENCE LTD.
   Uniscience House, 8 Jesus Lane, Cambridge, CB5 8BA
UPJOHN
   Puurs, Lichtevstraat B2670, Belgium
VICKERS LTD., VICKERS INSTRUMENTS
   Haxby Road, York, YO3 7SD
VINDON SCIENTIFIC LTD.
   Ceramyl Works, Diggle, Oldham OL3 5YJ
VOLAC
   77-93 Tanner Street, Barking, Essex, IG11 8QD
WATSON-MARLOWE LTD.
   Falmouth, Cornwall, TR11 4RU
WILLIAM R. WARNER & CO.
   See GENERAL DIAGNOSTICS
ZEISS (CARL) (OBERKOCHEN) LTD.
   Degenhardt House, 31-36 Foley Street, London W1P 8AP

North America

ADVANCED INSTRUMENTS INC.
   1000 Highland Ave., Needham Heights, MA 02194
AIR PRODUCTS & CHEMICALS INC.
   P.O. Box 538, Allentown, PA 18105

ALDRICH CHEMICAL CO.
Milwaukee, WI
AMERICAN OPTICAL CORP. SCIENTIFIC INSTRUMENT DIVISION
Sugar & Eggert Roads, Buffalo, NY 14215
AMICON CORP., SCIENTIFIC SYSTEMS DIVISION
21 Hartwell Ave., Lexington, MA 02173
ARTEK SYSTEMS CORP.
170 Finn Court, Farmingdale, NY 11735
ASEPTIC-THERMO INDICATOR COMPANY
North Hollywood, CA
BAKER CO., INC.
Sanford Airport, Sanford, ME 04073
BECKMAN INSTRUMENTS, INC., ELECTRONIC INSTRUMENTS DIVISION
3900 N. River Road, Schiller Park, IL 60176
BECTON DICKINSON & CO.
Oxnard, CA 93030
BELLCO GLASS, INC.
340 Edrudo Road, Vineland, NJ 08360
BETHESDA RESEARCH LABORATORIES
411 N. Stonestreet Ave., Rockville, MD 20850
BIOFLUIDS INC.
Rockville, MD
BIOMEDICAL TECHNOLOGIES, INC.
22 Thorndike St., Cambridge, MA 02141
BIO-RAD LABORATORIES
32nd & Griffin Ave., Richmond, CA 94804
BOEHRINGER MANNHEIM BIOCHEMICALS
P.O. Box 50816, Indianapolis, IN 46250
BOLEX (USA)
250 Community Drive, Great Neck, NY 11020
BRAUN, N. AMERICA
55 Cambridge Parkway, Cambridge, MA 02142
J. BROOKS LABORATORIES
P.O. Box 37, Olivenhain, CA 92024
BUCK, H.J., CO.
10534 York Road, Cockeysville, MD 21030
CALBIOCHEM-BEHRING CORP.
American Hoechst Corp., P.O. Box 12087, San Diego, CA 92112
CEDANCO
P.O. Box 42, Wellesley, MA 02181
CHEMAP, DIV. A C BIOTECHNICS INC.
230 Crossways Park Drive, Woodbury, NY 11746
CLAY ADAMS, DIVISION BECTON DICKINSON & CO.
299 Webro Road, Parsippany, NJ 07054
COLE-PARMER INSTRUMENT CO.
7425 N. Oak Park Ave., Chicago, IL 60648
COLLABORATIVE RESEARCH, INC.
Two Oak Park, Bedford, MA 01730
COLLAGEN CORPORATION
2455 Faber Place, Palo Alto, CA 94303
CONNAUGHT MEDICAL RESEARCH LABORATORIES
Toronto, Canada
CONTAMINATION CONTROL, INC.
P.O. Box 316, Kulpsville, PA 19443
COOPER BIOMEDICAL
Malvern, PA

CORNING MEDICAL, CORNING GLASS WORKS
Medfield, MA 02052
COULTER ELECTRONICS INC.
590 W. 20 Street, Hialeah, FL 33010
CRYO-MED
49659 Leona Drive, Mt. Clements, MI 48043
DAMON/IEC
300 Second Ave., Needham Heights, MA 02194
DEXTRAN PRODUCTS LTD.
P.O. Box 1360, Princeton, NJ 08542
DIFCO LABORATORIES
P.O. Box 1058A, Detroit, MI 48232
DOW CORNING CORP.
P.O. Box 1767, Midland, MI 48640
DYNATECH LABORATORIES, INC., DYNATECH COR.
900 Slaters Lane, Alexandria, VA 22314
EASTMAN KODAK CO.
343 Slate Street, Rochester, NY 14650
FALCON PLASTICS
(See BECTON DICKINSON)
FISHER SCIENTIFIC CO.
711 Forbes Ave., Pittsburgh, PA 15219
FLOW LABORATORIES, INC.
1710 Chapman Ave., Rockville, MD 20852
FMC CORP.
Rockland, ME
FORMA SCIENTIFIC
P.O. Box 649, Marietta, OH 45750
FUJI PHOTO FILM (U.S.A.)
350 Fifth Ave., New York, NY 10001
GELMAN SCIENCES, INC.
600 S. Wagner Road, Ann Arbor, MI 48106
GERMFREE LABORATORIES, INC.
2600 S.W. 28th Lane, Miami, FL 33133
GIBCO LABORATORIES, GRAND ISLAND BIOLOGICAL CO.
3175 Staley Road, Grand Island, NY 14072
GILFORD INSTRUMENT LABORATORIES, INC.
132 Artino Street, Oberlin, OH 44074
GILSON MEDICAL ELECTRONICS, INC.
P.O. Box 27, Middleton, WI 53562
HANA MEDIA, INC. (HANA BIOLOGICS)
629 Bancroft Way, Berkeley, CA 94710
HEINICKE INSTRUMENTS CO.
3000 Taft St., Hollywood, FL 33021
HELENA LABS
P.O. Box 752, 1530 Lindberg Drive, Beaumont, TX 77704
ILFORD
West 70 Century Road, Paramus, NJ 07652
INTERNATIONAL EQUIPMENT CO., (DAMON)
300, 2nd Ave., Needham Has, MA 02194
KAHLENBERG-GLOBE EQUIPMENT CO.
Sarasota, FL
K C BIOLOGICALS (CORNING GLASS WORKS)
P.O. Box 14848, Lenexa, KS 66215
KELVINATOR COMMERCIAL INC.
Wisconsin
LAB-TEK DIVISION, MILES LABORATORIES
475 North Aurora Rd., Naperville, IL 60540

LKB INSTRUMENTS INC.
12221 Parklawn Drive, Rockville, MD 20852
LAB-LINE INSTRUMENTS, INC.
15 & Bloomingdale Ave., Melrose Park, IL 60160
LEITZ, E., INC.
Link Drive, Rockleigh, NJ 07647
LUX SCIENTIFIC CORP.
1157 Tourmaline Drive, Newbury Park, CA 91320
L & W PHOTOOPTICAL LTD.
Van Nuys, CA
M.A. BIOPRODUCTS
East Coast, Building 100, Biggs Ford Rd., Walkersville, MD 21793, West Coast, 11841 Mississippi Ave., Los Angeles, CA 90025
MERCK CHEMICAL DIVISION, MERCK & CO., INC.
P.O. Box 2000, Rahway, NJ 07065
MICROBIOLOGICAL ASSOCIATES
5221 River Road, Bethesda, MD 20016
MILES SCIENTIFIC
Naperville, IL
MILLIPORE CORP.
Ashby Road, Bedford, MA 01730
MINNESOTA VALLEY ENGINEERING
407 7th Street, N.W., New Prague, MN 56071
MORGAN SHEET METAL CO.
Sarasota, FL
NAPCO
10855 S.W. Greenburg Rd., Portland, OR 97223
NEW BRUNSWICK SCIENTIFIC CO., INC.
44 Talmadge Road, Edison, NJ 08817
NEW ENGLAND NUCLEAR (NEN)
549 Albany Street, Boston, MA 02118
NIKON, NIPPON KOGAKU K.K.
Fuji Bldg., 2-3 Marunouchi 3-chome, Chiyoda-Ku, Tokyo, Japan
NUCLEPORE CORP.
7035 Commerce Circle, Pleasanton, CA 94566
OLYMPUS CORPORATION OF AMERICA
4 Nevada Drive, New Hyde Park, NY 11042
ORTHO INSTRUMENTS
376 University Ave., Westwood, MA 02090
OXOID U.S.A. INC.
9017 Red Branch Road, Columbia, MD 21045
PACKARD INSTRUMENT CO., INC.
2200 Warrenville Road, Downers Grove, IL 60515
PALL CORP.
30 Sea Cliff Ave., Glen Cove, NY 11542
PARTICLE DATA, INC.
Box 265 (111 Hahn), Elmhurst, IL 60126

PHARMACIA FINE CHEMICALS, DIVISION OF PHARMACIA, INC.
800 Centennial Ave., Piscataway, NJ 08854
PIERCE CHEMICALS
P.O. Box 117, Rockford, IL 611050
POLAROID CORP.
549 Technology Square, Cambridge, MA 02139
PRECISION SCIENTIFIC CO.
3737 West Cortland St., Chicago, IL 60647
PROPPER MANUFACTURING CO., INC.
Long Island City, NY 11101
RAININ INSTRUMENT CO., INC.
Woburn, MA 01801
REVCO INC.
Aiken Road, Route 1, Box 275, Asheville, NC 28804
SAGE INSTRUMENTS
DIVISION OF ORION INDUSTRIES,
Cambridge, MA
SCHERING CORPORATION
Kenilworth, NJ 07033
SERAGEN, INC.
54 Clayton St., Boston, MA 021122
SCHWARTZ MANN, (*See* BECTON DICKINSON)
SHANDON SOUTHERN INSTRUMENTS, INC.
515 Broad Street, Sewickley, PA 15143
SIGMA CHEMICAL CO.
P.O. Box 14508, St. Louis, MO 63178
TEKMAR CO.
P.O. Box 37202, Cincinnati, OH 45222
UNION CARBIDE CORP.
Linde Division, P.O. Box 372, South Plainfield, NJ 07080
U.S. BIOCHEMICAL CORP.
Cleveland, OH
USHER (DEXTRAN PRODUCTS LTD.)
421 Comstock Road, Scarborough, Ontario, Canada, MIL 2H5
VALLEY FORGE INSTRUMENT CO., INC.
55 Buckwalter Road, Phoenixville, PA 19460
VENTREX
217 Read St., Portland, ME 04103
VERNITRON/BETTERBUILT
S. Empire Blvd., Carlstadt, NJ 07072
VICKERS INSTRUMENTS, INC.
300 Commercial Street, Malden, MA 02148
WHEATON SCIENTIFIC
1000 N. Tenth Street, Millville, NJ 08332
ZEISS, CARL, INC.
444 Fifth Ave., New York, NY 10018

SCIENTIFIC SOCIETIES WITH INTERESTS IN TISSUE CULTURE

AMERICAN TISSUE ASSOCIATION
Executive Director: William G. Momberger,
19110 Montgomery Village Avenue, Suite 300,
Gaithersburg, MD 20879

EUROPEAN TISSUE CULTURE SOCIETY (ETCS)
Secretary: Caroline Wigley, Department of Anatomy, Guy's
Hospital London. ETCS can also supply information regarding
National Tissue Culture Societies in Europe.

AMERICAN SOCIETY FOR CELL BIOLOGY
Executive Officer: Richard S. Young, 9650 Rockville Pike,
Bethesda, MD 20814

EUROPEAN CELL BIOLOGY ORGANISATION
(Federation: Membership via National Cell Biology Societies
in Europe) Secretary General: Dr. Michael Balls, Department
of Human Morphology, University of Nottingham.

EUROPEAN SOCIETY FOR ANIMAL CELL CULTURE
TECHNOLOGY (ESACT)
Secretary: Dr. Bryan Griffiths, C.A.M.R. Porton, Porton
Down, Salisbury, Wiltshire.

BRITISH SOCIETY FOR CELL BIOLOGY
Secretary: Nancy Lane, Department of Zoology, University of
Oxford, Oxford, England

INTERNATIONAL ASSOCIATION FOR CELL CULTURE
Secretary: Dr. Richard Ham, M.C.D.B., University of
Colorado, Boulder, CO 80302

LATE ADDITIONS AND ALTERATIONS TO TRADE INDEX

CLONETICS CORPORATION
1800 30th St., Boulder, CO 80301

Keratinocyte growth medium, keratinocyte cultures, growth
factors

J.R. SCIENTIFIC
P.O. Box 1862, Woodland, CA 95695

Growth factors — bFGF, PDGF, fibronectin

IMCERA BIOPRODUCTS, INC.
P.O. Box 207, Terre Haute, IN 47808

Growth factors — IGF-1 (Samatomedin C)

HAZELTON RESEARCH PRODUCE, INC.
P.O. Box 14848, Lenexa, KS 66215

Serum substitute

GEN-PROBE
9880 Campus Point Drive, San Diego, CA 92121

Mycoplasma detection by DNA probe (Available through Fisher
Scientific)

NALGENE LABWARE DEPT., NALGE CO.
P.O. Box 20365, Rochester, NY 14602

Plasticware, filters, cryogenic storage systems

BIONIQUE LABORATORIES, INC.
Bloomingdale Rd., Saranac Lake, NY 12983

Culture chambers for cytology, mycoplasma testing

FENWAL LABORATORIES
Division of Travenol Labs., Inc., Deerfield, IL 60015

Plastic culture bags

NOVO BIOLABS, NOVO INDUSTRI A/S NOVO ALLÉ
DK-2880 Bagsvaerd, Denmark

Monoclonal antibodies, diagnostic kits

CD MEDICAL, INC.
P.O. Box 9308, Miami Lakes, FL 33014

Perfused capillary culture

EARL CLAY LABORATORIES, INC.
890 Lamont Ave., Novato, CA 94947

Filter wells

DU PONT COMPANY, BIOTECHNOLOGY SYSTEMS
Barley Mill Plaza, P-24, Wilmington, DE 19898

Culture media, serum substitute, mycoplasma test kit, culture
bags

COLLABORATIVE RESEARCH INC.
Two Oak Park, Bedford, MA 01730

Growth factors, attachment factors, intracellular matrix
(Matrigel), serum substitute

HYCLONE LABORATORIES, INC.
1725 So. State Hwy. 89–91, Logan, UT 84321

Media, sera

INTERLAB; INTERMED
187 East Wilbur Rd., Thousand Oaks, CA 91360

Nunc tissue culture plastics

INVITRON CORPORATION
4649 Le Bourget Drive, St. Louis, MO 63134

Large scale culture systems

Glossary*

Adaptation. Induction or repression of synthesis of a macromolecule (usually a protein) in response to a stimulus, e.g., *enzyme* adaptation—an alteration in enzyme activity brought about by an inducer or repressor and involving an altered rate of enzyme synthesis or degradation.

Allograft. See homograft.

Amniocentesis. Prenatal sampling of the amniotic cavity.

Anchorage-dependent. Requiring attachment to a solid substrate for survival or growth.

Anemometer. An instrument for measuring air flow rate.

Aneuploid. Not an exact multiple of the haploid chromosome number (haploid = that number present in germ cells after meiosis; i.e., each chromosome represented once).

Aseptic. Free of microbial infection.

Balanced salt solution. An isotonic solution of inorganic salts present in approximately the correct physiological concentrations. May also contain glucose but usually free of other organic nutrients.

Cell culture. Growth of cells dissociated from the parent tissue by spontaneous migration or mechanical or enzymatic dispersal.

Cell fusion. Formation of single cell body by fusion of two other cells; either spontaneously or, more often, by induced fusion with inactivated Sendai virus or polyethylene glycol.

Cell hybridization. See *Hybrid cell.*

Cell line. A propagated culture after the first subculture.

Cell strain. A characterized cell line derived by selection or cloning.

Chemically defined. Used of medium to imply that it is made entirely from pure defined constituents. Distinct from "serum-free" where other poorly characterized constituents may be used to replace serum.

*[See also Schaeffer, 1984]

Clone. A population of cells derived from one cell.

Confluent. Where all the cells are in contact all round their periphery with other cells, and no available substrate is left uncovered.

Contact inhibition. Inhibition of cell membrane ruffling and cell motility when cells are in complete contact with other adjacent cells, as in a confluent culture. Often precedes cessation of cell proliferation but not necessarily causally related.

Continuous cell line or cell strain. One having the capacity for infinite survival. Previously known as "established" and often referred to as "immortal."

Cyclic growth. Growth from a low cell density to a high cell density with a regular subculture interval. A regular repetition of the growth cycle for maintenance purposes.

Deadaptation. Reversible loss of a specific property due to the absence of the appropriate inducer (not always defined).

Dedifferentiation. A term implying irreversible loss of the specialized properties that a cell would have expressed in *vivo*. As evidence accumulates that cultures "dedifferentiate" by a combination of selection of undifferentiated cells or stromal cells and deadaptation resulting from the absence of the appropriate inducers, this term is going out of favor. It is still correctly applied to progressive loss of differentiated morphology in histological observations of, for example, tumor tissue.

Density limitation of growth. Mitotic inhibition correlated with an increase in cell density.

Diploid. Each chromosome represented as a pair, identical in the autosomes and female sex chromosomes and non-identical in male sex chromosomes, and corresponding to the chromosome number and morphology of most somatic cells of the species from which the cells were derived.

Ectoderm. The outer germ layer of the embryo giving rise to the epithelium of the skin.

Endoderm. The innermost germ layer of the embryo giving rise to the epithelial component of organs such as the gut, liver, and lungs.

Endothelium. An epithelial-like cell layer lining spaces within mesodermally derived tissues, such as blood vessels, and derived from the mesoderm of the embryo.

Epithelial. Used of a culture to imply cells derived from epithelium but often used more loosely to describe any cells of a polygonal shape with clear sharp boundaries between cells. *Pavement-like.* More correctly this should be termed "epithelioid" or "epithelial-like."

Epithelium. A covering or lining of cells, as in the surface of the skin or lining of the gut, usually derived from the embryonic endoderm or ectoderm but exceptionally derived from mesoderm as with kidney tubules and mesothelium lining body cavities.

Euploid. Exact multiples of the haploid chromosome set. The correct morphology characteristic of each chromosome pair in the species from which the cells were derived is not implicit in the definition but is usually assumed to be the case. Otherwise it should be stated as "euploid but with some chromosomal aberrations."

Explant. A fragment of tissue transplanted from its original site and maintained in an artificial medium.

Fibroblast. A proliferating precursor cell of the mature differentiated fibrocyte.

Fibroblastic. Resembling fibroblasts, i.e., spindle shaped (bipolar) or stellate (multipolar); usually arranged in parallel arrays at confluence if contact inhibited. Often used indiscriminately for undifferentiated mesodermal cells regardless of their relationship to the fibrocyte lineage. Implies a migratory type of cell with processes exceeding the nuclear diameter by threefold or more.

Finite cell line. A culture which has been propogated by subculture but is only capable of a limited number of cell generations *in vitro* before dying out.

Generation number. The number of population doublings (estimated from dilution at subculture) that a culture has undergone since explantation. Necessarily contains an approximation of the number of generations in primary culture.

Generation time. The interval from one point in the cell division cycle to the same point in the cycle, one division later. Distinct from doubling time or population doubling time which is derived from the total cell count of a population and therefore averages different generation times including the effect of non-growing cells.

Genotype. The total genetic characteristics of a cell.

Glycocalyx. Glycosylated peptides, proteins, and lipids, and glycosaminoglycans attached to the surface of the cell.

Growth curve. A semi-log plot of cell number on a log scale against time on a linear scale in a proliferating cell culture. Usually divided into *lag phase,* before growth is initiated, *log phase,* the period of exponential growth, and *plateau,* a stable cell count achieved when the culture stops growing at a high cell density.

Growth cycle. Growth interval from subculture to the top of the log phase, ready for a further subculture.

Haploid. That chromosome number where each chromosome is represented once. In most higher animals it is the number present in the gametes and half of the number found in most somatic cells.

Heterokaryon. Genetically different nuclei in a common cytoplasm, usually derived by cell fusion.

Heteroploid. A term used to describe a culture (not a cell) where the cells comprising the culture have chromosome numbers other than diploid.

Histotypic. A culture resembling tissue-like morphology *in vivo.* It is usually implied that this is a three-dimensional culture recreated from dispersed cell culture which attempts to retain, by cell proliferation and multilayering or by reaggregation, the tissue-like structure. Organ cultures cannot be propagated whereas histotypic cultures can.

Homiothermic. Able to maintain a constant body temperature in spite of environmental fluctuation.

Homograft. (Allograft). A graft derived from a genetically different donor of the same species as the recipient.

Homokaryon. Genetically identical nuclei in a common cytoplasm, usually a product of cell fusion.

Hybrid cell. Mononucleate cell which results from the fusion of two different cells, leading to the formation of a synkaryon.

Ideogram. The arrangement of (in the case of genetic analysis of a cell) the chromosomes in order by size and morphology so that the karyotype may be studied.

Induction. An increase in effect produced by a given stimulus. *Embryonic induction.* The interaction of cells from two different germ layers, promoting differentiation, often reciprocal. *Enzyme induction.* An increase in synthesis of an enzyme produced by, for example, hormonal stimulation.

Isograft. (Syngraft). A graft derived from a genetically identical or nearly identical donor of the same species as the recipient.

Karyotype. The distinctive chromosomal complement of a cell.

Laminar flow. The flow of a fluid that closely follows the shape of a streamlined surface without turbulence. Used in connection with laminar air flow cabinets to imply a stable flow of air over the work area such as to minimize turbulence.

Laminar flow cabinet or hood. A work station with filtered air flowing in a laminar nonturbulent flow parallel to (horizontal laminar flow) or perpendicular to (vertical laminar flow) the work surface, such as to maintain the sterility of the work.

Log phase. See *Growth curve*.

Malignant. A term to describe a tumor which has become invasive or metastatic (i.e., colonizing other tissues). Usually progressive leading to destruction of host cells and ultimately death of the host.

Malignant transformation. The development of the ability to invade normal tissue without regulation in space or time. May also lead to metastatic growth (colonization of a distant site with subsequent unregulated invasive growth).

Manometer. A "U" shaped tube containing liquid, the levels of which in each limb of the "U" reflect the pressure difference between the ends.

Medium. A mixture of inorganic salts and other nutrients capable of sustaining cell survival *in vitro* for 24 hours. *Growth medium*. That medium which is used in routine culture such that the cell number increases with time. *Maintenance medium*. A medium which will retain cell survival without growth (cell proliferation), e.g., a low serum or serum-free medium used with serum dependent cells to maintain cell survival without cell proliferation.

Mesenchyme. Loose, often migratory, embryonic tissue derived from the mesoderm, giving rise to connective tissue, cartilage, muscle, hemopoietic cells, etc. in the adult.

Mesoderm. A germ layer in the embryo arising between the ectoderm and endoderm and giving rise to connective tissue etc. (as above for mesenchyme).

Monoclonal. Derived from a single clone of cells. *Monoclonal antibody*. Antibody produced by a clone of lymphoid cells either *in vivo* or *in vitro*. *In vitro* the clone is usually derived from a hybrid of a sensitized spleen cell and continuously growing myeloma cell.

Morphogenesis. The development of form and structure of an organism.

Myeloma. A tumor derived from myeloid cells. Used in monoclonal antibody production when the myeloma cell can produce immunoglobulin.

Neoplastic. A new, unnecessary, proliferation of cells giving rise to a tumor.

Neoplastic transformation. The conversion of a non-tumorigenic cell into a tumorigenic cell.

Organ culture. The maintenance or growth of organ primordia or the whole or parts of an organ *in vitro* in a way that may allow differentiation and preservation of the architecture and/or function.

Organogenesis. The development of organs.

Organotypic. Histotypic culture involving more than one cell type to create a model of the cellular interactions characteristic of the organ *in vivo*. A reconstruction from dissociated cells or fragments of tissue is implied as distinct from *organ culture* where the structural integrity of the explanted tissue is retained.

Passage. The transfer or subculture of cells from one culture vessel to another. Usually, but not necessarily, implies subdivision of a proliferating cell population enabling propagation of a cell line or cell strain. *Passage number*. The number of times a culture has been subcultured.

Phenotype. The aggregate of all the expressed properties of a cell, being the product of the interaction of the genotype with the regulatory environment.

Plating efficiency. The percentage of cells seeded at subculture giving rise to colonies. If each colony can be said to be derived from one cell this is synonomous with cloning efficiency. Sometimes used loosely to describe the number of cells surviving after subculture but this is better termed the "seeding efficiency."

Plateau. See *Growth Curve*.

Poikilothermic. Body temperature close to that of the environment and not regulated by metabolism.

Population density. The number of monolayer cells per unit area of substrate. For cells growing in suspension this term is identical to the cell concentration.

Population doubling time. The interval required for a cell population to double at the middle of the logarithmic phase of growth.

Primary culture. A culture started from cells, tissue, or organs taken directly from an organism and before the first subculture.

Pseudodiploid. Numerically diploid chromosome number but with chromosomal aberrations.

Quasidiploid. See pseudodiploid.

Saturation density. Maximum number of cells attainable per cm^2 (monolayer culture) or per ml (suspension culture) under specified culture conditions.

Seeding efficiency. The percentage of the inoculum which attaches to the substrate within a stated period of time (implying viability, or survival but not necessarily proliferative capacity).

Somatic cell genetics. The study of cell genetics by recombination and segregation of genes in somatic cells. Usually by cell fusion.

Split Ratio. The divisor of the dilution ratio of a cell culture at subculture, e.g., one flask divided into four or 100 ml up to 400 ml would be a split ratio of 4.

Subconfluent. Less than confluent. All of the available substrate is not covered.

Subculture. See *Passage.*

Substrate. The matrix or solid underlay upon which a monolayer culture grows.

Superconfluent. When a monolayer culture progresses beyond the state where all the cells are attached to the substrate and multilayering occurs.

Suspension culture. Where cells will multiply suspended in medium.

Synkaryon. A hybrid cell which results from the fusion of the nuclei it carries.

Tetraploid. Twice the diploid (four times the haploid) number of chromosomes.

Tissue culture. Properly, the maintenance of fragments of tissue *in vitro* but now commonly applied as a generic term to include tissue explant culture, organ culture, and dispersed cell culture, including the culture of propagated cell lines and cell strains.

Transformation. A permanent alteration of the cell phenotype presumed to occur via an irreversible genetic change. May be *spontaneous* as in the development of rapidly growing continuous cell lines from slow-growing early passage rodent cell lines or induced by chemical or viral action. Usually produces cell lines which have an increased growth rate, an infinite life span, a higher plating efficiency, and are often (but not necessarily) tumorigenic.

Transfection. The transfer, by artificial means, of genetic material from one cell to another. Implies transfer of less than the whole nucleus of the donor cell and is usually achieved by using isolated chromosomes, DNA, or cloned genes.

Viral transformation. A permanent phenotypic change induced by the genetic and heritable effects of a transforming virus.

Variant. A cell line expressing a stable phenotype which is different from the parental culture from which it was derived.

Xenograft. Transplantation of tissue to a different species from which it was derived. Often used to describe implantation of human tumors in athymic (nude), immune deprived, or immune suppressed mice.

References

Aaronson, S.A., Todaro, G.J., Freeman, A.E. (1970) Human sarcoma cells in culture, identification by colony-forming ability on monolayers of normal cells. Exp. Cell Res. *61*:1–5.

Abercrombie, M., Heaysman, J.E.M. (1954) Observations on the social behaviour of cells in tissue culture, II. "Monolayering" of fibroblasts. Exp. Cell Res. *6*:293–306.

Adams, D.O. (1979) Macrophages. In Jakoby, W.B., Pastan, I.H. (eds): "Methods of Enzymology, Vol. LVII, Cell Culture." New York, Academic Press, pp. 494–506.

Adams, R.L.P. (1980) In Work, T.S., Burdon, R.H. (eds): "Laboratory Techniques in Biochemistry and Molecular Biology. Cell Culture for Biochemists." Amsterdam, Elsevier/North Holland Biomedical Press.

Albrecht, A.M., Biedler, J.L., Hutchison, D.J. (1972) Two different species of dihydrofolate reductase in mammalian cells differentially resistant to amethopterin and methasquin. Cancer Res. *32*:1539–1546.

Allen, R.D., Allen, N.B. (1983) Video-enhanced microscopy with a computer frame memory. J. Microsc. *129*.

Ambrose, E.J., Dudgeon, J.A., Easty, D.M., Easty, G.C. (1961) The inhibition of tumor growth by enzymes in tissue culture. Exp. Cell Res. *24*:220–227.

Ames, B.N. (1980) Identifying environmental chemicals causing mutations and cancer. Science *204*:587–593.

Andersson, L.C., Nilsson, K. Gahmberg, C.G. (1979a) K562—a human erythroleukemic cell line. Int. J. Cancer *23*:143–147.

Andersson, L.C., Jokinen, M., Klein, G., Nilsson, K. (1979b) Presence of erythrocytic components in the K562 cell line. Int. J. Cancer *24*:514.

Andersson, L.C., Jokinen, M., Gahmberg, C.G. (1979c) Induction of erythroid differentiation in the human leukaemia cell line K562. Nature *278*:364–365.

Antoniades, H.N., Scher, C.D., Stiles, C.D. (1979) Purification of human platelet-derived growth factor. Proc. Natl. Acad. Sci. USA *76*:1809.

Arrighi, F.E., Hsu, T.C. (1974) Staining constitutive heterochromatin and Giemsa crossbands of mammalian chromosomes. In Yunis, J. (ed): "Human Chromosome Methodology, 2nd Ed." New York, Academic Press.

Au, A.M.-J., Varon, S. (1979) Neural cell sequestration on immunoaffinity columns. Exp. Cell Res. *120*:269.

Aub, J.C., Tieslau, C., Lankester, A. (1963) Reactions of normal and tumor cell surfaces to enzymes, I. Wheat-germ lipase and associated mucopolysaccharides. Proc. Natl. Acad. Sci. USA *50*:613–619.

Auerbach, R., Grobstein, C. (1958) Inductive interaction of embryonic tissues after dissociation and reaggregation. Exp. Cell Res. *15*:384–397.

Augeron, C., Laboisse, C.L. (1984) Emergence of permanently differentiated cell clones in a human colonic cancer cell line in culture after treatment with sodium butyrate. Cancer Res. *44*:3961–3969.

Augusti-Tocco, G., Sato, G. (1969) Establishment of functional clonal lines of neurons from mouse neuroblastoma. Proc. Natl. Acad. Sci. USA *64*:311–315.

Avrameas, S. (1970) Immunoenzyme techniques: Enzymes as markers for the localization of antigens and antibodies. In Bourne, G.H., Danielli, J.F. (eds): "International Review of Cytology." New York, Academic Press, pp. 349–385.

Balin, A.K., Goodman, B.P., Rasmussen, H., Cristofalo, V.J. (1976) The effect of oxygen tension on the growth and metabolism of WI-38 cells. J. Cell. Physiol. *89*:235–250.

Ballard, P.L. (1979) Glucocorticoids and differentiation. Glucocorticoid Horm. Action *12*:439–517.

Ballard, P.L., Tomkins, G.M. (1969) Dexamethasone and cell adhesion. Nature *244*:344–345.

Bard, D.R., Dickens, M.J., Smith, Audrey U., Sarek, J.M. (1972) Isolation of living cells from mature mammalian bone. Nature *236*:314–315.

Barde, Y.A., Lindsay, R.M., Monard, D., Thoenen, H. (1978) New factor released by cultured cells supporting survival and growth of sensory neurones. Nature *274*:818.

Barkley, W.E. (1979) Safety considerations in the cell culture laboratory. In Jacoby, W.B., Pastan, I. (eds.): "Methods of Enzymology, LVIII." Chap. 4. New York, Academic Press, pp. 36–43.

Barnes, D., Sato, G. (1980) Methods for growth of cultured cells in serum-free medium. Anal. Biochem. *102*:255–270.

Barnes, W.D. Sirbasku, D.A. Sato, G.H., eds. (1984a) "Cell Culture Methods for Molecular and Cell Biology, Vol. 1. Methods for Preparation of Media, Supplements, and Substrata for Serum-Free Animal Cell Culture." N.Y., Alan R. Liss Inc.

Barnes, W.D., Sirbasku, D.A., Sato, G.H., eds. (1984b) "Cell Culture Methods for Molecular and Cell Biology, Vol. 2, Methods for Serum-Free Culture of Cells of the Endocrine System." N.Y., Alan R. Liss Inc.

Barnes, W.D., Sirbasku, D.A. Sato, G.H., eds. (1984c) "Cell Culture Methods for Molecular and Cell Biology, Vol.3. Methods for Serum-Free Culture of Epithelial and Fibroblastic Cells." N.Y., Alan R. Liss Inc.

Barnes, W.D. Sirbasku, D.A., Sato, G.H., eds. (1984d) "Cell Culture Methods for Molecular and Cell Biology, Vol. 4, Methods for Serum-Free Culture of Neuronal and Lymphoid Cells." N.Y., Alan R. Liss Inc.

Barnstable, C. (1980) Monoclonal antibodies which recognize different cell types in the rat retina. Nature *286*:231–234.

Barski, G., Sorieul, S., Cornefert, F. (1960) Production dans les cultures *in vitro* de deux souches cellulaires en association de cellules de caractère "hybride". C.R. Acad Sci. [D] Paris *251*:1825.

Baserga, R. (1962) A study of nucleic acid synthesis in ascites tumor cells by two-emulsion autoradiography. J. Cell Biol. *12*:633–637.

Bateman, A.E., Peckham, M.J., Steel, G.G. (1979) Assays of drug sensitivity for cells from human tumours: *In vitro* and *in vivo* tests on a xenografted tumour. Br. J. Cancer *40*:81.

Bell, E., Ivarsson, B., Merrill, C. (1979) Production of a tissue-like structure by contraction of collagen lattices by human fibroblasts of different proliferative potential *in vitro*. Proc. Natl. Acad. Sci. USA *76:*1274–1279.

Benda, P., Lightbody, J., Sato, G., Levine, L., Sweet, W. (1968) Differentiated rat glial cell strain in tissue culture. Science *161:*370.

Benya, P.D., Padilla, S.R., Nimni, M.E. (1978) Independent regulation of collagen types by chondrocytes during the loss of differentiated function in culture. Cell *15:*1313–1321.

Berger, S.L. (1979) Lymphocytes as resting cells. In Jakoby, W.B., Pastan, I.H. (eds): "Methods in Enzymology, Vol. LVII, Cell Culture." New York, Academic Press, pp. 486–494.

Bergerat, J.P., Barlogie, B., Drewinko, B. (1979) Effects of cisdichloro-diammineplatinum (II) on human colon carcinoma cells *in vitro*. Cancer Res. *39:*1334.

Berky, J.J., Sherrod, P.C., eds (1977) "Short Term *in vitro* Testing for Carcinogenesis, Mutagenesis and Toxicity." Philadelphia, The Franklin Institute Press.

Berry, M.N., Friend, D.S. (1969) High yield preparation of isolated rat liver parenchymal cells. A biochemical and fine structural study. J. Cell Biol. *43:*506–520.

Berry, R.J., Laing, A.H., Wells, J. (1975) Fresh explant cultures of human tumours *in vitro* and the assessment of sensitivity to cytotoxic chemotherapy. Br. J. Cancer *31:*218–227.

Biedler, J.L. (1976) Chromosome abnormalities in human tumour cells in culture. In Fogh, J. (ed): "Human Tumour Cells in Vitro." New York, Academic Press.

Biedler, J.L., Albrecht, A.M., Hutchinson, D.J., Spengler, B.A. (1972). Drug response dihydrofolate reductase, and cytogenetics of amethopterin-resistant Chinese hamster cells *in vitro*. Cancer Res. *32:*151–161.

Biggers, J.D., Gwatkin, R.B.C., Heyner, S. (1961) Growth of embryonic avian and mammalian tibiae on a relatively simple chemically defined medium. Exp. Cell Res. *25:*41.

Bignami, A., Dahl, D., Rueger, D.G. (1980) Glial fibrillary acidic GFA protein in normal neural cells and in pathological conditions. In Federoff, S., Hertz, L. (eds): "Advances in Cellular Neurobiology, Vol.I." New York, Academic Press.

Birch, J.R., Pirt, S.J. (1970) Improvements in a chemically-defined medium for the growth of mouse cells (strain LS) in suspension. J. Cell Sci. *7:*661–670.

Birch, J.R., Pirt, S.J. (1971) The quantitative glucose and mineral nutrient requirements of mouse LS (suspension) cells in chemically-defined medium. J. Cell Sci. *8:*693–700.

Birnie, G.D., Simons, P.J. (1967) The incorporation of ^{3}H-thymidine and ^{3}H-uridine into chick and mouse embryo cells cultured on stainless steel. Exp. Cell Res. *46:*355–366.

Bjerkvig, R., Laerum, O.D., Mella, O. (1986b) Glioma cell interactions with fetal rat brain aggregates in vitro, and with brain tissue in vivo. Cancer Res. *46:*4071–4079.

Bjerkvig, R., Steinsvag, S.K., Laerum, O.D. (1986a) Reaggregation of fetal rat brain cells in a stationary culture system I.: Methodology and cell identification. In Vitro *22:*180–192.

Blaker, G.J., Birch, J.R., Pirt, S.J. (1971) The glucose, insulin and glutamine requirements of suspension cultures of HeLa cells in a defined culture medium. J. Cell Sci. *9:*529–537.

Bobrow, M., Madan, J., Pearson, P.L. (1972) Staining of some specific regions on human chromosomes, particularly the secondary constriction of no. 9. Nature *238:*122–124.

Bohnert, A., Hornung, J., Mackenzie, I.C., and Fusenig, N.E. (1986) Epithelial-mesenchymal interactions control basement membrane production and differentiation in cultured and transplanted mouse keratinocytes. Cell Tissue Res. *244:*413–429.

Booyse, F.M., Sedlak, B.J., Rafelson, M.E. (1975) Culture of arterial endothelial cells. Characterization and growth of bovine aortic

cells. Thromb. Diathes. Ahemorrh. *34:*825–839.

Bornstein, M.B., Murray, M.R. (1958) Serial observations on patterns of growth, myelin formation, maintenance and degeneration in cultures of newborn rat and kitten cerebellum. J. Biophys. Biochem. Cytol. *4:*499.

Boukamp, P., Dzarlieva-Petrusevska, R.T., Breitkreutz, D., Hornung, J., and Fusenig, N.E. (1987) Spontaneous transformation in vitro of human skin keratinocytes with preserved potential of normal differentiation. (submitted for publication).

Bowman, P.D., Betz, A.L., AR, D., Wolinsky, J.S., Penney, J.B., Shivers, R.R., Goldstein, G. (1981) Primary culture of capillary endothelium from rat brain. In Vitro *17:*353–362.

Boyum, A. (1968a) Isolation of leucocytes from human blood. A two–phase system for removal of red cells with methylcellulose as erythrocyte aggregative agent. Scand. J. Clin Lab. Invest. (Suppl 97)*21:*9–29.

Boyum, A (1968b) Isolation of leucocytes from human blood. Further observations. Methylcellulose, dextran and Ficoll as erythrocyte aggregating agents. Scand. J. Clin. Lab. Invest. (Suppl. 97) *31:*50.

Braaten, J.T., Lee, M.J., Schewk, A., Mintz, D.H. (1974) Removal of fibroblastoid cells from primary monolayer cultures of rat neonatal endocrine pancreas by dosium ethylmercurithiosalicylate. Biochem. Biophys. Res. Comm. *61:*476–482.

Bradford, M. (1976) A rapid and sensitive method for the quantitation of microgram quantities of protein using the principle of protein-dye binding. Anal. Biochem. *72:*248–254.

Bradley, N.J., Bloom, H.J.G., Davies, A.J.S., Swift, S.M. (1978) Growth of human gliomas in immune-deficient mice: A possible model for pre-clinical therapy studies. Br. J. Cancer *38:*263.

Bradley, T.R., Metcalf, D. (1966) The growth of mouse bone marrow cells *in vitro*. Aust. J. Biol. Med. *44:*287–300.

Breen, G.A.M., De Vellis, J. (1974) Regulation of glycerol phosphate dehydrogenase by hydrocortisone in dissociated rat cerebral cell cultures. Dev. Biol. *41:*255–266.

Brockes, J.P., Fields, K.L., Raff, M.C. (1979) Studies on cultured rat Schwann cells. I. Establishment of purified populations from cultures of peripheral nerve. Brain Res. *165:*105.

Brouty-Boyé, D. Gresser, I., Baldwin, C. (1979) Reversion of the transformed phenotype to the parental phenotype by subcultivation of x-ray transformed C$_3$H/10T1/2 at low cell density. Int. J. Cancer *2:*253–260.

Brouty-Boyé, D., Tucker, R.W., Folkman, J. (1980) Transformed and neoplastic phenotype: Reversibility during culture by cell density and cell shape. Int. J. Cancer *26:*501–507.

Brower, M., Carney, D.N., Oie, H.K., Gazdar, A.F., Minna, J.D. (1986) Growth of cell lines and clinical specimens of human non-small cell lung cancer in a serum-free defined medium. Cancer Res. *46:*798–806.

Bruland, Ø., Fodstad, Ø., and Pihl, A. (1985) The use of multicellular spheroids in establishing human sarcoma cell lines in vitro. Int. J. Cancer *35:*793–798.

Brunk, C.F., Jones, K.C., James, T.W. (1979) Assay for nanogram quantities of DNA in cellular homogenates. Anal. Biochem. *92:*497–500.

Brysk, M.M., Snider, J.M., Smith, E.B. (1981) Separation of newborn rat epidermal cells on discontinuous isokinetic gradients of percoll. J. Invest. Dermatol. *77:*205–209.

Buehring, G.C. (1972) Culture of human mammary epithelial cells. Keeping abreast of a new method. J. Natl. Cancer Inst. *49:*1433–1434.

Buick, R.N., Stanisic, T.H., Fry, S.E., Salmon, S.E., Trent, J.M., Krosovich, P. (1979) Development of an agar-methyl cellulose clonogenic assay for cells of transitional cell carcinoma of the human bladder. Cancer Res. *39:*5051–5056.

Buonassisi, V., Sato, G., Cohen, A.I. (1962) Hormone-producing cultures of adrenal and pituitary tumor origin. Proc. Natl. Acad. Sci. USA *48*:1184–1190.

Burgess, A.W., Metcalf, D. (1980) The nature and action of granulocyte-macrophage colony stimulating factors. Blood *56*:947–958.

Burgess, A.W., Nicola, N.A. (1983) Growth factors and stem cells. London: Academic Press Inc. 355pp.

Burke, J.M., Ross, R. (1977) Collagen synthesis by monkey arterial smooth muscle cells during proliferation and quiescence in culture. Exp. Cell Res. *107*:387–395.

Burwen, S.J., Pitelka, D.R. (1980) Secretory function of lactating moose mammary epithelial cells cultured on collagen gels. Exp. Cell Res. *126*:249–262.

Cahn, R.D., Lasher, R. (1967) Simultaneous synthesis of DNA and specialized cellular products by differentiating cartilage cells *in vitro*. Proc. Natl. Acad. Sci. USA *58*:1131–1138.

Cahn, R.D., Cooh, H.G., Cahn, M.B. (1967) In Wilt, F.H., Wessells, N.K. (eds): "Methods in Developmental Biology." New York, Thomas Y. Crowell, pp. 493.

Carlsson J., Gabel, D., Larsson, E., Westermark, B. (1979) Protein-coated agarose surfaces for attachment of cells. In Vitro *15*:844–50.

Carney, D.N., Bunn, P.A., Gazdar, A.F., Pagan, J.A., Minna, J.D. (1981) Selective growth in serum-free hormone-supplemented medium of tumor cells obtained by biopsy from patients with small cell carcinoma of lung. Proc. Natl. Acad. Sci. USA *78*:3185–3189.

Carpenter, G., Cohen, S. (1977) Epidermal growth factor. In Acton, R.T., Lynn, J.D. (eds): "Cell Culture and Its Application." New York, Academic Press, pp. 83–105.

Carrel, A. (1912) On the permanent life of tissues outside the organism. J. Exp. Med. *15*:516–528.

Caspersson, T., Farber, S., Foley, G.E., Kudynowski, J., Modest, E.J., Simonsson, E., Wagh, U., Zech, L. (1968) Chemical differentiation along metaphase chromosomes. Exp. Cell Res. *49*:219–222.

Catsimpoolas, N., Griffith, A.L., Skrabut, E.M., Valeri, C.R. (1978) An alternate method for the preparative velocity sedimentation of cells at unit gravity. Anal. Biochem. *87*:243–248.

Center for Disease Control, Office of Biosafety, Atlanta, GA 00333, USA (1985) "Proposed Biosafety Guidelines for Microbiological and Bacteriological Laboratories." Publications Dept., DHHS, Public Health Service.

Ceriani, R.L., Taylor-Papadimitriou, J., Peterson, J.A., Brown, P. (1979) Characterization of cells cultured from early lactation milks. In Vitro *15*:356–362.

Chambard, M., Verrier, B., Gabrion, J., Mauchamp, J., Bugeia, J.C, Pelassy, C., Mercier, B. (1983) Polarization of thyroid cells in culture; evidence for the basolateral localization of the iodide "pump" and of the thyroid-stimulating hormone receptor-adenyl cyclase complex. J. Cell Biol. *96*:1172–1177.

Chang, S.E., Taylor-Papadimitriou, J. (1983) Cell Diff. *12*:143–154.

Chang, S.E., Keen, J., Lane, E.B., Taylor-Papadimitriou, J. (1982) Cancer Res. *42*:2040–2053.

Chaproniere, D.M., McKeehan, W.L. (1986) Serial culture of adult human prostatic epithelial cells in serum-free medium containing low calcium and a new growth factor from bovine brain. Cancer Res., *46*:819–824.

Chen, T.R. (1977) In situ detection of mycoplasm contamination in cell cultures by fluorescent Hoechst 33258 stain. Exp. Cell Res. *104*:255.

Choi, K.W., Bloom, A.D. (1970) Cloning human lymphocytes *in vitro*. Nature *227*:171–173.

Christensen, B., Kieler, J., Villien, M., Don, P., Wang, C.Y., Wolf, H. (1984) A classification of human urothelial cells propagated in vitro. Anticancer Res. *4*:319–338.

Chung, Y.S., Song, I.S., Erickson, R.H., Sleisinger, M.H., Kim, Y.S. (1985) Effect of growth and sodium butyrate on brush border membrane-associated hydrolases in human colorectal cancer cell lines. Cancer Res. *45*:2976–2982.

Clark, J., Hirtenstein, M., Gebb, C. (1979) Critical parameters in the microcarrier culture of animal cells. Presented at the Third General Meeting of the European Society of Animal Cell Technology, Oxford, October 2–5, 1979.

Clark, J.M., Pateman, J.A. (1978) Long-term culture of Chinese hamster Kupffer cell lines isolated by a primary cloning step. Exp. Cell Res. *112*:207–217.

Clarke, G.D., Ryan, P.J. (1980) Tranquilizers can block mitogenesis in 3T3 cells and induce differentiation in Friend cells. Nature *287*:160–161.

Clayton, R.M., Bower, D.J., Clayton, P.R., Patek, C.E., Randall, F.E., Sime, C., Wainwright, N.R., Zehir, A. (1980) Cell culture in the investigation of normal and abnormal differentiation of eye tissues. In Richards, R.J., Rajan, K.T. (eds): "Tissue Culture in Medical Research (II)." Oxford, Pergamon Press.

Cohen, J., Balazs, R., Hojos, F., Currie, D.N., Dutton, G.R. (1978) Separation of cell types from the developing cerebellum. Brain Res. *148*:313–331.

Cohen, S. (1962) Isolation of a mouse submaxillary gland protein accelerating incisor eruption and eyelid opening in the new-born animal. J. Biol. Chem. *237*:1555–1562.

Cole, R.J., Paul, J. (1966) The effects of erythropoietin on haem synthesis in mouse yolk sac and cultured foetal liver cells. J. Embryol. Exp. Morphol. *15*:245–260.

Collins, S.J., Gallo, R.C., Gallagher, R.E. (1977) Continuous growth and differentiation of human myeloid leukaemic cells in suspension culture. Nature *270*:347–349.

Committee for a Standardized Karyotype of *Rattus Norvegicus* (1973) Standard karyotype of the Norway rat, *Rattus Norvegicus*. Cytogenet. Cell Genet. *12*:199–205.

Committee on Standardized Genetic Nomenclature for Mice. (1972) Standard karyotype of the mouse *Mus musculis*. J. Hered. 63:69.

Fraenkel-Conrat, H., Robert R. Wagner, R.R., eds. (1979) "Comprehensive Virology." New York and London, Plenum Press.

Conkie, D., Affara, N., Harrison, P.R., Paul, J., Jones, K., (1974) *In situ* localization of globin messenger RNA formation. II. After treatment of Friend virus-transformed mouse cells with dimethyl sulphoxide. J. Cell Biol. *63*:414–419.

Coon, H.G., Cahn, R.D. (1966) Differentiation *in vitro*: Effects of sephadex fractions of chick embryo extract. Science *153*:1116–1119.

Coons, A.H., Kaplan, M.M. (1950) Localization of antigen in tissue cells. II. Improvements in a method for the detection of antigen by means of fluorescent antibody. J. Exp. Med. *91*:1–13.

Cooper, G.W. (1965) Induction of somite chondrogenesis by cartilage and notochord: A correlation between inductive activity and specific stages of cytodifferentiation. Dev. Biol. *12*:185–212.

Cooper, P.D., Burt, A.M., Wilson, J.N. (1958) Critical effect of oxygen tension on rate of growth of animal cells in continuous suspended culture. Nature *182*:1508–1509.

Cour, I., Maxwell, G., Hay, R.J. (1979) Tests for bacterial and fungal contaminants in cell cultures as applied at the ATCC. In Evans, V.J., Perry, V.P., Vincent, M.M. (eds): "Manual of the American Tissue Culture Association." *5*:1157–1160.

Courtenay, V.D., Selby, P.J. Smith, I.E., Mills, J., Peckham, M.J. (1978) Growth of human tumor cell colonies from biopsies using two soft-agar techniques. Br. J. Cancer *38*:77–81.

Cox, R.P., ed (1974) "Cell Communication." New York, John Wiley & Sons.

Crabb, J.W., Armes, L.G., Johnson, C.M., McKeehan, W.L. (1986) Characterization of multiple forms of prostatropin (prostate epithelial growth factor) from bovine brain. Biochem. Biophys. Res. Commun. 136:1155–1161.

Creasey, A.A., Smith, H.S., Hackett, A.J., Fukuyama, K., Epstein, W.L., Madin, S.H. (1979) Biological properties of human melanoma cells in culture. In Vitro 15:342.

Crissman, H.A., Mullaney, P.F., Steinkamp, J.A. (1975) Methods and applications of flow systems for analysis and sorting of mammalian cells. In Prescott, D.M. (ed): "Methods in Cell Biology, Vol. IX." New York, Academic Press.

Cunha, G.R. (1984) Androgenic effects upon prostatic epithelium are mediated via tropic influences from stroma. In "New Approaches to the Study of Benign Prostatic Hyperplasia." New York, Alan R. Liss, Inc. pp.81–102.

Curtis, A.S.G., Seehar, G.M. (1978) The control of cell division by tension of diffusion. Nature 274:52–53.

Cuttitta, F., Carney, D.N., Mulshine, J., Moody, T.W., Fedorko, J., Fischler, A., Minna, J.D. (1985) Bombesin-like peptides can function as autocrine growth factors in human small-cell lung cancer. Nature 316:823.

Das, S.K., Stanley, E.R. (1982) Structure-function studies of a colony stimulating factor (CSF-1). J. Biol. Chem. 257:13679–13684.

Davison, P.M., Bensch, K., Karasek, M.A. (1983) Isolation and long-term serial cultivation of endothelial cells from microvessels of the adult human dermis. In Vitro 19:937–945.

Defendi, V., ed. (1964) "Retention of Functional Differentiation in Cultured Cells." Philadelphia, the Wistar Institute Press.

Defendi, V. (1976) In Nelson, D.S. (ed): "Immunobiology of the Macrophage." New York, Academic Press, pp. 275–286.

DeLeij, L., Poppema, S., Nulend, J.K., Haar, A.T., Schwander, E., Ebbens, F., Postmus, P.E., Hauw The, T. (1985) Neuroendocrine differentiation antigen on huam lung carcinoma and Kulchitski cells. Cancer Res. 45:2192–2200.

Del Vecchio, P., Smith, J.R. (1981) Expression of angiotensin converting enzyme activity in cultured pulmonary artery endothelial cells. J. Cell Physiol. 108:337–345.

DeMars, R. (1957) The inhibition of glutamine of glutamyl transferase formation in cultures of human cells. Biochim. Biophys. Acta 27:435–436.

Dendy, P.P. (1976) Some problems in the use of short-term cultures of human tumours for in vitro screening of cytotoxic drugs. Chemotherapy 7:341–350.

De Ridder, L., Mareel, M. (1978) Morphology and ^{125}I-concentration of embryonic chick thyroids cultured in an atmosphere of oxygen. Cell Biol. Int. Rep. 2:189–194.

De Vitry, F., Camier, M., Czernichow, P., Benda, Ph., Cohen, P., Tixier-Vidal, A. (1974) Establishment of a clone of mouse hypothalamic neurosecretory cells synthesizing neurophysin and vasopressin. Proc. Natl Acad. Sci. USA 71:3575–3579.

De Vonne, T.L., Mouray, H (1978) Human α_2–macroglobulin and its antitrypsin and antithrombin activities in serum and plasma. Clin. Chim. Acta 90:83–85.

Dexter, D.L., Barbosa, J.A., Calabresi, P. (1979) N,N-dimethylformamide-induced alteration of cell culture characteristics and loss of tumorigenicity in cultured human colon carcinoma cells. Cancer Res. 39:1020–1025.

Dexter, T.M., Allen, T.D., Lajtha, L.G. (1977) Conditions controlling the proliferation of haemopoietic stem cells in vitro. J. Cell Physiol. 91:335–345.

Dexter, T.M. Allen, T.D., Scott, D., Teich, N.M. (1979) Isolation and characterisation of a bipotential haematopoietic cell line. Nature 277:417–474.

Dexter, T.M., Spooncer, E., Simmons, P., Allen, T.D. (1984) Long-term marrow culture: An overview of technique and experience. p. 57–96 in: "Long-term Bone Marrow Culture." Kroc Foundation Series 18, In Wright, D.G., Greenberger,J.S.(eds): publ New York, Alan R. Liss, Inc.

Dickson, J.A., Suzangar, M. (1976) The in vitro response of human tumours to cytotoxic drugs and hyperthermia (42°) and its relevance to clinical oncology. In Balls, M., Monnickendam, M. (eds): "Organ Culture and Biomedical Research." Cambridge, Cambridge University Press, pp. 417–446.

Dickson, J.D., Flanigan, T.P., Kemshead, J.T. (1983) Monoclonal antibodies reacting specifically with the cell surface of human astrocytes in culture. Biochem. Soc. Trans. 11:208.

DiPaolo, J.A. (1965) In vitro test systems for cancer chemotherapy. III. Preliminary studies of spontaneous mammary tumors in mice. Cancer Chemother. Rep. 44:19–24.

Douglas, W.H.J., Moorman, G.W., Teel, R.W. (1976) The formation of histotypic structures from monodispersed rat lung cells cultured on a three-dimensional substrate. In Vitro 12:373–381.

Douglas, W.H.J., McAteer, J.A., Dell'Orco, R.T., Phelps, D. (1980) Visualization of cellular aggregates cultured on a three-dimensional collagen sponge matrix. In Vitro 16:306–312.

Drejer, J., Larsson, O.M., Schousboe, A. (1983) Characterization of uptake and release processes for D- and L-aspartate in primary cultures of astrocytes and cerebellar granule cells. Neurochem. Res. 8:231–243.

Duksin, D., Maoz, A., Fuchs, S. (1975) Differential cytotoxic activity of anticollagen serum on rat osteoblasts and fibroblasts in tissue culture. Cell 5:83–86.

Dulak, N.C., Temin, H.M. (1973a) A partially purified rat liver cell conditioned medium with multiplication-stimulating activity for embryo fibroblasts. J. Cell Physiol. 81:153–160.

Dulak, N.C., Temin, H.M. (1973b) Multiplication-stimulating activity for chicken embryo fibroblasts from rat liver cell conditioned medium: A family of small peptides. J. Cell Physiol. 81:161–170.

Dulbecco, R., Vogt, M. (1954) Plaque formation and isolation of pure cell lines with poliomyelitis viruses. J. Exp. Med. 199:167–182.

Dulbecco, R., Freeman, G. (1959) Plaque formation by the polyoma virus. Virology 8:396–397.

Dulbecco, R., Elkington, J. (1973) Conditions limiting multiplication of fibroblastic and epithelial cells in dense cultures. Nature 246:197–199.

Eagle, H. (1955) The specific amino acid requirements of mammalian cells (stain L) in tissue culture. J. Biol. Chem. 214:839.

Eagle, H. (1959) Amino acid metabolism in mammalian cell cultures. Science 130:432.

Eagle, H. (1973) The effect of environmental pH on the growth of normal and malignant cells. J. Cell Physiol. 82:1–8.

Eagle, H., Foley, G.E., Koprowski, H., Lazarus, H., Levine, E.M., Adams, R.A. (1970) Growth characteristics of virus-transformed cells. J. Exp. Med. 131:863–879.

Earle, W.R., Schilling, E.L., Stark, T.H., Straus, N.P., Brown, M.F., Shelton, E. (1943) Production of malignancy in vitro. IV. The mouse fibroblast cultures and changes seen in the living cells. J. Natl. Cancer Inst. 4:165–212.

Easton, T.G., Valinsky, J.E., Reich, E. (1978) M540 as a fluorescent probe of membranes: Staining of electrically excitable cells. Cell 13:476–486.

Easty, D.M., Easty, G.C. (1974) Measurement of the ability of cells to infiltrate normal tissues in vitro. Br. J. Cancer 29:36–49.

Easty, G.C., Easty, D.M., Ambrose, E.J. (1960) Studies of cellular adhesiveness. Exp. Cell Res. 19:539–548.

Ebendal, T. (1976) The relative roles of contact inhibition and contact guidance in orientation of axons extending on aligned collagen fibrils in vitro. Exp. Cell Res. 98:159–169.

Ebendal, T. (1979) Stage-dependent stimulation of neurite outgrowth exerted by nerve growth factor and chick heart in cultured embryonic ganglia. Dev. Biol. 72:276.

Ebendal, T., Jacobson, C.O. (1977) Tissue explants affecting extension and orientation of axons in cultured chick embryo ganglia. Exp. Cell Res. *105:*379–387.

Edelman, G.M. (1973) Nonenzymatic dissociations. B. Specific cell fractionation of chemically derivatized surfaces. In Kruse, P.F., Jr., Patterson, M.K., Jr. (eds): "Tissue Culture Methods and Applications." New York, Academic Press, pp. 29–36.

Edwards, P.A.W., Easty, D.M., Foster, C.S. (1980) Selective culture of epithelioid cells from a human squamous carcinoma using a monoclonal antibody to kill fibroblasts. Cell Biol. Int. Rep. *4:*917–922.

Eisen, H., Bach, R., Emery, R. (1977) Induction of spectrin in Friend erythroleukaemic cells. Proc. Natl. Acad. Sci. USA. *74:*3898–4002.

Eisinger, M., Lee, J.S., Hefton, J.M., Darzykiewicz, A., Chiao, J.W., Deharven, E. (1979) Human epidermal cell cultures—growth and differentiation in the absence of dermal components or medium supplements. Proc. Natl. Acad. Sci. U.S.A. *76:*5340.

Elsdale, T., Bard, J. (1972) Collagen substrata for studies on cell behaviour. J. Cell Biol. *54:*626–637.

Elvin, P., Wong, V., Evans, C.W. (1985) A study of the adhesive, locomotory and invasive behaviour of Walker 256 carcinosarcoma cells. Exp. Cell Biol. *53:*9–18.

Eng, L.F., Bigbee, J.W. (1979) Immunochemistry of nervous-system specific antigens. In Aprison (ed): "Advances in Neurochemistry." New York, Plenum Press, pp. 43–98.

Espmark, J.A., Ahlqvist-Roth, L. (1978) Tissue typing of cells in cultures. I. Distinction between cells lines by the various patterns produced in mixed haemabsorption with selected multiparous sera. J. Immunol. Methods *24:*141–153.

Evans, V.J., Bryant, J.C. (1965) Advances in tissue culture at the National Cancer Institute in the United States of America. In: Ramakrishnan, C.V. (ed.): "Tissue Culture." The Hague, W. Junk, pp. 145–167.

Evans, V.J., Bryant, J.C., Fioramonti, M.C.,McQuilkin, W.T., Sanford, K.K., Earle, W.R. (1956) Studies of nutrient media for tissue C cells in vitro. I. A protein-free chemically defined medium for cultivation of strain L cells. Cancer Res. *16:*77.

Federoff, S. (1975) In Evans, V.J., Perry, V.P., Vincent, M.M. (eds.): "Manual of the Tissue Culture Association" *1:*53–57.

Fell, H.B. (1953) Recent advances in organ culture. Sci. Prog. *162:*212.

Fergusson, R.J., Carmichael, J., Smyth, J.F. (1986) Human tumour xenografts growing in immunodeficient mice: A useful model for assessing chemotherapeutic agents in bronchial carcinoma. Thorax *41:*376–380.

Ferrone, S. and Dierch, M.P., ed. (1985) "Handbook of Monoclonal Antibodies: Applications in Biology and Medicine." Noyes Publications.

Finbow, M.E., Pitts, J.D. (1981) Permeability of junctions between animal cells. Exp. Cell Res. *131:*1–13.

Fisher, M., Solursh, M. (1979) The influence of the substratum on mesenchyme spreading *in vitro*. Exp. Cell Res. *123:*

Fisher, H.W., Puck, T.T., Sato, G. (1958) Molecular growth requirements of single mammalian cells: The action of fetuin in promoting cell attachment of glass. Proc. Natl. Acad. Sci. USA *44:*4–10.

Flynn, D., Yang, J., Nandi, S. (1982) Growth and differentiation of primary cultures of mouse mammary epithelium embedded in collagen gel. Different. *22:*191.

Fogh, J. (1973) "Contamination in Tissue Culture." New York: Academic Press.

Foley, J.F., Aftonomos, B. Th. (1973) Pronase. In Kruse, P.F., Jr., Patterson, M.K. Jr. (eds): "Tissue Culture Methods and Applications." New York: Academic Press, pp. 185–188.

Folkman, J. (1985) Tumor angiogenesis. Adv. Cancer Res. *43:*175–203.

Folkman, J. (1986) How is blood vessel growth regulated in normal and neoplastic tissue?—G.H.A. Clowes Memorial Award Lecture. Cancer Res. *46:*467–473.

Folkman, J., Haudenschild, C. (1980) Angiogenesis *in vitro*. Nature *288:*551–556.

Folkman, J., Moscona, A. (1978) Role of cell shape in growth control. Nature *273:*345–349.

Folkman, J., Tucker, R.W. (1980) Cell configuration, substrate and growth control. In Subtellny, S., Wessells, N.K. (eds): "Cell Surface, Mediator of Developmental Processes." New York, Academic Press.

Folkman, J., Haudenschild, C. C., Zetter, B.R. (1979) Long-term culture of capillary endothelial cells. Proc. Natl. Acad. Sci. USA *76:*5217–5221.

Fontana, A., Hengarner, H., de Tribolet, N., Weber, E. (1984) Glioblastoma cells release interleukin 1 and factors inhibiting interleukin 2-mediated effects. J. Immunol. *132(4):*1837–1844.

Foreman, J., Pegg, D.E. (1979) Cell preservation in a programmed cooling machine: The effect of variations in supercooling. Cryobiology *16:*315–321.

Frame, M., Freshney, R.I., Shaw, R., Graham, D.I. (1980) Markers of differentiation in glial cells. Cell Biol. Int. Rep. *4:*732.

Fraser, C.M., Venter, J.C. (1980) The synthesis of β-adrenergic receptors in cultured human lung cells: Induction by glucocorticoids. Biochem. Biophys. Res. Commun. *94:*390–398.

Fredin, B.L., Seiffert, S.C., Gelehrter, T.D. (1979) Dexamethasone-induced adhesion in hepatoma cells: The role of plasminogen activator. Nature *277:*312–313.

Freedman, V.H., Shin, S. (1974) Cellular tumorigenicity in nude mice: Correlation with cell growth in semi-solid medium. Cell *3:*355–359.

Freeman, A.E., Igel, H.J., Herrman, B.J., Kleinfeld, K.L. (1976) Growth and characterisation of human skin epithelial cultures. In Vitro *12:*352–62.

Freshney, R.I. (1972) Tumour cells disaggregated in collagenase. Lancet *2:*488–489.

Freshney, R.I. (1976a) Separation of cultured cells by isopycnic centrifugation in metrizamide gradients. In Rickwood, D. (ed): "Biological Separations." London and Washington, Information Retrieval, Ltd., pp. 123–130.

Freshney, R.I. (1976b) Some observations on assay of anticancer drugs in culture. In Dendy, P.P. (ed.): "Human Tumours in Short Term Culture." New York, Academic Press, pp. 150–158.

Freshney, R.I. (1978) Use of tissue culture in predictive testing of drug sensitivity. Cancer Topics *1:*5–7.

Freshney, R.I. (1980) Culture of glioma of the brain. In Thomas, D.G.T., Graham, D.I. (eds): "Brain Tumours, Scientific Basic, Clinical Investigation and Current Therapy." London: Butterworths, pp. 21–50.

Freshney, R.I. (1986) "Animal Cell Culture, A Practical Approach." Oxford: IRL Press.

Freshney, R.I., Hart, E. (1982) Clonogenicity of human glia in suspension. Br. J. Cancer *46:*463.

Freshney, R.I., Paul, J., Kane, I.M. (1975) Assay of anti-cancer drugs in tissue culture: Conditions affecting their ability to incorporate ^{3}H-leucine after drug treatment. Br. J. Cancer *31:*89–99.

Freshney, R.I., Morgan, D., Hassanzadah, M., Shaw, R., Frame, M. (1980a) Glucocorticoids, proliferation and the cell surface. In Richards, R.J., Rajan, K.T. (eds): "Tissue Culture in Medical Research (II)." Oxford, Pergamon Press, pp. 125–132.

Freshney, R.I., Sherry, A., Hassanzadah, M., Freshney, M., Crilly, P., Morgan, D. (1980b) Control of cell proliferation in human glioma by glucocorticoids. Br. J. Cancer *41:*857–866.

Freshney, R.I., Celik, F., Morgan, D. (1982) Analysis of cytotoxic and cytostatic effects. In Davis, W., Malvoni, C., Tanneberger,

St. (eds): "The Control of Tumor Growth and Its Biological Base." Fortschritte in der Onkologie, Band 10. Berlin, Akademie-Verlag, pp. 349–358.

Freshney, R.I., Hart, E., Russell, J.M. (1982) Isolation and purification of cell cultures from human tumours. In Reid, E., Cook, G.M.W., Morre, D.J. (eds): "Cancer Cell Organelles. Methodological Surveys (B): Biochemistry, Vol. II." Chichester, England, Horwood, pp. 97–110.

Freyer, J.P. and Sutherland, R.M. (1980) Selective dissociation and characterization of cells from different regions of multicell tumour spheroids. Cancer Res. 40:3956–3965.

Friend, C., Patuleia, M.C., Nelson, J.B. (1966) Antibiotic effect of tylosine on a mycoplasma contaminant in a tissue culture leukemia cell line. Proc. Soc. Exp. Biol. Med. 121:1009.

Friend, C., Scher, W., Holland, J.G., Sato, T. (1971) Hemoglobin synthesis in murine virus-induced leukemic cells in vitro. 2. Stimulation of erythroid differentiation by dimethyl sulfoxide. Proc. Natl. Acad. Sci. USA 68:378–382.

Friend, K.K., Dorman, B.P., Kucherlapati, R.S., Ruddle, F.H. (1976) Detection of interspecific translocations in mouse-human hybrids by alkaline Giemsa staining. Exp. Cell Res. 99:31–36.

Fritz, G.R., Knobil, E. (1964) Amino acid transport and protein synthesis in muscle. Action of insulin. Proc. Exp. Biol. Med. 116:873–875.

Fry, J., Bridges, J.W. (1979) The effect of phenobarbitone on adult rat liver cells and primary cell lines. Toxicol. Lett. 4;295–301.

Fusenig, N.E. (1986) Mammalian epidermal cells in culture. In: Biology of the Integument, Vol. 2 Vertebrates. Eds. J. Bereiter-Hahn, A.G. Matoltsy, K.S. Richards, Springer Verlag, pp. 409–442.

Fusenig, N.E. and Worst, P.K.M. (1975) Mouse epidermal cell cultures. II. Isolation, characterization and cultivation of epidermal cells from perinatal mouse skin. Exp. Cell Res. 93:443–457.

Gabrilove, J.L., Welte, K., Harris, P., Platzer, E., Lu, L. et al. (1986) Pluripoietin alpha: A second hematopoietic colony stimulating factor produced by the human bladder carcinoma cell line 5637. Proc. Natl. Acad. Sci. USA 83:2478–2482.

Gartler, S.M. (1967) "Genetic Markers as Tracers in Cell Culture." Second Bicennial Review Conference on Cell, Tissue and Organ Culture. NCI Monographs, pp. 167–195.

Gartner, S. and Kaplan, H.S. (1980) Long-term culture of human bone marrow cells. Proc. Natl. Acad. Sci (USA) 77:4756–4759.

Gaush, C.R., Hard, W.L., Smith, T.F. (1966) Characterization of an established line of canine kidney cells (MDCK). Proc. Soc. Exp. Biol. Med. 122:931–933.

Geppert, E.F., Williams, M.C., Mason, R.J. (1980) Primary culture of rat alveolar type II cells on floating collagen membranes. Exp. Cell Res. 128:363–374.

Gey, G.O., Coffman, W.D., Kubicek, M.T. (1952) Tissue culture studies of the proliferative capacity of cervical carcinoma and normal epithelium. Cancer Res. 12:364–365.

Gilbert, S.F., Migeon, B.R. (1975) D-valine as a selective agent for normal human and rodent epithelial cells in culture. Cell 5:11–17.

Gilbert, S.F., Migeon, B.R. (1977) Renal enzymes in kidney cells selected by D-Valine medium. J. Cell Physiol. 92:161–168.

Gilchrest, B.A., Nemore, R.E., Maciag, T. (1980) Growth of human keratinocytes on fibronectin-coated plates. Cell Biol. Int. Rep. 4:1009–1016.

Gilchrest, B.A., Vrabel, M.A., Flynn, E., Szabo, G. (1984) Selective cultivation of human melanocytes from newborn and adult epidermis. J. Invest. Dermatol. 83:370.

Gilchrist, B.A., Albert, L.S., Karassik, R.L., Yaar, M. (1985) Substrate influences human epidermal melanocyte attachment and spreading in vitro. In Vitro, 21:114.

Gilden, D.H., Wroblewska, Z., Eng, L.F., Rorke, L.B. (1976) Human brain in tissue culture. Part 5. Identification of glial cells by immunofluorescence. J. Neurol. Sci. 29:177–184.

Gillis, S., Watson, J. (1981) Interleukin-2 dependent culture of cytolytic T cell lines. Immunol. Rev. 54:81–109.

Gimbrone, M.A., Jr., Cotran, R.S., Folkman, J. (1974) Human vascular endothelial cells in culture, growth and DNA synthesis. J. Cell Biol. 60:673–684.

Giovanella, B.C., Stehlin, J.S., Williams, L.J. (1974) Heterotransplantation of human malignant tumors in "nude" mice. II. Malignant tumors induced by injection of cell cultures derived from human solid tumors. J. Natl. Cancer. Inst. 52:921.

Goldberg, B. (1977) Collagen synthesis as a marker for cell type in mouse 3T3 lines. Cell 11:169–172.

Golde, D.W. (1984) "Hematopoiesis." New York, Churchill Livingstone Inc, p. 361.

Golde, D.W., Cline, M.J. (1973) Cultivation of normal and neoplastic human bone marrow leucocytes in liquid suspension. In: "Proc. 7th Leucocyte Culture Conference." New York: Academic Press.

Good, N.E., Winget, G.D., Winter, W., Connolly, T.N., Izawa, S., Singh, R.M.M. (1966) Hydrogen ion buffers and biological research. Biochemistry 5:467–477.

Goodwin, G., Shaper, J.H. Abezoff, M.D., Mendelsohn, G., Baylin, S.B. (1983) Analysis of cell surface proteins delineates a differentiation pathway linking endocrine and non endocrine human lung cancers. Proc. Natl. Acad. Sci. 80:3807–3811.

Gorham, L.W., Waymouth, C. (1965) Differentiation in vitro of embryonic cartilage and bone in a chemically defined medium. Proc. Soc. Exp. Biol. Med. 119:287–290.

Gospodarowicz, D. (1974) Localization of fibroblast growth factor and its effect alone and with hydrocortisone on 3T3 cell growth. Nature 249:123–127.

Gospodarowicz, D., Mescher, A.L. (1977) A comparison of the responses of cultured myoblasts and chondrocytes to fibroblast and epidermal growth factors. J. Cell. Physiol. 93:117–128.

Gospodarowicz, D., Moran, J. (1974) Growth factors in mammalian cell cultures. Ann. Rev. Biochem. 45:531–558.

Gospodarowicz, D., Moran, J., Braun, D., Birdwell, C. (1976) Clonal growth of bovine vascular endothelial cells: Fibroblast growth factor as a survival agent. Proc. Natl. Acad. Sci. USA 73:4120–4124.

Gospodarowicz, D., Moran, J.S., Braun, D.L. (1977) Control of proliferation of bovine vascular endothelial cells. J. Cell Physiol. 91:377–386.

Gospodarowicz, D., Greenburg, G., Bialecki, H., Zetter, B.R. (1978) Factors involved in the modulation of cell proliferation in vivo and in vitro: The role of fibroblast and epidermal growth factors in the proliferative response of mammalian cells. In Vitro 14:85–118.

Gospodarowicz, D., Greenburg, G., Birdwell, C.R. (1978) Determination of cell shape by the extra cellular matrix and its correlation with the control of cellular growth. Cancer Res. 38:4155–4171.

Gospodarowicz, D., Delgado, D., Vlodavsky, I. (1980) Permissive effect of the extracellular matrix on cell proliferation in vitro. Proc. Natl. Acad. Sci. 77:4094–4098.

Gough, N.M., Gough, J., Metcalf, D., Kelso, A., Grail, A. et al. (1984) Molecular cloning of cDNA encoding a murine haemopoietic growth regulator granulocyte/macrophage colony stimulating factor. Nature 309:763–767.

Graham, F.L., Van der Eb (1973) A new technique for the assay of infectivity of human adenovirus 5 DNA. Virology 52:456–461.

Granner, D.K., Hayashi, S., Thompson, E.B., Tomkins, G.M. (1968) Stimulation of tyrosine aminotransferase synthesis by dexamethasone phosphate in cell culture. J. Mol. Biol. 35:291–301.

Green H. (1977) Terminal differentiation of cultured human epidermal cells. Cell 11:405–416.

Green, A.E., Athreya, B., Lehr, H.B., Coriell, L.L. (1967) Viability of cell cultures following extended preservation in liquid nitrogen. Proc. Soc. Exp. Biol. Med. 124:1302–1307.

Green, C.L., Pretlow, T.P., Tucker, K.A., Bradley, E.L. Jr., Cook, W.J., Pitts, A.M., Pretlow II, T.G. (1980) Large-capacity separation of malignant cells and lymphocytes from the Furth mast cell tumor in a reorienting zonal rotor. Cancer Res. 40:1791–1796.

Green, H., Thomas, J. (1978) Pattern formation by cultured human epidermal cells: Development of curved ridges resembling dermatoglyphs. Science 200:1385–1388.

Green, H., Kehinde, O., Thomas, J. (1979) Growth of cultured human epidermal cells into multiple epithelia suitable for grafting. Proc. Natl. Acad. Sci. USA 76:5665–5668.

Greenberger J.S. (1980) Self-renewal of factor-dependent haemopoietic progenitor cell lines derived from long-term bone marrow cultures demonstrate significant mouse strain genotypic variation. J. Supramol. Struct. 13:501–511.

Greenleaf, R.D., Mason, R.J., Williams, M.C. (1979) Isolation of alveolar type II cells by centrifugal elutriation. In Vitro 15:673.

Griffiths, J.B., Pirt, G.J. (1967) The uptake of amino acids by mouse cells (Strain LS) during growth in batch culture and chemostat culture: The influence of cell growth rate. Proc. R. Soc. Biol. 168:421–438.

Grobstein, C. (1953) Epithelio-mesenchymal specificity in the morphogenesis of mouse submandibular rudiments in vitro. J. Exp. Zool. 124:383.

Grobstein, C. (1953) Morphogenetic interaction between embryonic mouse tissues separated by a membrane filter. Nature 4384:869–871.

Gross, M., Goldwasser, E. (1971) On the mechanism of erythropoietin-induced differentiation. J. Biol. Chem. 246:2480–2486.

Guguen-Guillouzo, C., and Guillozo, A. (1986). In Guillozo, A., and Guguen-Guillouzo (eds): "Isolated and Culture Hepatocytes." Paris, ISERM, and London, John Libbey Eurotext, pp. 1–12.

Guguen-Guillouzo, C., Campion, J.P., Brissot, P. et al. (1982). High yield preparation of isolated human adult hepatocytes by enzymatic profusion of the liver. Cell Biol. Int. Rep. 6:625–628.

Guguen-Guillouzo, C., Clement, B., Baffet, G. et al. (1983). Maintenance and reversibility of active albumin secretion by adult rat hepatocytes co-cultured with another liver epithelial cell type. Exp. Cell Res. 143:47–54.

Guilbert, L.J., Iscove, N.N. (1976) Partial replacement of serum by selenite, transferrin, albumin and lecithin in haemopoietic cell cultures. Nature 263:594–595.

Guillouzo, A., Guguen-Guillouzo, C., Bourel, M. (1981) Hepatocytes in culture: Expression of differentiated functions and their application to the study of metabolism. Triangle (Sandoz J. Med. Sci.) 20:121–128.

Gullino, P.M., Knazak, R.A. (1979) Tissue culture on artificial capillaries. In Jakoby, W.B., Pastan, I. (eds): "Methods in Enzymology, Vol. LVIII. Cell Culture." New York, Academic Press, pp. 178–184.

Guner, M., Freshney, R.I., Morgan, D., Freshney, M.G., Thomas, D.G.T., Graham, D.I. (1977) Effects of dexamethasone and betamethasone on in vitro: cultures from human astrocytoma. Br. J. Cancer 35:439–47.

Gwatkin, R.B.L. (1973) Pronase. In Kruse, P.F., Jr., Patterson, M.K., Jr. (eds): "Tissue Culture Methods and Applications." New York, Academic Press, pp. 3–5.

Habel, K., Salzman, N.P., eds. (1969) "Fundamental Techniques in Virology." New York, Academic Press.

Halton, D.M., Peterson, W.D. Jr., Hukku, B. (1983) Cell culture quality control by rapid isoenzymatic characteristics. In Vitro 19:16–24.

Ham, R.G. (1963) An improved nutrient solution for diploid Chinese hamster and human cell lines. Exp. Cell Res. 29:515.

Ham, R.G. (1965) Clonal growth of mammalian cells in a chemically defined synthetic medium. Proc. Natl. Acad. Sci. USA 53:288.

Ham, R.G., McKeehan, W.L. (1978) Development of improved media and culture conditions for clonal growth of normal diploid cells. In Vitro 14:11–22.

Ham, R.G., McKeehan, W.L. (1979) Media and growth requirements. In Jakoby, W.B., Pastan, I.H. (eds): "Methods in Enzymology, Volume LVIII, Cell Culture." Academic Press, New York, pp. 44–93.

Hamburger, A.W., Salmon, S.E. (1977) Primary bioassay of human tumor stem cells. Science 197:461–463.

Hamilton, W.G., Ham, R.G. (1977) Clonal growth of Chinese hamster cell lines in protein-free media. In Vitro 13:537–547.

Hanks, J.H., Wallace, R.E. (1949) Relation of oxygen and temperature in the preservation of tissues by refrigeration. Proc. Exp. Biol. Med. 71:196.

Hapel, A.J., Fung, M.C., Johnson, R.M., Young, I.G., Johnson, G., Metcalf, D. (1985) Biologic properties of molecularly cloned and expressed murine Interleukin-3. Blood 65: 1453–1459.

Harrington, W.N., Godman, G.C. (1980) A selective inhibitor of cell proliferation from normal serum. Proc. Natl. Acad. Sci. USA. Biol. Sci. 77:423–427.

Harris, H., Hopkinson, D.A. (1976) "Handbook of Enzyme Electrophoresis in Human Genetics." New York, American Elsevier.

Harris, H., Watkins, J.F. (1965) Hybrid cells from mouse and man: Artificial heterokaryons of mammalian cells from different species. Nature 205:640–646.

Harris, L.W., Griffiths, J.B. (1977) Relative effects of cooling and warming rates on mammalian cells during the freeze-thaw cycle. Cryobiology 14:662–669.

Harrison, P.R., Conkie, D., Affara, N., Paul, J. (1974) In situ localization of globin messenger RNA formation. I. During mouse foetal liver development. J. Cell Biol. 63:402–413.

Harrison, R.G. (1907) Observations on the living developing nerve fiber. Proc. Soc. Exp. Biol. Med. 4:140–143.

Hart, I.R., Fidler, I.J. (1978) An in vitro quantitative assay for tumor cell invasion. Cancer Res. 38:3218–3224.

Hassbroek, F.J., Neggle, J.C., Fleming, A.L. (1962) High-resolution auto-radiography without loss of water-soluble ions. Nature 195:615–616.

Hauschka, S.D., Konigsberg, I.R. (1966) The influence of collagen on the development of muscle clones. Proc. Natl. Acad. Sci. USA 55:119–126.

Hay, R.J. (1979) Identification, separation and culture of mammalian tissue cells. In Reid, E. (ed.): "Methodological Surveys in Biochemistry. Vol. 8, Cell Populations." London, Ellis Horwood, Ltd., pp. 143–160.

Hay, R.J., Strehler, B.L. (1967) The limited growth span of cell strains isolated from the chick embryo. Exp. Gerontol. 2:123.

Hay, R.J., Kern, J., Caputo, J. (1979) Testing for the presence of viruses in cultured cell lines. In: "Manual of American Tissue Culture Association." 5:1127–1130.

Hayashi, I., Sato, G.H. (1976) Replacement of serum by hormones permits growth of cells in a defined medium. Nature 259:132–134.

Hayflick, L., Moorhead, P.S. (1961) The serial cultivation of human diploid cell strains. Exp. Cell Res. 25:585–621.

Helden, C.H., Westermark, B., Wasteson, A. (1979) Platelet-derived growth factor: purification and partial characterization. Proc. Natl. Acad. Sci. USA 76:3722–3726.

Hemstreet, G.P., Enoch, P.G., Pretlow, T.G. (1980) Tissue disaggregation of human renal cell carcinoma with further isopyknic and isokinetic gradient purification. Cancer Res 40:1043–1049.

Hennings, H., Michael, D., Cheng, C., Steinert, P., Holbrook, K., Yuspa, S.H. (1980) Calcium regulation of growth and differentiation of mouse epidermal cells in culture. Cell 19:245–254.

Herzenberg, L.A., Sweet, R.G., Herzenberg, L.A. (1976) Fluorescence-activated cell sorting. Sci. Am. *234:*108–117.

Heyderman, E., Steele, K., Ormerod, M.G. (1979) A new antigen on the epithelial membrane: Its immunoperoxidase localisation in normal and neoplastic tissue. J. Clin. Pathol. *32:*35–39.

Higgins, P.J., Darzynkiewicz, Z., Melamed, M.R. (1983) Secretion of albumin and alpha-foetoprotein by dimethylsulphoxide-stimulated hepatocellular carcinoma cells. Br. J. Cancer *48:*485–493.

Higuchi, K. (1977) Cultivation of mammalian cell lines in serum-free chemically defined medium. Methods Cell Biol. *14:*131.

Hill, B.T. (1983) An overview of clonogenic assays for human tumour biopsies. In Dendy, P.P., Hill, B.T. (eds): "Human Tumour Drug Sensitivity Testing In Vitro." Academic Press, pp. 91–102.

Hilwig, I., Gropp, A. (1972) Staining of constitutive heterochromatin in mammalian chromosomes with a new fluorochrome. Exp. Cell Res. *75:*122–126.

Hokin, L.E., Hokin, M.R. (1963) Biological transport. Ann Rev Biochem. *32:*533–577.

Holbrock, K.A., and Hennings, H. (1983) Phenotypic expression of epidermal cells in vitro: a review. J. Invest. Dermatol. *81:*11s–24s.

Holden, H.T., Lichter, W., Sigel, M.M. (1973) Quantitative methods for measuring cell growth and death. In Kruse, P.F., Jr., Patterson, M.K., Jr (eds): "Tissue Culture Methods and Applications." New York, Academic Press, pp. 408–412.

Hollenberg, M.D., Cuatrecasas, P. (1973) Epidermal growth factor: Receptors in human fibroblasts and modulation of action by cholera toxin. Proc. Natl. Acad. Sci. *70:*2964–2968.

Holley, R.W., Armour, R., Baldwin, J.H. (1978) Density-dependent regulation of growth of BSC-1 cells in cell culture: Growth inhibitors formed by the cells. Proc. Natl. Acad. Sci. USA *75:*1864–1866.

Honn, K.V., Singley, J.A., Chavin, W. (1975) Fetal bovine serum: A multivariate standard (38805). Proc. Soc. Exptl. Biol. Med. *149:*344–347.

Horibata, K., Harris, A.W. (1970) Mouse myelomas and lymphomas in culture. Exp. Cell Res *60:*61–77.

Horita, A., Weber, L.J. (1964) Skin penetrating property of drugs dissolved in dimethylsulfoxide (DMSO) and other vehicles. Life Sci. *3:*1389–1395.

Hosokawa, M., Phillips, P.D., Cristofalo, V.J. (1986) The effect of dexamethasone on epidermal growth factor binding and stimulation of proliferation in young and senscent W138 cells. Exp. Cell Res. *164(2):*408.

House, W. (1973) Bulk culture of cell monolayers. In Kruse, P.F., Jr., Patterson, M.K., Jr. (eds): "Tissue Culture Methods and Applications." New York, Academic Press, pp. 338–344.

House, W., Shearer, M., Maroudas, N.G. (1972) Method for bulk culture of animal cells on plastic film. Exp. Cell Res. *71:*293–296.

Howard, M., Kessler, S., Chused, T., Paul, W.E. (1981) Long term culture of normal mouse B lymphocytes. Proc. Natl. Acad. Sci. USA *78:*5788–5792.

Howard, B.V., Macarak, E.J., Gunson, D., Kefalides, N.A. (1976) Characterization of the collagen synthesized by endothelial cells in culture. Proc. Natl. Acad. Sci. USA *73:*2361–2364.

Howie Report (1978) "Code of Practice for Prevention of Infection in Clinical Laboratories and Post-Mortem Rooms." London, H.M. Stationary Office.

Hume, D.A., Weidemann, M.J. (1980) Mitogenic lymphocyte transformation. Amsterdam, Elsevier/North Holland Biomedical Press.

Hurrell, J.G.R., ed. (1982) "Monoclonal Hybridoma Antibodies: Techniques and Applications." CRC Press, Boca Raton, Florida.

Hynes, R.O. (1973) Alteration of cell-surface proteins by viral transformation and by proteolysis. Proc. Natl. Acad. Sci. USA *70:*3170–3174.

Hynes, R.O. (1974) Role of cell surface alterations in cell transformation. The importance of proteases and cell surface proteins. Cell *1:*147–156.

Hynes, R.O. (1976) Cell surface proteins and malignant transformation. Biochim. Biophys. Acta *458:*73–107.

Ilsie, A.W., Puck, T.T. (1971) Morphological transformation of Chinese hamster cells by dibutyryl adenosine cycline 3':5'-monophosphate and testosterone. Proc. Natl. Acad. Sci. USA. *2:*358–361.

An international system for human cytogenetic nomenclature. (1978) Report of the Standing Committee on Human Cytogenetic Nomenclature. The National Foundation—March of Dimes.

Iscove, N., Melchers, F. (1978) Complete replacement of serum by albumin, transferrin and soybean lipid in cultures of lipopolysaccharide-reactive B lymphocytes. J. Exp. Med. *147:*923–933.

Iscove, N.N., Guilbert, L.W., Weyman, C. (1980) Complete replacement of serum in primary cultures of erythropoitin-dependent red cell precursors (CFU-E) by albumin, transferrin, iron, unsaturated fatty acid, lecithin and cholesterol. Exp. Cell Res. *126:*121–126.

Itagaki, A., Kimura, G. (1974) TES and HEPES buffers in mammalial cell cultures and viral studies: Problems of carbon dioxide requirements. Exp. Cell Res. *83:*351–360.

Jacobs, K., Shoemaker, C., Rudersdorf, R., Neill, S.D., Kaufman, R.J., et al. (1985) Isolation and characterisation of genomic and cDNA clones of human erythropoietin. Nature *313:*806–810.

Jaffe, E.A., Nachman, R.L., Becker, G.C., Minick, C.R. (1973) Culture of human endothelial cells derived from umbilical veins. J. Clin. Invest. *52:*2745–2744.

Jetten, A.M., Smets, H. (1985) Regulation and differentiation of tracheal epithelial cells by retinoids. In: "Retinoids: Differentiation and Disease." Ciba Foundation Symposium. London. Pitman, pp.61–76.

Jones, T.L., Haskill, J.S. (1973) Polyacrylamide: An improved surface for cloning of primary tumors containing fibroblasts. J. Natl. Cancer Inst. *51:*1575–1580.

Jones, T.L., Haskill, J.S. (1976) Use of polyacrylamide for cloning of primary tumors. Methods Cell Biol. *14:*195.

Kahn, P., Shin, S.-L. (1979) Cellular tumorigenicity in nude mice. Test of association among loss of cell-surface fibronectin, anchorage independence, and tumor-forming ability. J. Cell Biol. *82:*1.

Kaltenbach, J.P., Kaltenbach, M.H., Lyons, W.B. (1958) Nigrosin as a dye for differentiating live and dead ascites cells. Exp. Cell Res. *15:*112–117.

Kao, F.-T., Puck, T.T. (1968) Genetics of somatic mammalian cells. VII. Induction and isolation of nutritional mutants in Chinese hamster cells. Proc. Natl. Acad. Sci. USA *60:*1275–1281.

Kao, F.T., Chasin, L., Puck, T.T. (1969) Genetics of somatic mammalian cells, X. Complementation analysis of glycine-requiring auxotrophs. Proc. Natl. Acad. Sci., USA *64:*1284–1291.

Karasek, M.A. (1983) Culture of human keratinocytes in liquid medium. J. Invest. Dermatol. *81:*21s–28s.

Katsuta, H., ed. (1978) "Nutritional Requirements of Cultured Cells." Tokyo, Japan Scientific Societies Press/Baltimore, University Park Press.

Kaufmann, M., Klinga, K., Runnebaum, B, Kubli, F. (1980) *In vitro* adriamycin sensitivity test and hormonal receptors in primary breast cancer. Eur. J. Cancer *16:*1609–1613.

Kawamura, A., Jr., ed. (1969) "Fluorescent Antibody Techniques and Their Application." Tokyo, University of Tokyo Press.

Kawasaki, E.S., Ladner, M.B., Wang, A.M., van Arsdell, J., Waffen, M.K., et. al. (1985) Molecular cloning of a complementary DNA encoding human macrophage-specific colony-stimulating factor (SCF-1). Science *230:*291–296.

Kelley, D.S., Becker, J.E., Potter, V.R. (1978) Effect of insulin, dexamethasone, and glucagon on the amino acid transport ability of four rat hepatoma cell lines and rat hepatocytes in culture. Cancer Res. *38:*4591–4601.

Kempner, E.S., Miller, J.H. (1962) Autoradiographic resolution of doubly labeled compounds. Science *135:*1063–1064.

Kern, P.A., Knedler, A., Eckel, R.H. (1983) Isolation and culture of microvascular endothelium from human adipose tissue. J. Clin. Invest. *71:*1822–1829.

Kim, Y.S., Whitehead, J.S., Perdomo, J. (1979) Glycoproteins of cultured epithelial cells from human colonic adenocarcinoma and fetal intestine. Eur. J. Cancer *15:*725–735.

Kindler, V., Thorens, B., de Kossodo, S., Allet, B., Eliason, J.F. (1986) Stimulation of hematopoiesis *in vivo* by recombinant bacterial murine interleukin 3. Proc. Natl. Acad. Sci. USA *83:*1001–1005.

Kingsbury, A., Gallo, V., Woodhams, P.L., Balazs, R. (1985) Survival, morphology and adhesion properties of cerebellar interneurons cultured in chemically defined and serum-supplemented medium. Dev. Brain Res. *17:*17–25.

Kinlough, M.A., Robson, H.N. (1961) Chromosome preparations obtained directly from peripheral blood. Nature *192:*684.

Kissane, J.M., Robbins, E. (1958) The fluorometric measurement of deoxyribonucleic acid in animal tissues with specific reference to the central nervous system. J. Biol. Chem. *233:*184–188.

Kitos, P.A., Sinclair, R., Waymouth, C. (1962) Glutamine metabolism by animal cells growing in a synthetic medium. Exp. Cell Res. *27:*307–316.

Klann, R.C., Marchok, A.C. (1982) Effects of retinoic acid on cell proliferation and cell differentiation in a rat tracheal epithelial cell line. Cell Tissue Kinet. *15:*473–482.

Kleinman, H.K., McGoodwin, E.B., Rennard, S.I., Martin, G.R. (1981) Preparation of collagen substrates for cell attachment: Effect of collagen concentration and phosphate buffer. Anal. Biochem. *94:*308.

Klevjer-Anderson, P., Buehring, G.C. (1980) Effect of hormones on growth rates of malignant and nonmalignant human mammary epithelia in cell culture. In Vitro *16:*491–501.

Knazek, R.A. (1974) Solid tissue masses formed *in vitro* from cells cultured on artificial capillaries. Fed. Proc. *33:*1978–1981.

Knazek, R.A., Gullino, P., Kohler, P.O., Dedrick, R. (1972) Cell culture on artificial capillaries. An approach to tissue growth *in vitro.* Science *178:*65–67.

Knazek, R.A., Kohler, P.O., Gullino, P.M. (1974) Hormone production by cells grown *in vitro* on artificial capillaries. Exp. Cell Res. *84:*251.

Knox, P., Wells, P. (1979) Cell adhesion and proteoglycans. I. The effect of exogenous proteoglycans on the attachment of chick embryo fibroblasts to tissue cultured plastic and collagen. J. Cell Sci. *40:*77–88.

Kohler, G., Milstein, C. (1975) Continuous cultures of fused cells secreting antibody of predefined specificity. Nature *256:*495–497.

Konigsberg, I.R. (1979) Skeletal myoblasts in culture. In Jakoby, W.B., Pastan, I.H. (eds): "Methods in Enzymology, Vol. LVII, Cell Culture." New York, Academic Press, pp. 511–527.

Kopriwa, B.M. (1963) A model dark room unit for radioautography. J. Histochem. Cytochem. *11:*553–555.

Kosher, R.A., Church, R.L. (1975) Stimulation of *in vitro* somite chondrogenesis by procollagen and collagen. Nature *258:*327–330.

Kreisberg, J.L., Sachs, G., Pretlow, T.G.E., McGuire, R.A. (1977) Separation of proximal tubule cells from suspensions of rat kidney cell by free-flow electrophoresis. J. Cell Physiol. *93:*169–172.

Kreth, W., Herzenberg, L.A. (1974) Fluorescence-activated cell sorting of human T and B lymphocytes. Cell. Immunol. *12:*396–406.

Krog, H.H. (1976) Identification of inbred strains of mice, *Mus musculus.* I. Genetic control of mice using starch gel electropho-

resis. Biochem. Genet. *14:*319–326.

Kruse, P.F., Jr., Miedema, E. (1965) Production and characterization of multiple-layered populations of animal cells. J. Cell Biol. *27:*273.

Kruse, P.F., Jr., Keen, L.N., Whittle, W.L. (1970) Some distinctive characteristics of high density perfusion cultures of diverse cell types. In Vitro *6:*75–78.

Kuchler, R.J,. ed. (1974) "Animal Cell Culture and Virology." Stroudsburg, Pa., Dowden, Hutchinson & Ross, Inc.

Kuchler, R.J. (1977) "Biochemical Methods in Cell Culture and Virology." New York, Academic Press.

Kuriharcuch, W., Green, H. (1978) Adipose conversion of 3T3 cells depends on a serum factor. Proc. Natl. Acad. Sci. *75:*6107–6110.

Kurtz, J.W., Wells, W.W. (1979) Automated fluorometric analysis of DNA, protein, and enzyme activities: Application of methods in cell culture. Anal. Biochem. *94:*166.

Labarca, C., Paigen, K. (1980) A simple, rapid, and sensitive DNA assay procedure. Anal. Biochem. *102:*344–352.

Lan, S., Smith, H.S., Stampfer, M.R. (1981) Clonal growth of normal and malignant human breast epithelia. J. Surg. Oncol. *18:*317–322.

Langer, P.R., Waldrop, A.A., Ward, D.C. (1981) Enzymatic synthesis of biotin-labelled polynucleotides: Novel nucleic acid affinity probes. Proc. Natl. Acad. Sci., USA *78:*6633.

Langer-Safer, P.R., Levine, M., Ward, D.C. (1982) Immunological methods for mapping genes on *Drosophila* polytene chromosomes. Proc. Natl. Acad. Sci., USA *79:*4381.

Lasfargues, E.Y. (1973) Human mammary tumors. In Kruse, P, Patterson, M.K. (eds): "Tissue Culture Methods and Applications." New York, Academic Press, pp. 45–50.

Lasnitzki, I., Mizuno, T. (1979) Role of mesenchyme in the induction of the rat prostate gland by androgens in organ culture. J. Endocrinol. *82:*171.

Laug, W.E., Tokes, Z.A., Benedict, W.F., Sorgente, N. (1980) Anchorage independent growth and plasminogen activator production by bovine endothelial cells. J. Cell Biol. *84:*281–293.

Lebeau, M.M., Rowley, J.D. (1984) Heritable fragile sites in cancer. Nature *308:*607–608.

Lechner, J.F., LaVeck, M.A. (1985) A serum free method for culturing normal human bronchial epithelial cells at clonal density. J. Tissue Cult. Methods *9:*43–48.

Lechner, J.F. Haugen, A., Autrup, H., McClendon, I.A., Trump, B.F., Harris, C.C. (1981) Clonal growth of epithelial cells from normal adult human bronchus. Cancer Res. *41:*2294–2304.

Leder, A., Leder, P. (1975) Butyric acid, a potent inducer of erythroid differentiation in cultured erythroleukemic cells. Cell *5:*319–322.

Leibo, S.P., Mazur, P. (1971) The role of cooling rates in low-temperature preservation. Cryobiology *8:*447–452.

Leibovitz, A. (1963) The growth and maintenance of tissue cell cultures in free gas exchange with the atmosphere. Am. J. Hyg. *78:*173–183.

Leighton, J. (1951) A sponge matrix method for tissue culture. Formation or organized aggregates of cells *in vitro.* J. Natl. Cancer Inst. *12:*545–561.

Leighton, J., Mark, R., Rush, G. (1968) Patterns of three-dimensional growth in collagen coated cellulose sponge: Carcinomas and embryonic tissues. Cancer Res. *28:*286–296.

Lesser, B., Brent, T.P. (1970) Cold storage as a method for accumulating mitotic HeLa cells without impairing subsequent synchronous growth. Exp. Cell Res. *62:*470–473.

Levi-Montalcini, R. (1964) Growth control of nerve cells by a protein factor and its antiserum. Science *143:*105–110.

Levi-Montalcini, R.C.P. (1979) The nerve-growth factor. Sci. Am. *240:*68.

Levine, E.M., Becker, B.G. (1977) Biochemical methods for detecting mycoplasma contamination. In McGarrity, G.T., Murphy, D.G., Nichols, W.W. (eds): "Mycoplasma Infection of Cell Cultures." New York, Plenum Press, pp. 87–104.

Ley, K.D., Tobey, R.A. (1970) Regulation of initiation of DNA synthesis in Chinese hamster cells. II. Induction of DNA synthesis and cell division by isoleucine and glutamine in G_1-arrested cells in suspension culture. J. Cell Biol. 47:453–459.

Lieber, M., Mazzetta, J., Nelson-Rees, W., Kaplan, M., Todaro, G. (1975) establishment of a continuous tumor-cell line (PANC-1) from a human carcinoma of the exocrine pancreas. Int. J. Cancer 15:741–747.

Liebermann, D., Sachs, L. (1978) Nuclear control of neurite induction in neuroblastoma cells. Exp. Cell Res. 113:383–390.

Lillie, J.H., MacCallum, D.K., Jepsen, A. (1980) Fine structure of subcultivated stratified squamous epithelium grown on collagen rafts. Exp. Cell Res. 125:153–165.

Lim, R., Mitsunobu, K. (1975) Partial purification of morphological transforming factor from pig brain. Biochem. Biophys. Acta 400:200–207.

Limburg, H., Hekcmann, U. (1968) Chemotherapy in the treatment of advanced pelvic malignant disease with special reference to ovarian cancer. J. Obstet. Gynaecol. Br. Cwlth. 75:1246–1255.

Lin, C.C., Uchida, I.A. (1973) Fluorescent banding of chromosomes (Q-bands). In Kruse, P.F., Patterson, M.K. (eds): "Tissue Culture Methods and Applications" New York, Academic Press, pp. 778–781.

Lin, M.A., Latt, S.A., Davidson, R.L. (1974) Identification of human and mouse chromosomes in human-mouse hybrids by centromere fluorescence. Exp. Cell Res. 87:429–433.

Lindgren, A., Westermark, B., Ponten, J. (1975) Serum stimulation of stationary human glia and glioma cells in culture. Exp. Cell Res. 95:311–319.

Lindsay, R.M. (1979) Adult rat brain astrocytes support survival of both NGF-dependent and NGF-insensitive neurones. Nature 282:80.

Linser, P., Moscona, A.A. (1980) Induction of glutamine synthetase in embryonic neural retina-localization in Muller fibers and dependence on cell interaction. Proc. Natl. Acad. Sci. USA 76:6476–6481.

Littauer, U.Z., Giovanni, M.Y., Glick, M.C. (1979) Differentiation of human neuroblastoma cells in culture. Biochem. Biophy. Res. Commun. 88:933–939.

Littlefield, J.W. (1964) Selection of hybrids from matings of fibroblasts in vitro and their presumed recombinants. Science 145:709–710.

Litwin, J. (1973) Titanium disks. In Kruse, P.F., Patterson, M.K. (eds): "Tissue Culture Methods and Applications." New York, Academic Press, pp. 383–387.

Lloyd, K.O., Travassos, L.R., Takahashi, T., Old, L.J. (1979) Cell surface glycoproteins of human tumor cell lines: Unusual characteristics of malignant melanoma. J. Natl Cancer Inst. 63:623.

Lord, B.I., Molineux, G., Testa, N.G., Kelly, M., Spooncer, E., Dexter, T.M. (1986) The kinetic response of haemopoietic precursor cells, in vivo, to highly purified recombinant interleukin-3. Lymphokine Res. 5:97–104.

Lovelock, J.E., Bishop, M.W.H. (1959) Prevention of freezing damage to living cells by dimethyl sulphoxide. Nature 183:1394–1395.

Lowry, O.N., Rosebrough, N.J., Farr, A.L., Randall, R.J. (1951) Protein measurement with the folin phenol reagent. J. Biol. Chem. 193:265–275.

MacDonald, C.M., Freshney, R.I., Hart, E., Graham, D.I. (1985) Selective control of human glioma cell proliferation by specific cell interaction. Exp. Cell Biol. 53:130–137.

Maciag, T., Cerondolo, J., Ilsley, S., Kelley, P.R., Forand, R. (1979) Endothelial cell growth factor from bovine hypothalamus-identi-fication and partial characterization. Proc. Natl. Acad. Sci. USA 76:5674–5678.

Macieira-Coelho, A. (1973) Cell cycle analysis. A. Mammalian cells. In Kruse, P.F., Patterson, M.K. (eds): "Tissue Culture Methods and Applications." New York, Academic Press, pp. 412–422.

Macklis, J.D., Sidman, R.L., Shine, H.D. (1985) Cross-linked collagen surface for cell culture that is stable, uniform, and optically superior to conventional surfaces. In Vitro 21:189–194.

Macpherson, I. (1973) Soft agar technique. In Kruse, P.F., Patterson, M.K. (eds): "Tissue Culture Methods and Applications." New York, Academic Press, pp. 276–280.

Macpherson, I., Bryden, A. (1971) Mitomycin C treated cells as feeders. Exp. Cell Res. 69:240–241.

Macpherson, I., Montagnier, L. (1964) Agar suspension culture for the selective assay of cells transformed by polyoma virus. Virology 23:291–294.

Macpherson, I., Stoker, M. (1962) Polyoma transformation of hamster cell clones-an investigation of genetic factors affecting cell competence. Virology 16:147.

Macy, M. (1978) Identification of cell line species by isoenzyme analysis. Man. Am. Tissue Cult. Assoc. 4:833–836.

Mahdavi, V., Hynes, R.O. (1979) Proteolytic enzymes in normal and transformed cells. Biochim. Biophys. Acta 583:167–178.

Malan-Shibley, L., Lype, P.T. (1981) The influence of culture conditions on cell morphology and tyrosine aminotransferase levels in rat liver epithelial cell lines. Exp. Cell Res. 131:363–371.

Malinin, T.I., Perry, V.P. (1967) A review of tissue and organ viability assay. Cryobiology 4:104–115.

Maltese, W.A., Volpe, J.J. (1979) Induction of an oligodendrogilial enzyme in C-6 glioma cells maintained at high density or in serum-free medium. J. Cell Physiol. 101:459–470.

Maramorosch, K. (1976) "Invertebrate Tissue Culture." New York, Academic Press.

Marcus, M., Lavi, U., Nattenberg, A., Ruttem, S., Markowitz, O. (1980) Selective killing of mycoplasmas from contaminated cells in cell cultures. Nature 285:659–660.

Mardh, P.H. (1975) Elimination of mycoplasmas from cell cultures with sodium polyanethol sulphonate. Nature 254:515–516.

Mareel, M., Kint, J., Meyvisch, C. (1979) Methods of study of the invasion of malignant C3H-mouse fibroblasts into embryonic chick heart in vitro. Virchows Arch. B. Cell Pathol. 30:95–111.

Mareel, M.M., Bruynell, E., Storme, G. (1980) Attachment of mouse fibrosarcoma cells to precultured fragments of embryonic chick heart. Virchows Arch. B. Cell Pathol. 34:85–97.

Mark J. (1971) Chromosomal characteristics of neurogenic tumours in adults. Hereditas 68:61–100.

Martin, G.R. (1975) Teratocarcinomas as a model system for the study of embryogenesis and neoplasia. Cell 5:229–243.

Martin, G.R. (1978) Advantages and limitations of teratocarcinoma stem cells as models of development. In Johnson, M.H. (ed): "Development in Mammals, Vol. 3." Amsterdam, North-Holland Publishing Co., p. 225.

Martin, G.R., Evans, M.J. (1974) The morphology and growth of a pluripotent teratocarcinoma cell line and its derivatives in tissue culture. Cell 2:163–172.

Mather, J. (1979) Testicular cells in defined medium. In Jakoby, W.B., Pastan, I.H. (eds): "Methods in Enzymology, Vol. LVII, Cell Culture." New York, Academic Press, p. 103.

Mather, J.P. Sato, G.H. (1979a) The growth of mouse melanoma cells in hormone supplemented, serum-free medium. Exp. Cell Res. 120:191.

Mather, J.P., Sato, G.H. (1979b) The use of hormone supplemented serum free media in primary cultures. Exp. Cell Res. 124:215.

Matsamura, T., Nitta, K., Yoshikawa, M., Takaoka, T., Katsuta, H. (1975) Action of bacterial protease on the dispersion of mammalian cells in tissue culture. Jpn. J. Exp. Med. *45:*383–392.

Maurer, H.R. (1986) Towards chemically-defined, serum-free media for mammalian cell culture. In Freshney, R.I. (ed): "Animal Cell Culture—a Practical Approach." Oxford, IRL Press, pp. 13–30.

Mazur, P., Leibo, S.P., Farrant, J., Chu, E.H.Y., Hanna, M.G., Jr., Smith, C.H. (1970) Interactions of cooling rate, warming rate and protective additive on the survival of frozen mammalian cells. In Wolstenholme, G.E.W., O'Conor, M (eds): "The Frozen Cell." CIBA Foundation Symposium. London, J.A. Churchill, pp. 69–85.

McCall, E., Povey, J., Dumonde, D.C. (1981) The culture of vascular endothelial cells on microporous membranes. Thromb. Res. *24:*417–431.

McCool, D., Miller, R.J., Painter, R.H., Bruch, W.R. (1970) Erythropoietin sensitivity of rat bone marrow cells separated by velocity sedimentation. Cell Tissue Kinet. *3:*55–66.

McCoy, T.A., Maxwell, M., Kruse, P.F. (1959) Amino acid requirements of the Novikoff hepatoma in vitro. Proc. Soc. Exp. Biol. Med. *100:*115–118.

McGarrity, G.J. (1982) Detection of mycoplasmic infection of cell cultures. In Maramorosch, K. (ed): "Advances in Cell Culture, Vol. 2." New York, Academic Press, pp. 99–131.

McKay, I., Taylor-Papadimitriou, J. (1981) Junctional communication pattern of cells cultured from human milk. Exp. Cell Res. *134:*465–470.

McKeehan, W.L. (1977) The effect of temperature during trypsin treatment on viability and multiplication potential of single normal human and chicken fibroblasts. Cell Biol. Int. Rep. *1:*335–343.

McKeehan, W.L., Ham, R.G. (1976) Stimulation of clonal growth of normal fibroblasts with substrata coated with basic polymers. J.Cell Biol. *71:*727–734.

McKeehan, W.L., McKeehan, K.A. (1979) Oxocarboxylic acids, pyridine nucleotide-linked oxidoreductases and serum factors in regulation of cell proliferation. J. Cell Physiol. *101:*9–16.

McKeehan, W.L., Hamilton, W.G., Ham, R.G. (1976) Selenium is an essential trace nutrient for growth of WI-38 diploid human fibroblasts. Proc. Natl. Acad. Sci. USA *73:*2023–2027.

McKeehan, W.L., McKeehan, K.A. Hammond, S.L., Ham, R.G. (1977) Improved medium for clonal growth of human diploid cells at low concentrations of serum protein. In Vitro *13:*399–416.

McKeehan, W.L., Adams, P.S., Rosser, M.P. (1984) Direct mitogenic effects of insulin, epidermal growth factor, cholera toxin, unknown pituitary factors and possibly prolactin, but not androgen, on normal rat prostate epithelial cells in serum-free primary cell culture. Cancer Res. *44:*1998–2010.

McLean, J.S., Frame, M.C., Freshney, R.I., Vaughan, P.F.T., Mackie, A.E. (1986) Phenotypic modification of human glioma and non-small cell lung carcinoma by glucocorticoids and other agents. Anticancer Res. *6:*1101–1106.

McLimans, W.F. (1979) Mass culture of mammalian cells. In Jakoby, W.B., Pastan, I.H. (eds): "Methods in Enzymology, Vol. LVII, Cell Culture." New York, Academic Press, pp. 194–211.

Meera Khan, P. (1971) Enzyme electrophoresis on cellulose acetate gel: Zymogram patterns in man-mouse and man-Chinese hamster somatic cell hybrids. Arch. Biochem. Biophys. *145:*470–483.

Meier, S., Hay, E.D. (1974) Control of corneal differentiation by extracellular materials. Collagen as a promoter and stabilizer of epithelial stroma production. Dev. Biol. *38:*249–270.

Meier, S., Hay, E.D. (1975) Stimulation of corneal differentiation by interaction between cell surface and extracellular matrix. J. Cell Biol. *66:*275–291.

Meistrich, M.L., Meyn, R.E., Barlogie, B. (1977a) Synchronization of mouse L-P59 cells by centrifugal elutriation separation. Exp. Cell Res *105:*169.

Meistrich, M.L., Gordina, D.J. Meyn, R.E., Barlogie, B. (1977b)

Separation of cells from mouse solid tumors by centrifugal elutriation. Cancer Res. *37:*4291–4296.

Melamad, M.R., Mullaney, P.F., Mendelsohn, M.L. (1979) "Flow Cytometry and Sorting." New York, John Wiley & Sons.

Melera, P.W., Wolgemuth, D., Biedler, J.L., Hession, C. (1980) Antifolate-resistant Chinese hamster cells. Evidence from independently derived sublines for the overproduction of two dihydrofolate reductases encoded by different mRNAs. J. Biol. Chem. *255:*319–322.

Messer, A. (1977) The maintenance and identification of mouse cerebellar granule cells in monolayer culture. Brain Res. *130:*1–12.

Messing, E.M., Fahey, J.L., deKernion, J.B., Bhuta, S.M., Bubbers, J.E. (1982) Serum-free medium for the in vitro growth of normal and malignant urinary bladder epithelial cells. Cancer Res. *42:*2392–2397.

Metcalf, D. (1970) Studies on colony formation *in vitro* by mouse bone marrow cells. J. Cell Physiol. *76:*89–100.

Metcalf, D. (1985a) Haemopoietic colonies *in vitro*. In: "Recent Results in Cancer Research 61." Berlin, Springer-Verlag, p. 227.

Metcalf, D. (1985b) The granulocyte-macrophage colony stimulating factors. Science *229:*16–22.

Metcalf, D. (1986) The molecular biology and functions of the granulocyte macrophage colony-stimulating factors. Blood, *67:*257–267.

Metcalf, D., Begley, C.G., Nicola, N.A., Lopez, A.F., Williamson, D.J. (1986) Effects of purified bacterially synthesized murine multi-CSF (IL-3) on hematopoiesis in normal adult mice. Blood *68:*46–57.

Michalopoulos, G., Pitot, H.C. (1975) Primary culture of parenchymal liver cells on collagen membranes. Fed.Proc. *34:*826.

Miller, D.R., Hamby, K.M., Allison, D.P., Fischer, S.M., Slaga T.J. (1980) The maintenance of a differentiated state in cultured mouse epidermal cells. Exp. Cell Res. *129:*63–71.

Miller, G.G., Walker, G.W.R., Giblack, R.E. (1972) A rapid method to determine the mammalian cell cycle. Exp. Cell Res. *72:*533–538.

Miller, R.G., Phillips, R.A. (1969) Separation of cells by velocity sedimentation. J. Cell. Physiol. *73:*191–201.

Milo, G.E., Ackerman, G.A., Noyes, I. (1980) Growth and ultrastructural characterization of proliferating human keratinocytes *in vitro* without added extrinsic factors. In Vitro *16:*20–30.

Milstein, C., Galfre, G., Secher, D.S., Springer, T. (1979)Mini review, monoclonal antibodies and cells surface antigens. Cell Biol. Internat. Rep. *3:*1–16.

Minna, J., Gilman, A. (1973) Genetic analysis of the nervous system using somatic cell hybrids. In Davidson, R.L., de la Cruz, F.F. (eds): "Somatic Cell Hybridization." New York, Raven Press pp. 191–196.

Minna, J., Glazer, D., Nirenberg, M. (1972) Genetic dissection of neural properties using somatic cells hybrids. Nature New Biol. *235:*225–231.

Minna J.D., Carney, D.N., Cuttitta, F., Gazdar, A.F. (1983) The biology of lung cancer. In Chabner, B. (ed): "Rational Basis for Chemotherapy." New York, Alan R. Liss, Inc.

Moll, R., Franks, W.W. Schiller, D.L. (1982) The catalog of human cytokeratins: Patterns of expression in normal epithelia, tumours and cultured cells. Cell *31:*11–24.

Montagnier, L. (1968) Corrélation entre la transformation des cellule BHK21 et leur résistance aux polysaccharides acides en milieu gélifié. C.R. Acad. Sci. D. *267:*921–924.

Montes de Oca, F., Macy, M.L., Shannon, J.E. (1969) Isoenzyme characterization of animal cell cultures. Proc. Soc. Exp. Biol. Med. *132:*462–469.

Moore, G.E., Gerner, R.E. Franklin, H.A. (1967) Culture of normal human leukocytes. J. Am. Med. Assoc. *199:*519–524.

Morasca, L., Etba, E. (1986) Flow cytometry. In Freshney, R.I.(ed): "Animal Cell Culture—a Practical Approach." Oxford, IRL Press, pp. 125–148.

Morgan, J.G., Morton, H.J., Parker, R.C. (1950) Nutrition of animal cells in tissue culture. I. Initial studies on a synthetic medium. Proc Soc. Exp. Biol Med. 73:1.

Morton, H.J. (1970) A survey of commercially available tissue culture media. In Vitro 6:89–108.

Moscona, A.A. (1952) Cell suspension from organ rudiments of chick embryos. Exp. Cell Res. 3:535.

Moscona, A.A., Piddington, R. (1966) Stimulation by hydrocortisone of premature changes in the developmental pattern of glutamine synthetase in embryonic retina. Biochim. Biophys. Acta 121:409–411.

Moss, P.S., Strohman, R.C. (1976) Myosin synthesis by fusion-arrested chick embryo myoblasts in cell culture. Dev. Biol. 48:431–437.

Moss, P.S., Spector, D.H., Glass, C.A., Strohman, R.C. (1984) Streptomycin retards the phenotypic maturation of chick myogenic cells. In Vitro 20:473–478.

Munthe-Kaas, A.C., Seglen, P.O. (1974) The use of metrizamide as a gradient medium for isopycnic separation of rat liver cells. FEBS Lett. 43:252–256.

Nagy, B., Ban, K., Bradar, B. (1977) Fibrinolysis associated with human neoplasia: Production of plasminogen activator by human tumors. Int. J. Cancer 19:614–620.

Nardone, R.M., Todd, G., Gonzalezx, P., Gaffney, E.V. (1965) Nucleoside incorporation into strain L cells: Inhibition by pleuropneumonia-like organisms. Science 149:1100–1101.

Nelson, P.G., Lieberman, M. (1981) "Excitable Cells in Tissue Culture." New York, Plenum.

Nelson-Rees, W., Flandermeyer, R.R. (1977) Inter- and intraspecies contamination of human breast tumor cell lines HBC and BrCa5 and other cell cultures. Science 195:1343–1344.

Neugut, A.I., Weinstein, I.B. (1979) Use of agarose in the determination of anchorage-independent growth. In Vitro 15:351.

Newton, A.A., Wildy, P. (1959) Parasynchronous division of HeLa cells. Exp. Cell Res. 16:624–635.

Nicola, N.A., Begley, C.A., Metcalf, D. (1985) Identification of the human analogue of a regulator that induces differentiation in murine leukamic cells. Nature 314:625–628.

Nicola, N.A., Metcalf, D., Matsumoto, M., Johnson, G.R. (1983) Purification of a factor inducing differentiation in murine myelomonocytic leukaemic cells. J. Biol. Chem. 258:9017–9023.

Nichols, E.A., Ruddle, F.H. (1973) A review of enzyme polymorphism, linkage and electrophoretic conditions for mouse and somatic cell hybrids in starch gels. J. Histochem. Cytochem. 21:1066–1081.

Nicolson, G.L. (1976) Trans-membrane control of the receptors on normal and tumor cells. II. Surface changes associated with transformation and malignancy. Biochim. Biophys. Acta 458:1–72.

Nicosia, R.F., Leighton, J. (1981) Angiogenesis in vitro: Light microscopic, radioautographic and ultrastructural studies of rat aorta in histophysiological gradient cluture. In Vitro 17:204.

Nicosia, R.F., Tchao, R., Leighton, J. (1983) Angiogenesis-dependent tumor spread in reinforced fibrin clot culture. Cancer Res. 43:2159–2166.

Nilos, R.M., Makarski, J.S. (1978) Control of melanogenesis in mouse melanoma cells of varying metastatic potential. J. Natl. Cancer Inst. 61:523–526.

Nishihira, T., Kasai, M., Hayashi, Y., Kimura, M., Matsumura, Y., Akaishi, T., Ishiguro, S., Kataoka, S., Watanabe, H., Miura, Y., Sato, H. (1981) Experimental studies on differentiation of cells originated from human neural crest tumours in vitro and in vivo. Cell Molec. Biol. 27:181–196.

Noble, P.B., Levine, M.D. (1986) Computer-assisted analyses of cell locomotion and chemotaxis. Boca Raton, Florida, CRC Press.

Noguchi, P., Wallace, R., Johnson, J., Earley, E.M., O'Brien, S., Ferrone, S., Pellegrino, M.A., Milstein, J., Needy, C., Browne, W., Petricciani, J. (1979) Characterization of WiDr: A human colon carcinoma cell line. In Vitro 15:401.

Norwood, T.H., Zeigler, C.J., Martin, G.M. (1976) Dimethyl sulphoxide enhances polyethylene glycol-mediated somatic cell fusion. Somatic Cell Genet. 2:263–270.

Novak, J. (1962) A high-resolution autoradiographic apposition method for water-soluble tracers and tissue constituents. Int. J. Appl. Radiat. Isot. 13:187–190.

O'Brien, S.J., Kleiner, G., Olson, R., Shannon, J.E. (1977) Enzyme polymorphisms as genetic signatures in human cells cultures. Science 195:1345–1348.

O'Brien, S.J., Shannon, J.E., Gail, M.H. (1980) Molecular approach to the identification and individualization of human and animal cells in culture: Isozyme and allozyme genetic signatures. In Vitro 16:119–135.

O'Farrell, P.H. (1975) High resolution two-dimensional electrophoresis of proteins. J. Biol. Chem. 250:4007–4021.

O'Garra, A., Warren, D.J., Holman, M., Popham, A.M., Sanderson, C.J. et al. (1986) Interleukin 4 (B-Cell growth factor II/eosinophil differentiation factor) is a mitogen and differentiation factor for preactivated murine B lymphocytes. Proc. Natl. Acad. Sci. USA 83:5228–5232.

O'Hare, M.J., Ellison, M.L., Neville, A.M. (1978) Tissue culture in endocrine research: Perspectives, pitfalls, and potentials. Curr. Top. Exp. Endocrinol. 3:1–56.

Okazaki, K., Holtzer, H. (1966) Myogenesis: Fusion, myosin synthesis and the mitotic cycle. Proc. Acad. Sci. USA 56:1484–1489.

Olmsted, C.A. (1967) A physico-chemical study of fetal calf sera used as tissue culture nutrient correlated with biological tests for toxicity. Cell Res. 48:283–299.

Olsson, I., Ologsson, T. (1981) Induction of differentiation in a human promyelocytic leukemic cell line (HL-60). Exp. Cell Res 131:225–230.

Oram, J.D., Downing, R.G., Akrigg, A., Dollery, A.A., Duggleby, C.J., Wilkinson, G.W.G., Greenaway, P.J. (1982) Use of recombinant plasmids to investigate the structure of the human cytomegalovirus genome. J. Gen. Virol. 59:111–129.

Orly, J., Sato, G., Erickson, G.F. (1980) Serum suppresses the expression of hormonally induced function in cultured granulosa cells. Cell 20:817–827.

Osborne, C.K., Hamilton, B., Tisus, G., Livingston, R.B. (1980) Epidermal growth factor stimulation of human breast cancer cells in culture. Cancer Res. 40:2361–2366.

Owens, R.B., Smith, H.S., Hackett, A.J. (1974) Epithelial cell culture from normal glandular tissue of mice. Mouse epithelial cultures enriched by selective trypsinisation. J. Natl. Cancer Inst. 53:261–269.

Pahlman, S., Ljungstedt-Pahlman, A., Sanderson, P.J., Ward, G.A., Hermon-Taylor, J. (1979) Isolation of plasma-membrane components from cultured human pancreatic cancer cells by immunoaffinity chromatography of anti-βM Sepharose 6MB. Br. J. Cancer 40:701.

Parenjpe, M.S., Boone, C.W., Ande Eaton, S.del. (1975) Selective growth of malignant cells by in vitro incubation on Teflon. Exp. Cell Res. 93:508–512.

Paris Conference (1971), Supplement (1975): Standardization in human cytogenetics. Cytogenet. Cell Genet. 15:201–238.

Parks, W.M., Gingrich, R.D., Dahle, C.E., Hoak, J.C. (1985) Identification and characterization of an endothelial, cell-specific antigen with a monoclonal antibody. Blood 66:816–823.

Pastan, I. (1979) Cell transformation. In Jakoby, W.B., Pastan, I.H.

(eds): "Methods in Enzymology, Vol. LVII, Cell Culture." New York, Academic Press, pp. 368–370.

Patueleia, M.C., Friend, C. (1967) Tissue culture studies on murine virus-induced leukemia cells: Isolation of single cells in agar-liquid medium. Cancer Res. 27:726–730.

Paul, J. (1975) "Cell and Tissue Culture." Edinburgh, Churchill Livingstone, pp. 172–184.

Paul, J., Conkie, D., Freshney, R.I. (1969) Erythropoietic cell population changes during the hepatic phase of erythropoiesis in the foetal mouse. Cell Tissue Kinet. 2:283–294.

Paul, J., Fottrell, P.F. (1961) Molecular variation in similar enzymes from different species. Ann. NY Acad. Sci. 94:668–677.

Paul, W.E.., Sredni, B., Schwartz, R. H. (1981) Long-term growth and cloning of non-transformed lymphocytes. Nature 294:697–699.

Pearse, A.G.E. (1968) "Histochemistry, Theoretical and Applied." Boston, Little, Brown & Co., pp. 255–264.

Peehl D. M., Ham, R.G. (1980) Clonal growth of human keratinocytes with small amounts of dialysed serum. In Vitro 16:526–540.

Pereira, M.E.A., Kabat, E.A. (1979) A versatile immunoadsorbent capable of binding lectins of various specificities and its use for the separation of cell populations. J Cell Biol. 82:185–194.

Perper, R.J., Zee, T.W., Mickelson, M.M. (1968) Purification of lymphocytes and platelets by gradient centrifugation. J. Lab. Clin. Med 72:842–868.

Pertoft, H., Laurent, T.C. (1977) Isopycnic separation of cells and cell organelles by centrifugation in modified colloidal silica gradients. In Catsimpoolas, N. (ed): "Methods of Cell Separation." New York, Plenum Press.

Petersen, D.F., Anderson, E.C., Tobey, R.A. (1968) Mitotic cells as a source of synchronized cultures. In Prescott, D.M. (ed): "Methods in Cell Physiology." New York, Academic Press, pp. 347–370.

Peterson, E.A., Evans, W.H. (1967) Separation of bone marrow cells by sedimentation at unit gravity. Nature 214:824–825.

Petricciani, J.C., Hoops, H.E., Chapple, P.J., eds (1979) "Cell Substrates: Their Use in the Production of Vaccines and Other Biologicals." New York, Plenum Press.

Phillips, P., Steward, J.K., Kumar, S. (1976) Tumor angiogenesis factor (TAF) in human and animal tumors. Intl. J. Cancer 17:549–558.

Phillips, P., Kumar, P., S., Waghe, M. (1979) Isolation and characterization of endothelial cells from adult rat brain white matter. J. Anat. 129:261.

Pike, B.L., Robinson, W.A. (1970) Human bone marrow colony growth in agar-gel. J. Cell Physiol. 76:77–84.

Pilot, H., Periano, C., Morse, P., Potter, V.R. (1964) Hepatomas in tissue culture compared with adapting liver in vitro. Natl. Cancer Inst. Monogr. 13:229–245.

Platsoucas, C.D., Good, R.A., Gupta, S. (1979) Separation of human lymphocyte-T subpopulations (T-mu, T-gamma) by density gradient electrophoresis. Proc. Natl. Acad. Sci. USA 76:1972.

Pluznik, D.H., Sachs, L. (1965) The cloning of normal "mast" cells in tissue culture. J. Cell. Comp. Physiol. 66:319–324.

Polinger, I.S. (1970) Separation of cell types in embryonic heart cell cultures. Exp. Cell Res. 63:78–82.

Pollack, M.S., Heagney, S.D., Livingston, P.O., Fogh, J. (1981) HLA-A, B, C & DR alloantigen expression on forty-six cultured human tumor cell lines. J. Natl. Cancer Inst. 66:1003–1012.

Pollack, R., ed. (1981) "Reading in Mammalian Cell Culture, 2nd Edition." Cold Spring Harbor, New York, Cold Spring Harbor Laboratory.

Pollock, M.F., Kenny, G.E. (1963) Mammalian cell cultures contaminated with pleuro-pneumonia-like organisms. III. Elimination of pleuro-pneumonia-like organisms with specific antiserum. Proc.

Soc. Exp. Biol. Med. 112:176–181.

Pontén, J. (1975) Neoplastic human glia cells in culture. In Fogh, J. (ed): "Human Tumor Cells in vitro. "New York, Plenum Publishing, pp. 175–206.

Pontén, J., Macintyre, E. (1968) Interaction between normal and transformed bovine fibroblasts in culture. II. Cells transformed by polyoma virus. J. Cell Sci. 3:603–668.

Pontén, J., Westermark, B. (1980) Cell generation and aging of nontransformed glial cells from adult humans. In Fedorof, S., Hertz, L. (eds): "Advances in Cellular Neurobiology, Vol. 1." New York, Academic Press, pp. 209–227.

Pontecorvo, G. (1975) Production of mammalian somatic cell hybrids by means of polyethylene glycol treatment. Somat. Cell Genet. 1:397–400.

Post, M., Floros, J., Smith, B.T. (1984) Inhibition of lung maturation by monoclonal antibodies against fibroblast-pneumocyte factor. Nature 308:284–286.

Povey, S., Hopkinson, D.A., Harris, H. and Franks, L.M. Characterization of human cell lines and differentiation from HeLa by enzyme typing. Nature 264:60–63, 1976.

Pretlow, T.G. (1971) Estimation of experimental conditions that permit cell separations by velocity sedimentation on isokinetic gradients of Ficoll in tissue culture medium. Anal. Biochem. 41:248–255.

Pretlow, T.P., Stinson, A.J., Pretlow, T.G., Glover, G.L. (1978) Cytologic appearance of cells dissociated from rat colon and their separation by isokinetic and isopyknic sedimentation in gradients of Ficoll. J. Natl. Cancer Inst. 61:1431–1437

Prince, G.A., Jenson, A.B., Billups, L.C., Notkins, A.L. (1978) Infection of human pancreatic beta cell cultures with mumps virus. Nature 271:158–161.

Prop, F.J.A., Wiepjes, G.J. (1973) Sequential enzyme treatment of mouse mammary gland. In Kruse, P.F., Patterson, M.K. (eds): "Tissue Culture Methods and Applications." New York, Academic Press, pp. 21–24.

Puck, T.T., Marcus, P.I. (1955) A rapid method for viable cell titration and clone production with HeLa cells in tissue culture: The use of X-irradiated cells to supply conditioning factors. Proc. Natl. Acad. Sci. USA 41:432–437.

Prunieras, M., Regnier, M., and Woodley, D. (1983) Methods for cultivation of keratinocytes with an air-liquid interface. J. Invest. Dermatol. 81:28s–33s.

Quarles, J.M., Morris, N.G., Leibovitz, A. (1980) Carcinoembryonic antigen production by human colorectal adenocarcinoma cells in matrix-perfusion culture. In Vitro 16:113–118.

Quastler, H. (1963) The analysis of cell population kinetics. In Lamerton, L.F., Fry, R.J.M. (eds): "Cell Proliferation." Philadelphia, Davis. pp. 18–34.

Quintanilla, M., Brown, K., Ramsden, M., Balmain, A. (1986) Carcinogen specific mutation and amplification of Ha-ras during mouse skin carcinogenesis. Nature 322:78–79.

Rabito, C.A., Tchao, R., Valentich, J., Leighton, J. (1980) Effect of cell substratum interaction of hemicyst formation by MDCK cells. In Vitro 16:461–468.

Raff, M.C., Fields, K.L., Hakomori, S.L., Minsky, R., Pruss, R.M., Winter, J. (1979) Cell-type-specific markers for distinguishing and studying neurons and the major classes of glial cells in culture. Brain Res. 174:283–309.

Raff, M.C., Abney, E., Brockes, J.P., Hornby-Smith, A. (1978) Schwann cell growth factors. Cell 15:813–822.

Reddy, J.K., Rao, M.S., Warren, J.R., Minnick, O.T. (1979) Concanavalin A agglutinability and surface microvilli of dissociated normal and neoplastic pancreatic acinar cells of the rat. Exp. Cell Res. 120:55–61.

Reel, J.R., Kenney, F.T. (1968) "Superinduction" of tyrosine transaminase in hepatoma cell cultures: Differential inhibition of synthesis and turnover by actinomycin D. Proc. Natl. Acad. Sci. USA *61*:200–206.

Reid, L.M., Rojkind, M. (1979) New techniques for culturing differentiated cells: Reconstituted basement membrane rafts. In Jakoby, W.B., Pastan, I.H. (eds.): "Methods in Enzymology, Vol. LVII, Cell Culture." New York, Academic Press, pp. 263–278.

Reitzer, L.J., Wice, B.M., Kennel, D. (1979) Evidence that glutamine, not sugar, is the major energy source for cultured HeLa cells. J. Biol. Chem. *254*:2669–2677.

Richler, C., Yaffe, D. (1970) The in vitro cultivation and differentiation capacities of myogenic cell lines. Dev. Biol. *23*:1–22.

Richmond, A., Lawson, D.H., Nixon, D.W., Chawla, R.K. (1985) Characterization of autostimulatory and transforming growth factors from human melanoma cells. Cancer Res. *45*:6390–6394.

Rickwood, D., Birnie, G.D. (1975) Metrizamide, a new density-gradient medium. FEBS Lett. *50*:102–110.

Riddle, P.N. (1979) Time-lapse cinemicroscopy. In Treherne, J.E., Rubery, P.H. (eds): "Biological Techniques Series." Academic Press.

Rifkin, D.B., Loeb, J.N., Moore, G., Reich, E. (1974) Properties of plasminogen activators formed by neoplastic human cell cultures. J. Exp. Med. *139*:1317–1328.

Rindler, M.J., Chuman, L.M., Shaffer, L., Saier, M.H., Jr. (1979) Retention of differentiated properties in an established dog kidney epithelial cell line (MDCK). J. Cell Biol. *81*:635–648.

Rockwell, G.A., Sato, G.H., McClure, D.B. (1980) The growth requirements of SV40 virus transformed Balb/c-3T3 cells in serum-free monolayer culture. J. Cell Physiol. *103*:323–331.

Rogers, A.W. (1979) "Techniques of Autoradiography (3rd Edition)." The Netherlands, Elsevier/North-Holland Biomedical Press.

Rojkind, et al. (1980) Connective tissue biomatrix: Its isolation and utilization for long term cultures of normal rat hepatocytes. J. Cell Biol. *87*:255–263.

Rosenberg, M.D. (1965) The culture of cells and tissues at the saline-fluorocarbon interface. In Ramakrishnan, C.V. (ed): "Tissue Culture." The Hague, W. Junk, pp. 93–107.

Ross, R. (1971) The smooth muscle cell. II. Growth of smooth muscle in culture and formation of elastic fibers. J. Cell Biol. *50*:172–186.

Rossi, G.B., Friend, C. (1967) Erythrocytic maturation of (Friend) virus-induced leukemic cells in spleen clones. Proc. Natl. Acad. Sci. USA *58*:1373–1380.

Rothfels, K. H., Siminovitch, L. (1958) An air drying technique for flattening chromosomes in mammalian cells growth in vitro. Stain Technol. *33*:73–77.

Rotman, B., Papermaster, B.W. (1966) Membrane properties of living mammalian cells as studied by enzymatic hydrolysis of fluorogenic esters. Proc. Natl. Acad. Sci. USA *55*:134–141.

Ruoff, N.M. and Hay, R.J. (1979) Metabolic and temporal studies on pancreatic exocrine cells in culture. Cell Tissue Res. *204*:243–252, 1979.

Rutzky, L.P., Tomita, J.T., Calenoff, M.A., Kahan, B.D. (1979) Human colon adenocarcinoma cells. III. In vitro organoid expression and carcino-embryonic antigen kinetics in hollow fiber culture. J. Natl. Cancer Inst. *63*:893.

Sachs, L. (1982) Normal development programmes in myeloid leukaemia: Regulatory proteins is the control of growth and differentiation. Cancer Surveys *1*:321–342.

Said, J.W., Nash, G., Sassoon, A.F., Shintaku, I.P., Banks-Schlegel, S. (1983) Involucrin in lung tumours. A specific marker for squamous differentiation. Lab Invest. *49*:563–568.

Salpeter, M.M. (1974) Electron microscope autoradiography: A personal assessment. In Wisse, E., Daems, W.Th., Molenaar, I., van Duijn, P. (eds): "Electron Microscopy and Cytochemistry." Amsterdam, North-Holland Publishing Company, pp. 315–326.

Sandberg, A.A. (1980) "The Chromosomes in Human Cancer and Leukaemia." Amsterdam, Elsevier/North-Holland Press.

Sandberg, A.A. (1982) Chromosomal changes in human cancers: Specificity and heterogeneity. In Owens, A.H., Coffey, D.S., Baylin, S.B. (eds): "Tumour Cell Heterogeneity." New York, Academic Press, pp. 367–397.

Sandström, B. (1965) Studies on cells from liver tissue cultivated in vitro. I. Influence of the culture method on cell morphology and growth pattern. Exp. Cell Res. *37*:552–568.

Sanford, K.K., Earle, W.R., Likely G.D. (1948) The growth in vitro of single isolated tissue cells. J. Natl. Cancer Inst. *9*:229.

Sanford, K.K., Earle, W.R., Evans, V.J., Waltz, H.K., Shannon, J.E. (1951) The measurement of proliferation in tissue cultures by enumeration of cell nuclei. J. Natl. Cancer Inst. *11*:773.

Sanford, K.K., Handieman, S.L., Jones, G.M. (1977) Morphology and serum dependence of cloned cell lines undergoing spontaneous malignant transformation in culture. Cancer Res. *37*:821–830.

Sato, G. (1979) The growth of cells in serum-free hormone-supplemented medium. In Jakoby, W.B., Pastan, I.H. (eds): "Methods in Enzymology." New York, Academic Press, pp. 94–109.

Sato, G., ed. (1981) "Functionally Differentiated Cell Lines." New York, Alan R. Liss, Inc.

Sato, G., Reid, L. (1978) Biochemical mode of action of hormones II. In Richenburg, H.V., (ed): "International Review of Biochemistry." Baltimore, University Park Press, pp. 219–251.

Sato, G.H., Yasumura, Y. (1966) Retention of differentiated function in dispersed cell culture. Trans. NY Acad. Sci. *28*:1063–1079.

Sato, G.H., Pardee, A.B., Sirbasku, D.A. (eds.) (1982) "Growth of Cells in Hormonally Defined Media." Cold Spring Harbor Conference on Cell Proliferation, 9. Cold Spring Harbor, Maine, Cold Spring Harbor Laboratory.

Sattler, C.A., Michalopoulos, G., Sattler, G.L., Pitot, H.C. (1978) Ultrastructure of adult rat hepatocytes cultured on floating collagen membranes. Cancer Res. *38*:1539–1549.

Savage, C.R., Jr., Bonney, R.J. (1978) Extended expression of differentiated function in primary cultures of adult liver parenchymal cells maintained on nitrocellulose filters. I. Induction of phosphoenol pyruvate carboxykinase and tryosine aminotransferase. Exp. Cell Res. *114*:307–315.

Schaeffer, W.I., (1984) Usage of vertebrate, invertebrate and plant cell tissue and organ culture terminology. In Vitro, *20*:19–24.

Schengrund, C.L., Repman, M.A. (1979) Differential enrichment of cells from embryonic rat cerebra by centrifugal elutriation. J. Neurochem. *33*:283.

Scher, W., Holland, J.G., Friend, C. (1971) Hemoglobin synthesis in murine virus-induced leukemic cells in vitro. I. Partial purification and identification of hemoglobins. Blood *37*:428–437.

Schimmelpfeng, L., Langenberg, U., Peters, J.M. (1968) Macrophages overcome mycoplasma infections of cells in vitro. Nature *285*:661.

Schleicher, J.B. (1973) Multisurface stacked plate propagators. In Kruse, P.F., Patterson, M.K. (eds): "Tissue Culture Methods and Applications." New York, Academic Press, pp. 333–338.

Schmidt, R., Reichert, U., Michel, S., Shrott, B., Boullier, M. (1985) Plasma membrane transglutaminase and cornified envelope competence in cultured human keratinocytes. FEBS Lett. *186*:201–204.

Schneider, E.L., Stanbridge, E.J. (1975) A simple biochemical technique for the detection of mycoplasma contamination of cultured cells. Methods Cell Biol. *10*:278–290.

Schneider, H., Muirhead, E.E., Zydeck, F.A. (1963) Some unusual observations of organoid tissues and blood elements in monolayer cultures. Exp. Cell Res. *30*:449–459.

Schnook, L.B., Otz, U., Lazary, S., De Week, A.L., Minowada, J., Odavic, R., Kniep, E.M., Edy, V. (1981) Lymphokine and monokine activities in supernatants from human lymphoid and myeloid cell lines. Lymphokines 2:1–19.

Schousboe, A., Thorbek, P., Hertz, L., Krogsgaard-Larsen, P. (1979) Effects of GABA analogues of restricted conformation on GABA transport in astrocytes and brain cortex slices and on GABA receptor binding. J. Neurochem. 33:181.

Schulman, H.M. (1968) The fractionation of rabbit reticulocytes in dextran density gradients. Biochim. Biophys. Acta 148:251–255.

Schwartz, S.M. (1978) Selection and characterization of bovine aortic endothelial cells. In Vitro 14:966.

Scotto, K.W., Biedler, J.L., Melera, P.W. (1986) Amplification and expression of genes associated with multidrug resistance in mammalian cells. Science 232:751–755.

Seeds, N.W. (1971) Biochemical differentiation in reaggregating brain cell culture. Proc. Natl. Acad. Sci. USA. 68:1858–1861.

Segal, S. (1964) Hormones, amino-acid transport and protein synthesis. Nature 203:17–19.

Seglen, P.O. (1975) Preparation of isolated rat liver cells. Methods Cell Biol. 13:29–83.

Selby, P.J., Thomas, M.J. Monaghan, P., Sloane, J., Peckham, M.J. (1980) Human tumour xenografts established and serially transplanted in mice immunologically deprived by thymectomy, cytosine arabinoside and whole-body irradiation. Br. J. Cancer 41:52.

Shall, S. (1973) Sedimentation in sucrose and Ficoll gradients of cells grown in suspension culture. In Kruse, P.F., Patterson, M.K., (eds): "Tissue Culture Methods and Applications." New York, Academic Press, pp. 198–204.

Shall, S., McClelland, A.J. (1971) Synchronization of mouse fibroblast LS cells grown in suspension culture. Nat. New Biol. 229:59–61.

Shows, T.B., Sakaguchi, A.Y. (1980) Gene transfer and gene mapping in mammalian cells in culture. In Vitro 16:55–76.

Sinclair, R., Morton, R.A. (1963) Variations in X-ray response during the division cycle of partially synchronized Chinese hamster cells in culture. Nature 199:1158–60.

Sirica, A.E., Hwand, C.G., Sattler, G.L., Pitot, H.C. (1980) Use of primary cultures of adult rat hepatocytes on collagen gel-nylon mesh to evaluate carcinogen-induced unscheduled DNA synthesis. Cancer Res. 40:3259–3267.

Sxumiel, I., Nias, A.H.W. (1980) Isobologram analysis of the combined effects of anti-tumour platinum complexes and ionizing radiation on mammalian cells. Br. J. Cancer 42:292.

Sladek, N.E. (1973) Bioassay and relative cytotoxic potency of cyclophosphamide metabolites generated in vitro and in vivo. Cancer Res. 33:1150–1158.

Smith, H.S., Lan, S., Ceriani, R., Hackett, A.J., Stampfer, M.R. (1981) Clonal proliferation of cultured non-malignant and malignant human breast epithelia. Cancer Res. 41:4637–4643.

Smith, H.S., Owens, R.B., Hiller, A.J., Nelson-Rees, W.A., Johnston J.O. (1976) The biology of human cells in tissue culture. I. Characterization of cells derived from osteogenic sarcomas. Int. J. Cancer 17:219–234.

Sorieul, S., Ephrussi, B. (1961) Karylogical demonstration of hybridization of mammalian cells in vitro. Nature 190:653–654.

Sorour, O., Raafat, M., El-Bolkainy, N., Mohamad, R. (1975) Infiltrative potentiality of brain tumors in organ culture. J. Neurosurg. 43:742–749.

Souza, L.M., Boone, T.C., Gabrilove, J., Lai, P.H., Zsebo, K.M. et al. (1986) Recombinant human granulocyte colony-stimulating factor: Effects on normal and leukemic myeloid cells. Science 232: 61–65.

Spandidos, D.A., Wilkie, N.M. (1984a) Malignant transformation of early passage rodent cells by a single mutated human oncogene. Nature 310:469–475.

Spandidos, D.A., Wilkie, N.M. (1984b) Expression of exogenous DNA in mammalian cells. In Hames, B.D., Higgins,S.J. (eds): "In Vitro Transcription and Translation—A Practical Approach." Oxford, IRL Press, pp. 1–48.

Splinter, T.A.W., Beudeker, M., Beek, A.V. (1978) Changes in cell density induced by isopaque. Exp. Cell Res. 111:245–251.

Spremulli, E.N., Dexter, D.L. (1984) Polar solents: A novel class of antineoplastic agents. J. Clin. Oncol. 2:227–241.

Sredni, B., Sieckmann, D.G., Kumagai, S.H., Green, I., Paul, W.E. (1981) Long term culture and cloning of non-transformed human B-lymphocytes. J. Exp. Med. 154:1500–1516.

Stampfer, M., Halcones, R.G., Hackett, A.J. (1980) Growth of normal human mammary cells in culture. In Vitro 16:415–425.

Stanbridge, E.J., Doersen, C.-J. (1978) Some effects that mycoplasmas have upon their injected host. In McGarrity, G.J., Murphy, D.G., Nichols, W.W. (eds.): "Mycoplasma Infection of Cell Cultures." New York, Plenum Press, pp. 119–134.

Stanley, E.R., Guilbert, J. (1981) Methods for the purification, assay, characterisation and target cell binding of a colony stimulating factor (CSF-1). J. Immunol. Methods 45:253–289.

Stanley, M.A., Parkinson, E.,. (1979) Growth requirements of human cervical epithelial cells in culture. Int. J. Cancer 24:407–414.

Stanners, C.P., Eliceri, G.L., Green, H. (1971) Two types of ribosome in mouse-hamster hybrid cells. Nat. New Biol. 230:52–54.

Steck, P.A., Voss, P.G., Wang. J.L. (1979) Growth control in cultured 3T3 fibroblasts. J. Cell Biol. 83:562–575.

Steele, V.E., Marchok, A.C., Nettesheim, P. (1978) Establishment of epithelial cell lines following exposure of culture tracheal epithelium to 12-0-tetradecanoylphorbol-13-acetate. Cancer Res. 38:3563–3565.

Stein, G.H. (1979) T98G: An anchorage-independent human tumor cell line that exhibits stationary phase G1 arrest in vitro. J. Cell Physiol. 99:43–54.

Stein, H.G., Yanishevsky, R. (1979) Autoradiography. In Jakoby, W.B., Pastan, I.H. (eds): "Methods in Enzymology, Vol. LVII. Cell Culture." New York, Academic Press, pp. 279–292.

Stephenson, J.R., Axelrad, A.A., McLeod, D.I., Schreeve, M.M. (1971) Induction of colonies of hemoglobin-synthesizing cells by erythropoietin in vitro. Proc. Natl. Acad. Sci. USA 68:1542–1546.

Sternberger, L.A. (1970) "Immunocytochemistry, 2nd Ed." Englewood Cliffs, N.J., Prentice Hall.

Stockdale, F.E., Topper, Y.J. (1966) The role of DNA synthesis and mitosis in hormone dependent differentiation. Proc. Natl. Acad. Sci. USA 56:1283–1289.

Stoker, M., O'Neill, C., Berryman, S., Waxman, B. (1968) Anchorage and growth regulation in normal and virus transformed cells. Int. J. Cancer 3:683–693.

Stoker, M., Perryman, M., Eeles, R. (1982) Clonal analysis of morphological phenotype in cultured mammary epithelial cells from human milk. Proc. R. Soc. Lond. Ser. B. 215:231–240.

Stoker, M.G.P. (1973) Role of diffusion boundary layer in contact inhibition of growth. Nature 246:200–203.

Stoker, M.G.P., Rubin, H. (1967) Density dependent inhibition of cell growth in culture. Nature 215:171–172.

Stoner, G.D., Harris, C.C., Myers, G.A., Trump, B.F., Connor, R.D. (1980) Putrescine stimulates growth of human bronchial epithelial cells in primary culture. In Vitro 16:399–406.

Stoner, G.D., Katoh, Y., Foidart, J-M., Trump, B.F., Steinert, P., Harris, C.C. (1981) Cultured human bronchial epithelial cells: Blood group antigens, keratin, collagens and fibronectin. In Vitro 17:577–587.

Strickland, S., Beers, W.H., (1976) Studies on the role of plasminogen activator in ovulation. In vitro response of granulosa cells to goandotropins, Cyclic nucleotides, and prostaglandins. J. Biol. Chem. *251:*5694–5702.

Stubblefield, E. (1968) Synchronization methods for mammalian cell cultures. In Prescott, D.M. (ed.): "Methods in Cell Physiology." New York, Academic Press, pp. 25–43.

Styles, J.A. (1977) A method for detecting carcinogenic organic chemicals using mammalian cells in culture. Br. J. Cancer *36:*558.

Sutherland, R.M., Carlsson, J., Durand, R., Yuhas, J. (1981) Spheroids in cancer research. Cancer Res. *41:*2980–2984.

Sykes, J.A., Whitescarver, J., Briggs, L., Anson, J.H. (1970) Separation of tumor cells from fibroblasts with use of discontinuous density gradients. J. Natl. Cancer Inst. *44:*855–864.

Taderera, J.V. (1967) Control of lung differentiation in vitro. Dev. Biol. *16:*489–512.

Takahashi, K., Okada, T.S. (1970) Analysis of the effect of "conditioned medium" upon the cell culture at low density. Dev. Growth Differ. *12:*65–77.

Tashjian, A.H., Jr. (1979) Clonal strains of hormone-producing pituitary cells. In Jakoby, W.B., Pastan, I.H. (eds): "Methods in Enzymology, Vol. LVII, Cell Culture." New York, Academic Press, pp. 527–535.

Tashjian, A.H., Jr., Yasumura, Y., Levine, L., Sato, G.H., Parker, M.C. (1968) Establishment of clonal strains of rat pituitary tumor cells that secrete growth hormone. Endocrinology *82:*342–368.

Taub, M., Saier, M.H., Jr. (1979) An established but differentiated kidney epithelial cell line (MDCK). In Jakoby, W.B., Pastan, I.H. (eds): "Methods in Enzymology, Vol. LVII, Cell Culture." New York, Academic Press, pp. 552–560.

Taylor, C.R. (1978) Immunoperoxidase techniques. Arch. Pathol. Lab. Med. *102:*113–121.

Taylor-Papadimitriou, J., Shearer, M., Tilly, R. (1977a) Some properties of cells cultured from early-lactation human milk. J. Natl. Cancer Inst. *58:*1563–1571.

Taylor-Papadimitriou, J., Shearer, M., Stoker, M.G.P. (1977b) Growth requirement of human mammary epithelial cells in culture. Int. J. Cancer *20:*903–908.

Taylor-Papadimitriou, J., Purkiss, P., Fentiman, I.S. (1980) Choleratoxin and analogues of cyclic AMP stimulate the growth of cultured human epithelial cells. J. Cell Physiol. *102:*317–322.

Taylor-Robinson, D. (1978) Cultural and serologic procedures for mycoplasmas in tissue culture. In McGarrity, G., Murphy, D.G., Nichols, W.W. (eds): "Mycoplasma Infection of Cell Cultures." New York, Plenum Press, pp. 47–56.

Temin, H.M. (1966) Studies on carcinogenesis by avian sarcoma viruses. III. The differential effect of serum and polyanions on multiplication of uninfected and converted cells. J. Natl. Cancer Inst. *37:*167–175.

Temin, H.M., Rubin H. (1958) Characteristics of an assay for Rous sarcoma virus and Rous sarcoma cells in tissue culture. Virology *6:*669–688.

Testa, N.G. (1985) Clonal assays for haemopoietic and lymphoid cells in vitro. In Potten, C.S., Hendry, J.H. (eds): "Cell Clones." Edinburgh, Churchill Livingstone Inc, pp. 27–43.

Thilly, W.G. Levine, D.W. (1979) Microcarrier culture: A homogenous environment for studies of cellular biochemistry. In Jakoby, W.B., Pastan, I.H., (eds): "Methods in Enzymology. Vol. LVII, Cell Culture." New York, Academic Press, pp. 184–194.

Thomas, J.A., ed. (1970) "Organ Culture." New York, Academic Press.

Thompson, L.H., Baker, R.M. (1973) Isolation of mutants of cultured mammalian cells. In Prescott, D. (ed) "Methods in Cell Biology, Vol. VI." New York, Academic Press, pp. 209–281.

Thornton, S.C., Mueller, S.N., Levine, E.M. (1983) Human endothelial cells: Use of heparin in cloning and long-term serial cultivation. Science *222:*623–625.

Thurston, J.M., Joftes, D.L. (1963) Stain compatible with dipping radioautography. Stain Technol. *38:*231–235.

Tobey, R.A., Anderson, E.C., Petersen, D.F. (1967) Effect of thymidine on duration of G1 in chinese hamster cells. J. Cell Biol. *35:*53–67.

Tobner, J., Watts, M.T., Fu, J.J.L. (1980) An *in vitro* and *in vivo* investigation on three surface active agents as modulators of cell proliferation. Cancer Res. *40:*1173–1180.

Todaro, G.J., DeLarco, JE. (1978) Growth factors produced by sarcoma virus-transformed cells. Cancer Res. *38:*4147–4154.

Todaro, G.J., Green, H. (1963) Quantitative studies of the growth of mouse embryo cells in culture and their development into established lines. J. Cell Biol. *17:*299–313.

Tom, B.H., Rutzky, L.P., Jakstys, M.M., Oyasu, R., Kaye, C.I., Kahan, B.D. (1976) Human colonic adenocarcinoma cells. I. Establishment and description of a new line. In Vitro *12:*180.

Toshiharu, M., Keiko, N., Masaaki, Y., Toshiko, T., Hajim, K. (1975) Action of bacterial neutral protease on the dispersion of mammalian caells in tissue culture. Jpn. J. Exp. Med. *45:*383–392.

Tozer, B.T., Pirt, S.J. (1964) Suspension culture of mammalian cells and macromolecular growth promoting fractions of calf serum. Nature *201:*375–378.

Traganos, F., Darzynkiewicz, Z., Sharpless, T., Melamed, M.R. (1977) Nucleic acid content and cell cycle distribution of five human bladder cell lines analyzed by flow cytofluorometry. Int. J. Cancer *20:*30–36.

Trapp, B.D., Honegger, P., Richelson, E., Webster, H. de F. (1981) Morphological differentiation of mechanically dissociated fetal rat brain in aggregating cell cultures. Brain Res. *160:*235–252.

Trowell, O.A. (1959) The culture of mature organs in a synthetic medium. Exp. Cell Res. *16:*118–147.

Tsao, M.C., Walthall, B.J., and Ham, R.G. (1982) Clonal growth of normal human epidermal keratinocytes in a defined medium. J. Cell Physiol. *110:*219–229.

Turner, R.W.A., Siminovitch, L., McCulloch, E.A., Till, J.E. (1967) Density gradient centrifugation of hemopoietic colony-forming cells. J. Cell Physiol. *69:*73–81.

Tveit, K.M., Pihl, A. (1981) Do cells lines in vitro reflect the properties of the tumours of origin? A study of lines derived from human melanoma xenografts. Br. J. Cancer *44:*775–786.

Twentyman, P.R. (1980) Response to chemotherapy of EMT6 spheroids as measured by growth delay and cell survival. Eur. J. Cancer *42:*297–304.

Uchida, I.A., Lin, C.C. (1974) Quinacrine fluorescent patterns. In Yunis, J. (ed): "Human Chromosome Methodology, 2nd Ed." New York, Academic Press, pp. 47–58.

Unkless, J., Dano, K., Kellerman, G., Reich, E. (1974) Fibrinolysis associated with oncogenic transformation. Partial purification and characterization of cell factor, a plasminogen activator. J. Biol. Chem. *249:*4295–4305.

Ursprung, H., ed. (1968) "The Stability of the Differentiated State." New York, Springer-Verlag.

Vago, C., ed. (1971) "Invertebrate Tissue Culture, Vol. 1." New York, Academic Press.

Vago, C., ed. (1972) "Invertebrate Tissue Culture, Vol. 2." New York, Academic Press.

Vaheri, A., Ruoslahti, E., Westermark, B., Ponten, J. (1976) A common cell-type specific surface antigen in cultured human glial cells and fibroblasts: Loss in malignant cells. J. Exp. Med. *143:*64–72.

Valenti, C. (1973) Diagnostic use of cell cultures initiated from amniocentesis. In Kruse, P.F., Patterson, M.K. (eds): "Tissue Culture Methods and Applications." New York, Academic Press, pp. 617–622.

Van Beek, W.P., Glimelius, B., Nilson, K., Emmelot, P. (1978) Changed cell surface glycoproteins in human glioma and osteosarcoma cells. Cancer Lett. *5:*311–317.

Van Someren, H., Van Hemegowyen, H.B., Los, W., Wurzer-Figurelli, E., Doppert, B., Yerylolt, M. Meera Khan, P. (1974) Enzyme electrophoresis on cellulose acetate gel II zymogram patterns in man-Chinese hamster cell hybrids. Humangenetik *25:*189–201.

Van Zoelen, E.J.J., Van Oostwaard, T.M.J., Van der Saag, P.T., De Laat, S.W. (1985) Phenotypic transformation of normal rat kidney cells in a growth-factor-defined medium: Induction by a neuroblastoma-derived transforming growth factor independently of the EGF receptor. J. Cell Physiol. *123:*151–160.

Van der Bosch, J., Masui, H., Sato, G. (1981) Growth characteristics of primary tissue cultures from heterotransplanted human colorectal carcinomas in serum-free medium. Cancer Res. *41:*611–618.

VanDiggelen, O., Shin, S., Phillips, D. (1977) Reduction in cellular tumorigenicity affter mycoplasma infection and elimination of mycoplasma from infected cultures by passage in nude mice. Cancer Res. *37:*2680–2687.

Van't Hof, J. (1968) In D.M. Prescott (ed): "Methods in Cell Physiology." New York, Academic Press, p. 95.

Van't Hof, J. (1973) Cell cycle analysis B. In Kruse, P., Patterson, M.K. (eds): "Tissue Culture Techniques and Application." New York, Academic Press, pp. 423–428.

Varon, S., Manthorpe, M. (1980) Separation of neurons and glial cells by affinity methods. In Fedoroff, S., Hertz, L. (eds): "Advances in Cellular Neurobiology, Vol. I." New York, Academic Press, pp. 405–442.

Venitt, S. (1985) "Mutagenicity Testing. A Practical Approach." Oxford, IRL Press, Ltd.

Vlodavsky, I., Lui, G.M., Gospodarowicz, D. (1980) Morphological appearance, growth behavior and migratory activity of human tumor cells maintained on extracellular matrix versus plastic. Cell *19:*607–617.

Voyta, J.C., Via, D.P., Butterfield, C.E., Zetter, B.R. (1984) Identification and isolation of endothelial cells based on their increased uptake of acetylated-low density lipoprotein. J. Cell Biol. *99:*2034–2040.

Vries, J.E., Benthem, M., Rumke, P. (1973) Separation of viable from nonviable tumor cells by flotation on a Ficoll-triosil mixture. Transplantation *15:*409–410.

Walker, C.R., Bandman, E., Strohman, R.C. (1979). Diazepam induces relaxation of chick embryo muscle fibers *in vitro* and inhibits myosin synthesis. Exp. Cell Res. *123:*285–291.

Wallace, D.H., Hegre, O.D. (1979) Development *in vitro* of epithelial-cell monolayers derived from fetal rat pancreas. In Vitro *15:*270.

Walter, H. (1975) Partition of cells in two-polymer aqueous phases: A method for separating cells and for obtaining information on their surface properties. In Prescott, D.M. (ed): "Methods in Cell Biology." New York, Academic Press, pp. 25–50.

Walter, H. (1977) Partition of cells in two-polymer aqueous phases: A surface affinity method for cell separation. In Catsimpoolas, N. (ed): "Methods of Cell Separation." New York, Plenum Press, pp. 307–354.

Wang, H.C., Fedoroff, S. (1972) Banding in human chromosomes treated with trypsin. Nat. New Biol. *235:*52–53.

Wang, H.C., Fedoroff, S. (1973) Karyology of cells in culture E. Trypsin technique to reveal G-bands. In Kruse, P.F., Patterson, M.J. (eds): "Tissue Culture Methods and Applications." New York, Academic Press, pp. 782–787.

Wang, R.J. (1976) Effect of room fluorescent light on the deterioration of tissue culture medium. In Vitro *12:*19–22.

Warren, L., Buck, C.A., Tuszynski, G.P. (1978) Glycopeptide changes and malignant transformation. A possible role for carbohydrate in malignant behavior. Biochim. Biophys. Acta. *516:*97.

Watret, G.E, Pringle, C.R., Elliott, R.M. (1985) Synthesis of Bunyavirus-specific proteins in a continuous cell line (XTC-2) derived from Xenopus laevis. J. Gen. Virol. *66:*473–482.

Waymouth, C. (1959) Rapid proliferation of sublines of NCTC clone 929 (Strain L) mouse cells in a simple chemically defined medium (MB752/1). J. Natl. Cancer Inst. *22:*1003.

Waymouth, C. (1970) Osmolality of mammalian blood and of media for culture of mammalian cells. In Vitro *6:*109–127.

Waymouth, C. (1974) To disaggregate or not to disaggregate. Injury and cell disaggregation, transient or permanent? In Vitro *10:*97–111.

Waymouth, C. (1977) In Evans, V.J., Perry, V., Vincent, M.M. (eds): "Manual of American Tissue Culture Association." *3:*521.

Waymouth, C. (1979) Autoclavable medium AM 77B. J. Cell Physiol. *100:*548–550.

Waymouth, C. (1984) Preparation and use of serum-free culture media. In Barnes, W.D. Sirbasku, D.A., Sato, G.H. (eds): "Cell Culture Methods for Molecular and Cell Biology, Vol. 1. Methods for Preparation of Media, Supplements, and Substrata for Serum-Free Animal Cell Culture." New York, Alan R. Liss, Inc. pp. 23–68.

Weibel, E.R., Palade, G.E. (1964) New cytoplasmic components in arterial endothelia. J. Cell Biol. *23:*101–102.

Weichselbaum, R., Epstein, J., Little, J.B. (1976) A technique for developing established cell lines from human osteosarcomas. In Vitro *12:*833–836.

Weiss, M.C., Green, H. (1967) Human-mouse hybrid cell lines containing partial complements of human chromosomes and functioning human genes. Proc. Natl. Acad. Sci. USA *58:*1104–1111.

Weiss, R., Teich, N., Varmus, H., Coffin, C., eds (1982, 1985) "Molecular and Biology of the Tumour Viruses." New York, Cold Spring Harbor Laboratory.

Westermark, B. (1974) The deficient density-dependent growth control of human malignant glioma cells and virus-transformed glia-like cells in culture. Int. J. Cancer *12:*438–451.

Westermark, B. (1978) Growth control in miniclones of human glial cells. Exp. Cell Res. *111:*295–299.

Westermark, B., Wasteson, A. (1975) The response of cultured human normal glial cells to growth factors. In Luft and Hall (eds): "Advances in Metabolic Disorders, Vol. 8." New York, Academic Press, pp. 85–100.

Westermark, B., Ponten, J., Hugosson, R. (1973) Determinants for the establishment of permanent tissue culture lines from human gliomas. Acta Pathol. Microbiol. Scand. Section A *81:*791–805.

Whei-Yang, K.W., Prockop, D.J. (1977) Proline analogue removes fibroblasts from cultured mixed cell populations, Nature *266:*63–64.

Whetton, A.D., Dexter, T.M. (1986) Haemopoietic growth factors. Trends in Biochemical Sciences *11:*207–211.

Whittle, W.L., Kruse, P.F. (1973) Replicate roller bottles. In Kruse, P.F., Patterson, M.K. (eds): "Tissue Culture Methods and Applications." New York, Academic Press, pp. 327–331.

Whur, P., Magudia, M., Boston, J., Lockwood, J., Williams, D.C. (1980) Plasminogen activator in cultured Lewis lung carcinoma cells measured by chromogenic substrate assay. Br. J. Cancer *42:*305–312.

Wiepjes, G.J., Prop, F.J.A. (1970) Improved method for preparation of single-cell suspensions from mammary glands of adult virgin mouse. Exp. Cell Res. *61:*451–454.

Wigler, M., Sweet, R., Sim, G.K., Wold, B., Pellicer, A., Lacy, E., Maniatis, T., Silverstein, S., Axel, R. (1979) Transformation of mammalian cells with genes from procaryotes and eucaryotes. Cell *16:*777–785.

Wilkes, P.R., Birnie, G.D., Old, R.W. (1978) Histone gene expression during the cell cycle studied by *in situ* hybridization. Exp. Cell Res. *115:*441–444.

Wilkins, L., Gilchrest, B.A., Szabo, G., Weinstein, R., Maciag, T. (1985) The stimulation of normal human melanocyte proliferation in vitro by melanocyte growth factor from bovine brain. J. Cell Physiol. *122:*350.

Wilkinson, P.C., Lackie, J.M. Allen, R.B. (1982) Methods for analysing leucocyte locomotion. In Catsimpoolas, N. (ed): "Cell Analysis, Vol. 1." New York, Plenum.

Willecke, K., Klomfass, M., Mierau, R., Dohner, J. (1979) Intraspecies transfer via total cellular DNA of the gene for hypoxanthine phosphoribosyltransferase into cultured mouse cells. Mol. Gen. Genet. *170:*179–185.

Willey, J.C., Saladino, A.J., Ozanne, C., Lechner, J.F., Harris, C.C. (1984) Acute effects of 12-0-tetradecanoylphorbol-13-acetate, teleocidin B, or 2,3,7,8-tetrachlorodibenzo-p-dioxin on cultured normal human bronchial cells. Carcinogenesis *5:*209–215.

Willingham, M.C., Pastan, I. (1975) Cyclic AMP modulates microvillus formation and agglutinability in transformed and normal mouse fibroblasts. Proc. Natl. Acad. Sci. USA *72:*1263–1267.

Wilmer, E.N., ed. (1965) "Cells and Tissues in Culture." London, Academic Press.

Wilkouski, J.A., Durbridge, M., Dubowitz, V. (1976) Growth of human muscle in tissue culture. In Vitro *12:*98–106.

Wolff, D.A., Pertoft, H. (1972) Separation of HeLa cells by colloidal silica density gradient centrifugation. J. Cell Biol. *55:*579.

Wolff, E., Wolff, E. (1952) La determination de la differentiation sexuelle de la syrinx du canard cultivé *in vitro*. Bull. Biol. *86:*325.

Worton, R.G., Duff, C. (1979) Karyotyping. In Jakoby, W.B., Pastan, I.H. (eds): "Methods in Enzymology, Vol. LVII, Cell Culture." New York, Academic Press, pp. 322–244.

Wright, J.E., Dendy, P.P. (1976) Identification of abnormal cells in short-term monolayer cultures of human tumor specimens. Acta Cytol. (Baltimore) *20:*328–334.

Wright, W.C., Daniels, W.P. and Fogh, J. Distinction of seventy-one cultured human tumor cell lines by polymorphic enzyme analysis. J. Natl. Cancer Inst. *66:*239–248, 1981.

Wu, A.M., Siminovitch, L., Till, J.E., McCulloch, E.A. (1968) Evidence for a relationship between mouse hemopoietic stem cells and cells forming colonies in culture. Proc. Natl. Acad. Sci. USA *59:*1209–1215.

Wu, R., Wu, M.M.J. (1986) Effects of retinoids on human bronchial epithelial cells: Differential regulation of hyaluronate synthesis and keratin protein synthesis. J. Cell Physiol. *127:*73–82.

Wu, Y.J., Parker, L.M., Binder, N.E., Beckett, M.A., Sinard, J.H., Griffiths, C.T., Rheinwald, J.G. (1982) The mesothelial keratins: A new family of cytoskeletal proteins identified in cultured mesothelial cells and nonkeratinizing epithelia. Cell *31:*693–703.

Wurster-Hill, D., Cannizzaro, L.A., Pettengill, O.S., Sorenson, G.D., Cate, C.C., Maurer, L.H. (1984) Cytogenetics of small cell carcinoma of the lung. Cancer Genet. Cytogenet. *13:*303–330.

Yaffe, D. (1968) Retention of differentiation potentialities during prolonged cultivation of myogenic cells. Proc. Natl. Acad. Sci. USA *61:*477–483.

Yaffe, D. (1971) Developmental changes preceding cell fusion during muscle cell differentiation in vitro. Exp. Cell Res. *66:*33–48.

Yang, J., Richards, J., Bowman, P., Guzman, R., Enami, J., McCormick, K., Hamamoto, S., Pitelka, D., Nandi, S. (1979) Sustained growth and 3-dimensional organization of primary mammary tumor epithelial cells embedded in collagen gels. Proc. Natl. Acad. Sci. USA *76:*3401.

Yang, J., Richards, J. Guzman, R., Imagawa, W. Nandi, S. (1980) Sustained growth in primary cultures of normal mammary epithelial cells embedded in collagen gels. Proc. Natl. Acad. Sci. USA *77:*2088–2092.

Yang, J., Elias, J.J., Petrakis, N.L., Wellings, S.R., Nandi, S. (1981) Effects of hormones and growth factors on human mammary epithelial cells in collagen gel culture. Cancer Res. *41:*1021–1027.

Yang, Y-C., Ciarletta, A.B., Temple, P.A., Chung, M.P., Kovacic, S., et al. (1986) Human IL3 (Multi-CSF): Identification by expression cloning of a novel hematopoietic growth factor related to murine IL-3. Cell *47:*3–10.

Yasin, R., Kundu, D., Thomson, E.J. (1981) Growth of adult human cells in culture at clonal densities. Cell Differ. *10:*131–137.

Yavin, Z., Yavin E. (1980) Survival and maturation of cerebral neurons on poly(L-lysine) surfaces in the absence of serum. Dev. Biol. *75:*454–460.

Yerganian, G., Leonard, M.J. (1961) Maintenance of normal in situ chromosomal features in long-term tissue cultures. Science *133:*1600–1601.

Yoshida, Y., Hilborn, V., Hassett, C., Mezfi, P., Byers, M.J., Freeman, A.G. (1980) Characterization of mouse fetal lung cells cultured on a pigskin substrate. In Vitro *16:*433–445.

Yuhas, J.M., Li, A.P., Martinex, A.O., Ladman, A.J. (1977) A simplified method for production and growth of multicellular tumour spheroids (MTS). Cancer Res. *37:*3639–3643.

Yunis, J. (1974) "Human Chromosome Methodology, 2nd Edition." New York, Academic Press.

Yuspa, S.H., Hawley-Nelson, P., Stanley, J.R., Hennings, H. (1980) Epidermal cell culture. Transplant. Proc. *12:*114–122.

Yuspa, S.H., Koehler, B., Kulesz-Martin, M., Hennings, H. (1981) Clonal growth of mouse epidermal cells in medium with reduced calcium concentration. J. Invest. Dermatol. *76:*144–146.

Zaroff, L., Sato, G.H., Mills, S.E. (1961) Single-cell platings from freshly isolated mammalian tissue. Exp. Cell Res. *23:*565–575.

Zawydiwski, R., Duncan, G.R. (1978) Spontaneous ^{51}Cr release by isolated rat hepatocytes: An indicator of membrane damage. In Vitro *14:*707–714.

Zetter, B.R. (1981) The endothelial cells of large and small blood vessels. Diabetes *30 suppl 2:*;24–28.

GENERAL TEXTBOOKS FOR FURTHER READING

Paul, J. (1975) "Cell and Tissue Culture." Edinburgh, Scotland, Livingstone. Good basic textbook.

Kruse, P., Patterson, M.K. (1973) "Tissue Culture, Techniques and Applications." New York, Academic Press. Good for specialized techniques but some duplication.

Jakoby, W.B., Pastan, I.H., eds. (1979) "Cell Culture." In Colowick, S.P., Kaplan N.D. (series eds.): "Methods in Enzymology, Vol. LVIII "New York, Academic Press. Very good for specialized techniques, matrix, serum-free media.

Pollack, R. ed. (1981) "Reading in Mammalian Cell Culture, 2nd Edition." Cold Spring Harbor, New York, Cold Spring Harbor Laboratory. Very good compilation of key papers in the field. Good tutorial and general interest. Good for teaching.

Crowe, R., Ozer, H. Rifkin, D. (1978) "Experiments with Normal and Transformed Cells." Cold Spring Harbor, New York, Cold Spring Harbor Laboratory. Laboratory exercises for senior undergraduate and graduate students.

Hall, D., Hawkins, S. (1975) "Laboratory Manual of Cell Biology." London, English Universities Press. Laboratory exercises for undergraduates.

Freshney, R.I. (1986) "Animal Cell Culture, a Practical Approach." Oxford, IRL Press. Invited chapters on specialized techniques.

Maramorosch, K. (1976) "Invertebrate Tissue Culture." New York, Academic Press. Quite useful, but restricted in species by the extent of work done.

Vago, E., ed. (1971, 1972) "Invertebrate Tissue Culture, Vols. 1 and 2. New York, Academic Press.

Dendy, P.P., ed. (1973) "Human Tumours in Short Term Culture." London, Academic Press. Characterization, drug and radiosensitivity. Good techniques review and some useful articles.

Federoff, S., Herz, L., eds. (1977) "Cell, Tissue and Organ Cultures in Neurobiology." New York, Academic Press. A good introduction to culturing neural cells.

Fogh, J. (1975) "Human Tumor Cells in Vitro." New York, Plenum. Some useful listings of human tumor types in culture.

Sato, G., Pardee, A.B., Sirbasku, D.A. (1982) "Growth of Cells in Hormonally Defined Media." Cold Spring Harbor Conferences on Cell Proliferation, Vol. 9. Cold Spring Harbor, New York, Cold Spring Harbor Laboratories. Excellent up to date review of serum-free culture.

Reinert, J., Yeoman, M.M. (1982) "Plant Cell and Tissue Culture. A Laboratory Manual." Berlin, Heidelberg, New York, Springer-Verlag.

Harris, C.C., Trump, B.F., Stoner, G.D. (1981) "Normal Human Tissue and Cell Culture." In Prescott, D.M. (series ed.): "Methods in Cell Biology." New York, Academic Press.

Barnes, D.W., Sirbasku, D.A., Sato, G.H. eds. (1984) "Cell Culture Methods for Molecular and Cell Biology." 4 volumes. Alan R. Liss, Inc.

USEFUL JOURNALS

Journal of Tissue Culture Methods
In Vitro. Journal of American Tissue Culture Association
Journal of Cellular Physiology
Experimental Cell Research
Cell
Cell Biology, International Reports
Journal of Cell Biology
Experimental Cell Biology
European Journal of Cell Biology
Anticancer Research
Cancer Research
International Journal of Cancer
British Journal of Cancer
Journal of the National Cancer Institute
Journal of Cell Science
Cellular Biology
European Journal of Cancer and Clinical Oncology

Index

The numbers in bold type indicate primary references.